The Rand McNally

Atlas of the Body and Mind

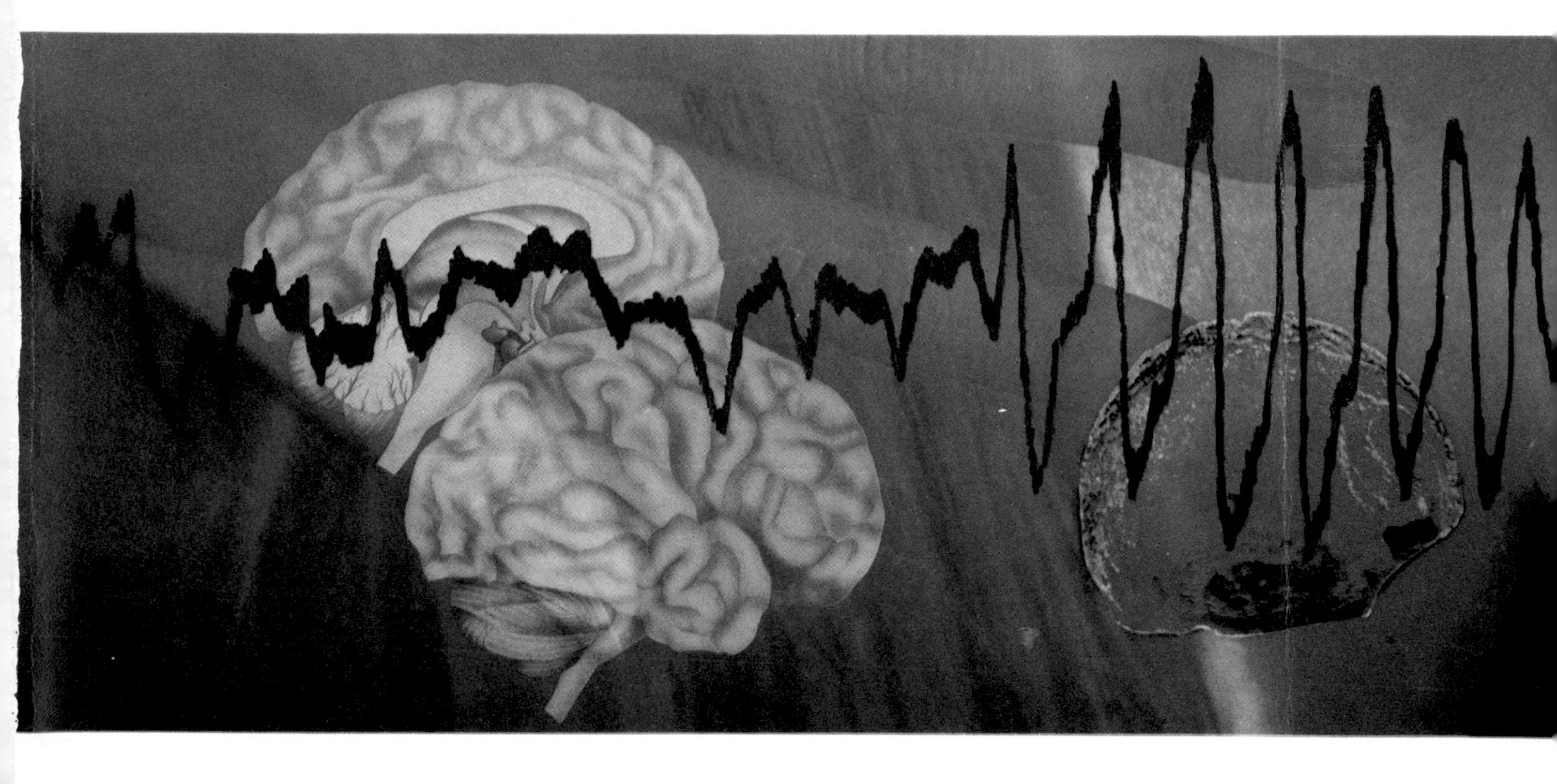

The Rand McNally

Atlas of the Body and Mind

Rand McNally and Company
New York Chicago San Francisco
in association with Mitchell Beazley Publishers Limited, London

Contents

The Rand McNally
Atlas of the Body and Mind
Edited and designed by
Mitchell Beazley Publishers Limited
87-89 Shaftesbury Avenue, London W1V 7AD

Library of Congress Catalog Card Number 75-41710

Published 1976 in the United States of America by
Rand McNally and Company,
P.O. Box 7600, Chicago,
Illinois 60680

Typesetting by Tradespools Ltd, Frome, England
Printed in The United States of America

Editors Ruth Binney, Michael Janson
Art Editor Peter Wrigley
Assistant Editors Lionel Bender, Nicholas Bevan, Marcia Lloyd, Sarina Turner, Sally Walters
Designers Javed Badar, Mary Ellis, Celia Welcomme
Picture Research Sue Pinkus
Editorial Assistants Bridget Alexander, Liz Whitaker
Production Barry Baker

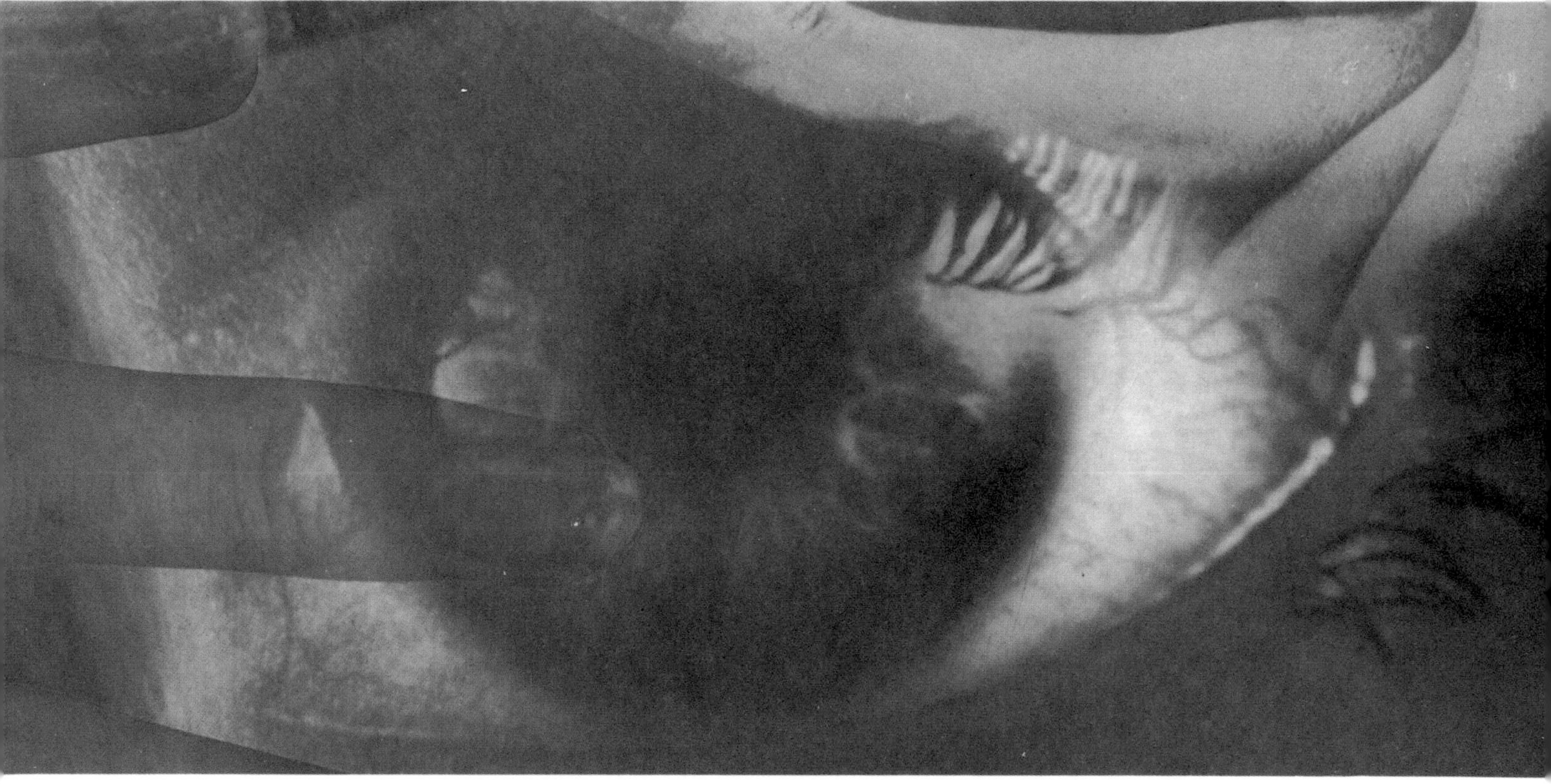

Contributing Editor
Claire Rayner

Contributors
Christopher Badcock
Jane Firbank
Paulette Pratt
John Rivers
David M. H. Williams
Clive Woods

Consultants
Peter Fenwick, M.B., B.Chir. (Cantab.), M.R.C.Psych.
John Reckless, M.B., M.R.C.P.

Foreword

The relationships between the body and the mind have intrigued many people for hundreds of years. About three centuries ago, Descartes, the French scientist and philosopher, set the scene for subsequent controversy by separating spatial substance, or matter, from thinking substance, or mind. This dualism, of mind and matter, tended to divide the physical sciences from philosophy and psychology, so that progress on these two fronts of human knowledge has been in parallel rather than convergent. In other words, the separateness of mind and body has been accepted as though each were a distinct entity without dependence of one upon the other. Yet it is obvious to even the most superficial inquiry that they are inextricably interdependent.

The body, especially if it is diseased, may affect the way you feel emotionally and may disturb rational thought, as is the case, for example, with a high fever. Drugs, physical matter, are used to alter mental states. It is a mental state which determines that you shall get up out of a chair, and yet that is a physical act of the body. What happens at the boundary between body and mind? Is there in fact a boundary at all? Is it just that certain very complex physical phenomena are labelled mental, and less complex ones are labelled as physical? This is the position of the materialistic philosophers. Both dualists and materialists bring philosophical problems upon themselves, but these are not the concern of this book.

Rather than starting from a defined philosophical standpoint, *The Atlas of the Body and Mind* unashamedly adopts the commonsense view that there is an interaction between body and mind, which is there for all to see, and then proceeds to explore what we know about this reciprocity in anatomical, physiological, medical and psychological terms.

The body is the substrate of mind. Apart from some doubtful phenomena we cannot scientifically be sure of the existence of mind without a body through which some of its manifestations are seen. The nervous system is the system which is most obviously concerned with mind, and it is the study of this which brings us most closely to the zone where mental phenomena may be indistinguishable from physical ones. It is here that this book scores so heavily, for a picture is worth a million words and wherever it is possible pictorial representation is used. This is sound psychology, for visual concepts are more easily grasped by most people than abstract ones. However, mental activity and subtle cellular reactions are not always reducible to simple visual terms and it is here that the text, written specially for ease of understanding, fills out the pictures. It is an admirable combination.

Man's special place in the world is due in large measure to his mental prowess based on the evolutionary development of his relatively big brain and complex nervous system. For our comfort and continued hold on a reasonable and reasoned existence we need all the understanding of ourselves that we can obtain. The place of the mind in nature and its limitations, and what imposes these, are vital to our comprehension. This is why this book is important, for the knowledge it contains ought to be widely known. By the simplicity of the presentation, without sacrifice of accuracy, a difficult area of study is made as plain as it could be. All who read and look carefully at this book will be enriched with deeper understanding of themselves.

Philip Rhodes, F.R.C.S., F.R.C.O.G.
Dean
Faculty of Medicine
University of Adelaide
South Australia

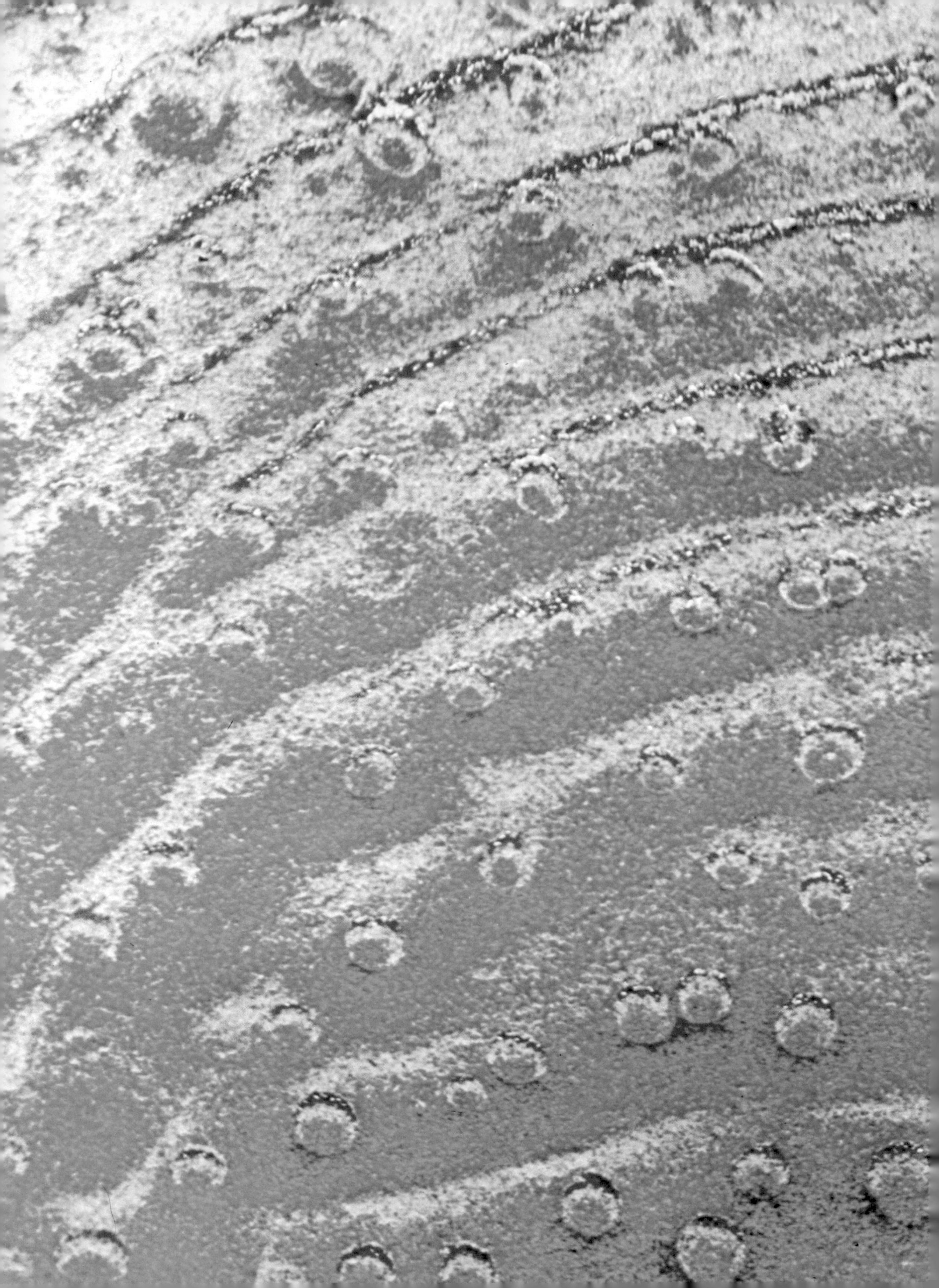

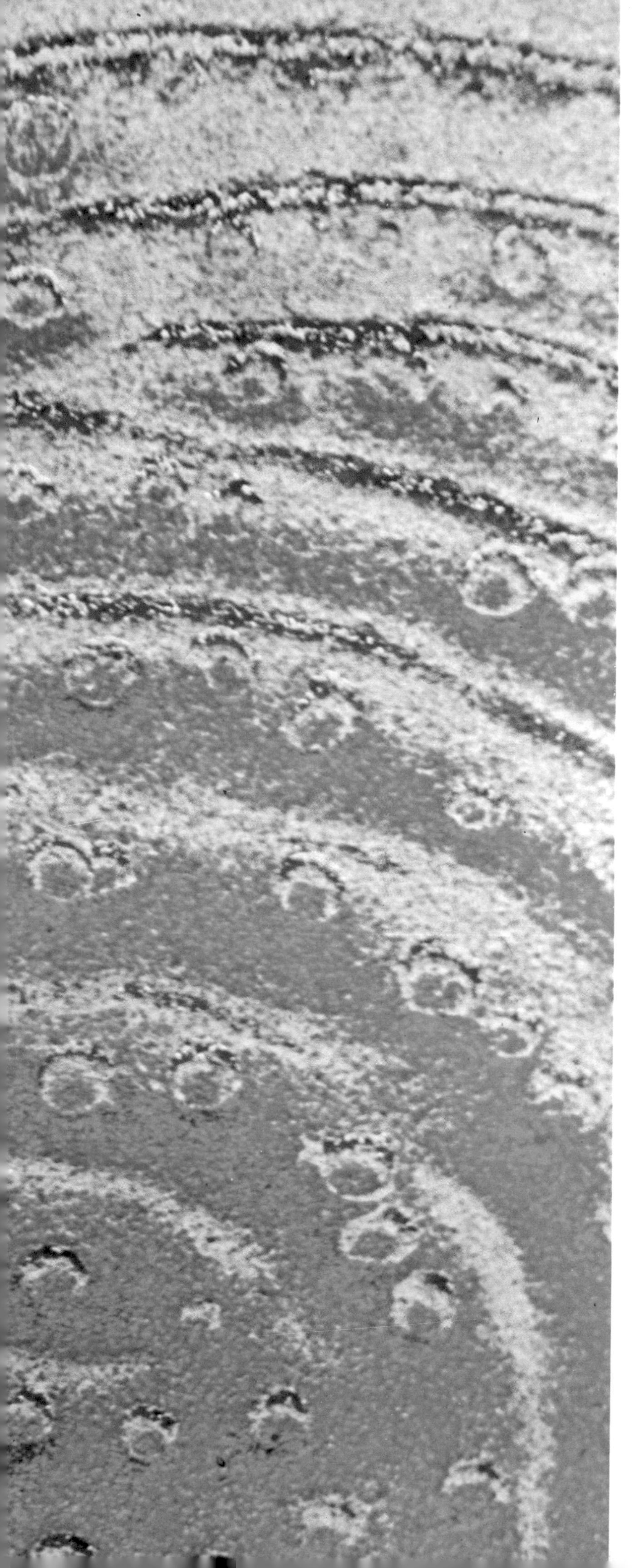

Evolution of man

As archaeologists sift inch by inch through the accumulated dusts, deposits and rocky debris of thousands of centuries, to uncover fossils and fossil impressions that tell the story of evolution, the complicated history of mankind is gradually being unraveled. Hardly a year goes by without a discovery that alters the views of anthropologists about the age of man himself and the relationships between the different cultures of our long-distant ancestors throughout the world.

In evolutionary terms man is a relative newcomer to the earth, but in this short time he has established himself as the most successful and dominant of species. As man's earliest ancestors left the forests to feed in the savannas and finally to form hunting societies on the open plains, so they changed physically, mentally and socially.

Physically, man is relatively weak, without the speed of the cheetah or the strength of the gorilla, but his bipedal gait has left his hands free to manipulate tools. This ability, along with the increase in brain size and, consequently, intelligence, has allowed him to develop superior technologies.

At the same time, as man climbed the evolutionary ladder, he developed a social structure which, molded and directed by the same forces of natural selection, led to the modern structured societies and material cultures of today.

Ridge patterns on a human thumb, seen here magnified more than one hundred and fifty times, are a universal attribute of mankind, but at the same time testify to the uniqueness of the individual, for no two people bear patterns that are exactly the same.

Man's place in evolution

Nobody is certain exactly when life began, but the first fossil sea animals date from about six hundred million years ago. It was another two hundred million years or so before vertebrates, animals with backbones, appeared on the land in the shape of the first amphibians, which were basically air-breathing fish which had begun to modify their fins into primitive legs. About 345 million years ago, the first reptiles appeared and with them the colonization of the earth by animals really began. But today the dominant forms of terrestrial life are mammals rather than reptiles, and of these mammals the most successful has been the only one who has evolved to the point of showing curiosity about his origins, his body and his mind—man.

The key to evolution is the concept of adaptation, the process by which, over many generations, animals and plants modify themselves to take advantage of their environment and so manage to survive. The better they adapt themselves, the better they survive and the more successful they become. If man is the most successful of animals, it is merely because he is the best adapted.

Man is a mammal, and the mammals succeeded the reptiles as the dominant form of animal life on the surface of the earth because they were significantly better adapted than reptiles. The first major characteristic of the mammals is warm-bloodedness. The importance of warm-bloodedness is that it maintains a relatively high and stable body temperature for the physiological processes going on in the cells of mammals and makes them independent of external sources of heat and able to flourish in many different environments. Along with warm-bloodedness goes the evolution of efficient insulation in the form of hair, fur or layers of fat deposits.

The second major characteristic of mammals is that they have developed differentiated teeth, rather than the rows of identical teeth found in reptiles, and this has greatly improved the efficiency of their mastication and the scope of their diet. Third, and even more important as far as man was concerned, mammals ceased to lay eggs, unlike all earlier forms of animal life, and instead gave birth to live young. This was a far more efficient mode of reproduction and made the mammals take much better care of their offspring than did the reptiles. It also enabled them to take advantage of their fourth, and perhaps, as far as man was concerned, their most important advance—the evolution of intelligent behavior. Mammals exhibit a plasticity of behavior and a curiosity and adaptability to the world which is unparalleled among animals. The mammals became capable of learning and of taking advantage of the opportunity to absorb some lessons from experience before being left to fend for themselves when the period of suckling and parental care came to an end. It is no accident that this last most important characteristic is nowhere shown more developed than among the primates, the main subgroup of the mammals, to which man belongs.

The first primates appeared about sixty-five million years ago. Essentially they were tree-living mammals. All primates have relatively large brains, well-developed eyes and good stereoscopic vision. This is because sight became much more important to them in their arboreal habitat than the sense of smell, on which most mammals rely. This led to a shortening of the muzzle and an increase in the size of the parts of the brain associated with the interpretation of visual sensations. It also led to a tendency to explore and to manipulate the surroundings with the limbs rather than with the muzzle, thus giving the primates an opportunity to develop enormously the intelligent response to phenomena which they shared with all other mammals.

A life in the trees also required generalized limbs with grasping, or prehensile, hands and feet. This capacity to grasp objects and, in particular, branches, has been lost in man in the case of the foot, but it has been compensated for by a greatly improved capacity for manipulation with the hands. Finally, a number of such minor features as the development of nails instead of claws, collarbones and only two nipples abnormally high on the body, were also evolved by primates and then bequeathed to man.

In many respects the primates as a whole continued important evolutionary trends begun by the mammals. Their arboreal life made them notably larger-brained, more intelligent and more curious about the world than most other mammals. Their tendency to have small litters and usually, among the higher primates and man, single births, continued the instinctive mammalian inclination to take good care of the young and to provide them with opportunities to acquire learned behavior before they have reached maturity.

The first primates of sixty-five million years ago looked a bit like modern rodents and were about the size of a domestic cat. Gradually, they diversified into four main subdivisions—the prosimians, monkeys, apes and man.

The prosimians are the smaller primates which frequently live on the outer, thinner branches of the trees and eat grubs, insects, small fruits and berries. Many are nocturnal. Their typical mode of locomotion is what is graphically described as vertical clinging and leaping, which results in proportionately longer legs than arms. Today they are represented by the bush babies, or galagos, tarsiers, lorises, lemurs and tree shrews.

The monkeys evolved independently in two families, the Old World and the New World monkeys. Generally larger than the prosimians, the monkeys have exploited life on the thicker, stronger branches of the trees and live on larger fruit, birds' eggs and grubs. The New World monkeys have evolved a prehensile tail, which has effectively become a fifth limb.

The apes are larger still. The gibbon, the smallest ape, and the least closely related to man, is a superb performer in the forest canopy, from which it seldom descends to the ground. The simians and orangutans are also essentially arboreal, but the gorillas and chimpanzees are now to a great extent terrestrial and spend most of their time on the ground, returning to the thick, inner boughs of the trees only at night.

Finally, man exists as the fourth and final subdivision of the primate family, now fully committed to a ground-living existence and showing certain unique adaptations of his own. These adaptations can all be seen as extensions of the basic trends in mammalian evolution already so successfully developed by the earlier primates.

In his capacity to adapt himself to life in any environment, man took warm-bloodedness to its logical conclusion. By becoming omnivorous, that is, able to eat anything, he developed to the extreme the greater dietary flexibility and efficiency of digestion first given to the mammals by their characteristic differentiated teeth. And by living in family groups and fostering culture, learning and curiosity about nature and himself, man took to its ultimate the behavioral adaptability and mental vivacity of the whole mammalian order.

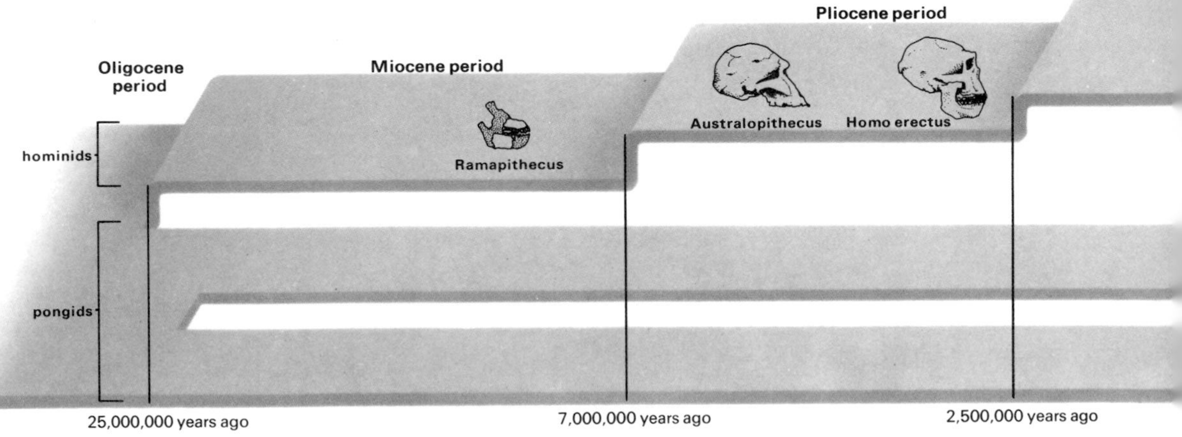

Man's evolution began about sixty-five million years ago. Until thirty-five million years ago different primate groups were evolving until one gave rise to the primitive stock from which the apes (pongids) and humans (hominids) evolved. The evolution of apes, including the chimpanzee and the gorilla, paralleled man's evolution. The earliest hominid remains, found in India and Africa, are those of Ramapithecus, which date back ten to twelve million years. Other remains have confirmed the existence of later links in the evolutionary chain—a more evolved hominid called Australopithecus, the first true man, Homo erectus and Homo sapiens, the species from which modern man finally evolved.

Man's ancestors, forty million years ago, were probably tree-dwelling primates (**1**) which resembled modern prosimians. From these evolved the anthropoid primates (**2**), which were able to swing and leap through the trees rather than crawl along the branches. Twenty-five to thirty million years ago, when the climate in many forested areas became cooler and there was more open grassland, some primates came down from the trees and adapted to the new habitat (**3**). These primates, which probably moved on all four limbs, but occasionally stood on their hind legs, preceded Australopithecus (**4**), who walked upright and lived in groups on the savanna.

The better an animal is adapted to its environment the more successful it becomes. Man, more than any other mammal, is able to adapt to and control most types of environment. (**A**) Man's technological advances have given him greater freedom of movement and independence of his environment than other animals. (**B**) By becoming omnivorous, using fire, developing agriculture and inventing methods of food storage, man has been able to use a wider range of foods and to develop a sophisticated method of feeding. (**C**) Although some other animals are able to use simple tools, man's greater intelligence has led to the creation and the use of complex objects which enhance his existence. (**D**) Because of a longer period of parental care, man learns more complex behavior patterns, which contribute to increasingly successful adaptation.

Pleistocene period

Homo sapiens

Rhodesia man — Neanderthal man — Solo man

Homo sapiens

Holocene period

Cro-Magnon man — modern human races

gibbon

gorilla

30,000 years ago

Cross-reference	pages
Brain and central nervous system	74-75
Early man	14-17
Primate locomotion	12-13
Reproduction	148-155

Man branches out

As each of the member species of the primate family began to adapt itself to its particular habitat in or around the trees, each of the main subgroups of the order came to manifest a characteristic locomotor behavior. The smaller prosimians became vertical clingers and leapers, specialized for a life among the thinner branches of trees and shrubs. The larger prosimians and the monkeys became, or remained, quadrupedal, that is they got about in the trees on all fours with the involvement of each limb being more or less equal. Thus the characteristic progression of a monkey in a tree is quadrupedally along a branch, which is held beneath it by its prehensile hands and feet, with the long tail acting as a balancing mechanism. Since this is obviously a good way of getting about on the ground as well, it is not surprising that some species of monkeys have become wholly terrestrial.

The smaller apes and one or two of the monkeys have evolved a means of locomotion which is called brachiation. This is quite different from vertical clinging and leaping or from quadrupedalism, and is best exemplified by the gibbons and simians, among the apes, and by the spider monkey. Essentially, it is a method of swinging through the trees by means of suspending the body below alternate arms. The legs and feet play no part at all in true brachiation, and in the case of the gibbon they are retracted up under the body during flight. Typically, brachiators have very long arms relative to their legs, and hooklike hands under which the body can be suspended.

Larger apes, including the chimpanzee and the gorilla, are frequently called semibrachiators. They, too, have long arms and short legs, but not to the same extent as does the gibbon. They are, in fact, too large to undertake true brachiation when they are adults and spend most of their time on the ground, where they progress by knuckle walking, walking with the weight of the body resting on the legs and the knuckles so that the body is held in a semierect posture.

If relative arm and leg lengths of primates are compared by means of what is known as intermembral index $\left(\frac{\text{arm length}}{\text{leg length}} \times 100\right)$, it is obvious that there is a smooth progression from vertical-clinging-and-leaping prosimians at one end, whose legs are longer than their arms, up through the monkeys, whose legs and arms are equal, or whose arms are a bit longer, to the brachiating apes and the gibbon, which has the highest index of all, with arms much longer than legs.

Man, however, is the exception here. In the intermembral index series he belongs among

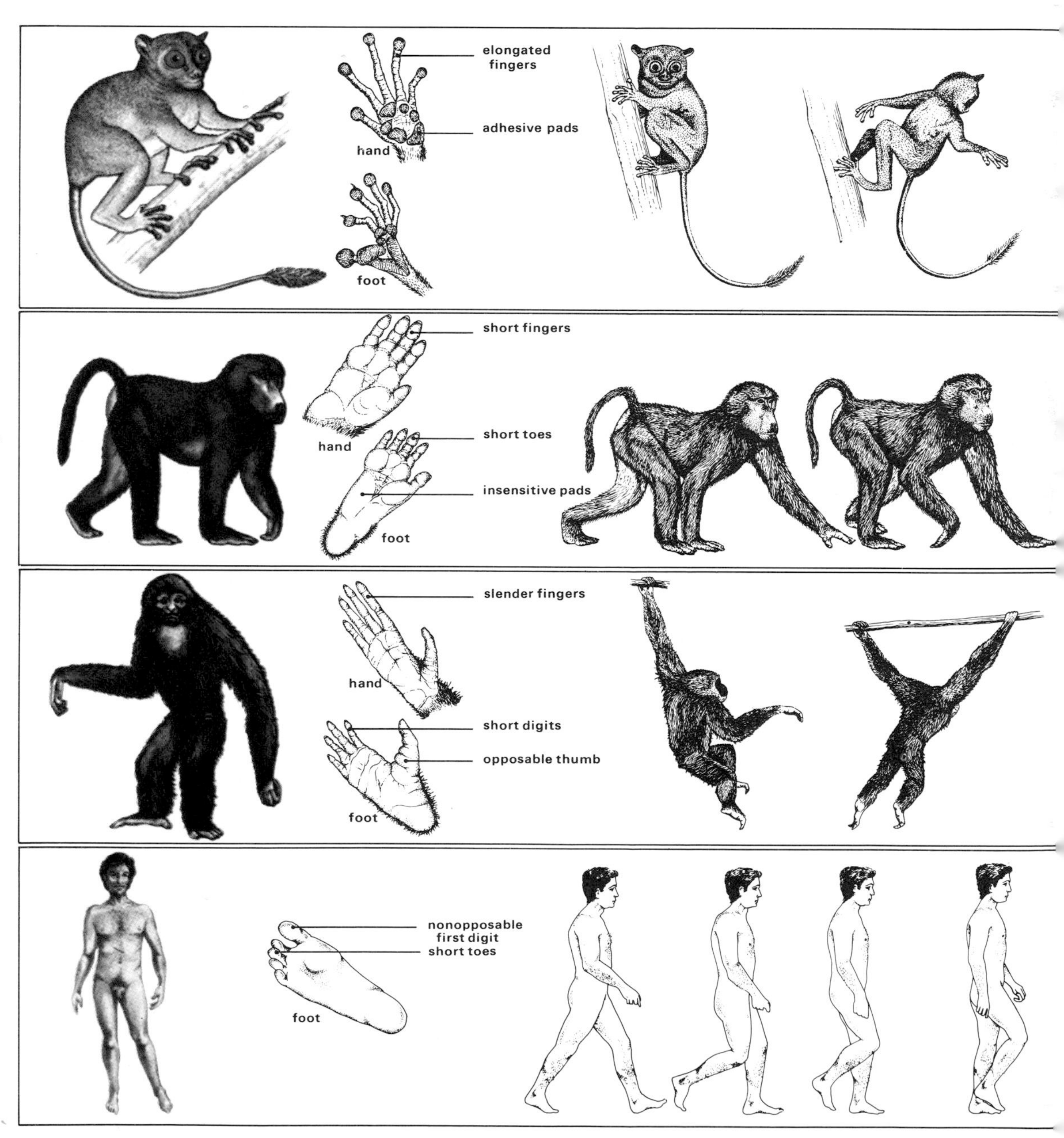

Vertical clinging and leaping is a specialized form of primate locomotion. Prosimians, such as the tarsier, move by suddenly extending their long hind limbs and leaping between vertical supports, usually branches or small tree trunks, and land feet first. The forelimbs play little part in movement, but are used to cling to the supports. The hand and foot are particularly adapted to vertical clinging and leaping. The fingers and toes are elongated. The palms, soles and tips of the digits are covered with adhesive pads.

Quadrupedalism is a generalized form of movement involving all four limbs. Among primates it goes along with a diversity of locomotory patterns, including running and jumping, arm swinging and leaping and ground running. The chacma baboon, for example, is a ground-living animal which uses a quadrupedal gait. It has short fingers and toes, and on the undersides of its hands and feet are tough, relatively insensitive pads.

Brachiation is an arm-swing mode of locomotion which represents an adaptation to arboreal life. The arms are fully extended above the head, suspending the body, which is propelled by hand-over-hand progression. Gibbons and simians, shown here, have evolved a true brachiating pattern. Orangutans, gorillas and chimpanzees, however, use their hind limbs as well as their arms to provide some support for the body. The brachiators have forelimbs far longer than hind limbs. The hand has strong, slender fingers and a short thumb, while the foot has short digits, but a large opposable thumb.

Bipedalism, exclusive to man, is the most highly evolved primate locomotory pattern. The body is propelled by alternate movements of the hind limbs. The foot has a nonopposable first digit and very short toes. The forelimbs are less important in this form of locomotion, but their appreciable length reflects man's ancestry.

the prosimians rather than the monkeys or his closest primate cousins, the apes. The reason for this is that man has evolved a unique mode of locomotion—the bipedal, or two-legged, striding gait. Other primates, and especially such apes as the chimpanzee or gorilla, which sometimes walk bipedally, cannot stride. Human stride walking requires that the legs should be long, hence man's apparently anomalous place in the intermembral index series, that they should be close together, which causes human females' obstetric problems which do not bother apes, and that the foot should be specially adapted to take the forces brought to bear on it in the course of the stride. Ape feet remain prehensile, with the big toes, the "thumb" of the foot, diverging sharply from the other digits. In striding bipedalism, however, the foot must be adapted so that the heel can come into contact with the ground at the beginning of the stride, and the weight of the body can be transferred forward along the outside and over the arch to the ball of the foot and from there to the big toe, which must be in line with the direction of motion in order to provide the pushoff into the next stride. Accompanying these changes in the foot are alterations to the pelvis, which becomes notably more narrow and is provided with characteristic musculature, the buttocks, which act to tilt and to rotate the pelvis in the most efficient way for two-legged walking.

These morphological changes could not have taken place overnight. A considerable lapse of time is required for such adaptations, and man must have branched out from his ape cousins many millions of years ago. In the intervening period things were happening to him which made these adaptations of prime importance.

Other features of human anatomy, such as man's shoulder blades and his broad rib cage, resemble those of the semibrachiating apes. This suggests that once, perhaps ten to twenty million years ago, man shared a common origin with them. Biochemical examination of human and primate proteins confirms that in terms of his body chemistry, man is most like the chimpanzee, slightly less like the gorilla and much less like the monkey. If this is true, then it suggests that one of man's most outstanding characteristics—his unique mode of locomotion—originated in the semibrachiation of the ancestors whom he probably shared with today's chimpanzees and gorillas.

Had the ancestors of modern man descended to a wholly terrestrial existence when they were quadrupeds like the baboons, they would probably never have evolved upright walking. Similarly, if they had first evolved thorough-going brachiation like the gibbon they might have missed the chance to succeed on the ground at all. It is most probable that man has evolved from an ancestor closely related to that of the modern chimpanzee, which had already begun to develop semibrachiation and the semiupright posture which that involves, as well as the characteristic rib cage and shoulders which man still retains. Man's ancestors then became wholly committed to a terrestrial existence and evolved from relatively inefficient knuckle walking to the highly efficient striding gait typical of man today.

It is probable, therefore, that while the smaller primates were colonizing the outer fringes of the trees and shrubs, and the larger primates, the monkeys and smaller apes, were adapting to a life in the trees, man was leaving the arboreal environment altogether and beginning to look for a life first on the ground between the trees and later completely away from them.

This was to be an existence in which he was to put his semibrachiating primate ancestry to good effect. The excellent sense of balance and advanced motor control of his limbs, which he retained from his arboreal predecessors, were preadaptive for the difficult art of walking balanced on two legs. The reliance on sight rather than on smell, which he shared with all the higher primates, stood him in good stead in an environment where his head was held high and in an ideal posture for putting his good primate eyesight to work for him in new surroundings. Finally, his prehensile forelimbs, now independent of any involvement in locomotion, could be devoted to other purposes, especially since they featured the generalized structure and refined manipulative capacity of all primate hands.

150
85
70
50
bipedalism
brachiation
quadrupedalism
vertical clinging and leaping

The intermembral index compares relative arm and leg lengths of primates. In the evolution of the larger primates, the arms become progressively longer as the legs become shorter, and the index therefore increases. The index is arrived at by dividing arm length by leg length and multiplying by 100. The index shows an increase from 50 to 150 as primate evolution advances. Following this trend, man's index should be more than 150, the index for the most advanced apes. Man, however, does not have longer arms than legs so his intermembral index is about 70.

Cross-reference	pages
Balance	**98-99**
Human skeleton	**32-33**
Motor control	**100-101**

The evolving intellect

However unique man's mode of walking may be, his feet and pelvis are not his most outstanding evolutionary specialization. His brain and its attendant behavioral specialization, intelligence, is what really makes man different from the other primates. Yet the development of high intelligence, which today is still the foundation of human culture and evolutionary success, did have something to do with bipedalism.

All the higher primates are notably intelligent when compared with most other mammals. This high intelligence is directly linked to the evolution of their large brains, associated with an adaptation to an arboreal existence, and the good vision, excellent motor control and other higher mental functions which that way of life demanded. The reduced reliance of the primates on the muzzle and on the sense of smell and their increased use of the limbs, especially the hands, to explore and to manipulate the environment, put an evolutionary premium on superior perceptual and manipulative ability. Once man became completely bipedal, this long-term trend in primate evolution came to fruition because his hands were now fully emancipated from participation in locomotion, his eyes were held well above the ground and he was now living in an environment which gave him sufficient scope to use his wits.

But for this to be possible, man had to be able to use his hands. Had his apelike ancestors first evolved by thoroughgoing brachiation, evolution would have bequeathed man a hand largely useless for the things to which he was subsequently to put it. A comparison of primate hands shows that those of such brachiators as the gibbon or orangutan feature long, hooked fingers with greatly reduced thumbs. Such hands are well adapted for brachiation, but they are not well suited for picking up objects, let alone for holding them. In contrast to these species, man has acquired relatively short fingers and a long, fully opposable thumb.

Opposability is the capacity to bring the thumb across the palm of the hand to meet the tips of the fingers. It is necessary for the manipulations upon which man relies for survival, and in particular for the precision grip and the power grip. Both of these grips are essential for the manufacture of stone or wooden tools which, typically, is done by holding the tool being made with a power grip in one hand, and the instrument used to fashion it in a precision grip in the other.

The earliest-known tools associated with man's ancestors have been linked with Australopithecus in Africa, who is believed to have emerged from more primitive and apelike antecedents about four to five million years ago. Australopithecus was a biped—probably not a fully evolved striding biped, but a biped all the same—and had a brain which was larger than his predecessors. He lived in savanna, grassy plains with scattered trees in tropical and subtropical regions, and he represents a point at which man's ancestors first became fully committed to a terrestrial existence, away from the trees in which they had originated.

Australopithecus was almost certainly a seedeater and may have lived a life comparable to today's gelada baboon, but unlike the gelada baboon he was able to make simple tools. This was doubtless because, unlike the baboons, Australopithecus was bipedal and so had hands free and available for tool using, and also because his brain was very much bigger.

The most common tools found which are associated with Australopithecus remains are

Australopithecus

Australopithecus was probably man's first fully terrestrial ancestor. Four to five million years ago he roamed the open, grassy savanna in small groups, living mostly on a diet of seeds. He fashioned crude pebble tools possibly as defense against such carnivores as the saber-toothed cat, for he was only about four feet tall and weighed less than one hundred pounds.

savanna • fossil sites

Man's early ancestor, Australopithecus, lived on open, grassy savanna. Fossil remains, mostly teeth and jaws, have been found primarily in the African rift valley, but others, found in such places as India and Java, indicate that Australopithecus had a widespread distribution.

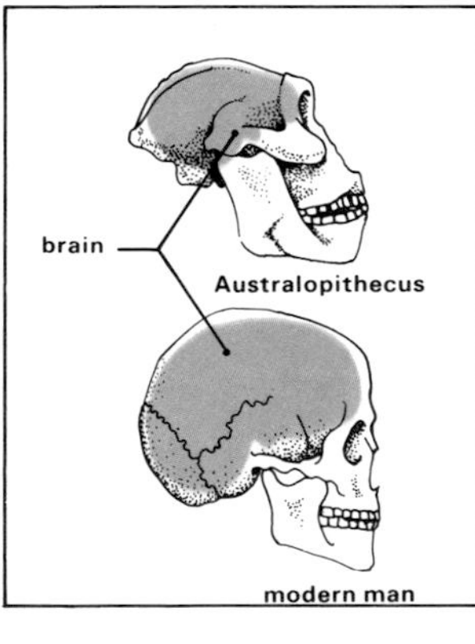

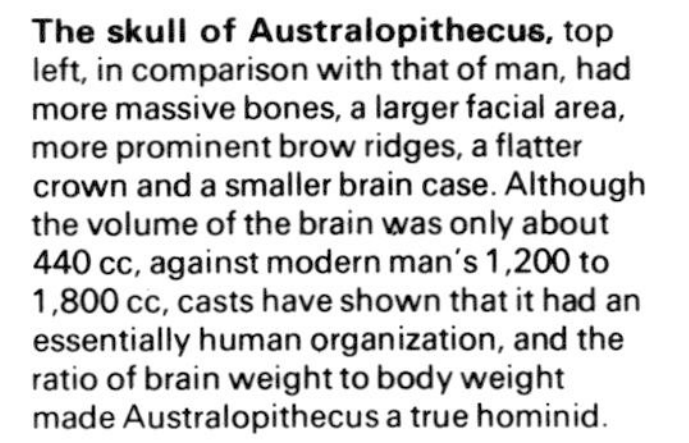

The skull of Australopithecus, top left, in comparison with that of man, had more massive bones, a larger facial area, more prominent brow ridges, a flatter crown and a smaller brain case. Although the volume of the brain was only about 440 cc, against modern man's 1,200 to 1,800 cc, casts have shown that it had an essentially human organization, and the ratio of brain weight to body weight made Australopithecus a true hominid.

The earliest and simplest tools used by man, more than two million years ago, were crude pebble choppers with irregular cutting edges.

so-called pebble tools. These are small, rounded stones which have been given an edge by chipping away flakes of stone on both sides. Most of them are about the size of a tennis ball. There is evidence that tools were also constructed from bones, pieces of wood and animal teeth.

Man's ancestors, as they wandered out onto the plains in the shape of Australopithecus, were puny, weak, four-foot-high primates who were not even equipped with the fanglike canine teeth of the baboons or chimpanzees. Lacking any means of natural self-defense, and living in an environment where predators were common, obviously there would have been a powerful stimulus for making up what was lacking in nature by the use of tools and weapons. First, perhaps, these were used only as a means of protection from other animals, but later they became a means of attack and a technological way to adapt nature to human uses. Therefore, what began simply as the throwing of sticks and stones could well have developed in time into the primitive tool industries which are associated with the remains of Australopithecus.

It is also clear, of course, that this could not have happened had not a progressively greater intelligence also evolved. Free hands and a need for weapons of attack and defense alone are not enough to produce material culture as we know it. An ability to put one's hands to good use is obviously of the first importance, and it was in this respect that the human brain was to make its great contribution to the early adaptive success of man.

A comparison of human and animal brains shows a steady progression in brain size in relation to body weight from the lowest animals to the reptiles, to the mammals and, finally, to the primates and to man. But not only is there an increase in size, there is also an even greater expansion in the areas of the cerebral cortex, the covering of the brain, and changes in the relative sizes of its parts. Compared to the brain of man's close primate cousins, the chimpanzee or the gorilla, the human brain shows greatest neurological development principally in the frontal, the temporal and the parietal lobes.

The frontal lobes are the part of the brain which lies immediately behind the forehead. Experiments with other primates and the results of surgery on this part of the human brain suggest that the frontal lobes are involved in the inhibition of instinctive behavior. This is clearly associated with intelligence, since the distinguishing characteristic of intelligent behavior is that it is acquired rather than innate and frequently demands the suppression of instinctual behavior. The "natural" response to a carnivore, for example, is to run away, not to throw stones at it.

The temporal lobes, which lie behind the temples on each side of the brain, are involved, among other functions, in information storage and retrieval. Obviously, memory is an essential ingredient in any response which, like intelligent behavior, needs to be learned. The parietal lobes, which lie toward the top of the brain, contain parts of the cerebral cortex which specialize in the linking up and synthesis of ideas and information, another important aspect of intelligence.

The notable advance which hominids like Australopithecus appear to have achieved in the story of man's evolution was, therefore, due not so much to one factor alone, but to the successful interplay resulting from man's acquisition of bipedalism, generalized hands and an advanced brain.

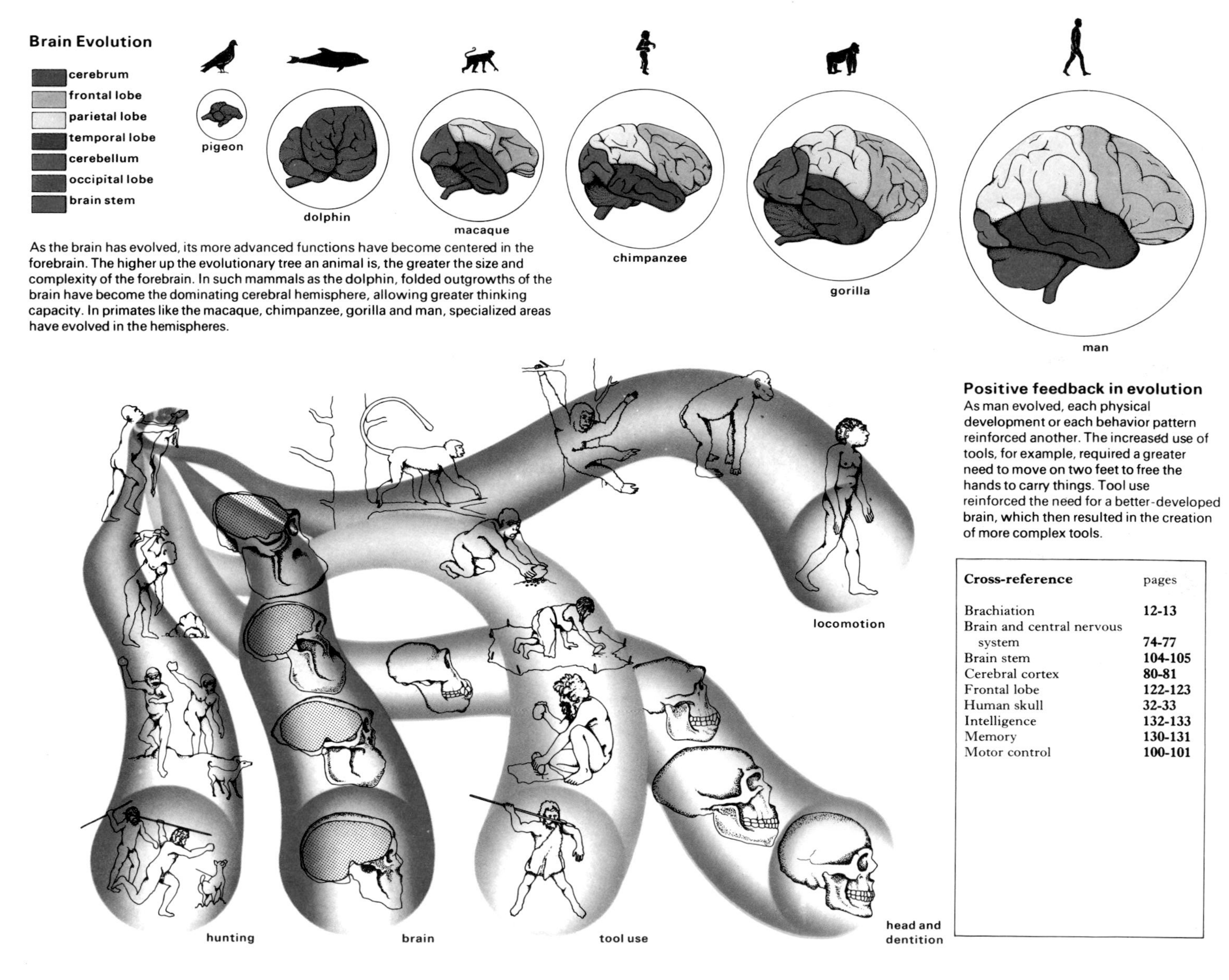

Brain Evolution

As the brain has evolved, its more advanced functions have become centered in the forebrain. The higher up the evolutionary tree an animal is, the greater the size and complexity of the forebrain. In such mammals as the dolphin, folded outgrowths of the brain have become the dominating cerebral hemisphere, allowing greater thinking capacity. In primates like the macaque, chimpanzee, gorilla and man, specialized areas have evolved in the hemispheres.

Positive feedback in evolution

As man evolved, each physical development or each behavior pattern reinforced another. The increased use of tools, for example, required a greater need to move on two feet to free the hands to carry things. Tool use reinforced the need for a better-developed brain, which then resulted in the creation of more complex tools.

Cross-reference	pages
Brachiation	12-13
Brain and central nervous system	74-77
Brain stem	104-105
Cerebral cortex	80-81
Frontal lobe	122-123
Human skull	32-33
Intelligence	132-133
Memory	130-131
Motor control	100-101

The ape who never grew up

Although the complexity of the human brain is certainly the principal factor in explaining man's acquisition of high intelligence and the capacity to produce culture, it is not the only factor. Of almost equal importance is the retardation of human maturity and the lengthening of childhood. This has come about as the result of the need to provide a long period of learning during which the immature human being can put that developed brain to good use in acquiring the skills and information necessary to an otherwise defenseless creature which relies almost entirely on its wits for survival.

Although the adult human brain is much larger than the adult chimpanzee brain, it is a fact of nature that all young mammals have large brains relative to their bodies. The brain of a newly born chimpanzee and the brain of an adult human being have about equal brain-body weight ratios. And in the infant chimpanzee brain, as in that of the adult human, the frontal, temporal and parietal lobes are very well developed. The only difference between the ape brain and the human brain is that while the human brain retains this immature form, this is soon lost in the ape and the rapid growth of the body after birth quickly results in an adverse brain-body weight ratio and the loss of the preponderance of the frontal, temporal and parietal lobes.

But the similarities between man and the immature ape do not end there. A comparison of an adult man with a fetal ape, such as a chimpanzee at three-quarters of the way through gestation, shows that there are many notable similarities. Both have large, globular heads with thin skulls and flat faces. Both are "naked" in that they lack body hair, and where human beings do acquire it is in the very places where it first appears in the ape fetus—on the head and the chin and around the lips.

The nonopposability of the human big toe is common to all primate fetuses, where, of course, it gives way to full opposability in adult monkeys and apes. In the case of the teeth, those of man resemble the milk teeth of apes more than the apes' milk teeth resemble their permanent dentition.

All these features suggest that human evolution has been dominated by an evolutionary process known as neoteny. Neoteny, or "remaining young," is not peculiar to man, but is found in other species, including the ostrich and a number of lower animals. Essentially, it means that man has evolved by retaining the fetal and immature characteristics of his ancestors. The probable principal reason for this is the paramount need for man to retain the large brain and period of educability found in immature and newborn primates. For not only does man retain the physiological characteristics of the fetal ape, he also retains the behavioral characteristics which go with them. Apart from the capacity to learn, these characteristics include a curiosity about the world and a playfulness and plasticity of response which is common to practically all young mammals, but is usually rapidly lost with the onset of maturity. Man "remains young" in this respect, too, since behaviorally he never loses the capacity to play and to be curious about things and to modify his reaction to them. Thus, neoteny has laid some of the most important foundations for human cultural achievement.

Advanced hominids

Seven hundred thousand years after the emergence of Homo erectus, the first true man, Neanderthal man appeared. He, in turn, was replaced by Cro-magnon man. Both of these advanced hominids had well-developed cultures, which included the use of sophisticated bone and stone tools and weapons. They both hunted for food and used animal skins for clothing.

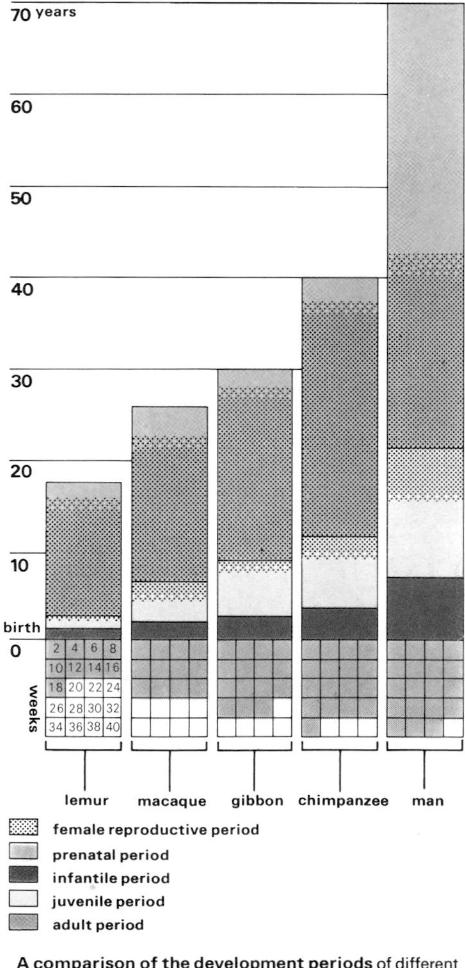

A comparison of the development periods of different primates shows that the higher up the evolutionary scale the animal is, the longer the development periods become. The juvenile period, for example, which is up to eighteen months in prosimians and between two and three years in monkeys, can last up to fourteen years in humans. As a result of a longer infant and juvenile period, the young have a greater exposure to adult behavior and cultural activities, which they begin to imitate. Their behavior, therefore, becomes more learned than instinctive.

However, neoteny also exacted its price. It required that human beings be born earlier—in their development if not in time—and remain helpless for far longer than any other primate. This naturally put, and still puts, a much greater burden on the human mother, who is forced to carry helpless babies for as long as two years before they can walk steadily on their own. She must also maintain care of children until the onset of maturity at puberty. By contrast, the only other savanna-living primate, the baboon, gives birth to young which can cling to the mother from the start and which can learn to walk and to look after themselves much more quickly.

Most of the specialized behavior and bodily adaptations which animals possess develop faster after birth than before. Indeed, at the embryonic stage of development the obvious close similarity between the embryos of different species results from this fact. But neoteny—which "keeps man young" in that it causes him to retain the fetal and immature characteristics of his ancestors—also inhibits him from developing the specializations of their maturity.

One of these specializations was almost certainly some form of semibrachiation. Neoteny, which has presumably been acting on man since he parted evolutionary company with the ancestors of today's apes, has been progressively despecializing him. Consequently, he has almost totally lost the bodily adaptations concerned with semibrachiation. Instead he has developed generalized limbs. But, most important as far as the dependence of the young is concerned, he has developed a generalized foot with a fetal appearance which was ideally suited for the one specialization which man did acquire—bipedalism.

Neoteny, therefore, which put the baby in the mother's arms, also freed those arms to carry the baby by enabling humans to develop a foot suited to bipedalism. In a habitat where humans lived a nomadic life this must have been of prime importance. Indeed, this interrelation between bipedalism and the helplessness of the young may well have been another factor in the complex interplay of bipedalism, intelligence and tool using, for that infantile helplessness and the capacity of the adult to transport and to care for the young were themselves aspects of the development of intelligence and culture.

The fossil record of human evolution shows a progressive despecialization of man's ancestors and the gradual acquisition of bipedalism, tool using and high intelligence. Ramapithecus, the earliest-known hominid, or forerunner of man, was closer to the ancestors he shared with today's apes than were his successors, and he was less "fetalized" than his successors were to become. Ramapithecus lived ten to twelve million years ago, probably on the fringes of the forests and he was at least in part arboreal. He may have been a knuckle walker like today's chimpanzees. Australopithecus appeared about four to five million years ago and seems to have been a savanna-dwelling biped with a larger brain, a flatter face and a generally more fetalized appearance.

Finally, the first true man, Homo erectus of about eight hundred thousand years ago, shows a further increase in brain size, tool use and cultural adaptation. He lived in what is now China, in caves, used fire, hunted big game and practiced cannibalism.

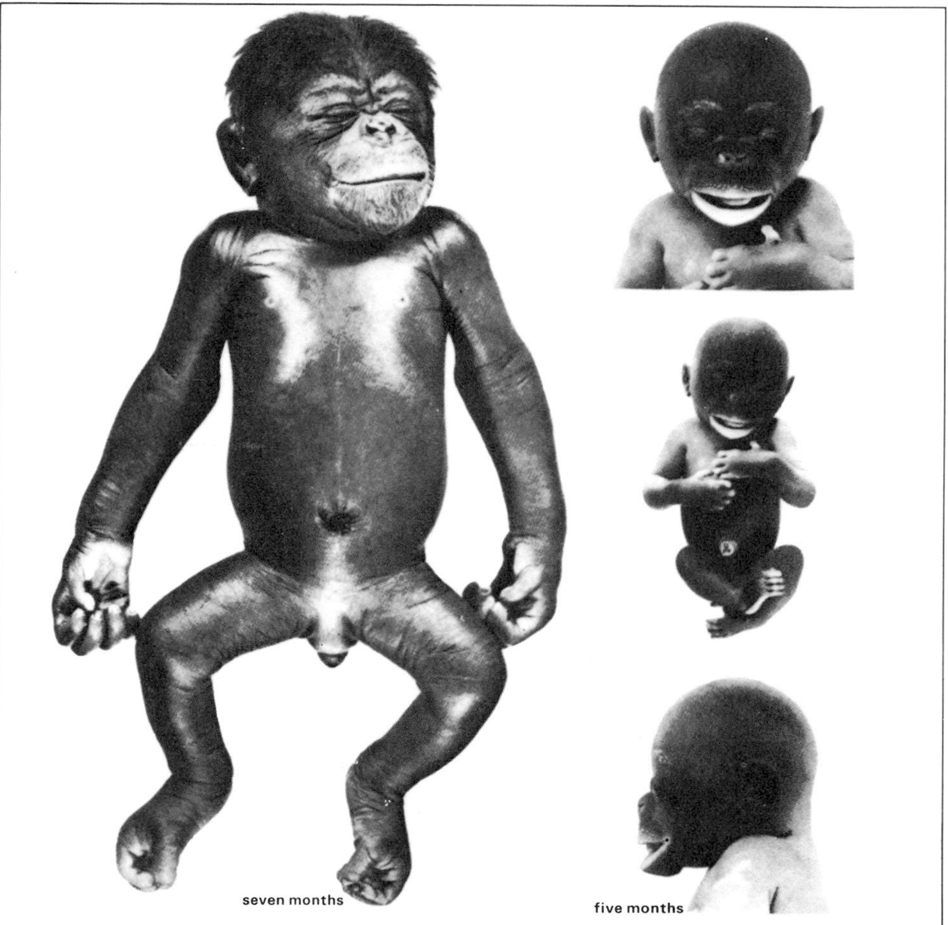

The chimpanzee fetus at five months and seven months is similar to a human fetus. After birth, however, humans retain such fetal characteristics as large brain weight to body size and the chimpanzee loses them as it becomes adult.

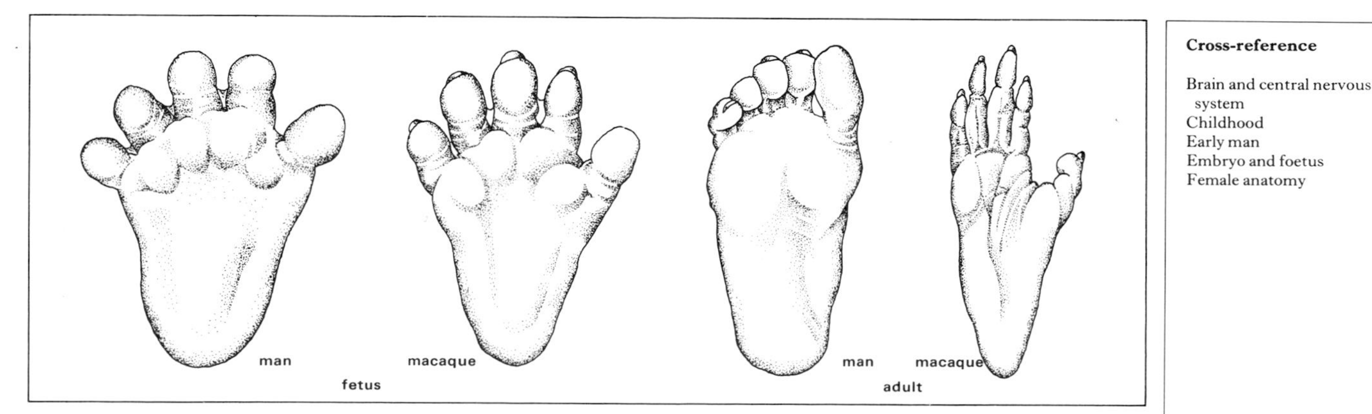

Evolution has progressively despecialized man and given him general limbs which are more suited to a terrestrial rather than to an arboreal existence. The feet of a human and a macaque monkey at correspondingly early fetal stages of development are similar. The human adult foot, however, keeps a fetal appearance and retains the short toes. In contrast, the adult macaque develops feet with four long toes, and a short opposable digit, which allows a firm grip on branches that is required for a life spent in the trees.

Cross-reference	pages
Brain and central nervous system	**74-75**
Childhood	**162-171**
Early man	**14-15**
Embryo and foetus	**152-155**
Female anatomy	**146-147**

The social animal

When the present condition of man is compared with the lower forms of life from which he originated, it is obvious that as the evolutionary ladder is ascended, leading from the earliest protoplasm to the amphibians and reptiles and on up to the mammals, the primates and man, a ladder of social evolution is also ascended. This is because an animal's social behavior is, like its bodily structure, subject to the forces of natural selection and functions principally as an adaptive mechanism.

Primate species—the monkeys, apes and man in particular—are, by and large, highly social in behavior. Part of the reason for this is that, like all mammals, primates suckle their young and need to look after them, sometimes, as is the case in man, for a considerable period of time. This creates what is perhaps the most basic of all social links—the relationship between mother and child. Among the higher primates the link is of special importance because of the length of childhood and the importance of learned behavior.

The second main social link is that which is created between a male and a female for the purposes of procreation. This clearly falls under the heading of sexual behavior. Finally, there are the nonmaternal, nonsexual links which animals establish between themselves and which usually can be seen as an adaptive response to available habitat and food supply.

In the evolution of sexual behavior, as in so many other respects, man has continued an important trend in primate evolution. In sexual behavior he has developed a greater reliance on sight rather than on smell and, because of this, on visual signals. The arboreal habitat of the original primates made complete reliance on scent as a means of attracting a mate no longer possible, because of distance and the wind blowing through the trees. Instead, the primates evolved elaborate visual stimuli, usually displayed by the female, in order to attract the attention of the male.

Like most mammals, the females of all the primates except humans have an estrous cycle. This is a regular and usually nonseasonal cycle of sexual receptivity, at the height of which the female ovulates and shows signs of greatly increased sexual excitability. Among many monkey species it is signaled by a change in the color and appearance of the sexual skin—the area surrounding the genitals of the female—and by the behavior of the female in question, which goes around soliciting copulation from males.

In humans, this trend has been continued, except that in the female the estrous cycle has become the menstrual cycle because the great development of the cerebral cortex has made human females largely unresponsive to changes in their hormone levels. These hormone levels, however, are broadly comparable to those of the female ape or monkey and still serve to trigger ovulation at about the middle of the cycle. But human females exhibit a permanent state of sexual receptivity which is even more dependent upon visual and social cues than is the sexual behavior of other primates.

Consequently, because human females show no external signs of readiness to copulate comparable to the sexual swelling of the female chimpanzee or the bright red sexual skin of some baboon species, there is a great emphasis on subtle signaling, which depends upon facial expression, at least at the initial stage of human sexual activity. Verbal and various socially determined sexual signals such as dancing together, or holding hands, also play an important part at first. It is not surprising, therefore, that humans have evolved eyebrows, prominent red lips and a luxuriant growth of head hair—and, in the case of the male, a beard—as facial adornments which greatly assist such sexual signaling. In providing women with lipstick and eye makeup, cosmetics manufacturers are only following a clear lead given by evolution.

Sexual adaptations also include the more curvaceous body outline, pubic hair and the development of prominent breasts. These are

Hominid evolution

From Ramapithecus, the forerunner of man, it has taken ten to twelve million years for the evolution of modern man. Four to five million years ago, Australopithecus was relatively intelligent, bipedal and used tools. The first true man, Homo erectus, used fire eight hundred thousand years ago and had a relatively well-developed culture. Neanderthal man and Cro-Magnon man were evolutionary members of the species Homo sapiens, today the most successful animal group on earth.

also secondary sexual characteristics evolved by a species habituated to standing upright and meeting face to face. This may also account for another peculiarity of human male sexual physiology—the large size of the penis. Given the permanent receptivity of females, man's penis, which in its erect state is far larger both relatively and absolutely than that of any other primate, may well have evolved as a means of signaling the male's sexual receptivity. This may also explain why among humans the man pursues the woman rather than the other way around, which is the case with the apes and monkeys.

In humans, maternal and sexual behavior are not entirely divorced, since human courtship, once it reaches the stage of bodily contact, proceeds to fulfill the function of all such behavior—that of breaking down the barriers between the individuals and facilitating their intimacy, by reproducing behavior otherwise only seen in the context of the mother–child relationship. Kissing is undoubtedly derived from sucking, and fondling, cuddling, holding hands, uttering affectionate sentiments and exchanging long, languid looks are all typical of nursing behavior.

The question of man's evolution of other social behavior is much less simple. He does, however, show some features of social behavior which resemble those of other primates. One of the most important of these and something which is seen in nearly all primates, man included, is dominance behavior. Dominance is the name given to the phenomenon of status ranking in animal societies. High-status animals take precedence over those with low status and usually get preferential access to food, to females and to rights of decision making. Rather like its human counterpart, the high-status chimpanzee can be recognized by the relaxed indifference with which it moves about and the way in which it can usually get whatever it wants. The low-status chimpanzee is, by contrast, nervous, everwatchful of what the dominants are doing and generally servile in its relations with them. Nevertheless, most primate societies, and especially that of the chimpanzee, are highly variable and often quite amorphous, since the membership of the groups is constantly changing.

If it is assumed that man's distant forest-dwelling, semibrachiating ancestors, who were so closely related to the chimpanzees, manifested a similar social behavior in conditions comparable to those of today's chimpanzees, then it can be concluded that at this stage man's ancestors (perhaps Ramapithecus) lived in loosely structured societies where, as among the chimpanzees, sexual promiscuity was the rule and dominance, although present, was not strongly developed.

As man's ancestors moved out onto the savannas, it is likely that they became organized into societies similar to those of today's baboons which, like the early hominids (for example Australopithecus), are terrestrial. Because they live in a dangerous environment, they show a more rigid social structure and strongly developed dominance hierarchies. It is likely that ultimately man's ancestors may have formed societies such as those of today's gelada baboon, which man closely resembles in the matter of his teeth (which suggests that we have become adapted for a similar seed-eating way of life) and in the fact that the gelada female, too, has secondary sexual characteristics on the chest.

The gelada baboons live in rigidly demarcated groups centered around dominant males and consisting of one such male and his harem of females and immature young. Subdominant males live together in looser, all-male groups. It is possible that such "bachelor groups" of subdominant males began the first systematic big game hunting, because what is known of other hunting animals suggests that a quite different social structure to that of the gelada one-male group is required and that something like the gelada bachelor groups might be just right. If this is the case, then the appearance of Homo erectus as a big game hunter some eight hundred thousand years ago may signal the changeover from seed-eating, gelada-type structures to fraternal hunting bands of the type which we can still find among the Bushmen and Australian aborigines.

Whatever the final truth about man's social origins, there can be no doubt that such behavioral adaptations as have led to the development of structured societies and advanced material culture have been dependent upon the mental and physical adaptations which have given man the highly developed body and mind that he possesses today.

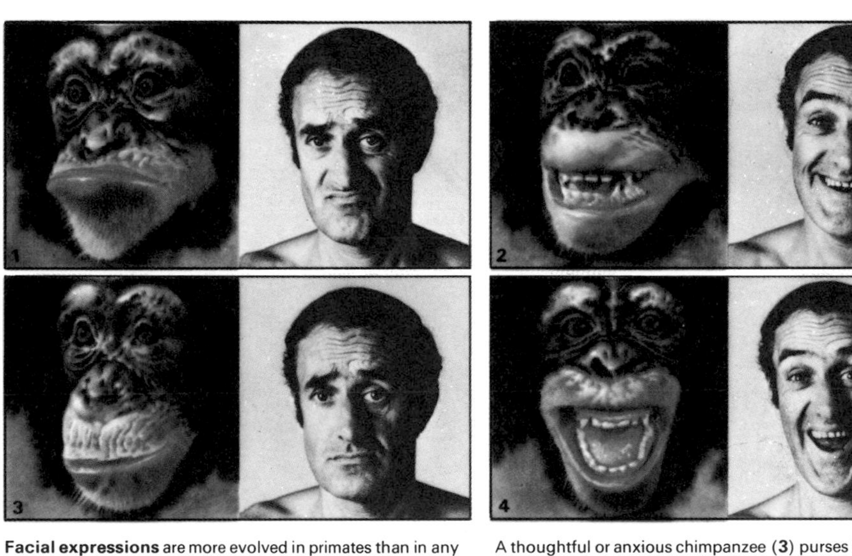

Facial expressions are more evolved in primates than in any other animals. Humans and chimpanzees have a range of expressions that are very similar. A chimpanzee's pout of curiosity (**1**) might be skepticism, or doubt, in a human. A smile (**2**) expresses pleasure in both chimpanzee and man. A thoughtful or anxious chimpanzee (**3**) purses its lips in an expression very much like that of human concentration. There is one disparity however. What appears to be outright laughter (**4**) in a chimpanzee is in fact an expression of annoyance or unhappiness.

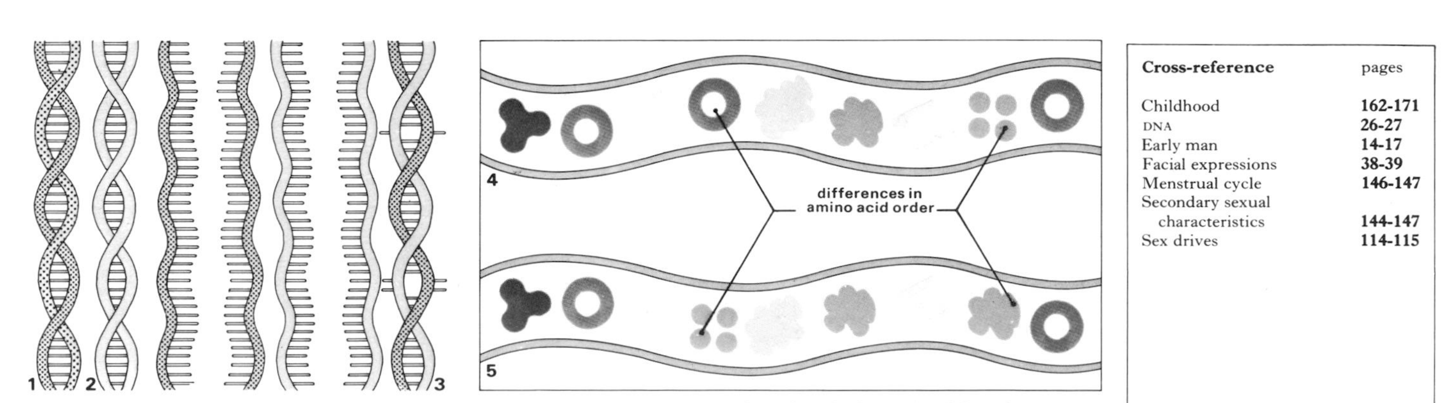

Evolutionary relationships between animal species can be determined by studying their genetic material, deoxyribose nucleic acid (DNA). By breaking chemical bonds between strands of gorilla DNA (**1**), as well as human DNA (**2**), the double helix unwinds. If single strands from each helix are combined they will connect (**3**), except where the links are chemically different. The more links formed the closer the species. The structure of the individual's proteins can also be compared, for the more alike the order of amino acids the closer the animals are related. Part of the blood protein hemoglobin from a man (**4**) and a gorilla (**5**), using symbols for six of the amino acids, shows a difference of only two amino acids and, therefore, a close relationship between the two species.

Cross-reference	pages
Childhood	162-171
DNA	26-27
Early man	14-17
Facial expressions	38-39
Menstrual cycle	146-147
Secondary sexual characteristics	144-147
Sex drives	114-115

Framework of the body

The amazing scheme of the human framework is to give support and protection while at the same time allowing great versatility of movement. The key to this versatility is the natural harmony and balance between the many and varied components of the framework and the ingenious way in which they are arranged.

The human skeleton is a structure which is flexible and yielding, yet firm enough to maintain the body in its characteristic upright position. The rigid elements are the more than two hundred separate bones of which it is made up. Bones protect soft, vulnerable internal parts, particularly the brain and the spinal cord, and act as mediators of movement through a variety of intricate joints and the attachment of contractile muscles. Other types of muscle form an important part of such organs as the heart and the digestive tract.

Other essential parts of the framework contribute to its vital plasticity. Connective tissue binds, packs and supports the internal organs. As cartilage it provides tough, friction-free surfaces at joints. As tendons and ligaments it creates links between muscles and bones, and as adipose tissue it forms an insulating and storage layer beneath the skin. The skin itself, the body's largest organ, is the tough, elastic outer fabric, resistant to water and the agents of disease.

The constituents of the body's framework, like every other tissue and organ, are composed of smaller units, the cells, the lowest denominators of all living matter. It is the specialization of cells—the process by which each one develops in a way suited to a particular task—which makes possible the whole range of body functions.

The rigid elements of the body's framework are the bones, which are composed of cylindrical units. In the illustration, left, the bone has been cut across, photographed in polarized light and magnified more than two hundred and fifty times, so that the active bone-forming cells are seen as dark, circular areas.

Cells: structure

Every living thing, whether plant or animal, is built of cells. Cells are the very units of all life as we know it, and it seems certain that the first creatures on our planet were primitive, single-celled organisms, probably living in the sea. Single-celled living things, right at the bottom of the scale of evolution, still exist today. Bacteria, everywhere around and even within us, consist of single cells. And there are many other microscopic creatures, so simple and unspecialized that it is a problem to decide whether to call them animals or plants.

Higher up the evolutionary scale are living things that consist of cells grouped together and working together, but each with a slightly different function. Some of the cells may be specialized to take in food, for example, and others to digest it. Others may be combined into structures that give the power of movement. At the top of the ladder are creatures such as ourselves, composed of literally thousands of different types of cells, each with its own particular place and function in the body. And yet, as complex as our adult bodies are—constructed of several hundred million million cells—we all started life as just one cell, the parent of all cells, a fertilized egg in the womb.

The body might be looked upon as a vast family or community of cells. Individual cells rely on the whole body for their livelihood, and the body relies on the sum of their contributions for its continued existence. Just as the members of a family have fundamental similarities, so all cells have the same basic makeup and structure. However, some cells have such specialized duties to fulfill that it is difficult to believe that they are even distant relations, so modified have they become in structure. A neuron, or nerve cell, for example, perhaps several feet long, bears very little resemblance to a tiny red corpuscle which has lost its nucleus and is circulating in the blood.

Each cell is a microscopically small unit, surrounded by a delicate membrane, the cell wall, some sixty-millionths of an inch thick. The wall is made up of a double layer of molecules of fat, sandwiched between two layers of protein. Its surface is dotted with minute cavities and sometimes has fingerlike projections. These increase the surface area of the cell and so help the exchange of materials that is constantly occurring between the cell's contents and the surrounding tissue fluids. Oxygen passes into each cell through the cell wall, dissolved into the fluid that bathes every tissue, and molecules of food material are also selectively absorbed. Simultaneously, carbon di-

A generalized human cell

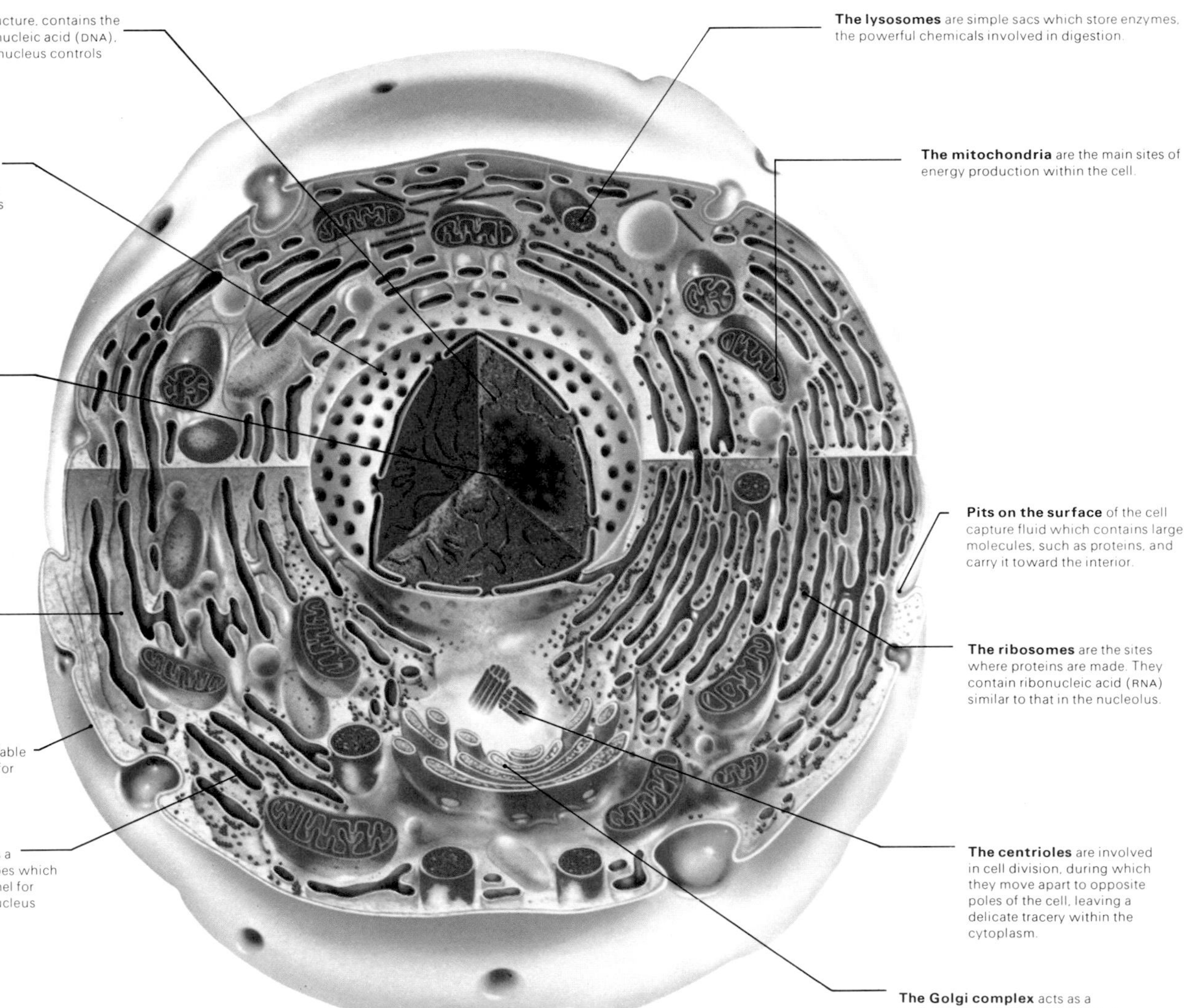

Cells are the microscopic building bricks of the body. Among the smallest human cells are some nerve cells, while the largest, the ovum, has a diameter of .005 of an inch—the size of a pinpoint. Each cell is a self-contained living unit, which becomes modified and specialized as it develops and is integrated into the body's systems. The cutaway view of a generalized cell—shown above magnified to six thousand times the average size—reveals the working parts concerned with systems of cell support, maintenance, repair and multiplication. These tiny structures are arranged in such a way that all parts of the cell are in close communication by way of a series of minute interconnecting channels running through the watery cytoplasm, from the nucleus to the cell membrane. Although this generalized cell is spherical, when actual functioning cells in tissues and organs are packed together in proximity, they take on a variety of shapes—muscle cells, for example, are elongated into fibers, liver cells are hexagonal and the cells which line the intestine are column shaped.

oxide and molecules of waste material, that are the result of the cell's activities, pass outward and are collected by the bloodstream for ultimate disposal by the excretory systems of the body.

Directing the activities of the cell is its nucleus. This is a densely packed globule made of the material called chromatin. Chromatin is the very material of heredity. When a cell divides, the chromatin becomes visible as the structures called chromosomes. Chromosomes carry the genes that are responsible for determining each and every characteristic of the living organism. Within the nucleus are found nucleoli. These are centers concerned with the formation of nucleic acids, which are necessary for the production of new protein material so that cells may grow and divide and repair themselves.

Surrounding the nucleus is cytoplasm, the material that forms the bulk of the cell. There is an extremely fine membrane separating the nucleus from the cytoplasm, but with numerous perforations that allow a continuous interchange of molecules between the two. Extending deep into the cytoplasm from the walls are irregular folds with, scattered on their outer surfaces, granular structures called ribosomes. The complicated folds make up the structure called the endoplasmic reticulum. Ribosomes are also found within the cytoplasm. They are made largely of the material called ribonucleic acid, and are the sites where proteins are built up from the molecules called amino acids.

Golgi bodies are gatherings of thin, flat, almost saucerlike structures within the cytoplasm. Their function was for long a puzzle, but it is now thought that they take up some of the protein that is synthesized by the nearby ribosomes, and attach carbohydrate units to it, forming the materials known as glycoproteins. Glycoproteins are found in mucus and in mucous membranes, and many Golgi bodies are found in the cells of mucus-secreting tissues. Microsomes are centers where new materials are accumulated in the neighborhood of the Golgi bodies, and forming part of them, ready for expulsion from the cytoplasm through the cell wall and into the surrounding fluid.

Under a high-powered microscope, it is also possible to make out structures that are sausage-shaped, and about one ten-thousandth of an inch in length, with an elaborately folded membrane within them. These are mitochondria, and are the sites where respiration takes place within the cell. All cells contain these mitochondria, but they are found in the greatest numbers in cells that are specially active, like muscle cells, and spermatozoa.

Lysosomes are small packets of powerful enzymes that bring about the breakdown of large molecules within the cell. From the units produced, other substances that are required, perhaps elsewhere in the cell, can then be built up. There are thought to be many thousands of different enzymes within the body, some in only certain types of tissue where their activity is most needed.

Also in the cytoplasm are minute globules of fat. These are found in the greatest concentration in cells that are specially adapted for fat storage, as in the adipose tissue that forms a layer under the skin and in the cells of the liver. In certain cells, granules of pigment, mainly the dark-colored pigment melanin, are found.

Where groups of cells that seem exactly similar in appearance and in function are gathered together, the result is called a tissue. Thus the lining layer of the digestive tract has cells that have essentially the same function, and, with variations from region to region, the same kind of appearance when seen under the microscope. This can be termed a tissue and given the general name mucous membrane. Similarly, bone cells together form a tissue, and the term could even be applied to the blood. From tissues, organs are built up, and the organs can be grouped together into the different systems that make up the body.

Epithelial cells are those that have the function of covering surfaces, and their structure varies according to their location in the body. The cells of the skin are found in layers that are more and more flattened the nearer the surface they are, forming a resistant covering which is constantly renewed from within. The cells that form the lining of the digestive tract make a soft and almost transparent epithelium, and some secrete mucus to lubricate the contents of the tracts.

Every minute, it is estimated, some three hundred million cells die in the body, most of them to be replaced immediately by the division of the cells that remain, so that the total number of cells in the adult body remains virtually constant throughout life.

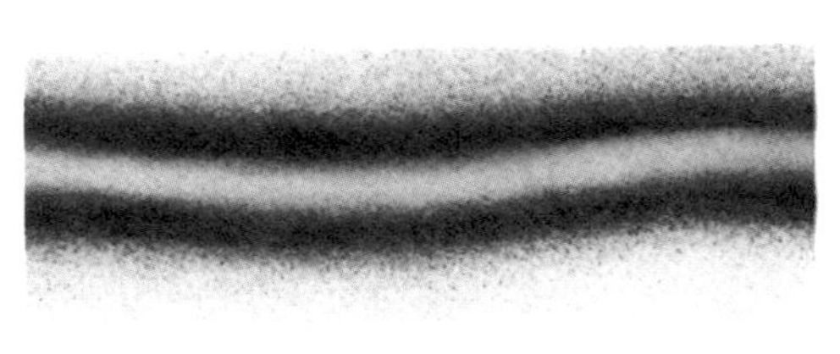

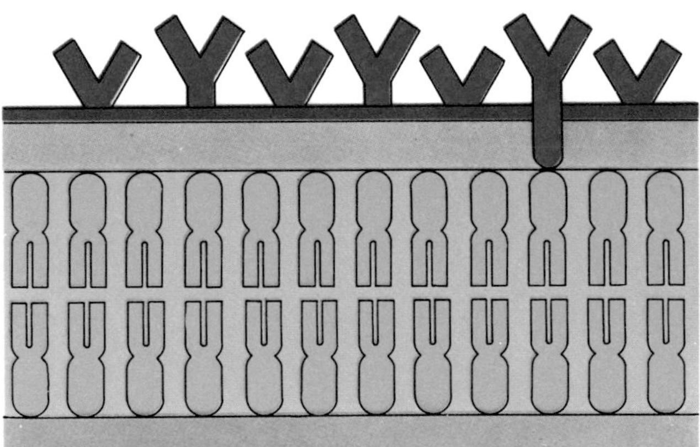

The cell membrane is a translucent pliable skin which controls the flow of materials into and out of the cell. The photograph, far left, is of the membrane as it appears under an electron microscope, magnified one million times. A highly complex unit, it is composed of a double layer of fat molecules sandwiched between two layers of protein molecules. The exact molecular structure of the cell membrane is still unknown, but the diagrammatic model, left, shows what is thought to be the relationship of carbohydrates (orange) with protein (yellow), water (light blue) and fat molecules (darker blue).

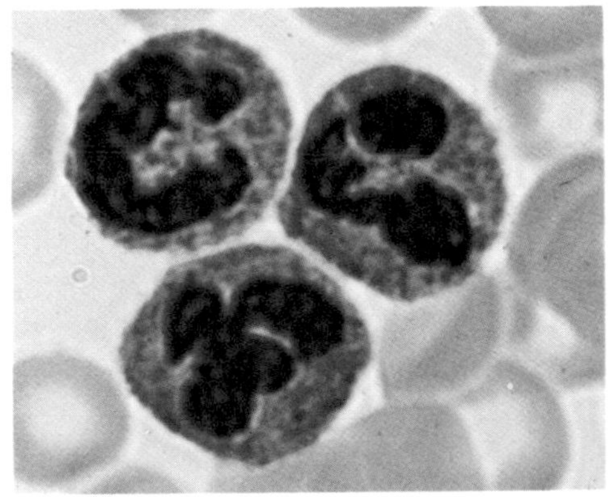

Polymorphs, cells with irregular nuclei, account for two-thirds of the white cells circulating in the blood. This microscopic view shows the translucent cells stained with a colored dye and magnified two and a half thousand times.

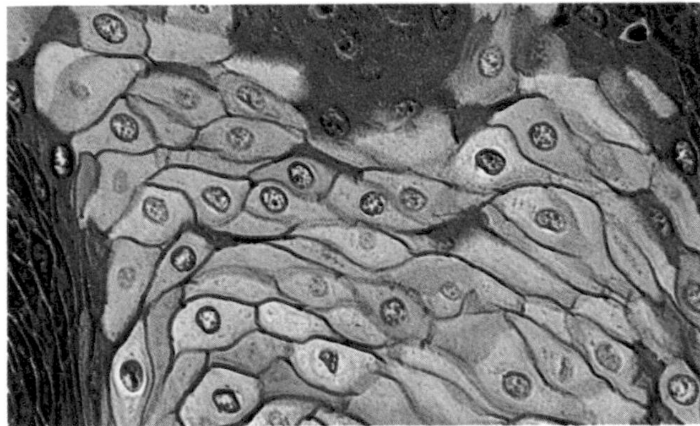

Epithelial cells, layered on the tongue, provide a tough, friction-resistant and virtually waterproof covering firmly attached to the tissue below. Here, the cells of the tongue, among the least specialized in the body, are seen under the microscope almost five hundred times their actual size.

Cross-reference

	pages
Chromosomes	26-27
DNA	26-27
Fertilized ovum	150-151
Functioning of cells	24-25
Heredity	158-159
Ovum	146-147
Sperm	144-145

Cells: function

Each of the countless millions of cells that together make up the human body has a life of its own. There are certain basic things that it must do simply to preserve that life. But at the same time every cell has a duty to the body as a whole, and no matter how specialized it may be in form or in function, it must cooperate with other cells if the whole body is to function smoothly. An analogy can be drawn between cells and the members of a well-organized community—each person, no matter what his occupation or trade, works not only for himself but also for the good of the community.

Life is based on a vast, continuous series of chemical changes that take place in and around our cells. The sum of all the chemical changes that take place in any living thing is called metabolism. Some of the changes lead to a breaking down of materials into smaller and simpler units, and these are called catabolic changes. Other chemical activities are directed to building up complex materials from simpler substances, and these are called anabolic changes. Anabolic and catabolic activities go hand in hand to release energy from the food that we take in, and to enable us to utilize the food materials for growth and repair of all the parts of our bodies.

To carry out its metabolic activities, each cell must obtain a constant supply of raw materials to work with. There must be a steady flow of organic materials—proteins, carbohydrates and fats—and of certain inorganic elements such as sodium, potassium and chloride. And as a medium in which all the chemical reactions can take place there must be water. These materials enter the cell through its surrounding skin, the cell membrane, which is highly selective in allowing substances to pass into and out of the cytoplasm. The substances that are taken in are different from cell to cell, and vary according to the cell's role in the body. Bone marrow cells take in iron, to incorporate in the hemoglobin of red blood corpuscles, and bone cells take in the calcium salts that go to make up bone.

Passing outward, to be carried elsewhere by the bloodstream, are the products of the cell's metabolism. Some of these may be enzymes, or hormones that the cell has manufactured, and some are merely waste products that cannot be allowed to accumulate. Some materials when they leave cells do not go into the blood. The protein keratin, for example, is built into hair and nails, and digestive enzymes are secreted directly into the alimentary canal.

Each cell can be regarded as a factory, equipped with and using appropriate machinery to process a particular range of raw materials. Certain of the materials will be used in the manufacture of products destined for distribution to the rest of the body, some will be used as fuel to keep the machinery moving and some will be used to repair and replace parts of the machinery itself.

Much of this so-called machinery is to be found in the minute structures known as mitochondria, scattered in the cytoplasm, and in the folded membrane called the endoplasmic reticulum, which is in places continuous with the outer cell membrane and in contact with the membrane that encloses the nucleus. The intricately folded reticulum has a large surface area on which all the molecules that take part in chemical reactions are arranged.

Maintaining chemical equilibrium in the cell

To function properly the cell must maintain a stable internal environment; therefore transfer of materials has to be achieved without an excessive buildup of chemicals. When particular molecules are needed—for example, glucose—the cell will take these in and discard other materials to preserve equilibrium.

Passive transport
As chemicals become concentrated outside the cell, a flow of small molecules takes place through the membrane until a balance exists. This process, known as diffusion, is the basis of the intake of digestive products by cells lining the intestines.

Active transport is the process, using chemical energy, by which the cell takes in larger molecules that would otherwise be unable to enter in sufficient quantities. Carrier molecules within the cell membrane bind themselves to incoming molecules, rotate around them and release them into the cell. This is the means by which the cell absorbs glucose. Active transport may upset stability, but the cell compensates by shedding other matter.

Osmosis
The movement of water through the cell membrane from areas of low to high chemical concentration is called osmosis. This alternative process provides for dilution of chemicals—unable to cross the membrane by diffusion—to maintain equilibrium within the cell.

The cell as a factory

Working around the clock, the human cell is comparable to a small factory. Here is a production unit, complete with its own computer—the cell nucleus—power plant, packaging and delivery systems. Like a factory, the cell takes in a continuous flow of raw materials for processing and packages its own chemical wastes ready for disposal.

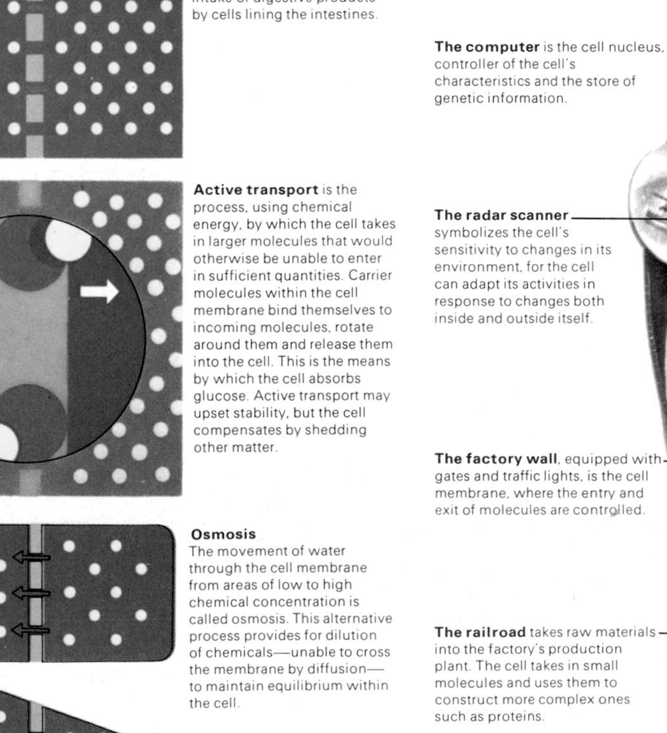

The production plant is the endoplasmic reticulum, where large molecules are built up from smaller ones.

The computer is the cell nucleus, controller of the cell's characteristics and the store of genetic information.

The radar scanner symbolizes the cell's sensitivity to changes in its environment, for the cell can adapt its activities in response to changes both inside and outside itself.

The factory wall, equipped with gates and traffic lights, is the cell membrane, where the entry and exit of molecules are controlled.

The power plant is the mitochondria, where essential energy is created by fuel breakdown. The fuel the cell uses comes from food.

The railroad takes raw materials into the factory's production plant. The cell takes in small molecules and uses them to construct more complex ones such as proteins.

One of the most important processes in the factory is the manufacture of protein. The raw materials for constructing large protein molecules are the units called amino acids. These are assembled together according to definite plans which are dictated by special controlling "messenger" molecules of the substance ribonucleic acid, or RNA. RNA molecules are produced within the nucleus of the cell, and pass into the cytoplasm. In the small, nearly spherical, bodies called ribosomes that are found encrusting the endoplasmic reticulum, the RNA molecules sort out the particular amino acids that are required to produce a specific protein and link them together. The proteins that the ribosomes have produced pass into the spaces in the reticulum and are there stored in enlarged vacuoles called cisternae. There are also ribosomes scattered through the cytoplasm. The materials that they produce are not stored in cisternae, but are used directly by the cell.

The proteins that are manufactured by the cell may be used for structure, for example to form a new endoplasmic reticulum. Or they may be enzymes which speed up the chemical reactions that are taking place, or they may be products, such as hormones, that the cell will "export" to other cells or to the rest of the body. Acting as the export-packing department of the factory is the Golgi apparatus. This receives material stored in the cisternae, and moves it, enclosed in membranous bags, to the wall of the cell, where the contents of the bags are released to the exterior.

The mitochondria in the cytoplasm are of fundamental importance. They contain the enzymes necessary for the complex process of cellular respiration, by which energy is released from the food material, mainly fatty acids and the simple sugar glucose, taken in by the cell. During respiration, oxygen is used up, and glucose and fatty acid molecules go through a series of chemical changes that results in the formation of an energy-rich substance called adenosine triphosphate (ATP). Carbon dioxide and water are produced as waste products. The important substance ATP, present throughout each cell, provides energy for all the biochemical processes that are taking place, including the manufacture of proteins.

Lysosomes in the cytoplasm are membranous bags containing enzymes that can break down large molecules into smaller ones. These must be kept separate from the rest of the cell's contents so that they cannot attack the cell itself. Food material that is taken into the cell is itself enclosed in a vacuole, and a vacuole fuses with a lysosome so that the breakdown of material can be carried out within the limits of the membranous bag. The smaller molecules which result can then diffuse through the membrane and be used wherever necessary in the cytoplasm.

In a commanding position within each cell is the nucleus, the "manager's office" of the factory. Here are stored the blueprints for the entire concern, so that others may be built to the same pattern when necessary, and minutely detailed instructions for the running of the factory. All of this information is in the form of molecular codes, built into the long, coiled molecules of deoxyribonucleic acid (DNA), which, associated with proteins, go to make up our chromosomes and are the basis of heredity. To act as messengers within the factory, molecules of RNA are produced in the nucleus, and these relay the coded instructions from the DNA blueprints to the sites of protein manufacture in the cytoplasm outside the nucleus. The nucleus is thus a memory bank for everything concerned with the cell's functions.

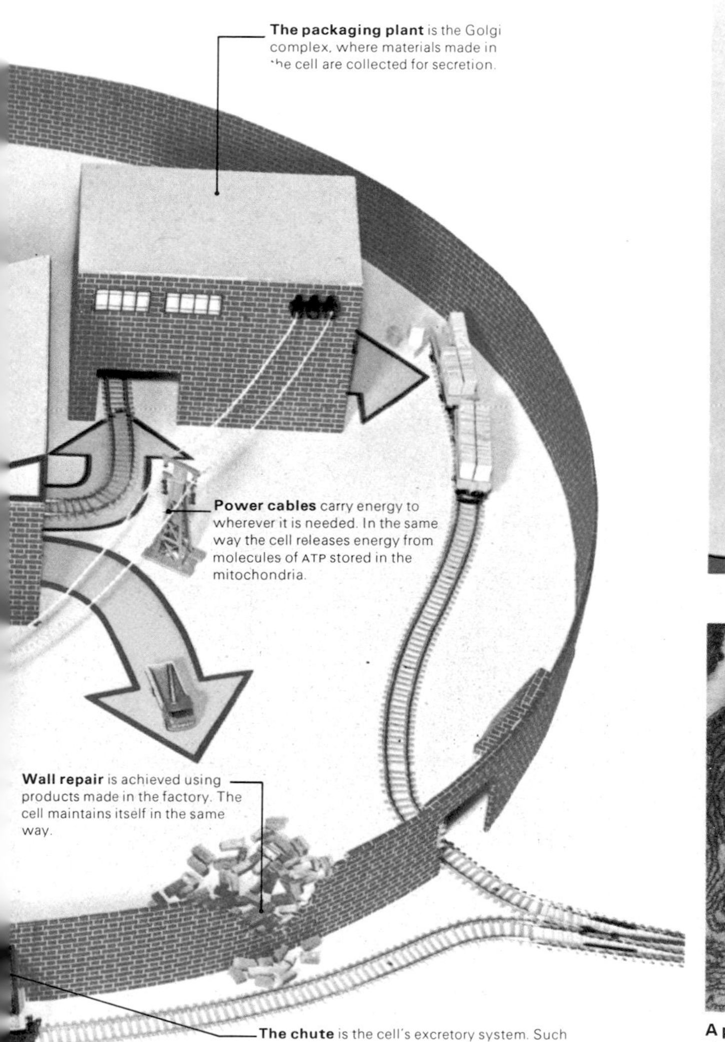

The packaging plant is the Golgi complex, where materials made in the cell are collected for secretion.

Power cables carry energy to wherever it is needed. In the same way the cell releases energy from molecules of ATP stored in the mitochondria.

Wall repair is achieved using products made in the factory. The cell maintains itself in the same way.

The chute is the cell's excretory system. Such metabolic wastes as carbon dioxide and ammonia are actively passed out across the cell membrane.

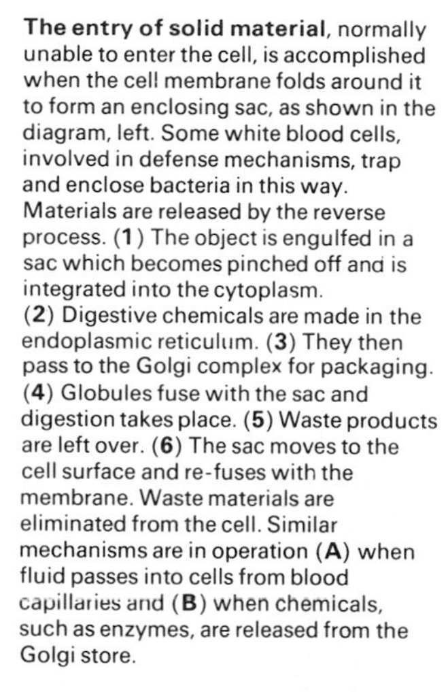

The entry of solid material, normally unable to enter the cell, is accomplished when the cell membrane folds around it to form an enclosing sac, as shown in the diagram, left. Some white blood cells, involved in defense mechanisms, trap and enclose bacteria in this way. Materials are released by the reverse process. (**1**) The object is engulfed in a sac which becomes pinched off and is integrated into the cytoplasm. (**2**) Digestive chemicals are made in the endoplasmic reticulum. (**3**) They then pass to the Golgi complex for packaging. (**4**) Globules fuse with the sac and digestion takes place. (**5**) Waste products are left over. (**6**) The sac moves to the cell surface and re-fuses with the membrane. Waste materials are eliminated from the cell. Similar mechanisms are in operation (**A**) when fluid passes into cells from blood capillaries and (**B**) when chemicals, such as enzymes, are released from the Golgi store.

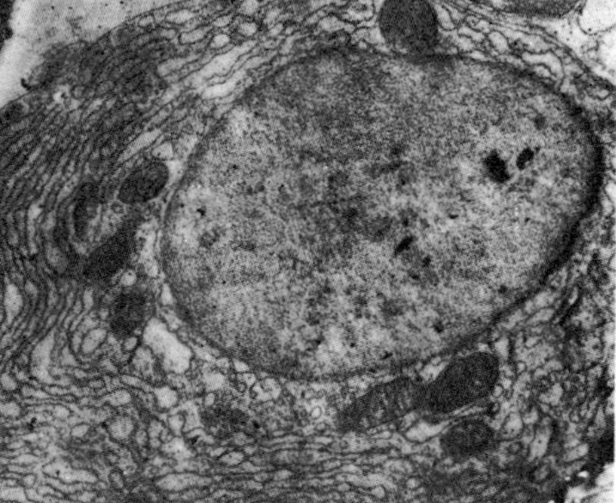

A plasma cell from human bone marrow, seen under the electron microscope at a magnification of thirteen thousand times, shows a large nucleus, seven mitochondria and a well-developed endoplasmic reticulum.

Cross-reference	pages
Blood	**58-61**
Diet	**44-45**
DNA and RNA	**26-27**
Hormones	**70-73**
Metabolism	**52-53**
Structure of cells	**22-23**

DNA: the code of life

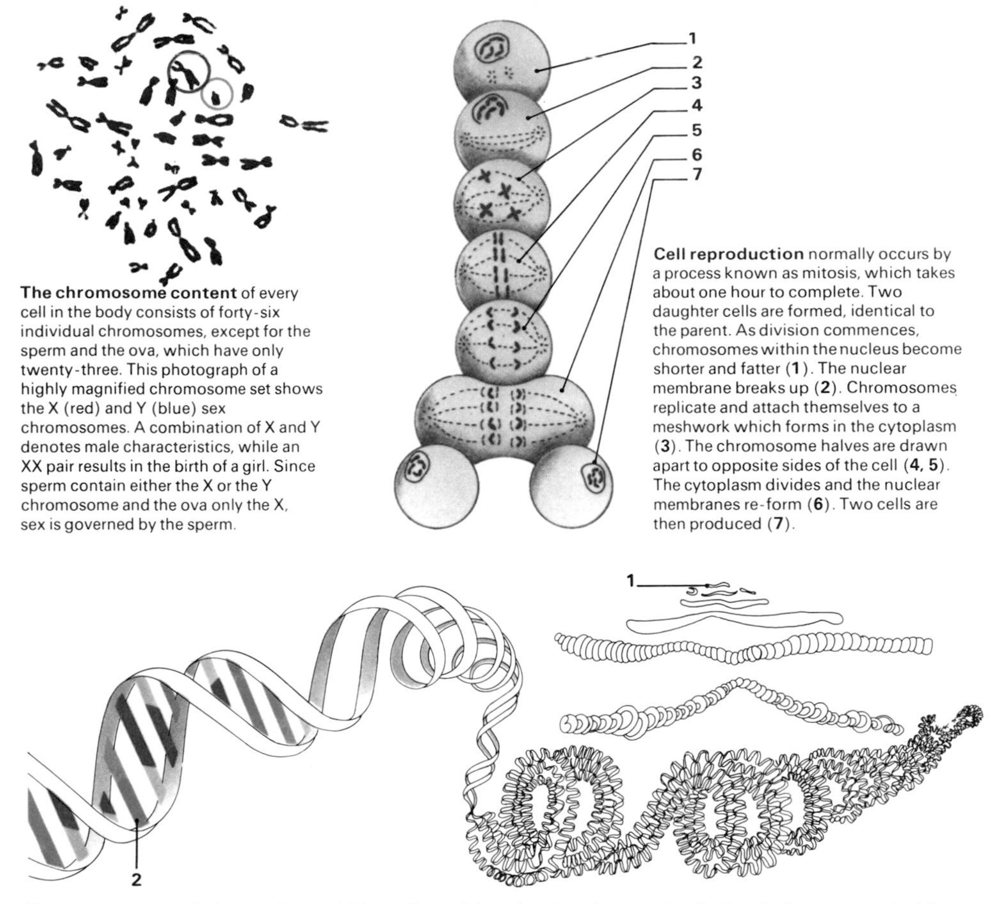

The chromosome content of every cell in the body consists of forty-six individual chromosomes, except for the sperm and the ova, which have only twenty-three. This photograph of a highly magnified chromosome set shows the X (red) and Y (blue) sex chromosomes. A combination of X and Y denotes male characteristics, while an XX pair results in the birth of a girl. Since sperm contain either the X or the Y chromosome and the ova only the X, sex is governed by the sperm.

Cell reproduction normally occurs by a process known as mitosis, which takes about one hour to complete. Two daughter cells are formed, identical to the parent. As division commences, chromosomes within the nucleus become shorter and fatter (**1**). The nuclear membrane breaks up (**2**). Chromosomes replicate and attach themselves to a meshwork which forms in the cytoplasm (**3**). The chromosome halves are drawn apart to opposite sides of the cell (**4, 5**). The cytoplasm divides and the nuclear membranes re-form (**6**). Two cells are then produced (**7**).

Chromosomes contain the genetic material deoxyribonucleic acid (DNA). In association with proteins, the twisted strands of DNA become coiled and folded to give chromosomes a beaded appearance. Each bead represents a gene—a length of DNA coding for the synthesis of one or more proteins. There are thought to be more than five hundred genes on each of the cell's forty-six chromosomes. Here a progression is shown from a chromosome, as it appears under the microscope (**1**), to the molecular architecture of DNA (**2**), shown diagrammatically to illustrate the double helix.

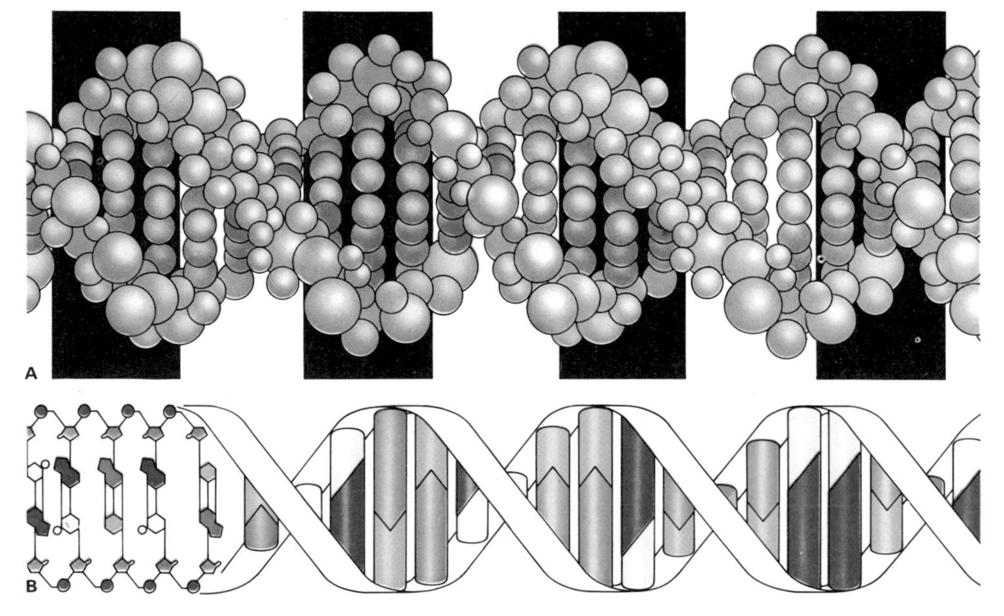

DNA acts as a code for the cell's life pattern. Its structure is customarily illustrated in two forms—actual molecular structure (**A**) and a diagrammatic representation (**B**). (**A**) Within each molecule, two interwoven, twisted strands (gray) are held together by cross-linkages (red) to form a spiral ladder—the double helix. (**B**) The diagram shows first that each strand is composed of alternating deoxyribose sugar (orange) and phosphate molecules (purple). Attached to each sugar are molecules called bases, of which there are only four types found in DNA (here designated as red, blue, green or yellow). The pairing of specific bases (red with yellow, blue with green) forms the "rungs" of the ladder. The DNA code consists of a sequence of triplets of bases on one strand. Each triplet codes for an amino acid, and during protein synthesis amino acids are joined together. Any change in the sequence of triplets changes the protein that is produced. As all chemical reactions within cells are controlled by the action of a group of proteins—the enzymes—the DNA code regulates cell activity.

Each of us began life as a solitary cell, a fertilized egg within the womb. It is hard to believe, considering the size and complexity of the fully grown body, that such a microscopic speck of matter could ever contain a complete set of instructions for building a new human being. Yet here, packed into the cell's nucleus, there are not only instructions to build a body conforming to the human plan, with living tissues, organs and systems intricately woven into a whole, but also instructions that refine and give an infinite number of individual variations to the familiar basic pattern. As human beings we are all similar, but as persons we are all different, and each of us has a unique array of physical and mental characteristics.

This vast amount of information—about everything from the length and color of our eyelashes and the size of our feet to how we should digest cheese and how fast our hearts should beat—is compressed into so small a space that only by thinking in terms of molecules can we begin to explain the mechanism. The "memory bank" of a computer consists of coded cards, disks or tapes; in the nucleus of a cell the information is stored in the winding molecules of different kinds of nucleic acids.

Deoxyribonucleic acid (DNA) holds the key to heredity and to the organization of all living things. Working in partnership with DNA is ribonucleic acid (RNA), which acts as a go-between in the control of life processes.

Microscopic examination of a cell nucleus usually gives little clue to the existence of any fine structure within it. There seems to be only a vague entanglement, meshwork, of material that can be stained with certain dyes. However, when a cell is in the process of dividing, quite definite, deeply stained, rod-shaped structures can be discerned. These are the chromosomes, relatively huge, linked molecules of DNA entwined in a double strand. Making up each chromosome, strung like beads on a thread, are its individual units, different DNA molecules—genes—that determine each characteristic of the living organism. Genes themselves cannot be seen, and it is only after years of observation and deduction that it has been possible to find out the locations of just a few of these hereditary units and relate them to specific characteristics. Human beings have forty-six chromosomes in twenty-three pairs, and one pair are sex chromosomes—XX for a female and XY for a male.

The process of cell division that enables an organism to grow is called mitosis. In the period between divisions, known as interphase, the DNA in the nucleus is in the form of long, extremely slender threads that cannot be seen with an ordinary microscope. It has been estimated that the total length of these threads in a human cell may be up to one yard, from which it can be imagined just how thin they must be. Although not much might appear to be happening during interphase, in fact the DNA molecules are at work all the time, producing smaller molecules of RNA with copies of parts of their own detailed structure. The RNA molecules pass out of the nucleus through pores in its membrane and move to the ribosomes in the cytoplasm, where they are necessary to direct the production of the many different proteins that the cell needs to function and maintain itself.

Also during interphase, but only when a cell is about to divide, the DNA molecules reproduce themselves by doubling and splitting length-

wise, so that when mitosis does occur there is enough DNA for distribution to the two new cells that will arise.

The first stage of mitosis is called prophase. The long threads of DNA, each now formed of two identical threads, become shorter and fatter and tightly coiled, and organize themselves into their visible chromosome shapes. The nuclear membrane disappears, so that the nuclear contents mingle with the cytoplasm. Each half of a chromosome, called a chromatid, remains linked to its twin at a region called the centromere.

The next stage is metaphase, when a spindle-shaped structure of fine threads made from cytoplasmic proteins appears in the cell. Around the center of this spindle the chromosomes congregate, with their centromeres aligned, and during anaphase the centromeres separate and the two chromatids that made up each chromosome part company and are pulled to each end of the cell. The final stage of mitosis is telophase, when new nuclear membranes form as the spindle disappears. Each set of chromatids is enclosed in a membrane and becomes the nucleus of a new cell. The rest of the cell separates into two halves, each of which contains one of the nuclei.

As cell division finishes, the chromosomes uncoil and become longer and thinner, disperse in each nucleus and finally are lost to sight in as mysterious a manner as they appeared. Each new cell has now exactly half of the genetic material that was present in the original cell, with identical DNA molecules.

The process of mitosis, if repeated rapidly, would lead to the production of four cells from two, eight from those four, then sixteen, thirty-two, sixty-four and so on. It has been estimated that forty cell divisions in succession are all that is required to produce from one cell the total number found in the adult body.

Some cells, for yet unknown reasons, become specialized, turning, for example, into brain cells, liver cells and heart muscle cells, and when a cell has become very specialized it loses the power to divide by mitosis. As growth progresses the total number of specialized cells in the body increases, so that the proportion of cells that actively divide regularly decreases. In most tissues of the body, however, there do remain cells that retain the ability to divide, and these few have the task of repairing the wear and tear in the body.

We have seen that cell division is a means of passing on, in the form of DNA blueprints, the instructions for the growth and functioning of the entire organism. But what exactly is the form of code used in DNA? The DNA molecule is a chain of chemical units called nucleotides, each of which contains three linked molecules—a sugar called deoxyribose, a phosphate group and one of four nitrogen-containing compounds called bases. Two of these bases, adenine and guanine, are in the chemical class known as purines, and two, thymine and cytosine, are pyrimidines.

It is now known that the DNA molecule is in the form of a double helix, two interlaced strands of a sugar and phosphate forming the sides of a structure like a twisted ladder. Acting as the rungs of the ladder are the bases, which extend across between the sugar-phosphate strands and meet in the middle. The base attached to one strand reaches out and forms a loose chemical bond with an opposite number on the other strand, and the rule is that a purine always links with a pyrimidine. Because of the shapes of the molecules, rungs cannot be built except by linking adenine (A) with thymine (T), and guanine (G) with cytosine (C). This gives rungs of equal length, and produces the ladder shape of the double DNA molecule.

Because of the rules about pairing, a particular sequence of nucleotides on one side of the ladder must be matched with a different complementary sequence on the other side. So a sequence of A-G-T on one side, for example, must be matched with T-C-A on the other side of the ladder, and T-T-A will form rungs only with A-A-T. This is the secret of the coding of the DNA molecule, and makes it possible to explain both how instructions for building specific proteins are written down, and how DNA molecules can duplicate themselves and give rise to "carbon copies" of the instructions for each new cell that is produced.

If the rungs of the ladder are all broken, and the sugar-phosphate strips allowed to unwind, section by section, then the bases left sticking out can attract to themselves new partners from the plentiful supply kept ready in the nucleus, a pool of waiting nucleotides. Each half-rung on each strip links up with a new opposite number to replace the one it has lost, and the sugar-phosphate ends of the new nucleotides then join up. Step by step, two new ladders are built up—the DNA has completely duplicated itself.

The sequence of bases in the DNA molecule is the recipe for making different proteins. Proteins are built up from sequences of amino acids linked together. There are twenty different amino acids, and the number, kind and order of these units determine the nature of each protein and its particular task. In the DNA molecule, groups of three adjacent bases, called codons, each form a code word for an amino acid in the recipe, and show the order in which the amino acids must join together. Thus the sequence A-C-A, C-A-C, A-C-A is the code for linking the amino acids cysteine-valine-cysteine together. However, proteins are manufactured, not in the nucleus, but in the ribosomes outside. This is where RNA comes in.

RNA is similar in structure to DNA, but has the sugar ribose instead of deoxyribose, and the base uracil (U) instead of thymine. RNA strands are built up from nucleotides that attach themselves to one side of a broken DNA ladder, join together and then peel away. The RNA now bears the recipe of the DNA molecule, but, of course, in complementary form: A-C-A is represented by U-G-U, and so on.

Now this "messenger" RNA, or mRNA, moves from the nucleus to the protein production lines, the ribosomes, with its pattern of codons. To build proteins the ribosomes need not just the recipe, but the ingredients—amino acids. These are supplied by molecules of another form of RNA, "transfer" RNA, or tRNA, each of which carries an amino acid of a particular kind. The mRNA passes through the ribosome, and as it does so tRNA molecules release the amino acid units corresponding to the sequence of codons, and these link together into a chain and form new protein. When it has yielded up its amino acid, a tRNA molecule can combine with another, and so repeat the process as long as there is a strand of mRNA and a ribosome waiting to translate the instructions from the nucleus of the cell.

The highly complex structure of DNA is apparent from the photograph, left, of a plastic model showing a tiny portion of a DNA molecule—a double helix.

Cross-reference	pages
Cells, structure and function	**22-25**
Embryo and fetus	**152-155**
Fertilization	**150-151**
Genetic engineering	**188-191**
Heredity	**158-159**
Ovum	**146-147**
Sperm	**144-145**

Skin, hair and nails

Tough, supple and self-renewing, human skin is the largest of the body's organs. In an adult the total area of skin is more than eighteen and a half square feet and weighs about six pounds, almost double the weight of the brain. Skin is a waterproof covering which acts as a first line of defense against injury or invasion by bacteria, viruses and parasites, and which, too, has other vital roles to play—as a sensory organ, as an agent of secretion and excretion and as a regulator of body temperature.

The two main layers of skin, the epidermis and the thicker underlying dermis, interlock tightly and work in close cooperation. On the surface there is a tough outer fabric of dead skin—spent cells which rub or flake away all the time. Deeper down in the epidermis are the living cells, which in turn will migrate outward, filling en route with keratin, the durable protein also found in hair, nails and, incidentally, animal horn.

It is the dermis, however, built around a network of protein fibers, which gives the skin its elasticity and strength. Here in this inner layer lies a complex collection of nerve endings, sweat glands, lymphatic and blood vessels and hair follicles. Deeper still, the subcutaneous space between skin and muscle contains fat cells which serve as an energy store, provide insulation for the body and help to give shape and substance to the human form.

Hair and the nails on our fingers and toes derive directly from skin: both are made up of modified epidermal cells filled with hard keratin. But whereas hair and nails are dead and nerveless and can be cut without pain, the skin as a whole is keenly responsive to such external sensations as pressure, temperature and pain. Even a trivial skin injury which exposes the sensitive dermis can give rise to severe pain.

The dermis, too, is the seat of hair growth. A tube of dead keratin is pushed up from the follicle to form the hair shaft. Supplied with a tiny erector muscle, each hair is capable of independent movement—standing on end, for example, in response to cold and alarm. At maturity the only parts of the body free of hair are the undersides of the fingers and toes, the palms of the hands, the soles of the feet and parts of the external genitalia. The scalp alone sprouts from one hundred thousand to two hundred thousand hairs, with an average daily growth rate of one-eightieth of an inch.

The color of the skin depends upon the amount of the dark pigment called melanin that is present. The pigmentation has a protective function, preventing the harmful ultraviolet component of sunlight from penetrating. Exposure to strong sunlight stimulates the growth of the epidermis, which becomes much thicker, as well as browner, due to an increased rate of production of melanin. In the condition called albinism, little or no melanin is produced in the skin, and albinos are therefore very liable to become painfully sunburned. In some people the distribution of melanin-producing cells in the skin is patchy, and exposure to sunlight causes the development of freckles rather than an even tan. A mole is a collection of cells that contain an unusually high concentration of melanin.

Hair color also depends upon the amount of melanin present in the individual. In the case of red hair there is an additional pigment. As people get older less pigment is formed and minute bubbles of air in the hair shaft make the hair look gray or white.

Closely linked with all types of body hair are the sebaceous glands, which empty a fatty secretion called sebum into the hair follicles. Besides oiling the hair and the surrounding skin, sebum also prevents excess evaporation of the sweat produced by some three million sweat glands which lie coiled in the dermis. Most of these are eccrine glands, yielding sweat which carries away sodium chloride and the waste products urea and lactic acid, as well as the breakdown products of such foodstuffs as garlic and spices. Apocrine sweat glands develop in association with the coarse hair found in the armpits and the pubic region. These glands are larger and secrete fluid which is much thicker than the fluid secreted by the eccrine glands. Bacteria on the skin surface break down this odorless fluid, yielding body odors. Modified sweat glands, ceruminous glands, in the ears produce cerumen, earwax.

Although skin is primarily a surface fabric, it does in fact continue into the body orifices in the modified form of mucous membrane. It is mucous membrane which forms the lining of various internal structures like the vagina, bladder and ureters, the lungs, intestines, nose and mouth. Lacking the heavily keratinized outer layer of skin proper, this membrane is rich in glands which secrete mucus to protect and to lubricate delicate surfaces.

Like surface skin, mucous membrane can also repel bacterial invasion—partly due to its own integrity and partly by providing a mildly acid barrier in which hostile organisms cannot thrive. Since the surface cells of skin and mucous membrane are constantly being renewed, both types of tissue are self-cleansing.

The protective coat

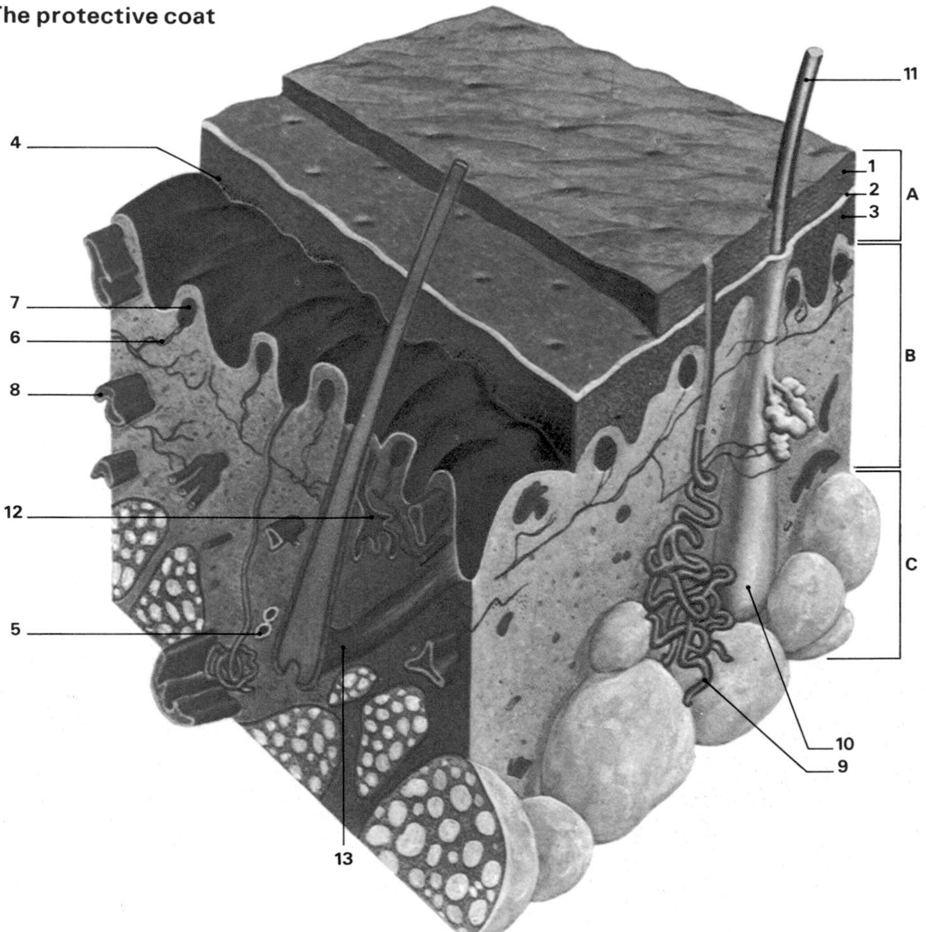

Skin is like a tiered cake with a thin layer, the epidermis (**A**), on top and a thicker layer of connective tissue, the dermis (**B**), underneath. This is underlaid by subcutaneous tissue, containing fat cells (**C**). The epidermis itself is composed of a waterproof layer of flattened, dead cells (**1**), a thinner layer of nondividing living cells (**2**) and an innermost germinative layer (**3**) of dividing cells. Just below this layer are the melanocytes (**4**), which produce the pigments that color skin. Embedded in the dermis are lymphatic vessels (**5**), nerve fibers (**6**) and sensory nerve endings (**7**), blood capillaries (**8**), sweat glands (**9**) and hair follicles (**10**). These follicles surround the shafts of hair (**11**) and each has a lubricating sebaceous gland (**12**) and a muscle which can erect the hair (**13**) whenever body temperature falls.

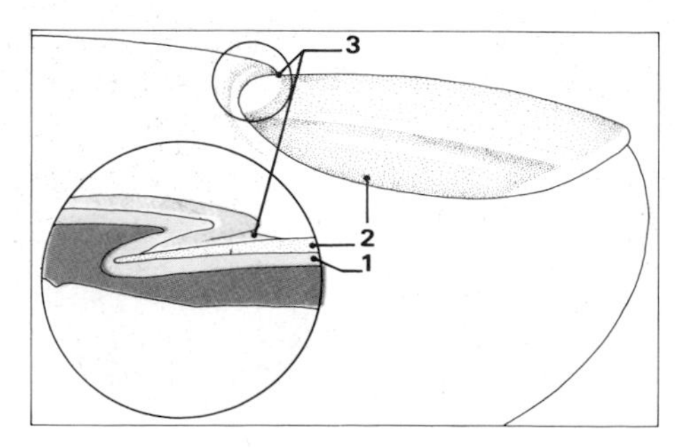

Growth of nail begins in the third month of fetal life, when cells of the epidermis, which form the nail bed, begin to divide (**1**). The upper cells become keratinized to form the nail plate (**2**), which is pushed out of the nail groove (**3**) over the epidermis.

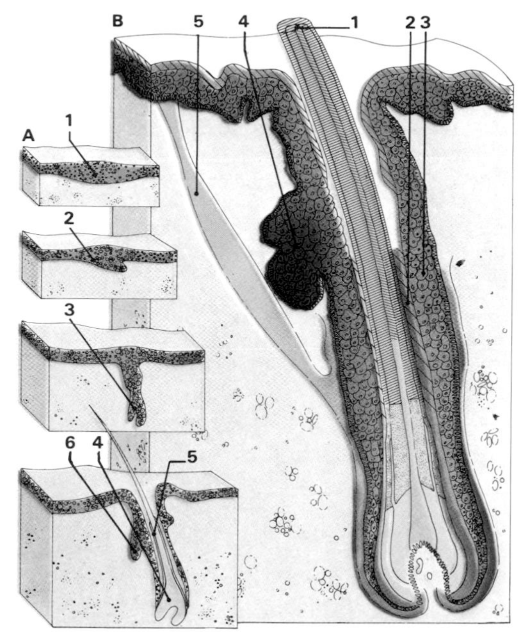

Hair formation begins in the third month of fetal life. Each hair grows in a follicle, which is made from epidermal cells that grow down into the dermis. Hair is created by the multiplication of special cells at the follicle base.
A. A hair develops.
The epidermis thickens (**1**) and cells begin to grow down into the dermis (**2**). This invading downgrowth forms a cap over some of the connective tissue to create a papilla (**3**). Cells of this papilla multiply to form the hair (**4**). As these cells are pushed up the central canal of the hair shaft and thus farther away from their source of nourishment, they become impregnated with the hard protein keratin. The cells of the papilla continue to multiply and are successively filled with keratin as the hair grows. Other cells of the papilla form an internal root sheath of keratinized cells (**5**). A sebaceous gland (**6**) develops from cells of the newly formed root sheath.
B. A formed hair follicle.
In a fully formed hair follicle the hair (**1**) lies within a two-layered hair shaft (**2, 3**). The sebacious gland (**4**) oils the hair and a small erector muscle (**5**) is able to make the hair stand on end.

Body temperature is modified by the skin. (**A**) In hot conditions, the body temperature rises, blood vessels (**1**) in the outer part of the dermis dilate, more blood flows through them and heat is radiated from the skin's surface. Sweat glands (**2**) become more active and the evaporation of sweat acts as a cooling system. (**B**) If the body temperature falls the blood vessels constrict and sweat production decreases. The hair-erecting muscles (**3**) contract and the hairs (**4**) stand on end, increasing the thickness of the layer of trapped air which insulates the skin's surface, rather like double glazing on a house window.

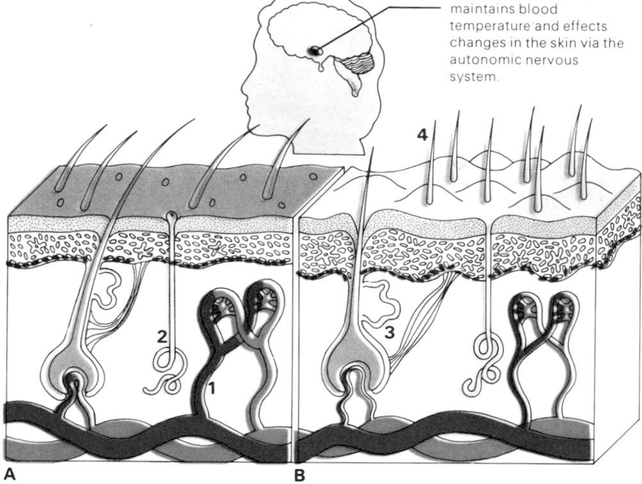

The hypothalamus maintains blood temperature and effects changes in the skin via the autonomic nervous system.

Skin color is created by production of the pigment melanin from cells called melanocytes (**1**) in the epidermis (**2**). All races have equal melanocyte numbers, but genetic differences control the amount of melanin injected into the cells which are produced in the germinative layer (**3**). Melanin production is stimulated by the ultraviolet rays of the sun. The melanin absorbs these harmful rays, and the skin becomes tanned and protected. Freckles appear when only some groups of melanocytes (**4**) become active and produce melanin.

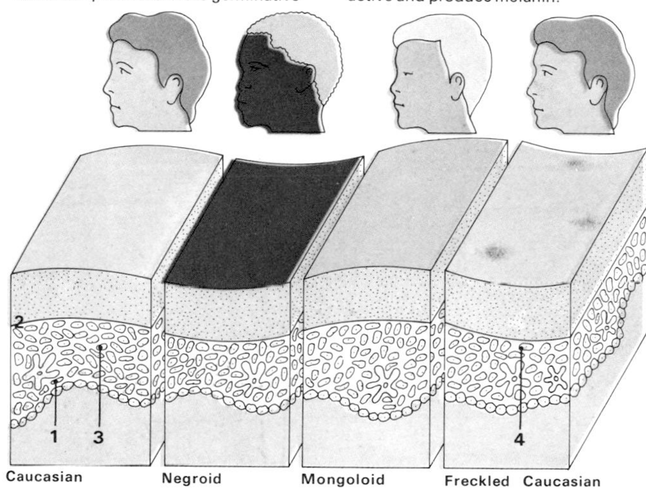

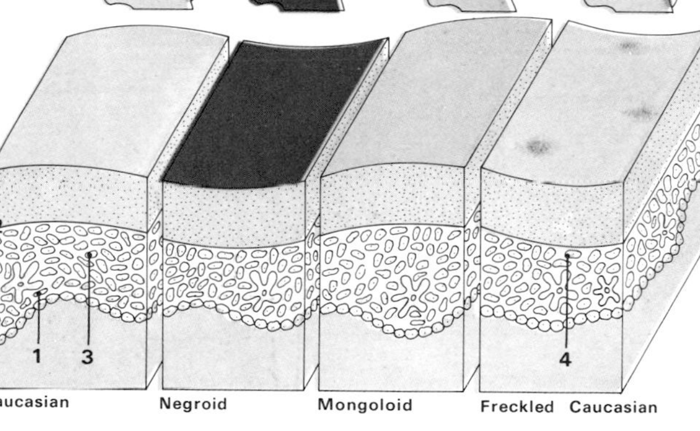

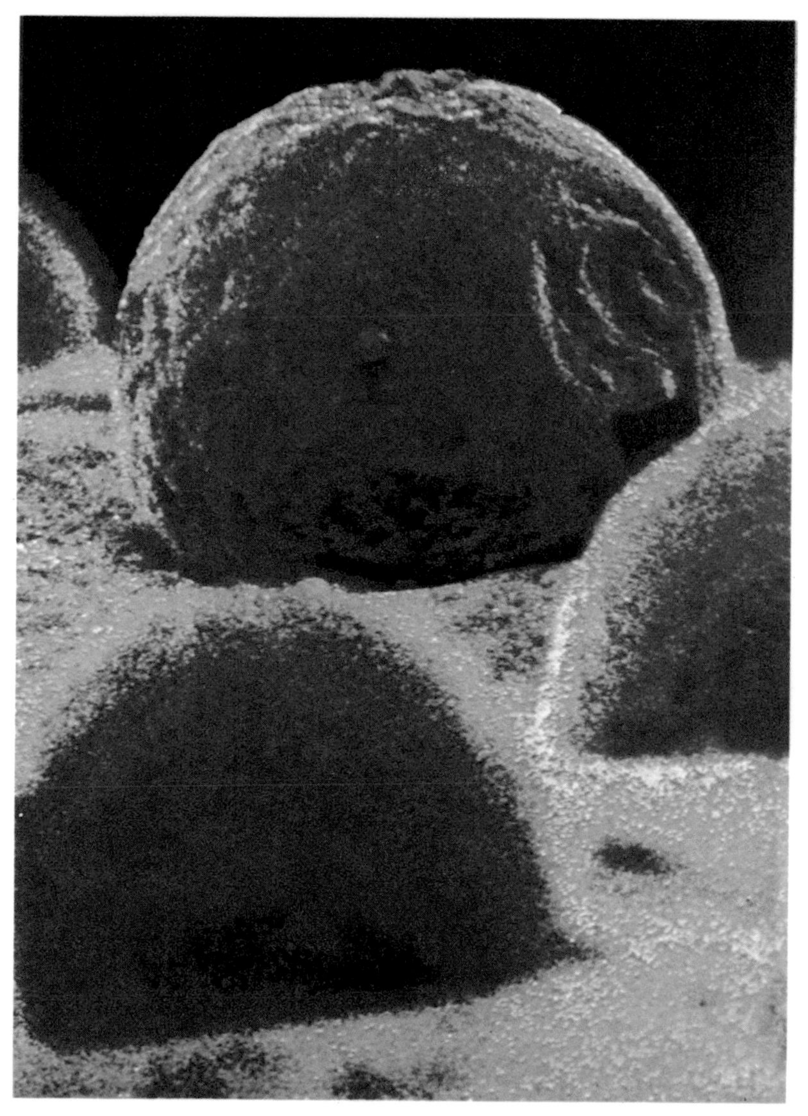

Perspiration acts as a cooling system for the body—heat is lost to the air as water evaporates from the surface of the skin. There are sweat glands over the entire surface of the skin. Droplets of sweat on the human thumb are seen through a scanning electron microscope, an instrument capable of producing three-dimensional pictures of objects normally invisible to the naked eye. An entire thumb magnified the same number of times as this picture would be larger than an average-sized man.

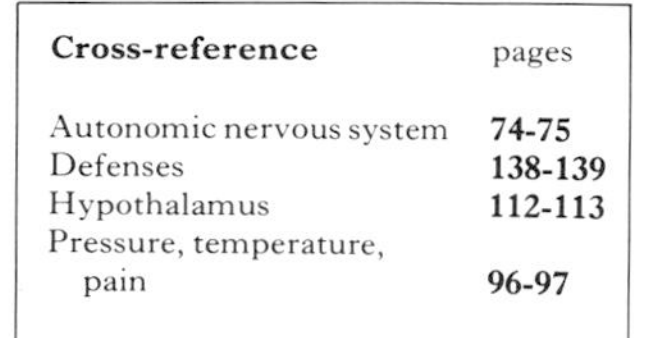

Cross-reference	pages
Autonomic nervous system	**74-75**
Defenses	**138-139**
Hypothalamus	**112-113**
Pressure, temperature, pain	**96-97**

Bone and connective tissue

Acting as the body's internal scaffolding is the skeleton—that assembly of blocks, girders, plates and struts of hard bone that forms a basis for head, body, arms and legs arranged in the human pattern. Arrayed around or within the framework are the softer structures: the vital organs and vessels, the muscles and fat, the skin and all the other tissues that in life fill out, clothe and enclose our bodies and give them their familiar outward form.

But what prevents this complex combination of living "hardware" and "software" from sagging into shapelessness? To be built into a framework that is stable yet movable, each bony unit of the skeleton needs firm linkage with another. Soft and mobile structures, such as arteries, intestines, liver and kidneys, must somehow be kept in proper shape and tethered in place. And muscles need tough, pliable cables to pull on the bony levers and to move joints. Throughout the body, supporting, packing and binding tissues are necessary if its assembled parts are not to distort and collapse. These are called connective tissue.

Different kinds of connective tissue have different properties suiting them to their own special tasks. In makeup all are variations on the same theme—a jellylike "ground substance" of water, salts, carbohydrates and proteins, in which cells and fibers are embedded. The cells are white cells, which play a defensive role, plasma cells, fat cells and fibroblasts, which are responsible for producing the fibers. The fibers themselves are of two sorts, those made of the tough white protein collagen, and those made of the yellowish, and more elastic, protein elastin. The properties of the particular tissue depend upon the proportions and the arrangement of all these ingredients.

Throughout the body there is loose, areolar tissue that forms sheaths around muscles, nerves and blood vessels, and acts as packing in and around such organs as the liver and the spleen. In this areolar tissue a meshwork of cells and both collagen and elastin fibers is arranged in soft and flexible sheets.

The tendons that act as cables linking muscle to bone are of much denser tissue, and have great tensile strength because they are composed mainly of tightly packed collagen fibers. The ligaments that link bone and bone and help to prevent dislocation of joints are also composed mainly of collagen. It is the combination of suppleness and great strength in collagen that enables our bodies to withstand not merely the normal everyday stresses on the skeleton of walking or bending down, but the deliberately severe stresses of such strenuous exertion as weight lifting.

Cartilage is another form of connective tissue. Dense, tough, pearly white and translucent, this is the material that forms a temporary skeleton for the unborn child. Most of the bones in your body began as rods or blocks of cartilage, which ossified, hardened into bone, as mineral materials were laid down in the growing tissue.

In the adult, cartilage has three main forms. Elastic cartilage contains collagen and a great deal of elastin, giving springiness, for example, to the external ear and the epiglottis in the throat. Fibrocartilage, with dense bundles of collagen and less elastin, makes up the resilient disks which separate the bones of the spine and links the pubic bones at the front. Hyaline cartilage, hard, clear and bluish, forms the articular cartilage that covers the ends of bones in joints, links ribs and breastbone, keeps open the larger airways and gives shape to the nose. When fully developed, cartilage, unlike looser connective tissues, does not contain nerves or blood vessels, but is penetrated by numerous tiny channels.

Bone, in developing from cartilage, becomes harder and more rigid, but keeps a degree of resilience because of the meshwork of pliable protein threads within it. As ossification proceeds, minute crystals of calcium salts are arranged, layer upon layer, forming tunnels of bone. The minerals are laid down by cells called osteoblasts that have invaded the cartilage. As they work, they wall themselves in and become immobile bone cells, or osteocytes, communicating with their neighbors only through tiny pores.

Seen in cross section under the microscope, the tunnels of bone form concentric circles that are known as Haversian systems. In the canal at the center of each system run the blood vessels that keep the living bone cells supplied with oxygen and food materials.

Growing bones retain bands of cartilage near their ends, separating, in long bones, the shaft, or diaphysis, from the caplike epiphyses, or ends. Here growth can take place without interfering with joint movement. In the late teens or early twenties growth ceases when even this cartilage ossifies.

There are two types of bone tissue, compact outer bone, dense and ivorylike, and cancellous bone, an inner corallike network which encloses cell-filled spaces. The combination gives strength and rigidity, and also lightness, to all the major bones in our bodies.

The surface enamel of teeth is the hardest of all human tissue—much harder than bone. Its composition is 96 percent mineral, notably small crystals of calcium phosphate and calcium fluoride, plus carbonate, magnesium, potassium, sodium and other ions contained in a fibrous protein network. It is this protein—similar to keratin in hair—which makes enamel so resistant to acid attack. Although prone to decay during life, after death the teeth resist decay longer than any other body tissue, including bone.

The structure of bone

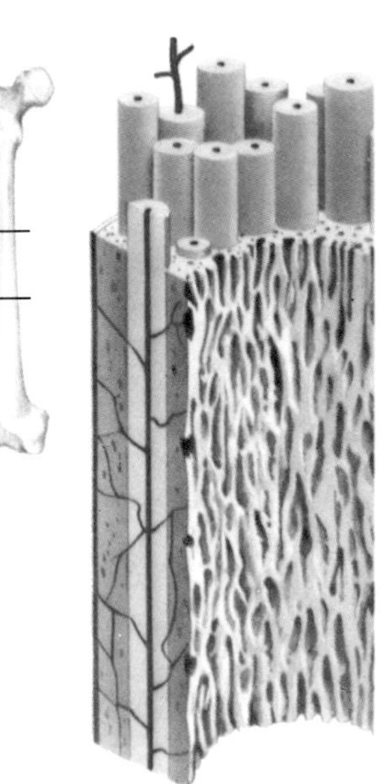

A section through the shaft of the femur, a long bone, above, shows a central mass of spongy bone surrounded by the branching, tubular Haversian systems. Blood vessels on the bone surface reach into the center of each Haversian system supplying the bone cells. In magnification, right, spongy bone, which is in the center, is seen as a honeycomb structure. Much heavier, the surrounding compact bone consists of concentric circles called lamellae, which make up the Haversian systems. Compact bone is actually connective tissue in which the ground substance has been hardened by a network of collagen fibers and mineral salts, such as those of calcium, laid down in lamellae. The bone cells, exchanging materials with the blood supply, lie in tissue fluid in spaces called lacunae within these rings.

Connective tissue

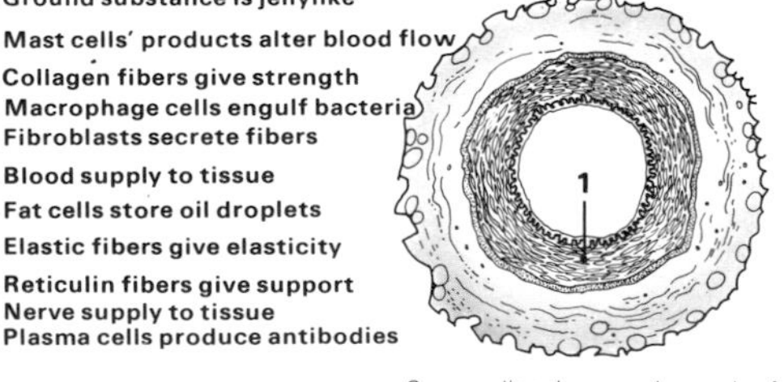

Surrounding the smooth muscle of an artery (**1**) is a protective sheath of loose areolar, connective tissue.

Connective tissue, above, is made up of cells and fibers in a jellylike "ground substance." Present throughout the body, connective tissue holds parts together and acts as packing. There are several kinds of connective tissue, which differ in the amount and type of cells and fibers in their makeup. Most common is loose areolar tissue, which contains most of the basic connective tissue components (listed above right). Widely distributed throughout the body, areolar tissue forms binding sheaths around muscles, nerves and blood vessels, and is present within organs as filling materials.

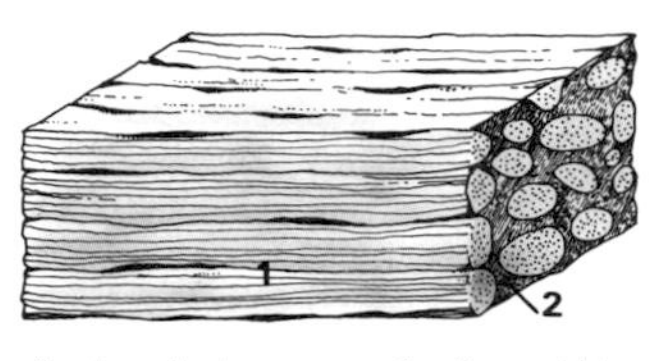

Tendons, the dense connective tissue which attaches muscle to bone, are bundles of inelastic collagen fibers (**1**) made by fibroblasts (**2**) in the ground substance.

Adipose tissue is composed mostly of fat cells. It cushions internal organs, forms an insulating layer under the skin and provides a fuel store.

Spongy or cancellous bone forms the center bulk of all our bones. Its honeycomb structure keeps bone light, in contrast to the heavier, compact bone, which gives it strength.

Bone cells, or osteocytes, are contained in spaces called lacunae. The minute projections of bone cells trail into adjoining channels known as canaliculi. Tissue fluid, which fills the lacunae, allows the transfer of materials between bone cells and capillaries.

Haversian systems, each about one-sixtieth of an inch wide, make up the structure of compact bone. Each system is formed by a series of rings called lamellae, which are deposits of mineral salts and collagen fibers.

Haversian canals, which run through the center of each Haversian system, contain the blood and lymphatic vessels supplying the bone. Here arteries carrying oxygenated blood are shown in red. Veins carrying deoxygenated blood are shown in blue. The lymphatic vessels are shown in white.

Three different kinds of cartilage

Cartilage, unlike other types of connective tissue, has no blood vessels. It is tough but pliant because the ground substance between the cells contains combinations of proteins and sugars. Cartilage is described as hyaline, fibrous or elastic, depending on the density and type of fibers present in its composition.

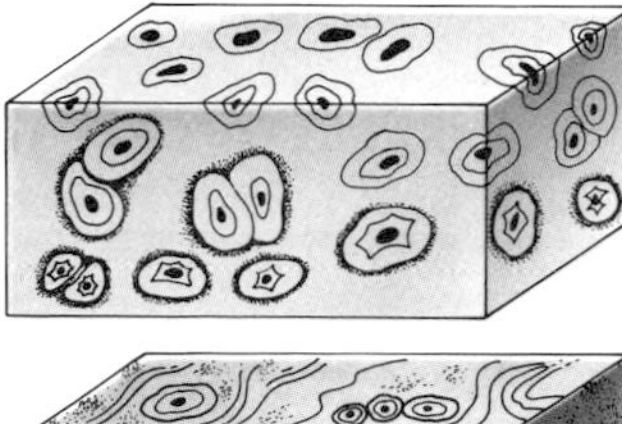

Hyaline, the most abundant type of cartilage, is clear and glassy, with few cells and fibers in the ground substance. Hyaline covers the ends of bones at the joints, and also forms the rings which keep the trachea open.

Fibrocartilage is made up of tightly packed bundles of collagen fibers, making it resilient and able to withstand compression. Fibrocartilage lies, for example, between vertebrae.

Elastic cartilage contains, in addition to collagen, fibers of the protein elastin. This makes it firm yet supple, giving support, for example, to the external ear and epiglottis.

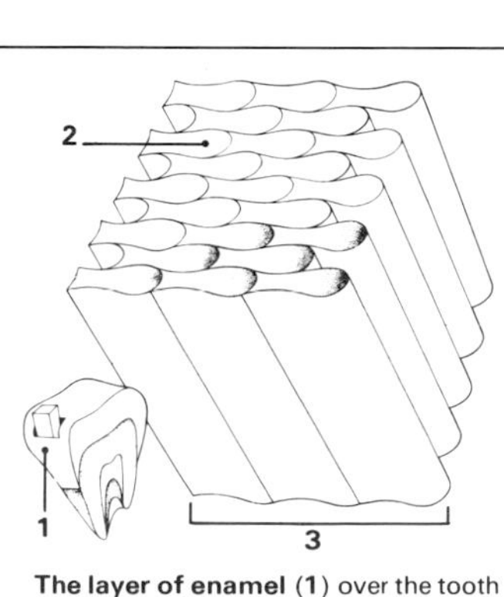

The layer of enamel (1) over the tooth crown is a substance similar to bone. A section through enamel shows that, like bone, it is composed of crystals of calcium phosphate (**2**) embedded in a ground substance. These crystals are built up into rods, or prisms (**3**) which make enamel the hardest substance in the body.

Cross-reference	pages
Blood	**58-61**
Defenses	**138-141**
Joints	**34-35**
Muscles	**36-41**
Skeleton	**32-33**

Joints: the skeleton in action

Wherever one bone meets another in the human body there is a joint of some kind. These linkages are of various designs, each one custom-built to fulfill a particular function. Some joints, for example those between the bony plates of the skull, do not move at all. At the other extreme are highly versatile joints like those in the hip and in the shoulder—masterpieces of engineering which give a phenomenal range of power and movement.

Freely movable joints are especially complex in structure because of the demands made by mobility. Where two bones meet and move there is friction and if, like the hip or knee, the joint is also load-bearing, there is the additional stress caused by such everyday movements as walking or climbing stairs. At the joint, therefore, each bone is capped with cartilage, which reduces friction and acts as a buffer.

To offset friction, the movable joint is contained in a tough, fibrous capsule lined with synovial membrane. This membrane secretes a runny, yellow-tinged substance, synovial fluid, which lubricates the working parts. All bony surfaces, tendons and ligaments involved in the joint are covered with synovial membrane, which extends to the edges of the cartilage layers. Beyond the capsule are ligaments, which bind and strengthen the joint, and the tendons of muscles, which are necessary to move its bones.

In large joints, like the knee, there are small, fluid-filled sacs, known as bursas, which act as shock absorbers. Also lined with synovial membrane, bursas are found between the movable parts (where bone meets tendon, muscle or ligament) lying outside the capsule. Sometimes undue stress on a joint causes the bursa to produce excessive fluid, giving rise to inflammation, pain and stiffness. This is a condition known as bursitis, and commonly called "tennis elbow" or "housemaid's knee."

At the elbow and knee, hinge joints combine strength with mobility. Here is a design yielding only bend-stretch movements, although the knee assembly, with crossed ligaments in the center uniting the femur and tibia, is somewhat more complex than the elbow. Besides the primary hinge movement between the upper and lower parts of the leg, the knee also contains a gliding mechanism where the patella, the kneecap, slides over the femur.

At the elbow, giving greater mobility to the strength of the associated hinge joint, is a pivot joint which enables the two bones of the forearm, the radius and the ulna, to twist.

The shoulders and hips are both served by the ball-and-socket-type joint—a freely moving joint which gives a wide range of play to the limbs. At the hip, the head of the femur fits into a deep socket, called the acetabulum, where the ilium, ischium and pubis meet. A rim of cartilage helps to grip the ball firmly, and strong

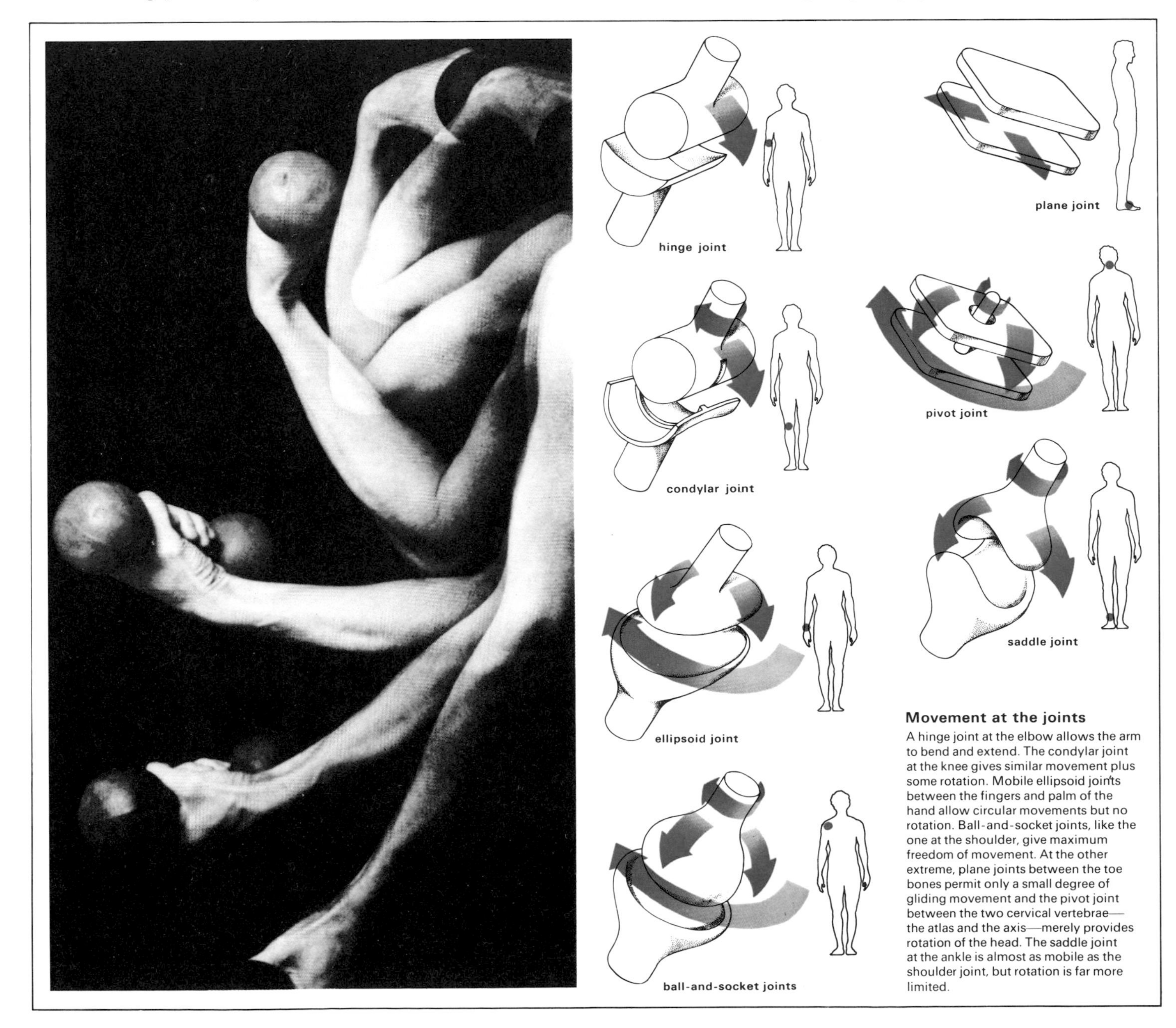

Movement at the joints
A hinge joint at the elbow allows the arm to bend and extend. The condylar joint at the knee gives similar movement plus some rotation. Mobile ellipsoid joints between the fingers and palm of the hand allow circular movements but no rotation. Ball-and-socket joints, like the one at the shoulder, give maximum freedom of movement. At the other extreme, plane joints between the toe bones permit only a small degree of gliding movement and the pivot joint between the two cervical vertebrae—the atlas and the axis—merely provides rotation of the head. The saddle joint at the ankle is almost as mobile as the shoulder joint, but rotation is far more limited.

ligaments and muscles give firm but flexible outside support. A special internal ligament called the ligamentum teres also unites the femur with the pelvic girdle, and also carries nerves and a blood supply to the head of the femur. At the shoulder, the head of the humerus fits into a much shallower socket, the glenoid cavity, on the scapula. Although it is this shallowness which gives the arm a greater range of movements than the leg, it also renders the joint more liable to dislocation.

At the top of the spine the first cervical vertebra, called the atlas, forms the kind of joint called a condylar joint with the occipital bone of the skull. Two rounded projections below the occipital bone rest upon shallow concavities in the atlas, and the joint allows nodding movement of the head. The joint between the atlas and the second cervical vertebra, which is called the axis, is another example of a pivot joint. A peglike projection from the body of the axis is encircled by the bony ring of the atlas, being kept in place by a ligament. This joint allows the head to turn from side to side and also to bend from side to side. Between them, these two joints give the head quite a wide range of movements.

There are several other kinds of joints, with different degrees of mobility. A gliding joint allows one surface to slide over another, the only restriction in movement being imposed by the surrounding ligaments and other structures. Gliding joints are found between the ribs and the thoracic vertebrae. They are between the carpal bones of the wrist as well as between the tarsal bones of the foot.

A saddle joint has surfaces so shaped that combinations of movements in two planes can be made, but one bone cannot rotate on the other. The joint at the base of the thumb, where its metacarpal meets the carpal bone, called the trapezium, is a good example. This allows the thumb to be moved across the palm to meet the other fingers—the movement called opposition.

Where not very much movement is needed between two bones there are the joints classified as either slightly movable or immovable. In the spine there are slightly movable joints between the vertebrae, which are united by disks of cartilage. Between the two pubic bones at the front of the body there is also tough cartilage, forming a joint which may become more movable during pregnancy, to allow expansion of the pelvis at childbirth.

Between the bones of the skull there are sutures, joints which are completely immovable. Here the bony plates are serrated, and their edges fit together rather like pieces of a jigsaw puzzle interlocking, with a firm bond of fibrous tissue.

Ranging in scope from the simple pad of cartilage between two bones to the complex units which give the human body grace and movement, our joints are as intricate as anything to be found in the sphere of mechanical engineering. Nevertheless, by copying the joints in metal or in plastic, biomedical engineers are beginning to provide a wide range of substitutes—spare parts to replace live structures missing or damaged through congenital defect, injury or disease.

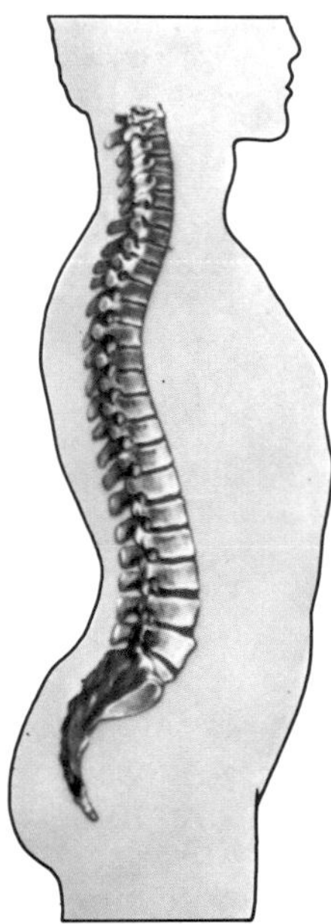

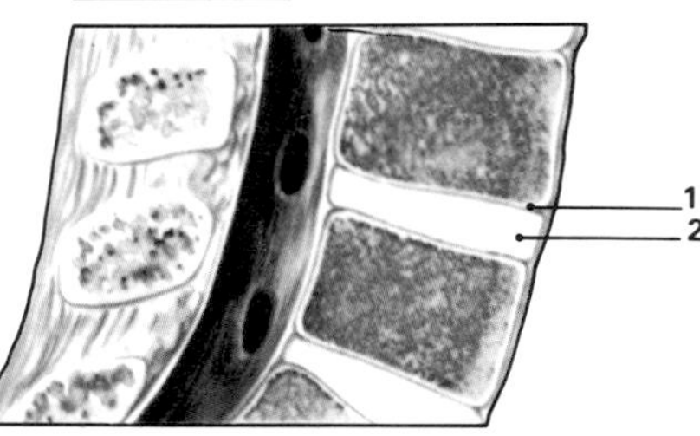

Cartilaginous joints are found between bones of the spine. A cutaway view of the spine shows that the vertebrae encircling the spinal cord are covered with the clear, glassy connective tissue known as hyaline cartilage (**1**). In between the vertebrae are the spinal disks—pads of a tougher, fiber-filled material called fibrocartilage (**2**). These disks allow the spine, which would otherwise be inflexible, to articulate freely in various directions. Able to withstand compression, the cartilage disks act as shock absorbers, cushioning the spine from injury.

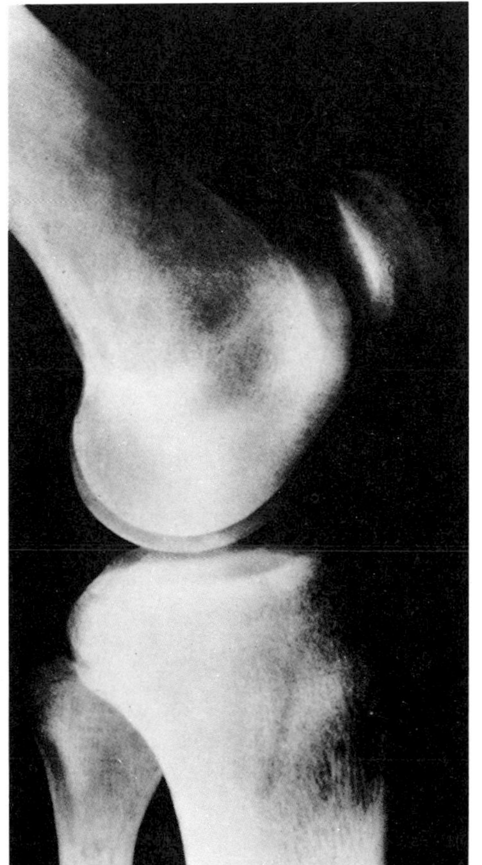

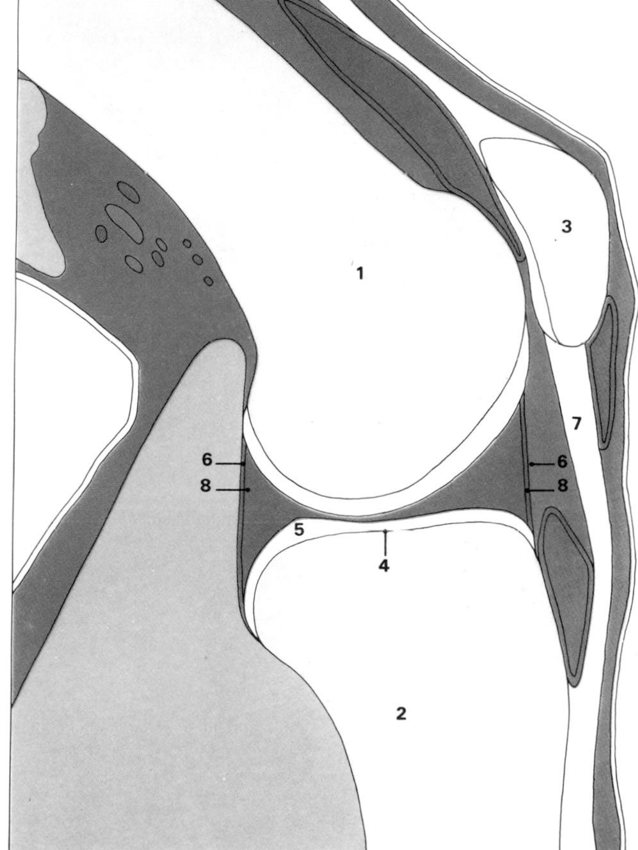

The body's largest joint is at the knee, where condylar joints link the thighbone, or femur (**1**), and the tibia (**2**). In addition, the kneecap, or patella (**3**), forms a plane joint with the femur. The femur fits into grooves, or condyles (**4**), in the tibia. To prevent friction, the bone surfaces are capped by smooth hyaline cartilage (**5**). Tough, inelastic ligaments hold the bone in place and form the joint capsule (**6**). The ligament holding the kneecap in place is an extension of the tendon (**7**) of the thigh muscle. The joint capsule is lined with synovial membrane (**8**), shown here in cross section so that it does not obscure the bones of the joint. Synovial fluid secreted by the membrane lubricates the joint and supplies nutrients to the hyaline cartilage.

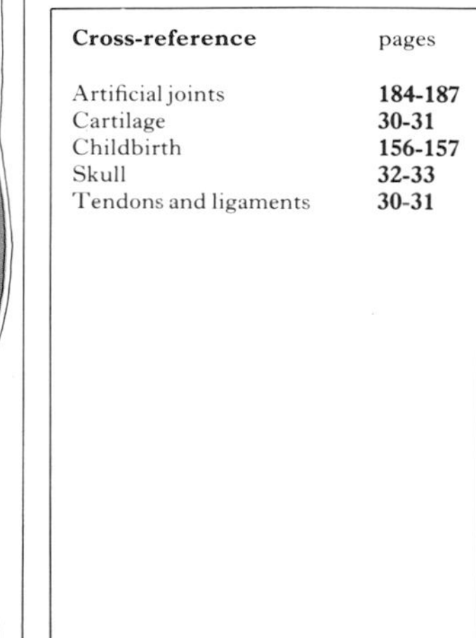

Cross-reference	pages
Artificial joints	184-187
Cartilage	30-31
Childbirth	156-157
Skull	32-33
Tendons and ligaments	30-31

Muscles: structure

Without muscles the human body would be floppy and immobile—as useless as a puppet without strings. Muscle is the raw material of movement, a contractile tissue which makes up from 35 percent to 45 percent of the body's weight, and which does everything from keeping the body's systems ticking over during sleep to making possible the great feats of strength and endurance seen, for example, at the Olympic Games.

With so many complex parts to move, the body contains three distinct types of muscle. Skeletal muscle is strongest and most abundant. Cardiac muscle is found only in the heart. Smooth muscle is found in the blood vessels, in the intestines and in some other internal organs. Skeletal muscle is capable of conscious control—acting at our will through the central nervous system—and is, therefore, also described as voluntary muscle.

Interlaced with blood vessels and nerve endings, voluntary muscle is built up in fibers, in length from a fraction of an inch to about one foot. These fibers are arranged in bundles, and each individual fiber itself consists of bundles of fibrils, or filaments. In turn, each fibril contains strands of protein, which overlap to give the characteristic striations, or striped pattern, seen under a microscope. The two proteins involved, actin and myosin, interact in response to nerve impulses and thus cause the muscle to contract.

When you wish to move a part of your body, nerve impulses travel from the brain along outgoing, or "motor," nerve pathways to the appropriate voluntary muscles. Although a nerve impulse acts rather like an electric current passing along a fine wire, in fact it is a ripple of chemical activity that travels along the nerve membrane, changing its molecular structure for an instant. At the very end of each motor nerve is an expanded part, called a motor end plate, in contact with the fibers of the muscle on which it acts. A nerve impulse arriving here causes the release of a special chemical "transmitter," which diffuses across the nerve-muscle junction and triggers off the molecular interaction between actin and myosin that causes shortening of the muscle fibrils. The transmitter substance for voluntary muscle is acetylcholine.

Impulses are delivered by nerves to bundles of muscle fibers known as motor units. A single motor unit may contain from five muscle fibers to two thousand. (The fewer the fibers to each nerve end plate, the more precise the action controlled by the motor unit.)

When muscles shorten, energy is dissipated, there is a buildup of heat and carbon dioxide and water and lactic acid are formed. Muscles used heavily for some time suffer from an accumulation of lactic acid because there is not enough oxygen available for its metabolism. In this case the result is an aching sensation and "heaviness" of the limbs. A cramp after strenuous exercise signals the buildup of lactic acid, accompanied by salt and water and/or oxygen deprivation. These shortages can also be caused by dehydration, by excessive fluid loss, due to illness or injury, or by interruption of the blood supply to the muscle.

The smooth muscle that serves the blood vessels, the bronchial tree, the alimentary tract, the uterus and urinary tract and the other internal systems is involuntary. This muscle, controlled by the autonomic part of the nervous system, provides slow, even contractions, which like all muscles are closely integrated with the actions of their neighbors. The transmitter substance at nerve-muscle junctions is in some cases acetylcholine and in others noradrenaline.

Unlike the skeletal or cardiac muscle, smooth muscle is not striated. Under the microscope it shows up as a complex of tiny spindle-shaped cells, with no sign of the actin/myosin interplay which is the essence of voluntary muscle contraction. It could be that, while actin is present in filaments in smooth muscle, myosin is contained elsewhere in the cells—invisible perhaps, but still on hand to work with its protein partner to produce movement.

Cardiac muscle, the force behind the heartbeat, is striated like skeletal muscle, but is involuntary, not subject to our conscious control. The fibers are branched. Each cardiac muscle fiber has its own spontaneous rhythm of contraction and relaxation, so in isolation the heart muscles would not all contract at the same time. The bulk of cardiac muscle has a built-in rhythm of thirty to forty contractions a minute. However, overruling and regulating this rhythm is a small mass of tissue called the pacemaker. From this, waves of contraction start and spread to all parts of the heart in turn, governing its rate of beating.

An exploded view of skeletal muscle

Skeletal muscle is composed of long cylindrical cells with many nuclei, called fibers, bound together by a membrane. These fibers consist of threadlike myofibrils containing two different proteins—myosin, which has knoblike projections, and actin, formed in a spiral configuration. These two proteins are arranged in parallel, creating dark bands along the myofibril where they overlap. Blood capillaries supply each muscle fiber with the oxygen which is essential for the energy-producing chemical reactions powering the muscle contractions. Stimulation by nerve impulses causes a chemical and mechanical link to occur between the actin and myosin filaments. This interaction results in the shortening of the fibers and the contraction of the whole muscle.

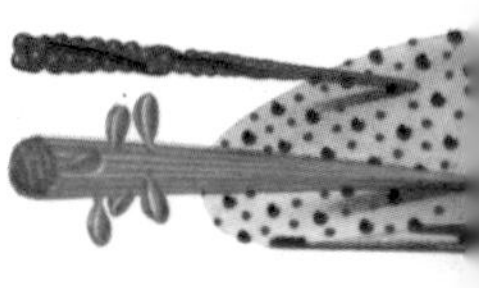

Actin filaments have a spiral configuration which couples them with neighboring myosin filaments to contract the muscle.

Three types of muscle structure

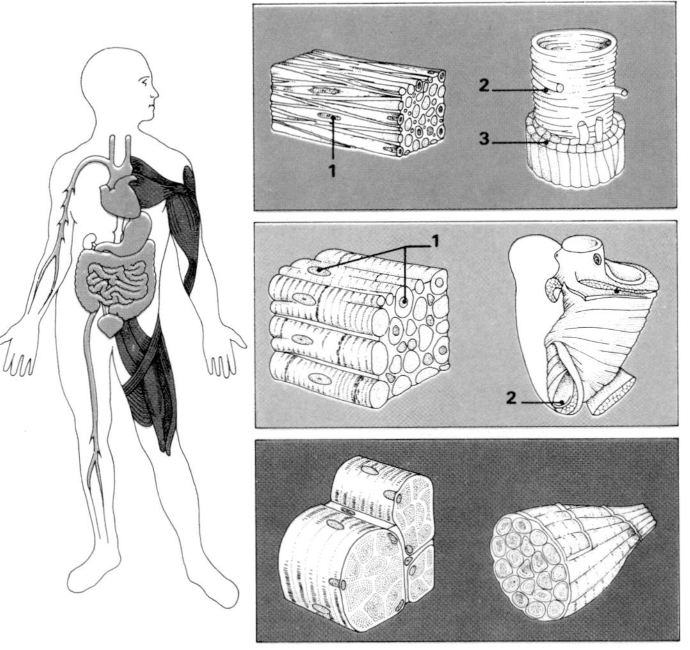

Smooth muscle is made up of elongated cells arranged in bundles, each with a single nucleus (**1**). Within the intestines an inner circular band (**2**) produces ringlike constrictions and an outer longitudinal band (**3**) produces wavelike motions. Controlled by the autonomic system, smooth muscle contraction is involuntary.

Cardiac muscle forming the heart is made up of branched cells containing many nuclei (**1**). It occurs as thick spiral bands (**2**) around the ventricles of the heart and performs rapid rhythmical contractions under the control of the heart's pacemaker.

Skeletal muscle is under conscious control and is therefore described as voluntary. It is composed of cells, elongated up to twelve inches in length, containing many minute myofibrils and nuclei. It can contract and relax rapidly, but is easily fatigued.

There are three types of muscle, each with a different structure and function. Skeletal muscle produces the most powerful contractions and is essentially involved in performing the body's movements. Smooth muscle is found in the arteries and in the digestive and urinary tracts, and performs the slow, long-term contractions needed to maintain these systems. Cardiac muscle, found exclusively in the heart, continuously generates very strong contractions and rarely tires in its task of pumping blood throughout the body. It has a built-in rhythm of contraction, but is regulated by the heart's pacemaker.

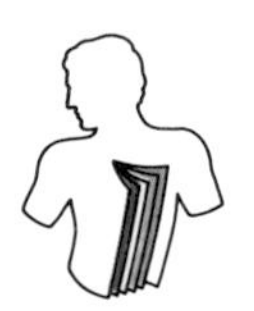

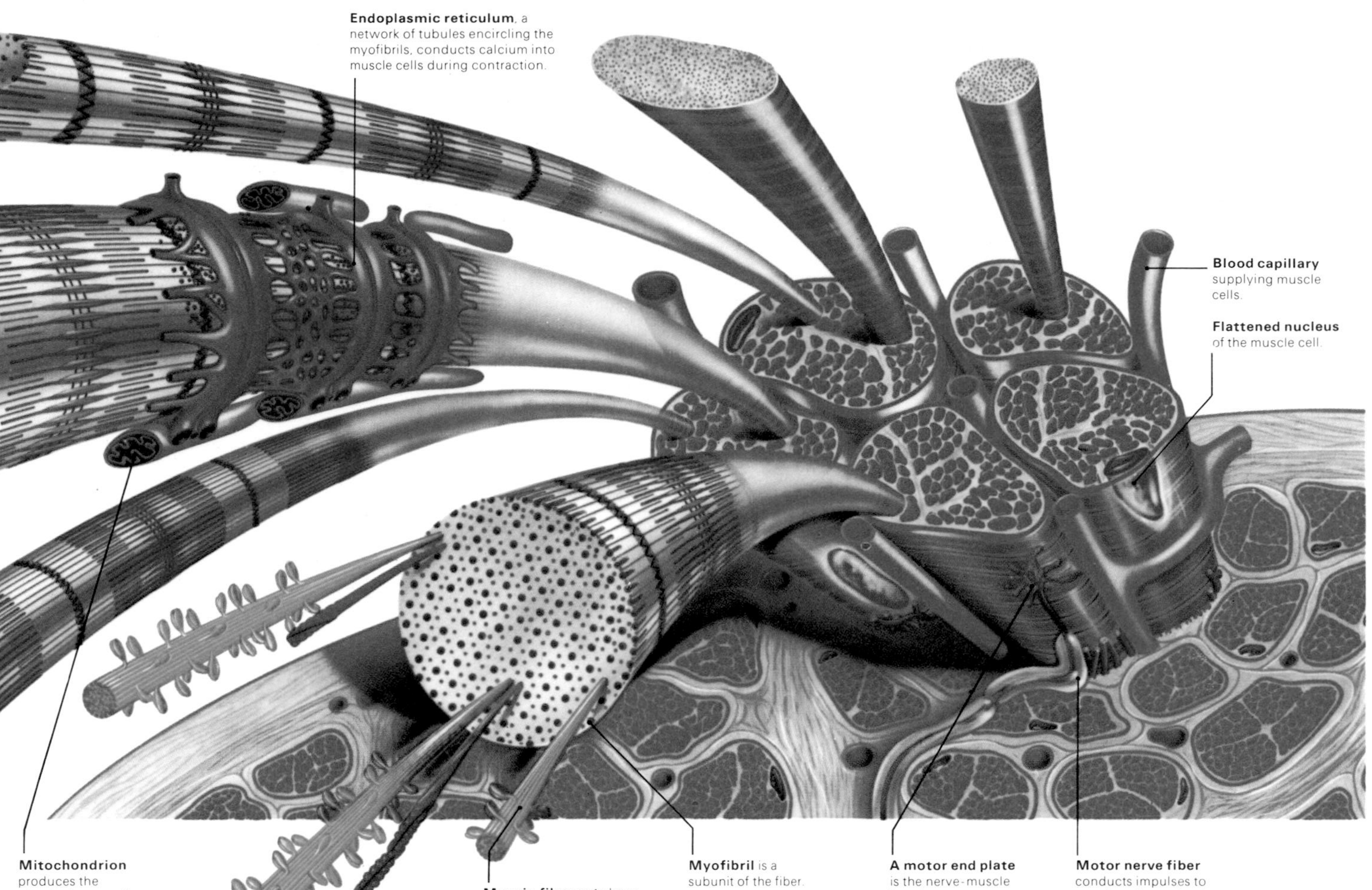

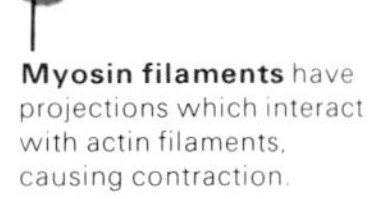

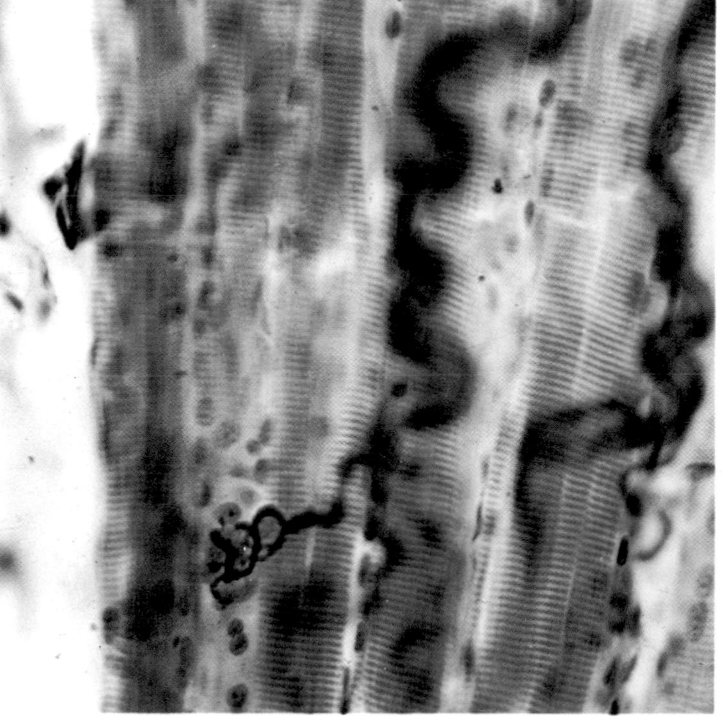

A biochemical ratchet acts when muscle contracts. Projections on the myosin filament (red) attach themselves successively along the spiral actin filament (olive green).

A nerve-muscle junction is shown, left, magnified more than one thousand times. The dark, threadlike nerve joins the muscle at the motor end plate, a point where many cell nuclei are present. At these neuromuscular junctions, the knoblike nerve endings act as chemical transmitters for impulses arriving from the central nervous system. The more nerve endings that are present, the more precise the movement a muscle can perform.

Cross-reference	pages
Brain and central nervous system	74-77
Heart	56-57
Involuntary muscles	40-41
Metabolism	52-53
Voluntary muscles	38-39

Muscles: voluntary movement

Skeletal muscles

The voluntary or skeletal muscle system consists of more than 650 muscles which produce body movement and help maintain an upright posture. Voluntary muscles are built onto the skeletal framework in several layers; the ones shown here are those directly under the skin. Facial muscles are involved in speech, eating and expression. Muscles across the chest and the top of the back function in respiration and are also responsible for shoulder and arm movements. Muscles in the lower back help to maintain an erect posture. Abdominal muscles protect the internal organs. Arm and leg muscles, among the most powerful in the body, produce limb movements. Acting like cables, muscles pull on bones to produce movement. This involves the contraction and shortening of individual muscle fibers. Most of the body's voluntary muscles are arranged in opposing groups—as one muscle contracts the opposing muscle relaxes. Even when the body is being maintained in a relaxed position the muscles contract partially, but do not shorten. This maintains tone.

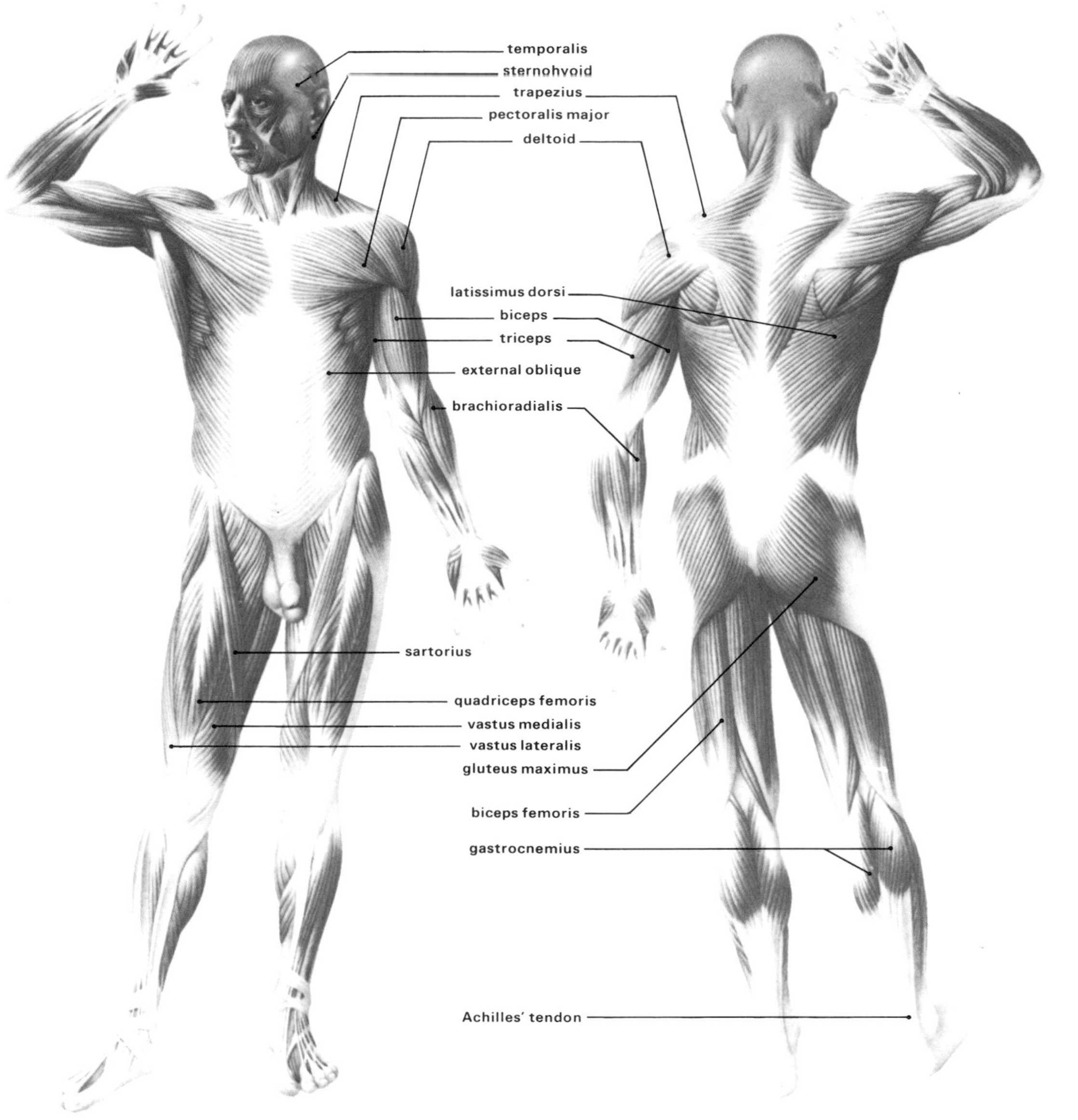

To enable us to move our bodies at will, we each have well over six hundred voluntary muscles clothing our skeletons. In a well-developed person the shape of many muscles can be discerned through the skin, but far more lie hidden within. They range in size from the bulky gluteus maximus, which gives shape to the curves of the buttock and is built of layers of many thousands of fibers, to the few delicate fibers of the tiny stapedius muscle inside the cavity of the middle ear.

Many muscles join one bone with another. They can be said to have an "origin" from one bone and an "insertion" into the other. At a muscle's origin, which may cover a large area, the fibers are attached directly to the connective tissue covering, or periosteum, of the bone. Some muscles have more than one origin. The biceps has two, the triceps three and the quadriceps has four separate origins. At the other end, a muscle usually tapers and its fibers are replaced by tough, inelastic connective tissue. This may be in the cablelike form of a tendon, sometimes many inches long, which is attached to a small area of bone, or it may be in the form of a flat sheet, or aponeurosis, attached to other structures. Sometimes a muscle is inserted into the fibrous sheath of another muscle, adding to its power. Yet other muscles are attached to the loose connective tissue within the skin—for example, the muscles that control facial expression.

Stimulated by impulses from motor nerves, voluntary muscles become shorter and fatter. In long muscles the fibers run parallel to one another and contract in line with the pull of the whole muscle. This arrangement gives the greatest shortening, but not necessarily the most power. In the most powerful muscles, an arrangement of fibers at an angle to the tendon and to the line of pull, like the vanes of a feather, gives strong contraction over a relatively short distance. Other muscles show a fan-shaped arrangement of fibers.

The movement produced by a muscle is magnified by the lever on which it pulls. Thus, when the brachialis muscle of the upper arm contracts fully, it may swing the fingertips through an arc of more than three feet, although it has moved its insertion in the forearm less than three inches. At the same time, of course, the power at the fingertips is reduced to one-twelfth of that at the insertion.

Without an opposing force, an individual muscle would be of little use to the body. Muscles contract to produce their actions. They pull, but cannot push to reverse these actions. To bend the wrist, for example, muscles on the front of the forearm are brought into play; to straighten it again, opposing muscles on the back of the forearm must be used. As one muscle, or group of muscles, contracts, so an opposing, antagonistic, muscle group allows itself to be gradually stretched. All movement is a matter of "give and take." When a part of the body is apparently immobile, and even during sleep, its muscles are nevertheless gently pulling against one another in response to a stream of signals from the brain. This gives muscles "tone"—a basic state of preparedness for action.

When we are merely standing still, a great deal of coordinated muscular activity is being carried out invisibly. We have to fight against the force of gravity, and this requires energy. The tension in all the muscles involved—those of the feet, legs, pelvis, spine, abdomen and even of the shoulders—is increased as they

Muscle architecture

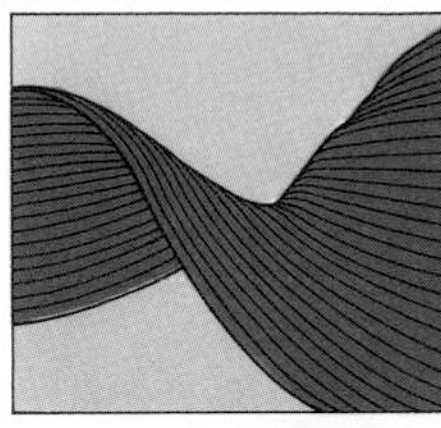

A spiral orientation of bundles of muscle fibers shapes their external appearance and determines their range of movement. This arrangement, as in the trapezius muscle of the back, produces a twisting action at joints.

Triangular bundles of muscle fibers, which appear fan-shaped, are arranged at an angle to the line of pull, as for example the temporalis muscles involved in chewing.

Straplike muscles, such as the sartorius of the leg and the sternohyoid of the neck, have a parallel arrangement of fiber bundles. They are capable of both subtle and powerful contractions.

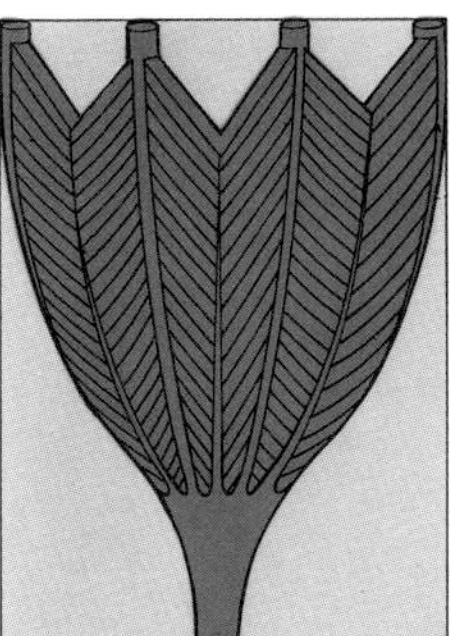

Multipennate muscles, like the deltoid of the arm, are composed of many short bundles of fibers. Despite a limited range of shortening, these muscles are powerful in their actions.

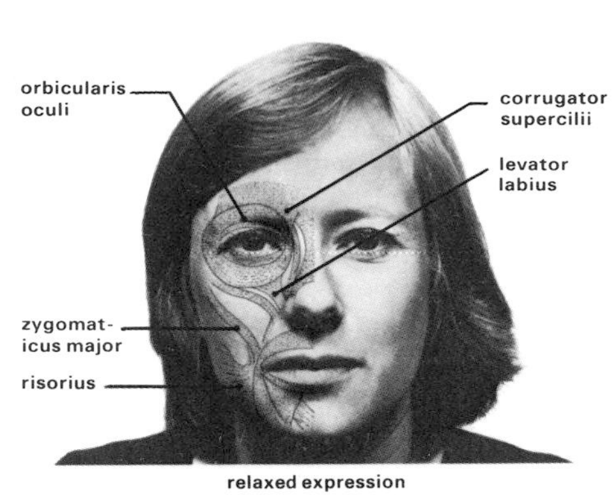

relaxed expression

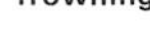

frowning

smiling

Facial expression

There are more than thirty facial muscles that are attached to the bones of the skull and inserted into the facial skin. These are the muscles which give our faces their mobility, moving under the skin to create the repertoire of expressions we use to match our conversation or mood. Even when a face appears expressionless, the muscles still maintain tone, giving tension to the skin. When we smile the zygomaticus and risorius muscles pull the mouth up and back and orbicularis oculi muscles contract to narrow the eyes. Frown lines and narrowed eyes, which characterize an expression of concern, are caused by contraction of the corrugator and orbicularis muscles.

Muscles move joints

Movement at joints is effected by the action of muscles that are generally arranged in opposing pairs. As one muscle contracts and shortens, the other relaxes and becomes longer. The biceps (**1**) and triceps (**2**) control movement of the forearm. To raise the forearm, the biceps contract and act as the prime mover, while the triceps, the antagonist, relaxes. When the elbow is straightened the reverse occurs.

constantly adjust and readjust the alignment of bones transmitting our weight to the ground. So, although we are apparently not doing anything, standing upright for any length of time eventually proves tiring.

The interplay of muscle groups is initiated by brain activity, but is not necessarily always ordained by conscious will. The passenger standing in a train, for example, swaying but remaining upright, does not have to think about the muscle actions needed to keep him on his feet. Constant adjustments are made automatically. The brain stem receives a flow of information from the eye, from the part of the ear devoted to balance and from special organs within the muscles, and sends appropriate messages to the muscle groups that are keeping the body erect. All this is achieved without true conscious control.

An outstanding feature of skeletal muscle is its versatility. Its activity enables us to perform, on the one hand, such feats of sheer strength as bending iron bars and, on the other, the delicate and precise action of threading a needle or the incredibly rapid and accurate fingerwork of the concert pianist.

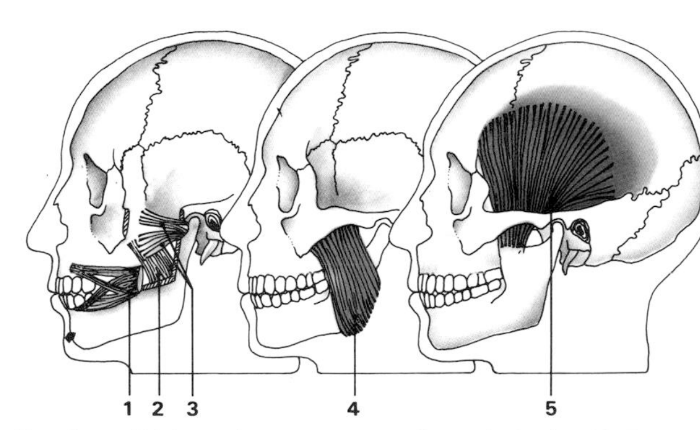

Chewing and biting actions are largely controlled by the movement of five muscles on either side of the head. Contraction of the buccinators (**1**) compresses the cheeks. These muscles, together with the tongue, are used to position food in the mouth. The internal pterygoids (**2**) are responsible for the side-to-side grinding movement of the jaws, while the external pterygoids (**3**) open the mouth and allow the jaw to protrude, creating back and forth grinding movements. The masseter muscles (**4**), stretching from the cheekbones to the base of the lower jaw, are responsible for raising and for lowering the jaw and are particularly important in biting. Contractions of the temporalis muscles (**5**), situated at the sides of the head, clench the teeth. They work in conjunction with the masseter muscles to coordinate the biting of food.

Cross-reference	pages
Brain and central nervous system	74-77
Brain stem	104-105
Connective tissue	30-31
Skeleton	32-33

Muscles: involuntary movement

It is fortunate that we do not have to direct, by conscious effort, the activities of all the muscles in our bodies. We do not, for example, have to remind our hearts to carry on beating, nor instruct our stomachs to churn up our food for digestion. Instead we have involuntary mechanisms working day and night without our knowledge—whole muscle systems governed automatically by impulses from the autonomic, or self-governing, division of the nervous system and by hormones circulating in the bloodstream.

Involuntary muscle is of two kinds. Smooth, or plain, muscle is distributed throughout the entire body, but cardiac muscle is to be found only in the heart. Fibers of smooth muscle in the walls of arteries control the amount of blood passing through the vessels. In the lungs, smooth muscle regulates the width of the small air passages. In the eye, tiny fibers in the iris make the size of the pupil greater or smaller, and in the skin, smooth muscle fibers attached to hair follicles can make hair "stand on end," and cause "gooseflesh." Sheets of smooth muscle form the walls of the bladder, and give the uterus its powers of contraction. But it is in the long, coiled and looped tube that forms the digestive tract that the greatest concentration of smooth muscle is to be found.

Almost from its beginning to its end, the digestive tract is composed of layers of smooth muscle lined with mucous membrane. There is an outer layer of fibers that runs longitudinally and, by contracting, it shortens sections of the tube. Within this there is a layer of circularly arranged fibers, whose contraction causes a narrowing of the tube. (In the wall of the stomach there is also a layer of fibers that runs obliquely, or spirally, to give a twisting movement.) By alternate contraction and relaxation, these muscle layers churn, mix and propel onward food that is being digested. Their coordinated rhythmic activity is called peristalsis.

The act of swallowing a mouthful of food that has been chewed is at first voluntary, but once the food mass, or bolus, enters the esophagus, we have no further conscious control over its progress—peristalsis takes over. From about halfway down, the esophagus has only smooth muscle in its walls, and from here waves of constriction automatically urge the bolus onward and into the stomach. These waves are powerful enough to overcome the pull of gravity if necessary; we can swallow food and drink quite efficiently, if uncomfortably, when standing on our heads.

Peristalsis is an interplay of contracting and relaxing muscles that occurs in the whole length of the alimentary canal. Muscle fibers in one section of the tube first relax to receive a food mass, and then contract behind and around it to squeeze it into an adjoining section. To ensure thorough mixing of the food with digestive juices, the movement may be temporarily reversed in short sections, but eventually food always moves onward.

Control of involuntary muscle

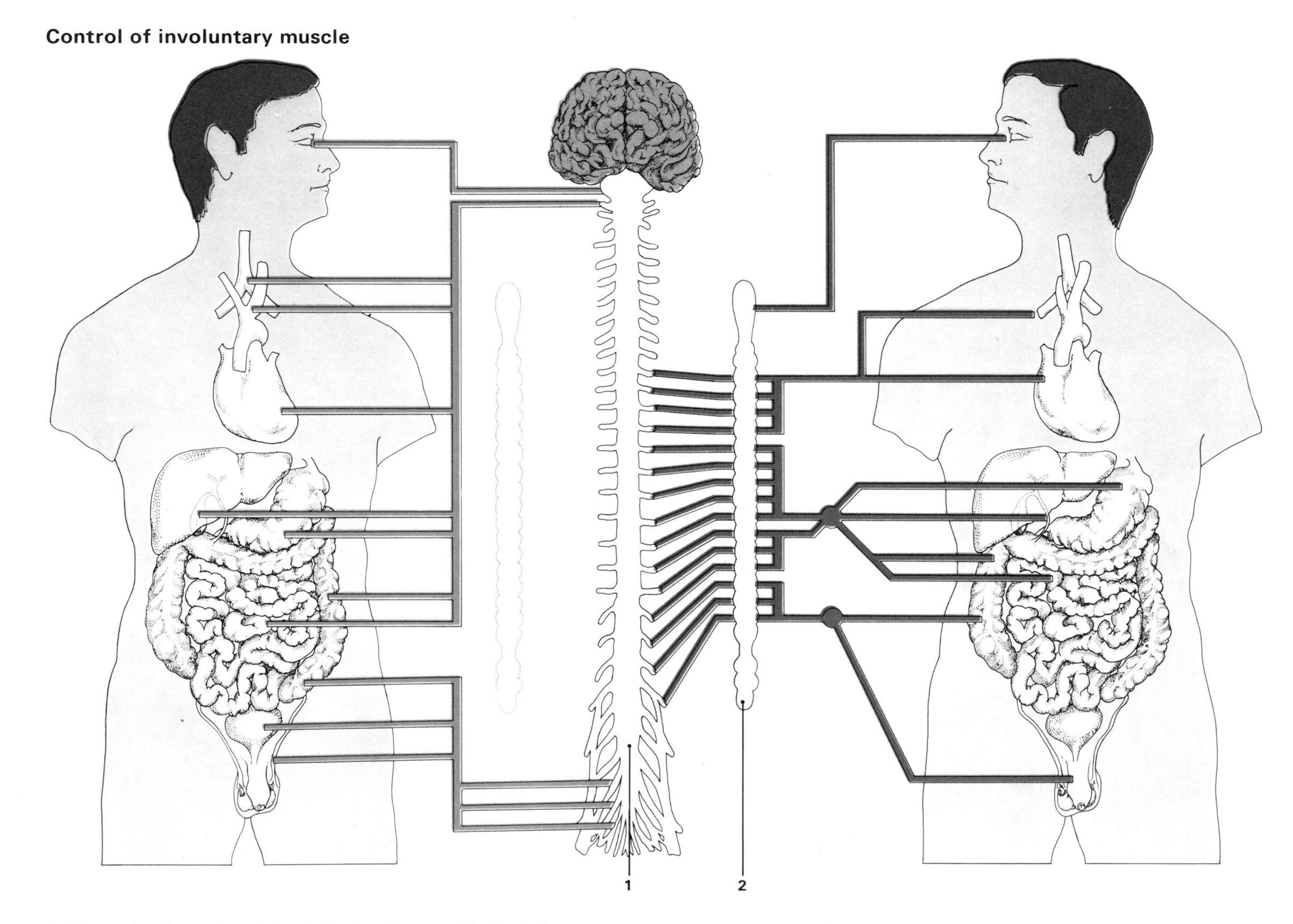

Cardiac muscle and smooth muscle function involuntarily—beyond our conscious control. There is smooth muscle in organs of the digestive, respiratory, circulatory and urinogenital systems, while cardiac muscle is found exclusively in the heart. These muscles are controlled by the autonomic nervous system, which is composed of two opposing divisions, the sympathetic (red lines) and the parasympathetic (blue lines). Sympathetic nerves leaving the spinal cord (**1**) lead first to nearby nerve chains (**2**) before connecting the body organs, where they stimulate muscle activity. In the parasympathetic division, nerves pass from the spinal cord directly to the organs and have the opposite effect—decreasing muscle activity. This dual mechanism provides a method for maintaining the activity of the muscles within controlled limits. The sympathetic system is dominant in situations requiring rapid action. When, for example, danger threatens there is an increase in heart rate, respiration rate, blood pressure and sweat gland activity. The parasympathetic system is dominant while the body is resting—usually during sleep, when the heart rate is slower and respiration is deeper and more regular.

The patterns of relaxation and contraction that produce peristalsis are regulated by trains of impulses passing to and fro in networks of autonomic nerve fibers within the intestinal wall. Autonomic nerves are of two kinds with, in general, opposing actions. Nerves of the sympathetic system release at their endings noradrenaline. This brings about relaxation of the muscle in the intestinal wall and constriction of the rings of muscle, or sphincters, such as that found at the exit from the stomach. Nerves of the parasympathetic system release acetylcholine, which has the opposite effect—contraction of muscle in the intestinal wall and relaxation of the sphincters. It is rhythmic alternation in the activity of the sympathetic and parasympathetic nerve networks that brings about peristalsis.

This system of dual control is found wherever smooth muscle is at work. Thus in the small blood vessels in the skin, sympathetic nerves cause constriction, and the skin becomes cold and pale, and parasympathetic nerves cause dilatation, resulting in reddening and warming. In the arteries that supply blood to skeletal muscles, the roles of the two systems are reversed, and adrenaline causes dilatation, while acetylcholine produces constriction. Again, in the eye, sympathetic nerve fibers stimulate the smooth muscle that dilates the pupil, and parasympathetic nerves act on the muscle fibers that constrict the pupil.

Although the activities of involuntary muscles are largely beyond our conscious control, in some cases we can override the instructions of the autonomic nervous system. In the bladder, for example, the buildup of a large volume of urine, causing a stretching of the muscular wall, would normally be a signal to bring the parasympathetic system into play. The muscle in the wall would contract, and the sphincter at the exit from the bladder relax, so that urine would be expelled. We have learned, however, to control by willpower the relaxation of the sphincter, and only allow it to release stored up urine when it is socially convenient. The same is true of bowel movements.

The involuntary muscle that is found in the heart, cardiac muscle, is striped like skeletal muscle, but has branching fibers. Cardiac muscle has a built-in rhythm of contraction and relaxation, and would continue to pulsate, after a fashion, without any nerve supply. To regulate and coordinate its contractions, however, the autonomic nervous system is constantly in action. The heart speeds up or slows down according to messages from the sympathetic and parasympathetic systems, so that its output adapts to circumstances.

If the blood pressure is too high, special sensors in the carotid arteries cause impulses to reach the heart through the brain and the vagus nerve (part of the parasympathetic system), and this slows down the heart and reduces blood pressure. If the blood pressure is too low, impulses from the sympathetic system increase the rate and force of contractions of the heart to bring the pressure up to normal again.

The functioning of involuntary muscle is also affected by circulating hormones, notably adrenaline, released into the bloodstream by the adrenal glands. Because adrenaline is one of the substances also released by sympathetic nerve fibers, the effects of the hormone largely parallel those produced by the action of sympathetic nerves.

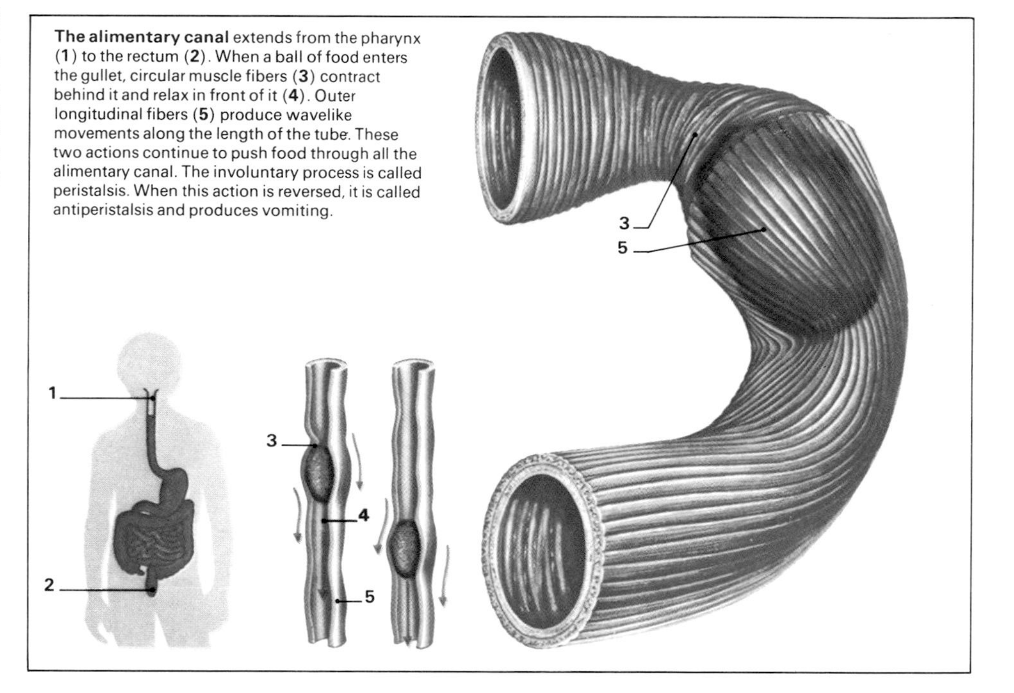

The alimentary canal extends from the pharynx (**1**) to the rectum (**2**). When a ball of food enters the gullet, circular muscle fibers (**3**) contract behind it and relax in front of it (**4**). Outer longitudinal fibers (**5**) produce wavelike movements along the length of the tube. These two actions continue to push food through all the alimentary canal. The involuntary process is called peristalsis. When this action is reversed, it is called antiperistalsis and produces vomiting.

Release of urine

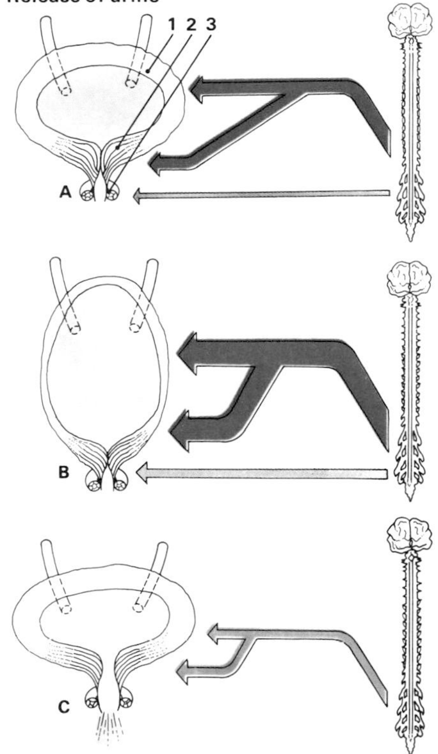

(**A**) The release of urine from the bladder is controlled by the autonomic nervous system, which acts on the muscular wall (**1**) of the bladder and an internal sphincter (**2**). Voluntary nerve impulses control the actions of an external sphincter (**3**), so that urine can be voided at convenient times. (**B**) As the bladder fills, sympathetic nerve impulses (red) increase, keeping the internal sphincter firmly closed. (**C**) When the bladder is nearly full the sympathetic impulses cease and the parasympathetic nerve output (blue) causes the bladder wall to contract and the internal sphincter is opened. As this occurs, controlled voluntary impulses may also cease and urine can then be released.

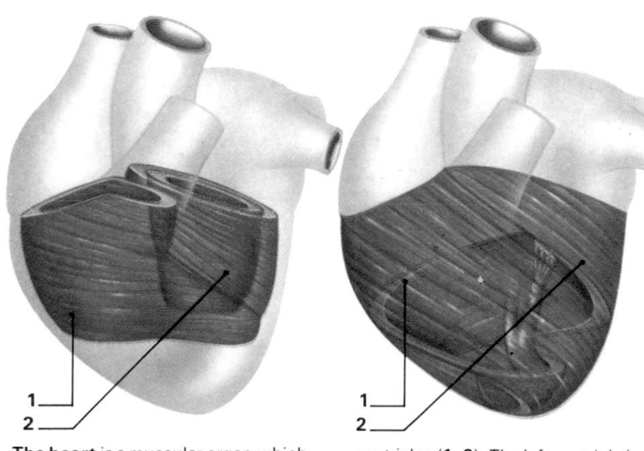

The heart is a muscular organ which pumps blood throughout the body. The continuous driving force is generated by the action of cardiac muscle—a specialized form of muscle. The muscle is arranged as whorls of tissue surrounding the two cavities of the ventricles (**1, 2**). The left ventricle is larger as it is required to pump blood around the general circulatory system. The right ventricle only pumps blood along the shorter circuit through the lungs. The autonomic nervous system controls changes in the rate of heartbeat.

Cross-reference	pages
Bladder	**74-77**
Brain and central nervous system	**66-67**
Cardiac muscle	**56-57**
Circulation	**54-55**
Eye	**82-83**
Hormones	**70-73**
Peristalsis	**46-47**

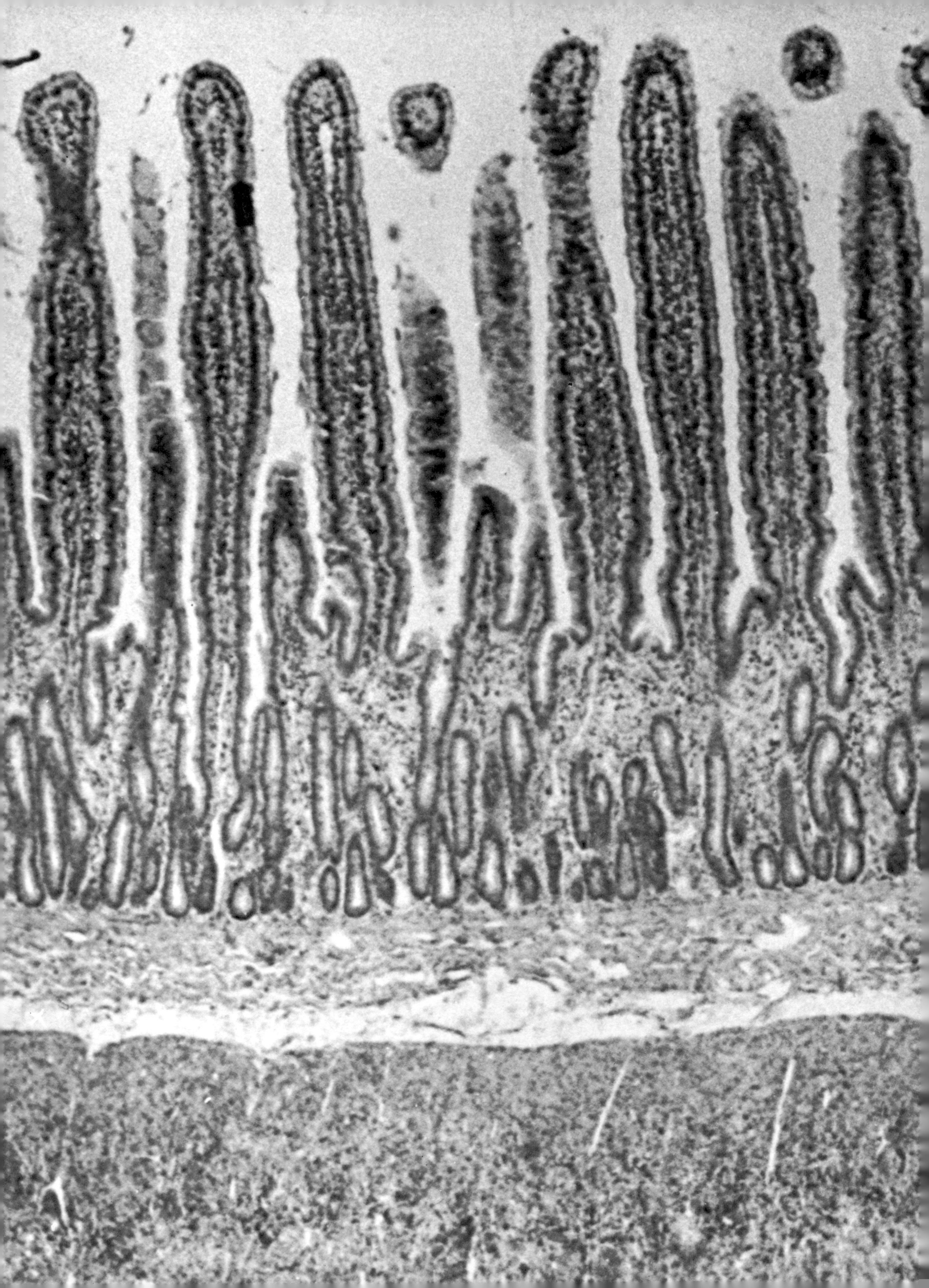

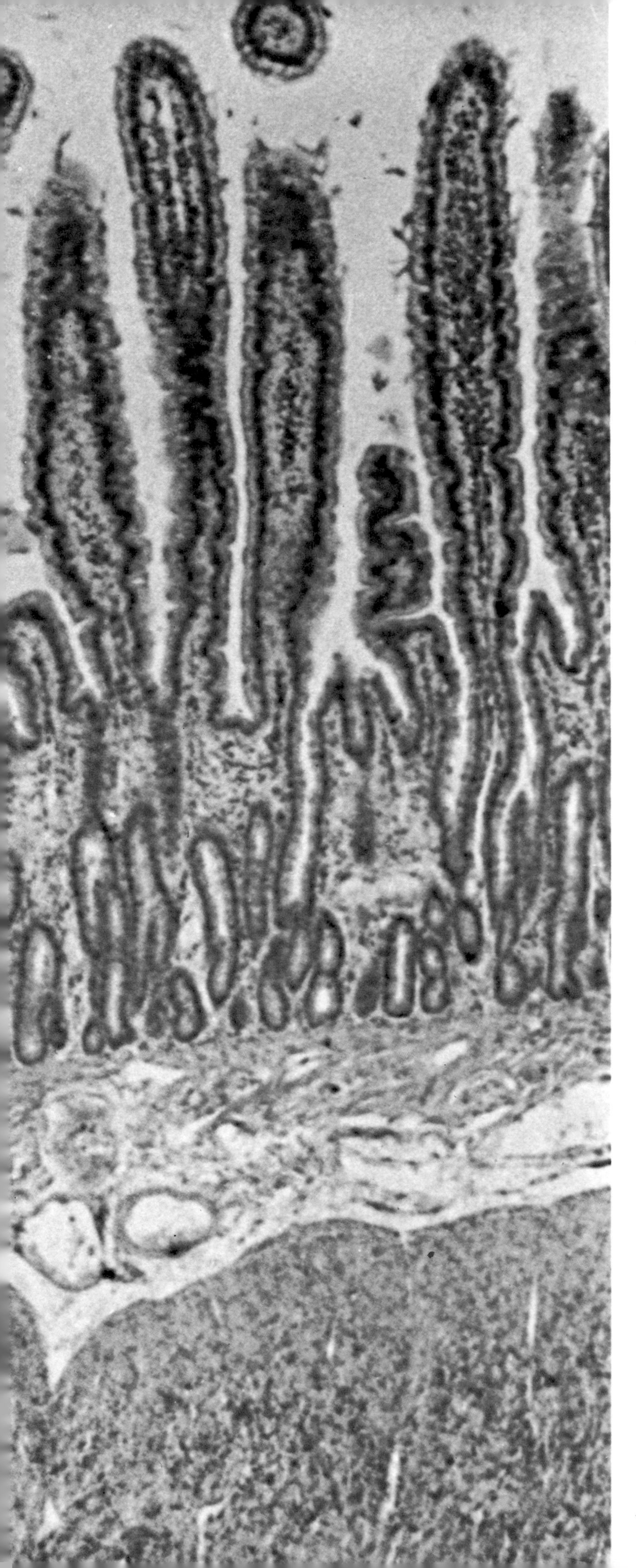

Energy for the machine

Activity of any kind, in any part of the universe, involves the same basic transaction—the exchange of energy between one structure and another. Every living organism has its own organization of energy input and usage, and the human organism is no exception. To serve our sophisticated energy requirement, we have within our bodies a series of interrelated systems which are involved in the essential functions of gathering in energy-containing materials, using them in the performance of the body's activities and disposing of unwanted leftovers.

The body has an unceasing demand for sources of energy, and these must be provided in such a way that each one can be used precisely as and when it is needed. Thus food, the prime source of energy, passes through the digestive system, where it is broken down into substances that are absorbed into the bloodstream and then circulated to all the cells of the body. Within the cells, the chemical energy stored in food fuels is released by the process of oxidation, for which the essential oxygen is gathered by the lungs and transported in the bloodstream. Chemical reactions take place all over the body as nutrients are passed from cell to cell, tissue to tissue and system to system. Vital to the transaction is the disposal of the waste products that are generated during the energy conversion process. They are carried away from cells by the blood, to be excreted by way of the kidneys and lungs.

An important feature of this energy cycle is the body's ability to store and to redistribute nutrients in accordance with periods of high or low demand. The materials essential to maintain life have markedly different storage times—the human body can survive for weeks without food, for days without water, but for only a few minutes without oxygen.

Food materials, the source of body energy, can only be used if they are absorbed into the bloodstream. This mainly occurs through the walls of the small intestine. To aid this uptake, the intestinal wall, seen left, magnified about one hundred and fifty times, is highly folded to create a greater area for absorption.

Diet: energy requirements

We are built from the food we eat. Since we are mostly water (about 60 percent of the weight of an average man), it is not surprising that water is also the largest single component of our diet. We consume about half a gallon of water a day—not just as the liquids we drink, but also in the food we eat. Both sources can fluctuate greatly, but, on average, nearly half the weight of our food is water. Most of the rest is a mixture of carbohydrate, fat and protein. The average daily Western diet includes about ten ounces of carbohydrate, five ounces of fat and four to five ounces of protein.

Carbohydrates, fats and proteins are the body's major fuels, and the energy they provide is measured as calories. One ounce of fats provides about 250 Calories, and an ounce of most carbohydrates provides 110 Calories. The exceptional carbohydrate is ethanol, or alcohol, which would provide almost 200 Calories an ounce if we could manage to drink it undiluted. In fact alcoholic drinks are dilute solutions, with even the strongest spirits containing only one-third alcohol, about 65 Calories an ounce. Protein provides about the same number of calories as carbohydrate and is also important in body structure.

There are two types of carbohydrate in the diet. There are those which are usually classed as sugars and are sweet-tasting, and there are those which are classed as starches, which are bland-tasting and are found in all cereals. One of the simplest sugars is glucose, and it is this carbohydrate that the body cells use. When we normally speak of sugar, we are referring to sucrose, a compound made of a molecule of glucose joined to another simple carbohydrate, fructose. Starches are also built of glucose units, joined together in long branching chains. During digestion, the complicated carbohydrates are split into the glucose or fructose from which they are made.

Proteins are also built of chains of simple building units called the amino acids. There are twenty different types of amino acid, which get their name from the amino, nitrogen-containing, group they include. The thousands of different proteins the body makes are composed of differing amounts of amino acids in differing sequences, and the right mixture must be present for the body to make proteins. Fortunately, twelve of the amino acids can be easily interconverted in the body, and it is only the amounts of the remaining eight which are critical in the diet. The lack of any one of these essential amino acids will prevent the synthesis of the protein which contains them and render the rest of the amino acids useless.

Fats are made of glycerol molecules (shaped like the letter E) linked to three molecules of fatty acids. Although we often distinguish between fats and oils, this distinction is entirely arbitrary—oils are merely fats that are liquid at room temperature. The temperature at which fats melt depends upon the type of fatty acid they contain. Unsaturated fatty acids melt at very low temperatures, and oils containing these are almost all liquid. Unsaturated fatty acids are found in many plant oils and in the tissues of those animals, including whales and fish, which live in very cold climates.

Cholesterol is one of the additional compounds of animal fat. The body is able to make its own supply of cholesterol, but any excess in the diet gets absorbed and has to be excreted. While cholesterol is an essential body component, it is believed that an excess amount of cholesterol taken in with animal products might be a factor leading to heart disease.

The energy that foods contain is needed by the body to provide heat and for work. If we eat more than we can get rid of, the excess energy is stored as triglycerides in adipose tissue. The only way to become fat is to eat too much "energy," and the only way to diet is to eat less or to expend more energy. Since the body can make fat from protein or carbohydrate, just eating less fat in the food is no shortcut to dieting.

Much is said about carbohydrates being fattening and proteins slimming, and although they contain the same number of calories, the theory appears to have some substance. Their different effects result because the food we do not eat does not make us fat. While we can eat relatively large amounts of carbohydrates, especially sugar, we are satisfied by far fewer calories if we restrict ourselves to foods that contain only protein and fat.

Many things affect the number of calories an individual needs, and, therefore, how much food can be eaten without gaining or losing weight. People doing a lot of physical work can use enormous amounts of energy, but mental work alone involves no great energy expenditure. If they are equally active, large people need more calories than small people, and men need more calories than women, but simple comparisons are made difficult by large variations in activity. Children, pregnant and lactating women all need extra energy for growth or for feeding the infant.

The major determinant of individual calorie requirement is the amount of energy the body uses to keep itself going. This is measured as the basal metabolic rate (BMR), the speed of metabolism in a resting person. The basal metabolic rate (adjusted for body size) is low in a newborn infant, but rapidly rises to a peak in the second year. It falls slowly thereafter, until adolescence, then remains constant into adulthood. It declines slowly in middle and old age, changes which parallel and may largely cause variations in the amount of food that is needed. In chronically starving people the basal metabolic rate falls by about 40 percent, helping adaptation to the lack of food.

The protein requirement is greatest in growing children, who need protein for growth. Adults need protein for tissue maintenance. About one ounce of good-quality

About ninety-five pounds of a one hundred and fifty pound man is water. Protein and stored fat account for a further fifty pounds and the rest is glycogen, minerals and other substances, such as nucleic acids.

water
protein
fat
minerals
glycogen
others

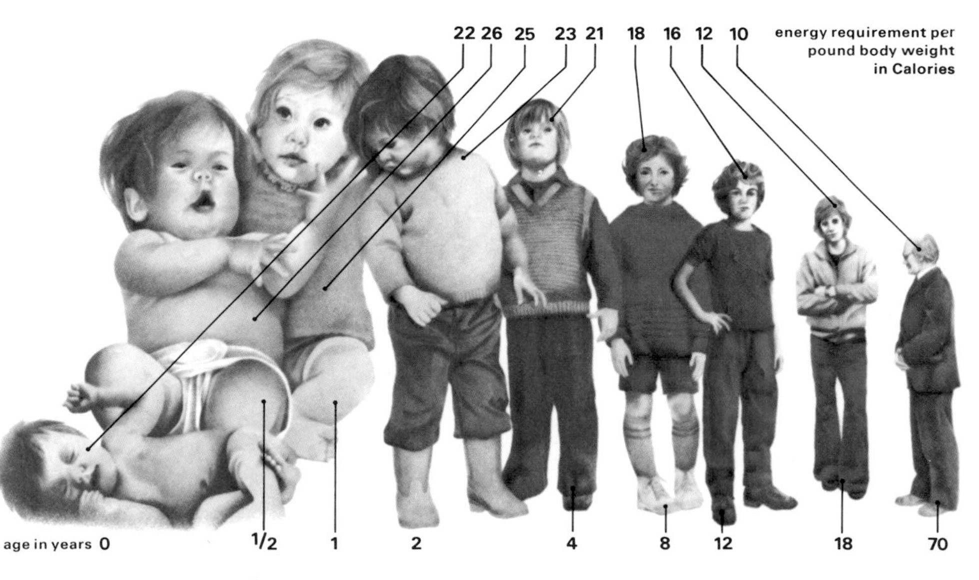

The resting, or basal, metabolic rate, the amount of energy in relation to body weight required to sustain such basic functions as the heartbeat, breathing and the maintenance of body heat, decreases after infancy. A newborn baby needs about twice the energy of an adult because it is growing and the proportion of such metabolically active tissues as the liver, brain and skeletal muscle is increasing. Energy requirement falls rapidly until adolescence, when it is only slightly higher than in adults of the same weight. Adult resting metabolic rate then declines every decade as people become less active.

protein is sufficient for this. Almost any balanced diet would provide it, since most foods contain surprising amounts of protein. Bread, for example, provides about as much protein as beefsteak, watercress or milk.

About fifteen highly complicated organic compounds are needed by the body in minute amounts to keep it in good repair. These compounds are called vitamins and are usually known by the letters given to them at the time of their discovery. Some vitamins, however, have specific names—biotin and folic acid, for example—while others have a collection of aliases. Pyridoxine, pyridoxal and pyridoxamine are all pseudonyms of what is more simply called vitamin B_6.

The different vitamins have a variety of different effects and not all their roles are understood. Some of them seem to be important in helping the body's enzymes work, while the function of the others is unknown. The difficulty in understanding what vitamins do must be due, in part at least, to the effects of vitamin deficiency, which can be bizarre and apparently unrelated. Lack of vitamin A, for example, leads to a failure to see in the dark, and eventually to blindness, blocking of the sweat glands, flaking skin, increased susceptibility to infection, and infertility.

Many vitamins are found in nearly all foods, and they are, in general, so ubiquitous that anyone eating a varied diet is never likely to run out. The occasional exceptions, such as folic acid deficiency in pregnancy or vitamin B_{12} deficiency in people lacking the ability to absorb it, are well recognized medical problems. So is the result of the overenthusiastic consumption of vitamin pills, which results in occasional deaths from poisoning by too much of vitamins A and D.

The other important group of nutrients is the minerals, or salts, of which we eat about one ounce each day. Most of this is sodium chloride, or common salt, a chemical not present in great amounts in foods, but which is added to them. Sodium chloride is found in all body tissues and a supply of this is needed to replace losses in sweat. Except for people living in tropical areas, however, the loss in sweat is very small, and our taste for salt is well in excess of any physiological need.

The need for some of the other minerals is obvious from their role in the body. Calcium and phosphorus are needed for bones and teeth and to help cells to work. Iron is needed for forming hemoglobin and iodine for thyroid hormones. But there are, too, a host of trace elements—minerals present in the diet or body in only small amounts. Their importance is not always fully understood because they are needed in such small amounts that almost every diet contains enough. For example, although our diet provides only one four-thousandth of an ounce of the element manganese a day, it seems to be enough to meet any need we might have, so it still is not known if it is important.

The final item of food, chance poisons aside, is only important because the body does not use it. The fiber in the diet is not digested and, therefore, never enters the body. It remains in the intestine and that is the reason it is important. It has long been known that fiber prevents constipation, but scientists now believe that its presence in the bowel helps the body get rid of such waste products as cholesterol and thereby remain healthy.

Vitamins

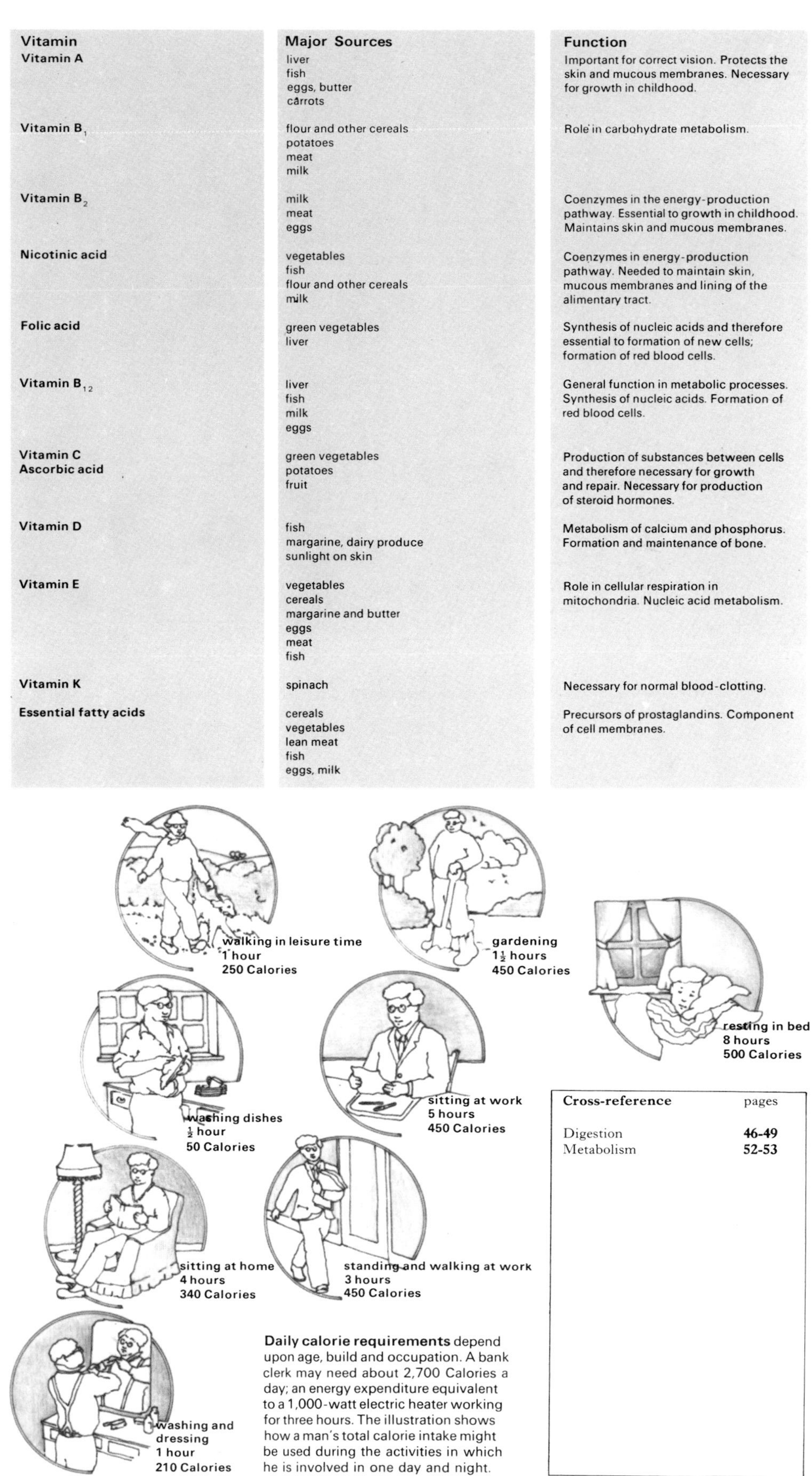

Vitamin	Major Sources	Function
Vitamin A	liver fish eggs, butter carrots	Important for correct vision. Protects the skin and mucous membranes. Necessary for growth in childhood.
Vitamin B_1	flour and other cereals potatoes meat milk	Role in carbohydrate metabolism.
Vitamin B_2	milk meat eggs	Coenzymes in the energy-production pathway. Essential to growth in childhood. Maintains skin and mucous membranes.
Nicotinic acid	vegetables fish flour and other cereals milk	Coenzymes in energy-production pathway. Needed to maintain skin, mucous membranes and lining of the alimentary tract.
Folic acid	green vegetables liver	Synthesis of nucleic acids and therefore essential to formation of new cells; formation of red blood cells.
Vitamin B_{12}	liver fish milk eggs	General function in metabolic processes. Synthesis of nucleic acids. Formation of red blood cells.
Vitamin C Ascorbic acid	green vegetables potatoes fruit	Production of substances between cells and therefore necessary for growth and repair. Necessary for production of steroid hormones.
Vitamin D	fish margarine, dairy produce sunlight on skin	Metabolism of calcium and phosphorus. Formation and maintenance of bone.
Vitamin E	vegetables cereals margarine and butter eggs meat fish	Role in cellular respiration in mitochondria. Nucleic acid metabolism.
Vitamin K	spinach	Necessary for normal blood-clotting.
Essential fatty acids	cereals vegetables lean meat fish eggs, milk	Precursors of prostaglandins. Component of cell membranes.

Daily calorie requirements depend upon age, build and occupation. A bank clerk may need about 2,700 Calories a day; an energy expenditure equivalent to a 1,000-watt electric heater working for three hours. The illustration shows how a man's total calorie intake might be used during the activities in which he is involved in one day and night.

Cross-reference	pages
Digestion	46-49
Metabolism	52-53

Digestion: the breakdown of food

Man is a machine built from, and fueled by, food. Like the food he eats, he is mostly water, with varying amounts of protein, fat, carbohydrate and salts. If food is to be utilized, however, it must first be broken down into molecules small enough to be absorbed. This is the duty of the digestive tract—a twenty-six-foot-long tube from mouth to anus in which food is subjected to a coordinated thermal, mechanical and, above all, chemical assault called digestion. Food enters the digestive system at the mouth, where it is warmed or cooled to the optimal temperature for digestion. Solid food is torn and ground by the teeth into small pieces, which are easier to swallow and to digest. Dry food is now mixed and lubricated with saliva to ease its passage down the esophagus. Each day the salivary glands produce about three pints of saliva which contains the starch-digesting enzyme ptyalin (amylase). This however, has little chance to act before food is swallowed.

Food leaves the mouth when we decide to swallow, a process which once begun is beyond conscious control. When the ball of food, the bolus, leaves the mouth, it passes into the narrow esophagus, where powerful waves of muscular contraction, called peristalsis, squeeze it down to the stomach.

The adult stomach is a distensible sac which can hold up to two and a half pints. It serves to store the food so that it can be slowly pushed on to the duodenum at the rate required, while the gastric juice it produces starts the breakdown of the proteins in the food and kills any contaminating bacteria. Like the rest of the digestive tract, the activities of the stomach are controlled by both hormones and nerves. The sight, the smell and even the thought of food will stimulate the production of gastric juice. The continuation of gastric juice production is controlled by the presence of food itself, which causes the stomach to release into the blood the hormone gastrin, the action of which further stimulates the glands that secrete this juice.

There are three major components of gastric juice—digestive enzymes, hydrochloric acid and mucus. Digestive enzymes are compounds which cause the chemical breakdown of foods. The one of most importance in the stomach is

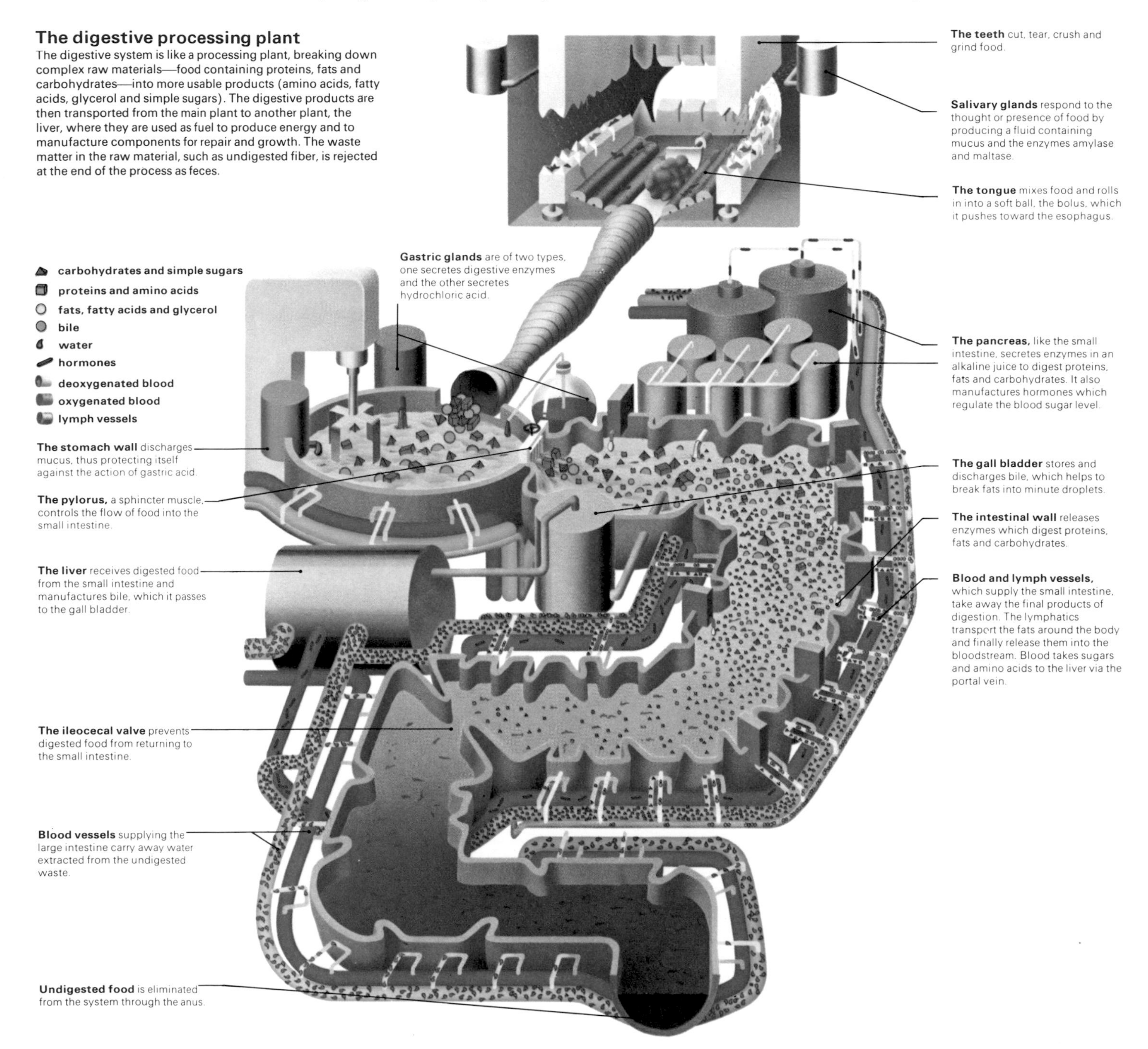

The digestive processing plant
The digestive system is like a processing plant, breaking down complex raw materials—food containing proteins, fats and carbohydrates—into more usable products (amino acids, fatty acids, glycerol and simple sugars). The digestive products are then transported from the main plant to another plant, the liver, where they are used as fuel to produce energy and to manufacture components for repair and growth. The waste matter in the raw material, such as undigested fiber, is rejected at the end of the process as feces.

pepsin, which begins the digestion of protein. The stomach secretes pepsinogen, from which pepsin is formed by the action of hydrochloric acid, which also aids its chemical action and kills most bacteria. Mucus has no role in digestion, but protects the gastric lining from any risk of self-digestion. At the same time as the stomach is secreting gastric juice, it is also rhythmically contracting, mixing food and digestive juice, and slowly squeezing its contents toward the exit to the duodenum—the first ten inches of the small intestine, where digestion really proceeds.

The rate at which food leaves the stomach is controlled by its chemical composition. Carbohydrate-rich food leaves in a few hours, protein-rich food is retained longer, while it may take many hours before fatty food is gone. This intricate control is achieved by the rate of contraction of the stomach being slowed, to varying degrees, by exactly what it ejects into the duodenum. Proteins and fats, which are more difficult to digest than carbohydrates, cause the duodenum to release into the blood varying amounts of the hormone enterogastrone, which reduces gastric activity, so that the duodenum receives food at the rate it can deal with it rather than at the rate the food is eaten.

Other hormones produced by the duodenum act on the glands which open into the small intestine. First, as soon as acid is detected in the duodenum, the hormone secretin is released. This stimulates the pancreas to secrete a watery alkaline juice that neutralizes the corrosive hydrochloric acid. The hormone pancreozymin then causes the pancreas to start the somewhat slower production of enzymes to break down proteins and start the digestion of the carbohydrates, the fats and the nucleic acids.

The presence of fats causes the duodenum to release yet another hormone, cholecystokinin, which makes the gall bladder contract. As a result, bile, a thick green alkaline solution of salts and pigments made in the liver and stored in the gall bladder, is poured out and mixes with the food. Bile salts are necessary for the efficient digestion of fats, which they emulsify so that they are more easily acted on by enzymes.

One end of each bile-salt molecule is soluble in fat, the other is soluble in water, so fats received by the small intestine for digestive breakdown link themselves to the fat-soluble ends of the bile salts. Lecithin and cholesterol then help to disperse the fat globules into small, stable particles called micelles. They present a vast surface area which is now exposed to the pancreatic enzyme lipase and are digested. The bile salts are also necessary for the absorption of some fat-soluble vitamins.

This complicated process of producing hormones in response to the type of food being digested, and using these hormones to control enzyme release and the movement of food, is the reason that the digestive system can, with apparent equanimity, cope with the bewildering variety of foods—from ice cream to caviar—that social custom or taste leads us to include in our diet.

By the time it leaves the duodenum almost anything we have eaten looks the same—a watery mixture of partly digested food, enzymes and other secretions. This mixture, called chyme, still needs further processing before it can be absorbed. Both of these processes occur in the next and longest part of the small intestine.

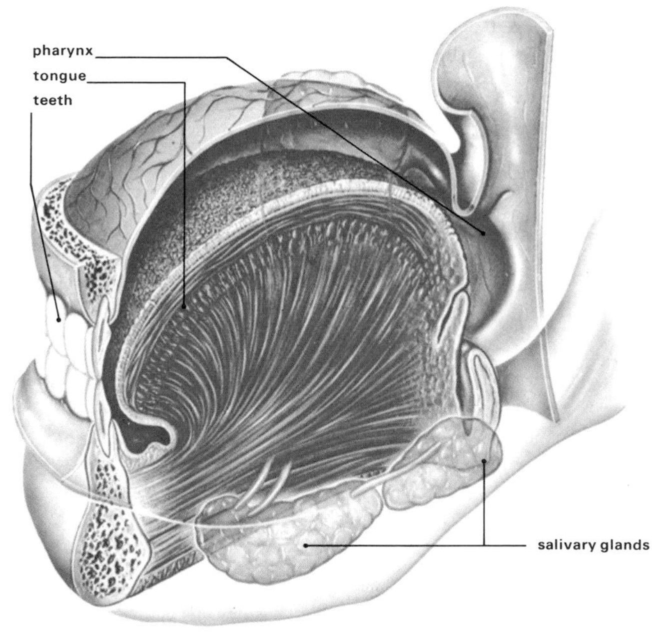

The mouth is the beginning of the digestive system. It is here that the first stages of food breakdown begin. When food enters the mouth, saliva, a fluid containing the digestive enzyme ptyalin and mucus, is secreted by the salivary glands. Saliva not only lubricates the food and begins its chemical digestion but it also helps to keep the mouth cavity moist and clean. The teeth function mechanically to break up the food into smaller, more readily swallowed and digestible pieces. The tongue and the muscular walls of the mouth shape the food into a moist ball, the bolus, which is pushed to the back of the mouth and into the pharynx to be swallowed.

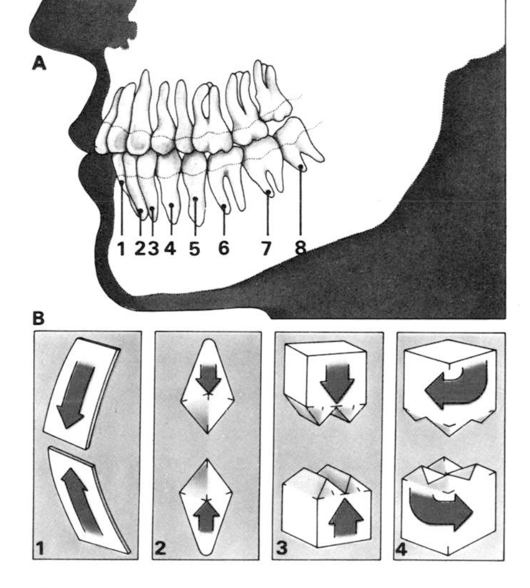

The permanent teeth of an adult (**A**) consist of the incisors (**1, 2**), the canines (**3**), premolars (**4, 5**) and molars (**6, 7**), including wisdom teeth (**8**)—which may not erupt until twenty-five years of age, if at all.
(**B**) Each type of tooth has a specific function in dealing with food and reducing it to conveniently sized small pieces for swallowing—incisors (**1**) bite and cut, canines (**2**) bite and tear, premolars (**3**) crush and the molars (**4**) crush and grind.

The throat is the pathway linking the mouth, nose, esophagus and trachea. When food is swallowed the soft palate closes off the postnasal cavity to prevent food from entering. At the same time a flap of elastic cartilage, the epiglottis, temporarily closes off the trachea to prevent food and liquid entering.

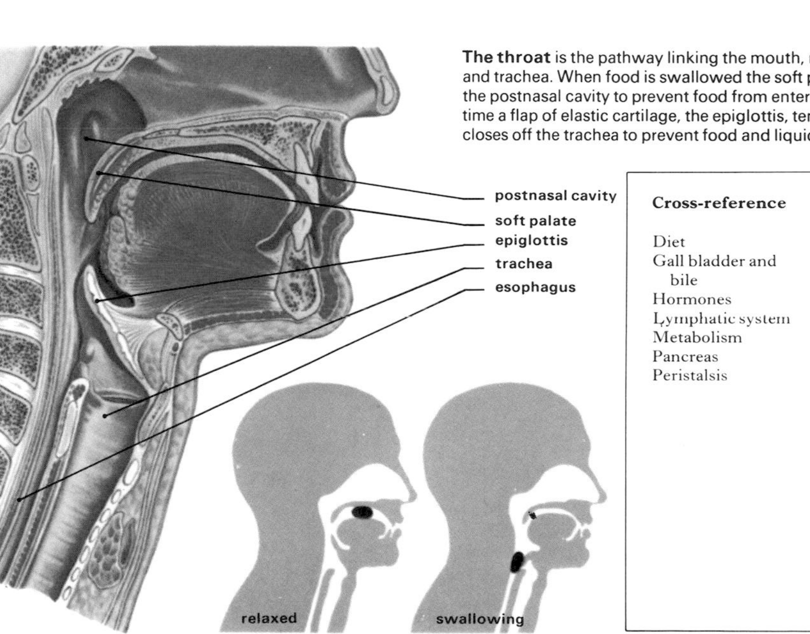

Cross-reference	pages
Diet	44-45
Gall bladder and bile	50-51
Hormones	70-73
Lymphatic system	54-55
Metabolism	52-53
Pancreas	72-73
Peristalsis	40-41

Stomach and intestines: the absorption of food

The logic of digestion and absorption is best understood when it is realized that a human being is a kind of fancy-shaped doughnut, whose digestive tract is the hole in the middle. Although we talk of our intestines as our "insides," food in the intestine is in fact still outside the body. The object of digestion is to reduce that food to particles small enough to pass across the "inner skin" of the intestinal wall and into the body. If absorption could occur without digestion, then, apart from the problem of the body chemicals leaking out, the large proteins and carbohydrates of food would trigger the immune system of the body, which is designed to protect it from infection.

When the chyme reaches the small intestine, although some food substances are ready to be absorbed, further processing is required for most. Food protein has already fragmented into chunks called peptones and these are now broken into the absorbable amino acids. Sugars and starches, similarly fragmented in the duodenum, are broken down into the individual monosaccharides from which they are built. Fats emulsified by the action of bile are split into particles which can be absorbed.

All this is achieved by the enzymes contained in the intestinal secretion, which is produced by the millions of cells that line the crypts of Lieberkühn in the intestinal wall. About five pints of this secretion are produced a day, but only in response to food, a control achieved by the hormone enterokinin, which is produced by the chyme-filled duodenum. The food is slowly pushed along the small intestine by rhythmic contractions, which also break up any remaining lumps.

As this final breakdown of food is occurring, the end products are being absorbed across the intestinal wall into the body. The small intestine is a narrow tube which is about twenty-one feet long, and the amount of absorption required is enormous. The total volume of food, drink and gastrointestinal secretions is some two and a half gallons per day—of which only about one-fifth of a pint is finally lost in the feces. To facilitate the absorption of nutrients as well as water, the wall of the small intestine is folded to increase its surface area. Not only do broad, spirally arranged folds create a series of permanent ridges down its length, but also these are in turn covered on the inner wall by minute fingerlike projections called villi.

The villi are just visible to the naked eye and are said to give the intestine a velvety appearance. The analogy is a good one. The villi increase the absorptive surface of the intestine by about five times what it would be if smooth, just as a piece of velvet has a much larger surface area than, say, a piece of smooth cotton cut to the same dimensions.

The villi themselves are further subdivided, their surface being covered in the microvilli, or brush border, so small that only the most powerful microscope can detect it. One square inch of intestinal wall is covered by something like twenty thousand villi and more than ten thousand million microvilli. This picture of folds on folds on folds gives a total surface area of the small intestine of 350 square yards, or two hundred times the surface area of the whole of the skin. Although this enormous surface makes the absorption of food highly efficient, it also increases the area of cells exposed to wear and tear. A phenomenal four ounces of cells may be shed every day from the intestinal wall, representing a potential loss of more than an ounce of protein alone. Fortunately, the small intestine is as efficient at digesting its own sloughed dead cells as it is with food, so that most of this protein is recovered.

The successful absorption of nutrients depends on their transport away from the cells of the villi. Each villus contains a network of capillaries which drain, via the mesenteric veins, into the hepatic portal system and small blind sacs called lacteals, which feed the lymphatic system. Most nutrients passing into the villus are carried away in the bloodstream, but fat and fat-soluble vitamins find their way chiefly into the lacteals. These are frequently squeezed out by rhythmic contractions of the villi, which provide the pump for the intestinal lymphatic system.

Once absorption is complete, a watery mix containing mainly fibrous waste, indigestible cellulose, some salts and unwanted breakdown products remains, and this now passes through the ileocecal valve into the large intestine. (Just below this valve is the hollow cul-de-sac of the

The small intestine

The small intestine, a twenty-one-foot-long tube, lies in coils in the abdomen. The first ten to twelve inches are called the duodenum, the next eight feet the jejunum and the remaining twelve feet the ileum. Its major function is to complete the digestion of food passing into it from the stomach and to absorb the final products of digestion.

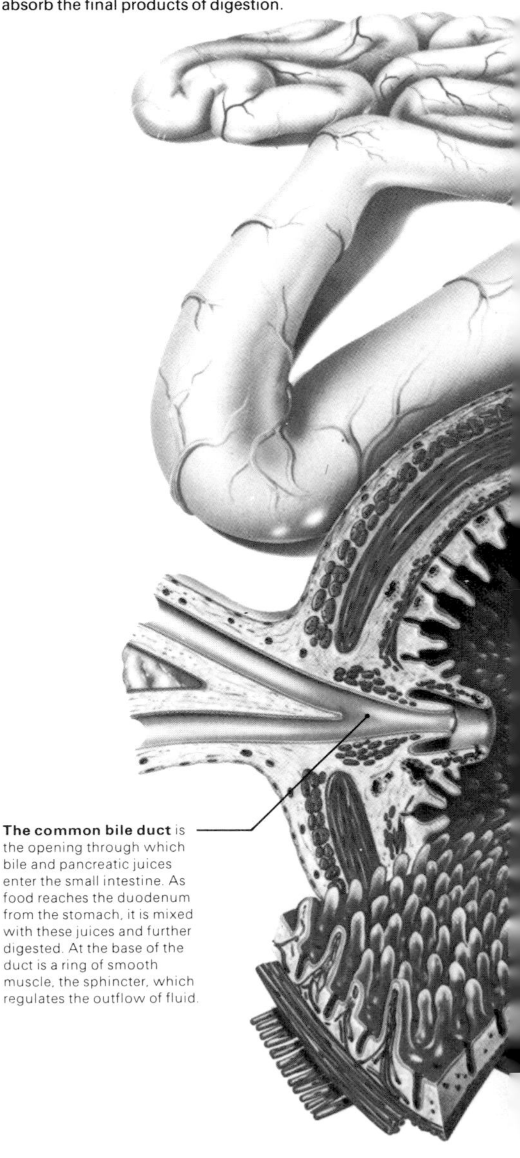

The common bile duct is the opening through which bile and pancreatic juices enter the small intestine. As food reaches the duodenum from the stomach, it is mixed with these juices and further digested. At the base of the duct is a ring of smooth muscle, the sphincter, which regulates the outflow of fluid.

The digestive system

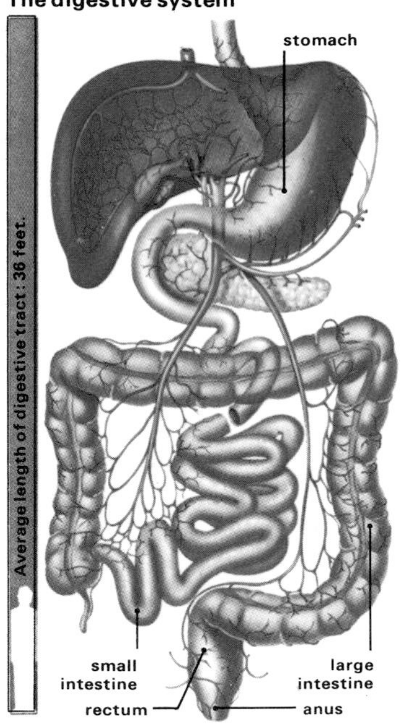

From the mouth, food passes down the throat into the stomach, where a churning action mixes it into a semiliquid mass. It is then forced into the small intestine, where the major part of digestion and absorption occurs. Undigested material passes into the large intestine, where most of the remaining water is absorbed by the bloodstream. Solid waste collects in the rectum and is expelled through the anus.

The stomach

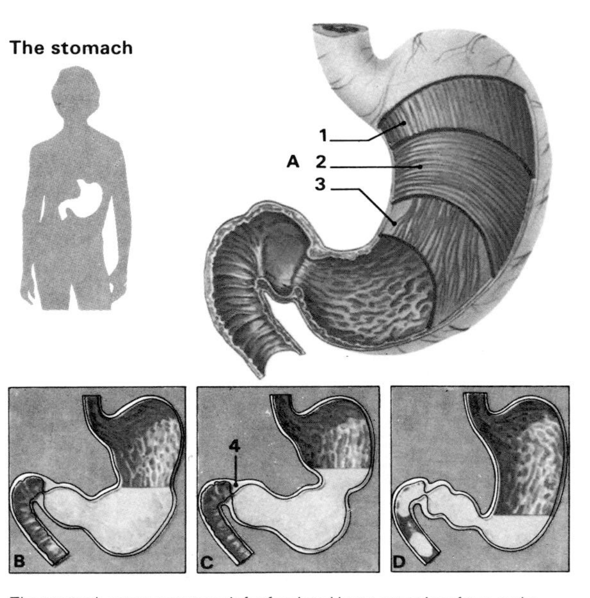

The stomach acts as a reservoir for food and has a capacity of two and a half pints. Within the stomach, solid food material is churned and kneaded for about three hours until it becomes a semiliquid mass known as chyme. The chyme is then forced into the small intestine, where the process of digestion is completed. The wall of the stomach (**A**) has three muscular layers, an outer longitudinal layer (**1**), a middle circular layer (**2**) and an inner oblique layer (**3**). As the stomach fills with food, wavelike contractions of the wall begin (**B**), and as these waves move along the stomach wall (**C**) some of the food is passed through the relaxed muscle valve at the base (**4**) and into the duodenum (**D**), the first part of the small intestine, where it is further digested before being absorbed into the body.

appendix, apparently useless and regarded as a relic of man's past evolution.) On a normal Western diet, about twelve ounces of chyme enters the cecum every day. The continuing reabsorption of water reduces its weight by two-thirds, forming the feces. These weights, however, are extremely variable—much reduced on a refined diet and considerably increased by the consumption of roughage.

Most of the large intestine is colonized, permanently and harmlessly, by bacteria which use the indigestible residues as nutrients and may synthesize some of the body's vitamin supplies. There is a continual loss of these bacteria into the feces.

Like the stomach the large intestine, particularly the descending colon and the rectum, has use as a store. Although the presence of increasing amounts of feces causes a reflex desire to defecate, the large intestine can store considerable amounts of fecal matter.

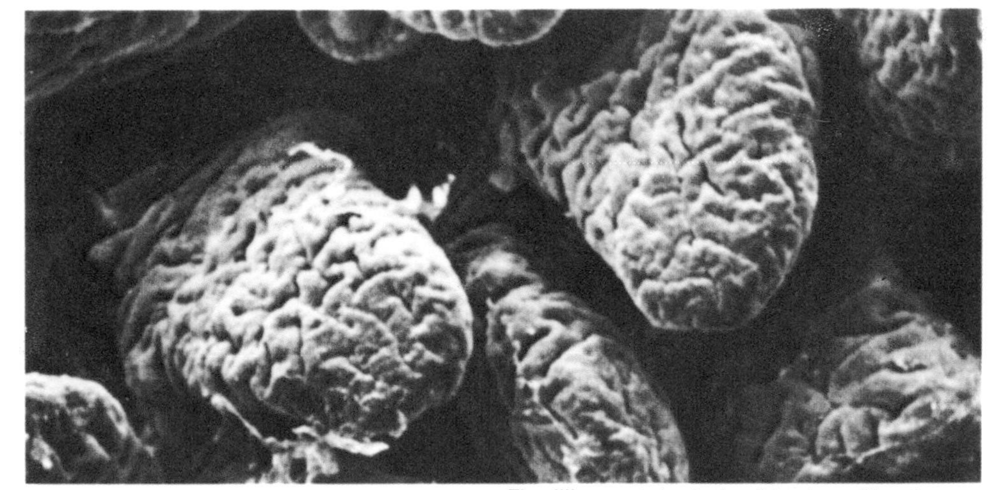

The villi, corrugated projections lining the small intestine, seen under a scanning electron microscope and magnified more than three hundred times. Villi create an enormous surface area which efficiently absorbs the end products of digestion.

The jejunum and the ileum are two subdivisions of the small intestine and connect the duodenum with the large intestine.

The muscle layers of the small intestine are arranged longitudinally on the outside and circularly on the inside. Between the layers are a series of nerve fibers which control the muscular movements that mix and move food along.

The duodenum is the short U-shaped section joining the stomach and jejunum.

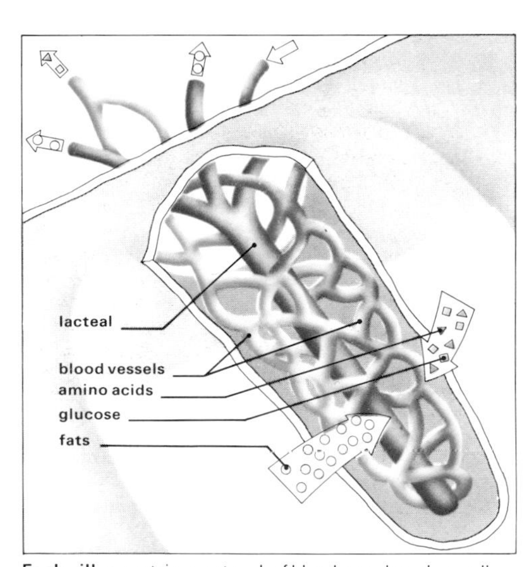

Each villus contains a network of blood vessels and a small lymphatic vessel, the lacteal. Some products of digestion—amino acids and glucose—pass into the blood vessels and then into the hepatic portal vein, which carries them to the liver. Most fatty acids and glycerol are recombined to become fats in the intestinal lining and are absorbed into the lacteals. The fluid in the lacteals passes into the lymphatic vessels and is emptied into the bloodstream.

Tributaries of the portal vein carry digested materials away from the small intestine to the liver.

An artery, one of several which supply the small intestine with blood.

Brunner's glands pour out an alkaline juice to neutralize the acid chyme reaching the small intestine and to create the necessary conditions for further digestion.

Crypts of Lieberkühn are numerous tubular glands which open into the intestine from between the villi. They secrete an alkaline fluid rich in digestive enzymes and mucus. The mucus protects the tissues of the small intestine from being digested by its own enzymes.

The villi are fingerlike projections lining the small intestine. Digested food is absorbed into the cells lining the villi.

Cross-reference	pages
Circulation	54-55
Digestion	46-47
Fat-soluble vitamins	52-53
Gall bladder and bile	50-51
Hormones	70-73
Metabolism	52-53
Muscles of digestive tract	40-41
Pancreas	72-73

The liver: processing, purification and storage

The liver is a central chemical-processing plant of staggering complexity. Weighing three pounds, it is the largest single gland in the body and with more than five hundred functions so far identified, it is easily the most versatile. Unlike other organs, the liver is served by two distinct blood supplies. The hepatic artery supplies it with fresh oxygenated blood, while the hepatic portal vein transports blood to it for processing. This portal vein supply comes from the stomach and intestines carrying absorbed nutrients which the liver extracts, processes and stores, and from other main abdominal organs—the spleen, the gall bladder and the pancreas.

When the portal vein enters the liver it branches repeatedly, its smaller branches, the interlobular veins, running with branches of the hepatic artery and bile ducts along the connective tissue-lined tubes that surround the liver lobules. The lobules are tiny filters of which the four lobes of the liver are composed.

The liver contains fifty to a hundred thousand lobules, each of which is, in effect, a chamber of hepatic cells honeycombed by an interconnecting maze of sinuses, or caverns, that lead to a central draining vein. The central veins, in turn, join, finally leading to the hepatic vein that takes blood from the liver. The cells of the lobules are washed by arterial blood, bringing them necessary oxygen, and portal blood, from which they remove nutrients, bacteria and old red blood cells. At the same time, into this cascade of blood passing through the microscopic sinuses, they secrete glucose, proteins, vitamins, fats and most of the other compounds required by the body.

Running between the individual cells that line the sinuses is yet another network of tunnels, much smaller than the sinuses, which drains outward, carrying bile toward the local branch of the bile duct, at the edge of the lobule.

Bile, a solution of organic compounds in water, is produced by the hepatic cells. Some of the contents of the bile, pigments that give it its green color, cholesterol and bile salts produced from cholesterol are, in part, waste products from the breakdown of old red blood cells. The economy of the body is such that the waste products are used as essential substances for the digestion and absorption of fats in the intestine. Part of the bile salts and cholesterol in the bowel is absorbed and recycled and part is lost in the "drain" of the intestine, with the pigments in the bile contributing to the color of the feces. Eventually, the pint or so of bile produced each day reaches the gall bladder, a small muscular sac underneath the liver, where it is stored until it is needed in digestion.

When hemoglobin is broken down in the liver much of the iron that it contains is reutilized. Excess amino acids from hemoglobin, or, more generally, from all the other proteins that are broken down, are either used by the versatile hepatic cell for building new proteins or are further broken down to provide energy. Poisonous ammonia produced during this process is converted to urea, a nontoxic compound, and returned to the blood for transport to the kidneys. From there it is excreted in the urine. Some amino acids thus broken down can be converted in the liver into the sugar—glucose—which, with other excess carbohydrates, is stored in the hepatic cells as glycogen, ready for turning back into glucose should the level in the blood start to fall.

This superficially rather curious procedure is part of an intricate mechanism in the body designed to ensure that a constant supply of glucose is available for use by the brain, which normally requires this fuel. Fats are also removed from portal blood by the hepatic cell, where they are either stored, modified into more useful fats and returned to the blood or attached to specially made proteins to form compounds called lipoproteins, by which most fats are transported to those organs that then use them as fuel.

The hepatic cell also acts as a bank account for vitamins A, B, D, E and K, excess amounts of which are stored for a rainy (or at least a low-vitamin) day, when the stores are released to keep up supplies to the tissues. The size of these liver stores can be so enormous that a well-nourished man can go for months without vitamin A and two to four years without vitamin B_{12} with no sign of deficiency. The liver stores smaller, but still important, amounts of some of the other B vitamins.

The chemistry of the hepatic cell enables the liver to act also as a center for protection against potential poisons. Because of this capacity, socially acceptable poisons, like alcohol, and potentially toxic medicines, like barbiturates, can be consumed as custom or physician requires. So-called poisons are compounds which damage body cells before the liver has dealt with them. Interestingly, the livers of strict carnivores, those who eat no food but other animals, have lost much of this detoxifying ability. They eat those animals whose livers have already filtered out potential poisons. Consequently, compounds that may be harmless to human beings, if eaten, for example, by cats, can have severe and bizarre effects.

Perhaps what is most surprising of all is that the site of this immense chemical plant—factory, warehouse and sewage works—is the simple hepatic cell. Since it is unlikely that in the foreseeable future it will be possible to build machines that can replace the liver, as they can, for example, the kidney, this lack of specialized cells is of immense benefit. For the liver has another property that makes it different from most of the rest of the body. If it is damaged, or part is removed, it can regrow to a large extent to continue its work as before.

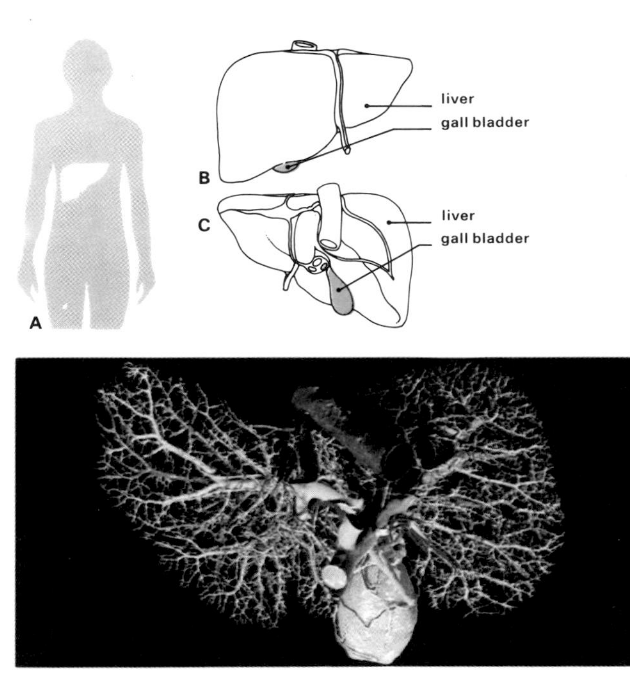

The liver and gall bladder are positioned close to each other high up in the abdomen (**A**). In the front view (**B**), the gall bladder cannot be clearly seen as it lies tucked underneath the liver. In the back view of the liver (**C**), however, the pear-shaped gall bladder is clearly visible. The liver produces bile juice, which passes down the hepatic duct to be stored in the gall bladder. Periodically, bile is released into the duodenum, where it aids the digestion of fats and neutralizes the acidity of chyme—food partly digested in the stomach.

Within the liver at any one moment is about one pint of blood—13 percent of the body's total blood. Seen in the resin cast of the liver, right, are the branches of the hepatic artery (red), hepatic vein (dark blue) and the portal vein (light blue), which form a dense network of blood vessels.

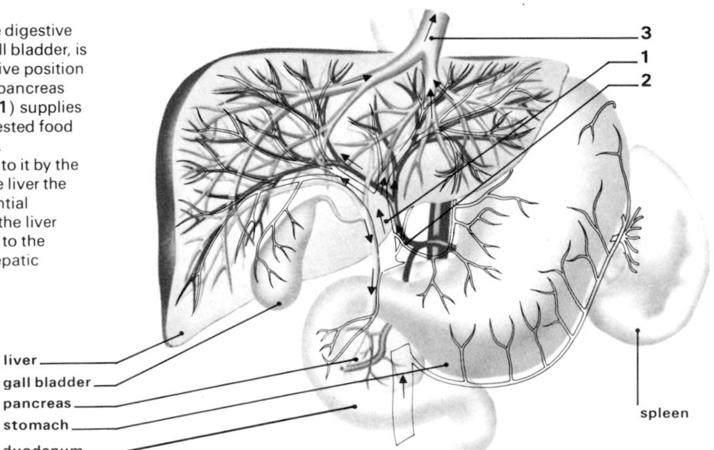

An accessory organ of the digestive system, the liver, with the gall bladder, is shown here, right, in its relative position to the stomach, duodenum, pancreas and spleen. The portal vein (**1**) supplies it with blood containing digested food materials from the intestines. Oxygenated blood is carried to it by the hepatic artery (**2**). Within the liver the blood is processed and essential substances are extracted by the liver cells. The blood then returns to the general circulation via the hepatic vein (**3**).

The architecture of the liver

The liver, a cutaway section of which is shown here, is the largest organ in the body. Liver tissue is composed of a compact mass of multisided units, the hepatic lobules. Each lobule consists of a central vein surrounded by plates of liver cells. The liver receives blood from two sources; 80 percent, which carries digested food materials, arrives from the intestines via the portal vein; the remaining 20 percent is oxygenated blood from the heart, which enters through the hepatic artery. An exchange of materials takes place between the liver cells and the blood, which then passes into the central veins and returns to the general body circulation via the hepatic vein. When an adult is at rest, about two and a half pints of blood flow through the liver each minute.

Hepatic lobules, structural units of the liver, consist of a central vein surrounded by plates of liver cells. A continuous stream of blood flows from the arterioles and venules around each lobule to the central veins.

The bile canaliculus carries bile juice from the liver.

A central vein, a tributary of the hepatic vein, conducts "processed" blood away from a lobule.

Sinusoids, spaces between plates of liver cells, are composed of tributaries of the hepatic artery and portal vein. Sinusoids conduct blood to the central vein.

Branches of the bile duct convey bile from the lobules to the gall bladder.

Hepatic arterioles carry oxygenated blood from the heart to the liver cells.

Portal venules conduct blood carrying digested food materials from the intestines to the liver.

Cross-reference	pages
Blood	**58-61**
Circulation	**54-55**
Digestion	**46-49**
Kidneys	**66-67**
Spleen	**138-139**
Stomach	**47-49**
Vitamins and diet	**44-45**

Metabolism: the fire of life

During his lifetime the average man eats about fifty tons of food and yet maintains a weight of about 160 pounds. Even the most obese person retains in his body only a minute fraction of what he eats. Most of the food we absorb into our bodies through the process of digestion must sooner or later pass out again.

Few nutrients leave the body unchanged. Most are modified or interconverted in one of the multitude of chemical pathways that are described as metabolism. The breakdown of nutrients is the prerequisite of life, since the energy required to keep the body going comes from this breakdown. In some ways the body is exactly like a flame—fuel (food) is burned, using oxygen from the air, releasing energy as heat or for work. Indeed, the fuels of food can be burned in an ordinary fire. In all the body, however, they are burned slowly and step by step, so that the energy of the fire of life is released a little at a time in a way that is the most useful.

This stepwise regulation of metabolism is achieved by the thousands of different kinds of giant molecules of protein, called enzymes, which are found in all the cells of the body. It is enzymes that make body chemistry work at all, since the fire of life not only burns slowly but also burns in the internal environment of the body, which is three-quarters water.

The fragments of food, nutrients, that are absorbed after digestion are the raw materials of metabolism. They are carried to the cells of the body tissues in the blood, passing into the cells and being used as a fuel or as a building material. Obviously, the body economy would be ill-served if nutrients were burned pell-mell; if, for example, the muscles of the foot were to run out of amino acids for repair because the muscles of the leg had burned them all. Just which fuels get put on the fire is, therefore, carefully controlled.

The first control is the liver, which sieves amino acids and carbohydrates out of the blood as it passes from the alimentary canal into the general circulation. The liver needs some of these nutrients itself, but essentially it acts as a careful controller of the body's nutrient supply. When carbohydrates are absorbed, the liver removes as much as it needs to keep its stores of carbohydrates as glycogen full, releasing them again as carbohydrates in the blood are used up. The liver, which has no protein store, removes amino acids from the blood to make plasma proteins, which it returns to the blood.

The amino acid economy of the body is controlled by various hormones—particularly growth hormone and insulin, which also regulates carbohydrates used. The level of these hormones in the blood controls the quantity of amino acids that such tissues as muscle can take from the blood, and, therefore, the amount that gets used up. Once amino acids enter, the cell uses what it needs for growth and repair and burns the rest as it does carbohydrates or fat.

Absorbed fat passes straight into the general circulation via the thoracic duct. Although the liver removes this fat and builds from it the phospholipids needed for body structures, it does so in competition with the demands of all the other tissues which use fat as fuel or, to a much lesser extent, as a building material.

This scheme of regulation is a careful one. It ensures that the brain is well supplied with glucose—the only fuel it normally uses—and that other tissues use fat or amino acids as fuel when glucose supplies are low.

All the cells of the body break down nutrients. This is as important for the nature of the fragments of food molecules produced as for the energy which they provide. As each type of nutrient is broken down, intermediates can be used for the synthesis of other important compounds, instead of being further fragmented. In this way, some of the amino acids of protein or the glycerol from fats, for example, can be partly broken down, then used to make necessary glucose if there is not

More than half a ton of food per year is consumed, digested and metabolized by the average person. Shown above is the amount of food consumed by an average English family of five in a year. It includes 224 pounds of sugar, 35 pounds of tea, 1,344 pounds of potatoes, 87 pounds of poultry, 92 pounds of biscuits, 1,235 pints of milk, 2,165 eggs, 62 pounds of cheese, 22 pounds of meat, 58 pounds of a wide variety of cakes and pastries, 800 pints of beer, 58 pounds of sausages and 112 pounds of apples.

enough present in the diet. Most broken-down compounds are finally reduced to the relatively simple food fragment derivative acetyl-coA, and from this point onward all foods are oxidized by the same route. Conversely, supplies of acetyl-coA in excess of the quantity that needs to be burned can be used as a simple building block for synthesizing a range of complex compounds.

Because of the continual diversion of breakdown products to synthetic pathways, and the diversity of the compounds that can be made, the body, despite the thousands of complicated chemicals that make it up, needs so very few from the diet. The body can, if necessary, make all carbohydrates and virtually all the fat it requires. While there is a major requirement for amino acids to build proteins, even that is remarkably small. Of the twenty amino acids found in protein, only eight cannot be made in the body. Only vitamins (with the exception of vitamin D) and minerals cannot be made in the chemical melting pot that is the fire of life.

As foods are broken down by a series of steps, the energy they contain is also released in a succession of small amounts. Much of this released energy is used to synthesize a compound called ATP, which is the form in which energy is stored in the cell and the way energy released by catabolism, the process of chemical breakdown, is harnessed for work—or building, by the processes called anabolism.

If we eat more food than we require to provide energy or for tissue growth and repair, the excess is stored in adipose tissue as fat. This fat acts as our energy store and as an outer insulating layer. The advantage of storing excess food as fat is that it is a compact, energy-rich compound and all that is needed for its synthesis is acetyl-coA and appropriate enzymes. Whatever nutrient is available in excess can, therefore, be converted to fat, and the acetyl-coA level is able to act like a buffer tank, available to the fat reserve for oxidation or conversion. However, the amount of food energy consumed each day is usually the same as the amount of energy expended and in most adults body weight varies only slightly on a day-to-day basis.

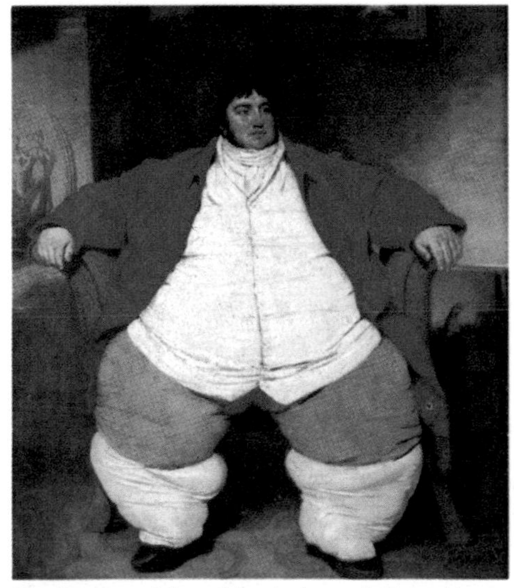

Obesity can result from a failure in the metabolic machinery, which allows most people to maintain a constant weight. An extreme case of obesity was Daniel Lambert, above, who weighed about 742 pounds.

The regulation of metabolism

The pituitary gland controls the metabolism of fats, carbohydrates and proteins by releasing growth hormone and by controlling hormone output of the thyroid gland.

The thyroid gland controls the metabolic rate of the body. It manufactures the hormone thyroxine, which affects the rate of chemical reactions in the cells.

The liver controls the supply of nutrients to the cells. In particular it sieves carbohydrates from the blood and regulates the amount of glucose in the circulation.

The pancreas secretes the hormones insulin and glucagon, which affect carbohydrate metabolism and, to a lesser extent, fat and protein metabolism.

The overall metabolism of the body is governed and constantly monitored by various organs of the body which are anatomically separate, but which function as a complete unit.

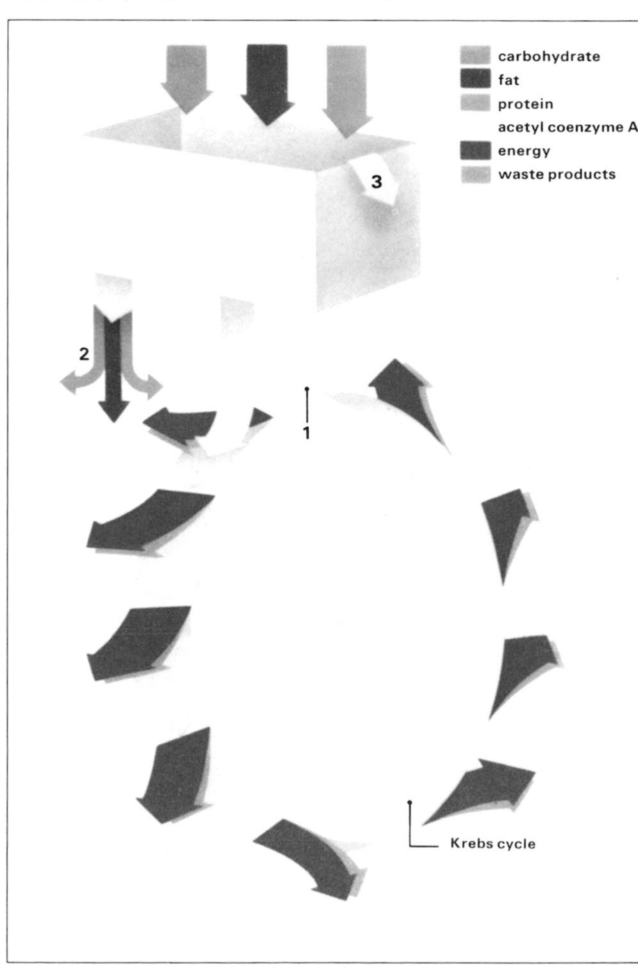

Metabolism involves chemical changes which convert food we eat into the components of the body, or burn food to provide energy. Within each body cell digested carbohydrate, fat and protein are finally reduced to a relatively simple substance called acetyl coenzyme A. The amount of acetyl-coA can be thought of in terms of a tank with an outlet and overflow, the reserve of which can be used in several ways. The acetyl-coA, for example, can be further broken down in a series of steps called the Krebs cycle (**1**). Each step produces energy which can be used to fuel the body and build up new molecules. Waste products from these reactions are removed from the cell. Acetyl-coA can also be used as a simple building block for the synthesis of new molecules (**2**). Excess acetyl-coA is converted into fat (**3**).

Cross-reference	pages
Circulation	54-55
Diet	44-45
Digestion	46-47
Hormones	70-73
Liver	50-51

The circulation of blood

No matter how simple an animal may be, it still needs some form of circulatory system within it to ensure that the oxygen and food materials that it absorbs reach every part of the organism and that the waste materials that it produces can be passed to the outside. Even in the single-celled amoeba there is a continuous circulatory movement in the jellylike cytoplasm. Fluid-filled vacuoles, some transporting food particles and others waste matter, circulate within the cell, and dissolved gases are carried around in the cytoplasm as it moves.

In our own bodies, composed of not one but countless millions of cells in solid blocks and sheets, the problem is correspondingly magnified. Every cell in the body, from those deep in the brain to those at the tips of our toes, needs oxygen and food to stay alive. It is clear that very few of these cells could hope to survive on materials diffusing into the body through the skin. Nor could all the waste matter be eliminated by diffusing in the opposite direction. We have a complex transport problem, and to cope with it Nature has evolved a complex system that relies on pumping and circulating around the body the versatile fluid called blood.

Blood has special properties that enable it to take up oxygen in chemical combination in regions where it is plentiful—those in contact with the outside air—and then, during circulation, to release it where oxygen is being used up and is not so plentiful—deep in the body's tissues. During circulation, blood can also transport carbon dioxide in the opposite direction, from the tissues where it is produced to the outside air, so that the result is an exchange of the two gases.

Because our bodies have not an adequate surface area to exchange all the oxygen and carbon dioxide necessary through our skins, we have developed a special internal surface—the lining of the lungs—through which such an exchange can take place readily. The total surface area provided by the lungs is about

The continuous circuit

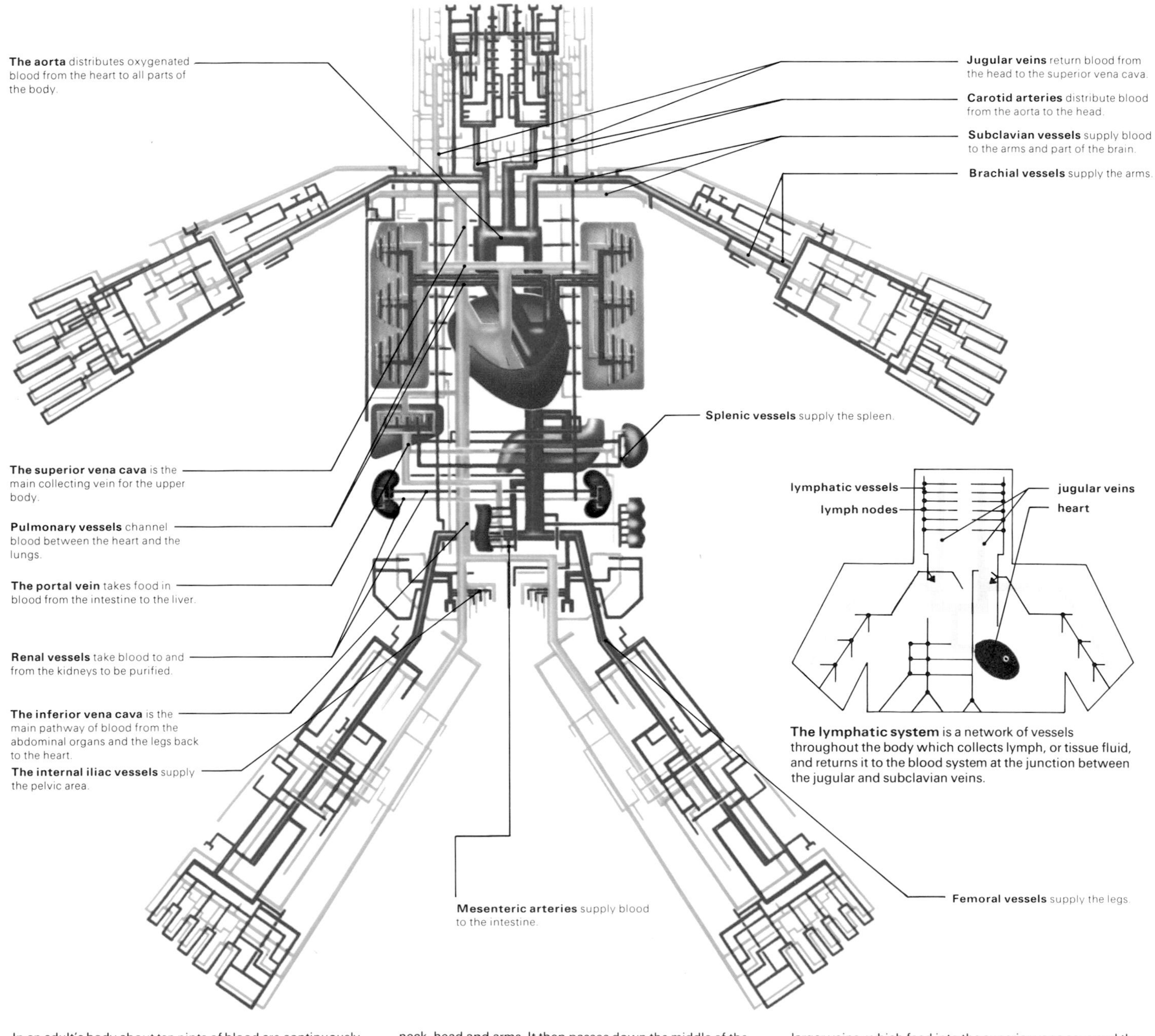

The lymphatic system is a network of vessels throughout the body which collects lymph, or tissue fluid, and returns it to the blood system at the junction between the jugular and subclavian veins.

In an adult's body about ten pints of blood are continuously being pumped by the heart through sixty thousand miles of blood vessels. Blood, which has acquired oxygen in the lungs, leaves the heart through the aorta, an inch-wide tube. The aorta branches into large arteries, which take the blood to the neck, head and arms. It then passes down the middle of the body carrying blood to the kidneys, liver, intestines and legs. The large arteries further divide into smaller arterioles and, finally, into minute capillaries, which surround the body cells. Blood then passes into venules, small veins, and then into larger veins, which feed into the superior vena cava and the inferior vena cava, which return the deoxygenated blood to the heart and then to the lung, where the blood gives up carbon dioxide and acquires a fresh supply of oxygen. In the illustration above, the arteries are indicated in red, the veins in blue.

forty times that of the skin. Circulating here, beneath the moist, thinner-than-paper lining, blood can come into very close contact with the air that we breathe, and carry out the exchange.

Again, we cannot absorb food from our surroundings directly through our skins. There has, therefore, developed another special internal surface—the lining of the digestive tract—through which simple food materials and water provided in the diet can be taken into the body. Blood circulating in the walls of the intestines picks up water and dissolved food substances that are destined for distribution and use as fuel and construction materials throughout the whole of the body.

Some, but very little, of the waste matter produced by living cells does in fact find its way out of the body through the skin, such as in sweat. The components of sweat, including water and salts, are derived from the bloodstream. By far the greater proportion of unwanted material leaves the body in the urine, a solution filtered by the kidneys from the blood circulating through them. This is the main route of excretion.

The essential function of the blood, then, is to transport substances as it circulates. But how is the circulation of blood around the body brought about? The answer is that the human body has a pump, the heart, and a vast network of blood-carrying vessels that penetrates to every corner of the body, servicing and bathing with fluid every ceil in every tissue. If all the tubes and tubules of the circulatory system were to be unraveled, the total length would be many thousands of miles. In a never-ending journey around the body are about ten pints of blood, impelled by the beating of the heart.

Arteries and veins are the largest tubes of the circulatory system, and the major routes by which blood travels to all parts of the body from the heart, and returns to it for recirculation. Arteries all carry blood away from the heart, and veins all carry blood toward the heart. The blood in arteries is under high pressure, and the vessel walls are appropriately thick, muscular and slightly elastic. The surge of pressure that follows each contraction of the heart causes the arterial walls to yield slightly, smoothing out the blood flow. This momentary expansion, echoing every heartbeat, is felt as a pulse where any large artery runs near the surface of the body.

The largest artery in the body is the aorta, nearly one inch wide, that arches, like the handle of a walking stick, up and backward from the left ventricle of the heart to pass down in front of the spine, giving off branches in its course to supply all parts of the body. Arteries branch into smaller vessels, and at their ends become arterioles, which form a network in the skin, in the muscles and in the organs that they supply. The muscle fibers in the walls of the arterioles enable them to contract and so regulate the blood flow in particular regions according to needs. Arterioles in turn divide into minute capillaries, tubules averaging only one twenty-fifth of an inch in length, and a hundredth of that in diameter. Capillaries have a wall only one cell thick. Through this wall materials pass into and out of the blood.

On the return side of the circulation, capillaries join up with the smallest branches of the veins, the venules, and the venules unite to form wider and wider collecting vessels until they become the major veins that return blood to the heart.

Veins have thinner walls than arteries and the pressure in them is lower. They expand and collapse to adjust to the blood volume within them. The massaging effect of nearby muscles, and the presence of simple cup-shaped valves in the larger veins, helps to prevent blood from collecting in the lower parts of the body.

The basic circulation, then, is from heart to artery, to arteriole, to capillary, to venule, to vein and back again to the heart. But, in fact, because blood must pass through the lungs to pick up oxygen and release the carbon dioxide it has gathered from tissues, there is a double circuit. Instead of being pumped directly around the body again, blood returning from the tissues is sent through the pulmonary artery, through capillaries in the lining of the lungs and taken back to the heart through the pulmonary veins. Now the circuit can start again.

The heart has four chambers that cope with the double circulation. While the left ventricle contracts to send blood, freshly oxygenated by the lungs, around the body, the right ventricle is contracting to send the "stale" blood through the lungs. The blood returning from the body is passed to the right ventricle by the right atrium, and the blood returning from the lungs is passed to the left ventricle by the left atrium.

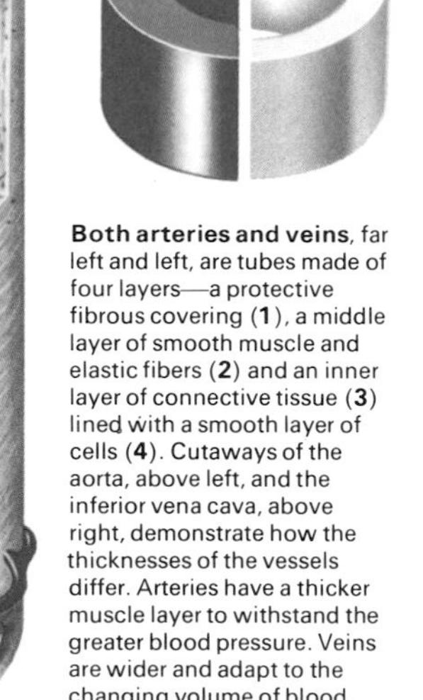

Both arteries and veins, far left and left, are tubes made of four layers—a protective fibrous covering (**1**), a middle layer of smooth muscle and elastic fibers (**2**) and an inner layer of connective tissue (**3**) lined with a smooth layer of cells (**4**). Cutaways of the aorta, above left, and the inferior vena cava, above right, demonstrate how the thicknesses of the vessels differ. Arteries have a thicker muscle layer to withstand the greater blood pressure. Veins are wider and adapt to the changing volume of blood.

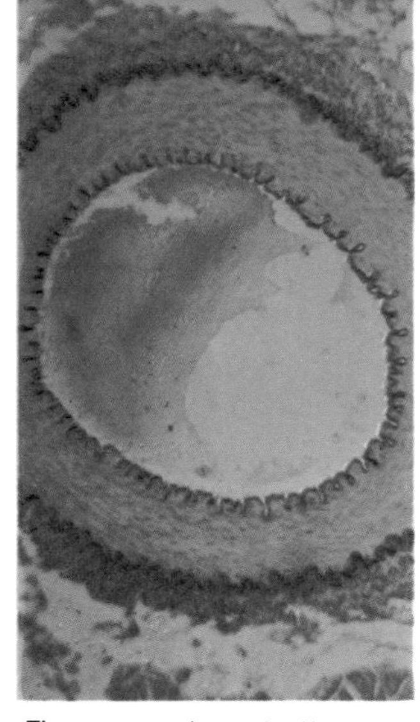

The aorta, seen here stained in cross section and twice life-size, is the body's largest artery. Where it leaves the heart it is one inch in diameter with stout, elastic and muscular walls. One and a half feet long, it narrows slightly and branches into smaller arteries, which distribute oxygenated blood all over the body.

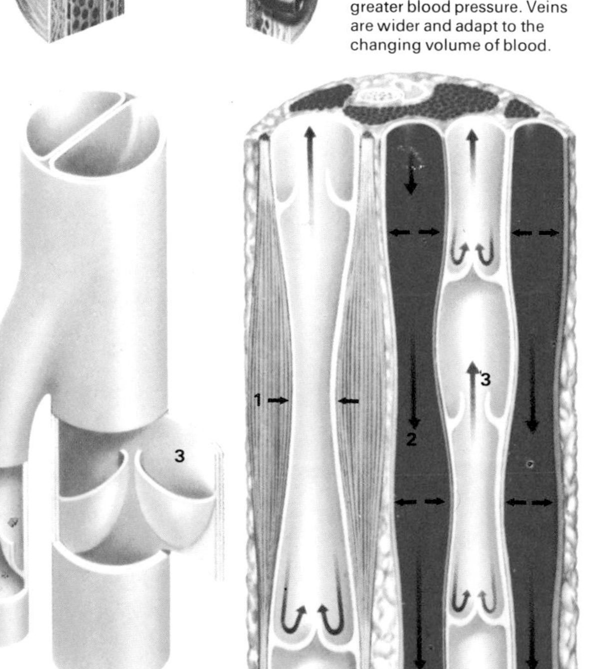

Blood flows slowly through veins because it has little pressure behind it. Blood flow in the veins relies on the rhythmic contraction of surrounding muscles (**1**) and the pumping action of nearby arteries (**2**) to return it to the heart. One way semilunar valves (**3**), which occur at intervals inside the veins, prevent any possible backflow of blood.

Cross-reference	pages
Blood	**58-61**
Digestion	**46-49**
Heart	**56-57**
Intestines	**48-49**
Kidneys	**66-67**
Lungs	**64-65**
Lymph	**58-59**
Lymphatic system	**138-139**

The heart: a muscular pump

Inside the heart

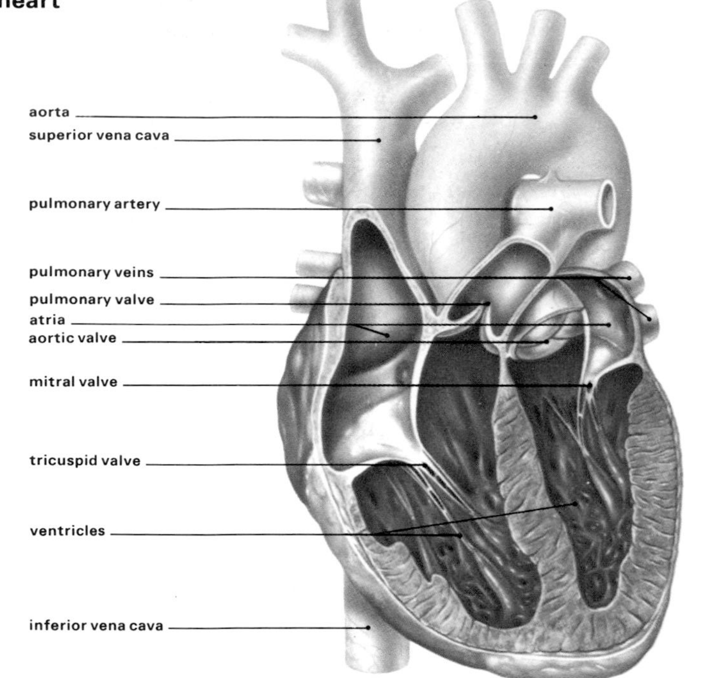

The heart is a muscular pumping organ which beats nonstop to circulate blood around the body. It functions as two halves, each consisting of a holding chamber, or atrium, and a pumping center, or ventricle. After circulating around the body, blood, now deoxygenated, returns to the right atrium through large veins, the superior vena cava and the inferior vena cava. When the atrium is full, blood is forced through the tricuspid valve into the right ventricle. It is then pumped to the lungs through the pulmonary valve into the pulmonary artery. The oxygenated blood returns via the pulmonary veins to the left atrium. After flowing through the mitral valve it is pumped out of the left ventricle into the aorta to return to the general circulation.

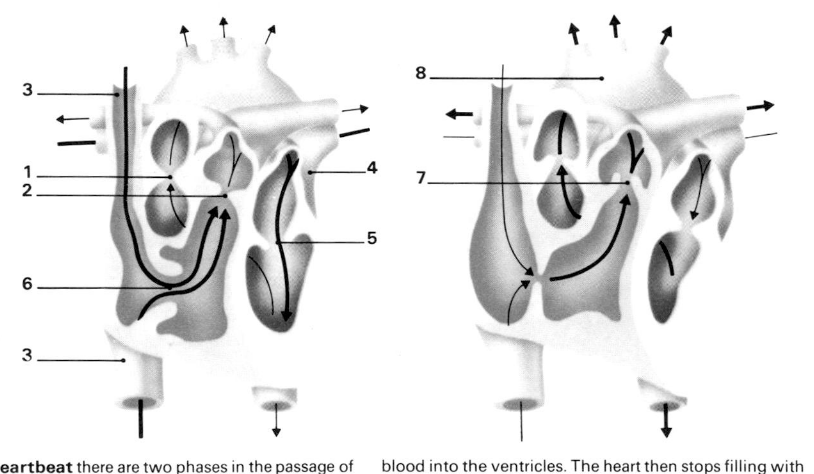

With each heartbeat there are two phases in the passage of oxygenated blood (red) and deoxygenated blood (blue) through the heart. In the phase of relaxation, or diastole (left), the heart fills with blood. As the ventricles relax, valves in the aorta (**1**) and in the pulmonary artery (**2**) close with a dup sound and blood pours into the two atria from the venae cavae (**3**) and the pulmonary veins (**4**). The mitral (**5**) and tricuspid (**6**) valves between the atria and ventricles open, allowing blood into the ventricles. The heart then stops filling with blood. In the phase of contraction, or systole (right), the heart empties and a lub sound is made by the closing of the mitral and tricuspid valves. This prevents a backflow of blood into the atria as it is pumped from the ventricles into the pulmonary artery (**7**) and aorta (**8**). Each heartbeat lasts for up to eight-tenths of a second, with systole lasting for four-tenths of a second and diastole occupying the remaining time.

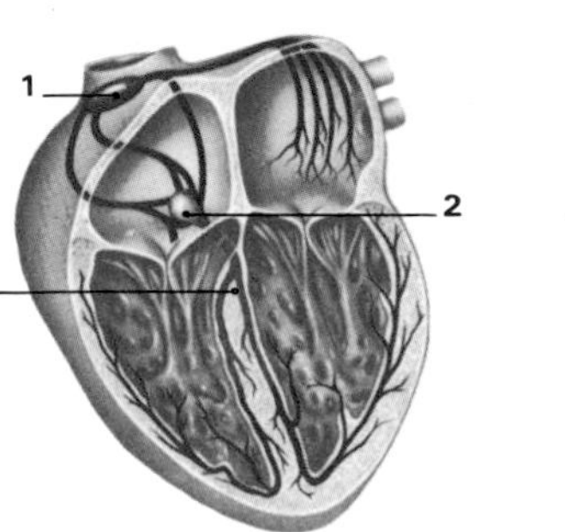

The pacemaker of the heart, a tiny area of specialized nervous tissue in the right atrium, sets the heart beating about seventy times a minute. Without it the heart would beat only forty times per minute, which is too slow for the body's needs. The pacemaker, or sinuatrial node (**1**), regularly sends out nerve impulses which spread through the two atria, causing them to contract. From the atrioventricular node (**2**) the contraction spreads down special conducting tissue, the bundle of His (**3**), causing the ventricles to contract and pump blood out of the heart.

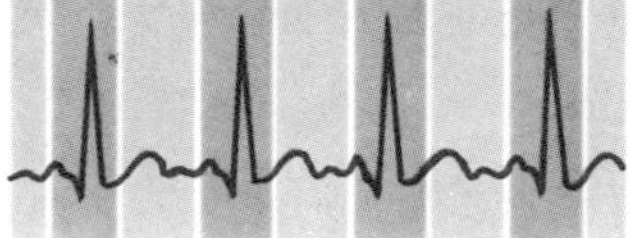

An electrocardiogram, or EKG, is a recording of the wave of electrical impulses passing through the heart muscle during each beat. The line traced on a moving roll of paper indicates contractions of the atrial muscle and the ventricular muscle followed by a period of relaxation.

Weighing about ten ounces, and not much larger than a clenched fist, the heart is a remarkable pump that works ceaselessly throughout life, forcing blood in an endless circuit through all the miles of living tubing that go to make up the circulatory system. During one day, something like two thousand gallons of blood pass through its chambers, carrying out the endless task of transporting oxygen and food materials for all the body's requirements.

Because there is a double circulation, blood in effect travels in a figure-eight circuit. The heart lies at the point of crossover, and can be regarded as two linked pumps. The left side of the heart impels oxygenated blood that it has received from the pulmonary veins into the aorta, and so to all the body's tissues, while simultaneously the right side impels deoxygenated blood that it has received from the body through the venae cavae and then into the pulmonary arteries.

The heart bears a rough resemblance to the valentine shape of tradition. The point of the cone, the apex of the heart, is directed downward and outward to the left. Its movement can be detected with the fingertips through the chest wall to the left of the sternum. This throb is called the apex beat. At the base of the cone, facing upward and to the right, the great vessels enter and leave the heart.

The walls of the heart are made of thick cardiac muscle and enclose four chambers, two of which receive blood returning to the heart. These are the atria, with thinner walls than the other two chambers, the more powerful ventricles, which are concerned with the output of blood into the systemic and pulmonary circulations. Because the systemic circulation is so large, the left ventricle has the greatest muscle mass; the apex beat is due to its forceful movement. Dividing the left atrium and ventricle from the right atrium and ventricle is a wall. The upper part of this wall is the interatrial septum, and the lower part is the interventricular septum. Within this tissue runs a branching band of special conducting tissue that carries impulses to regulate the heart's contractions.

Cardiac muscle has a built-in rhythm of contraction, but to ensure that the chambers contract in an orderly way, and cooperate in their activities, there is a specially developed area of tissue in the wall of the right atrium that takes charge. This is called the sinuatrial (SA) node, or more commonly the "pacemaker," and is the source of bursts of electrical activity that spread through the walls of the heart, like ripples spreading across the surface of a pond, and stimulate first the atria and then the ventricles to contract.

Both atria and both ventricles contract at the same time, in rapid succession. The period of ventricular contraction is called systole, and is followed by a relaxation or recovery period called diastole. Systole and diastole are both very short. The heart goes through about seventy cycles of systole and diastole every minute, but its rate can be more than doubled to keep up with the body's oxygen needs during violent exertion.

The electrical activity that accompanies each heartbeat can be detected on the surface of the body with electrodes, and a tracing, called an electrocardiogram (EKG), made to show the sequence of events. In an EKG tracing the P wave is produced by electrical activity

spreading over the atria, the QRS complex is the result of ventricular systole and the T wave is seen during diastole, before the heartbeat cycle starts again.

All the heart's activity would be of no use whatsoever if there were not some method of ensuring that the blood moved in one direction only, and did not simply surge back and forth between the heart chambers as they contracted and relaxed. It is the valves of the heart that turn it into a pump. They are one-way valves, which means that they shut automatically if blood attempts to move against the proper direction of flow.

Around the inflow veins, the venae cavae and the pulmonary veins, simple rings of muscle prevent backflow. Between the right atrium and the right ventricle is the tricuspid valve, and between the left atrium and left ventricle is the mitral valve. At the exits from the ventricles are the aortic valve of the left heart and the pulmonary valve of the right heart. These valves ensure that blood can only flow in the direction vein-atrium-ventricle-artery as the heart contracts and relaxes. Their closing gives rise to the "lub-dup" noise that can be heard through a stethoscope placed against the chest wall.

Being in constant action, the muscle of the heart walls must itself be supplied liberally with oxygenated blood, and this is brought about by the coronary arteries, which branch from the aorta just beyond the aortic valve and divide to spread over the surface of the heart in a network. Blood returns from the heart muscle into the heart through coronary veins, which empty directly into the right ventricle.

The pressure of the blood in the arteries is dependent upon both the force with which the heart contracts and the resistance that the blood meets as it passes through the small vessels in the body's tissues. The narrower the arterioles, the greater the blood pressure needed to overcome the resistance. The caliber of the arterioles is constantly changing under the influence of the autonomic nervous system, according to the body's needs, and the rate and force of the heartbeat are also adjusted by nervous control, and altered by hormones, particularly adrenaline, which circulates in the bloodstream.

The mitral valve

The mitral, or bicuspid, valve lies between the left atrium and ventricle. One of the heart's four valves, it is the only one with two, rather than three, flaps. This view from underneath an opened valve shows that attached to each valve flap there are tough, inelastic tendons, which in turn are anchored to papillary muscles, extensions of the ventricle wall. When blood is pumped into the ventricle from the atrium the papillary muscles relax, allowing the flaps to move apart and the valve to open. When the ventricle contracts, to force blood into the aorta, the attached papillary muscles also contract. This puts tension on the tendons, which pull the flaps together to prevent a backflow of blood.

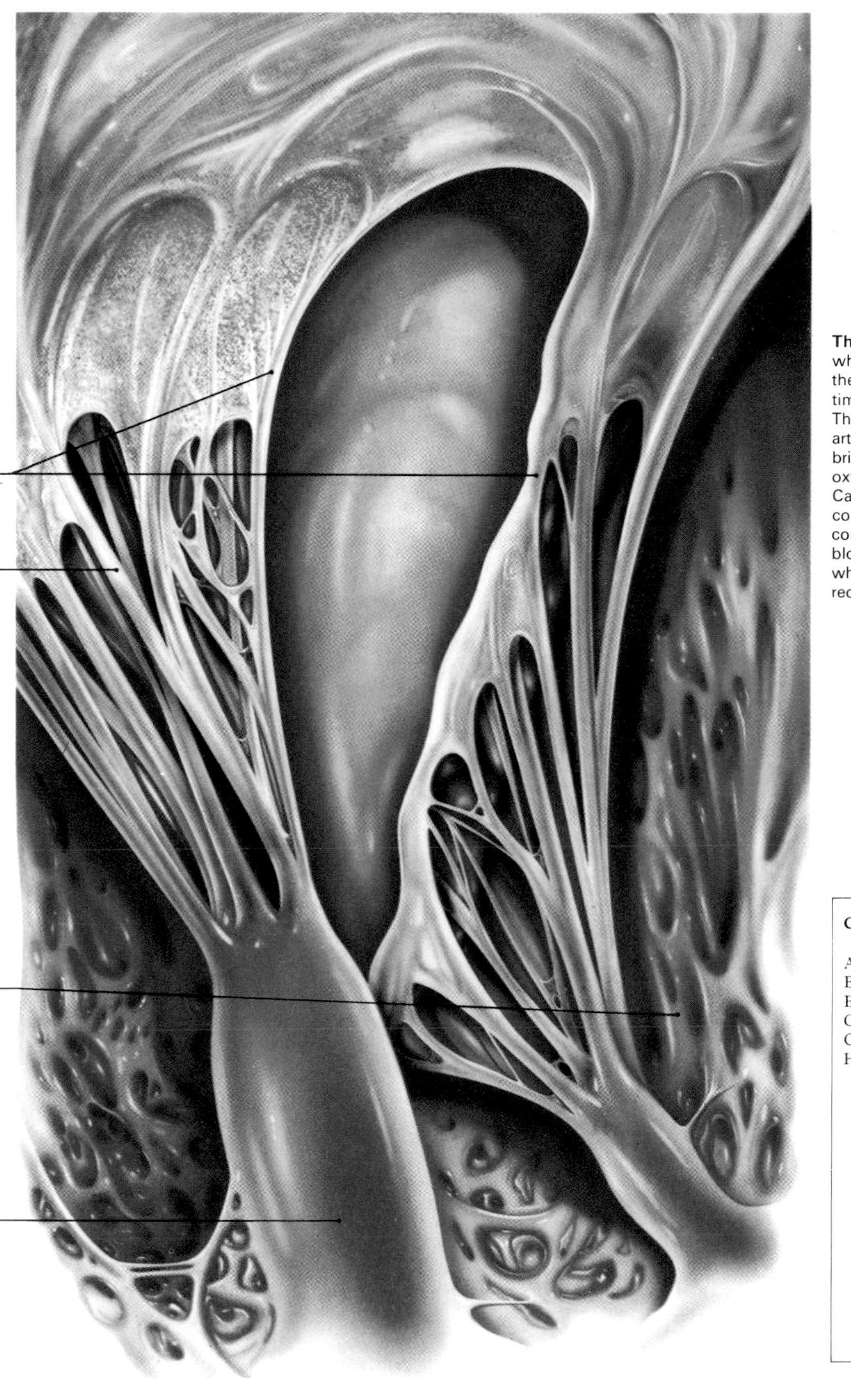

The valve flaps let blood into the ventricle and then close to prevent the blood from flowing back into the atrium.

The tendons are inelastic cords which hold the flaps in position.

The muscular walls of the ventricle contract to force blood into the aorta at the same time as the mitral valve closes.

The papillary muscles contract and relax with each heartbeat, altering the tension in the tendons.

3
2
1
4

The heart has its own blood supply which provides the energy needed for the heart muscle to pump about seventy times a minute when the body is at rest. The left (**1**) and right (**2**) coronary arteries, branches of the aorta (**3**), bring a constant supply of food and oxygen to the cardiac muscle cells. Carbon dioxide produced by muscle contraction is taken away by the coronary veins (**4**), which empty the blood into the right side of the heart, from where it is pumped to the lungs to be reoxygenated.

Cross-reference	pages
Autonomic nervous system	**74-75**
Blood	**58-61**
Blood pressure	**40-41**
Cardiac muscle	**40-41**
Circulation	**54-55**
Hormones	**70-73**

Blood: components

Surging through the complicated system of main road, branch road, bypasses and narrow streets that make up the circulatory system of the body flows an unending stream of traffic—the blood. Without this essential fluid our bodies would be rather like countries that were completely without supplies of food and water, without any means of waste disposal and, above all, without any air for the inhabitants to breathe. We need blood to stay alive.

Blood consists of plasma, a yellowish fluid in which are suspended cells, or corpuscles, and other small particles. The cells are of two main groups. There are red cells, also called erythrocytes, and several forms of white cells, or leukocytes. The other particles are called platelets. Each of these components has its own special role to play in the day-to-day working of each of the body's tissues and organs.

When it leaks from a cut or pinprick in the skin, blood is an opaque, slightly sticky red fluid. Normally, within a matter of minutes, it clots, darkens and solidifies as the body's defense mechanisms go to work to prevent further loss of the vital fluid and seal the gap.

When blood is taken in quantity in a laboratory and spun in a centrifuge, or specially treated and allowed to settle, the solid components form a reddish-brown sediment which makes up about 45 percent of the total volume. Above the sediment is the straw-colored plasma, which is a solution in water of salts and proteins. Plasma is similar in composition to the tissue fluid that lies between and bathes all of the body's cells, to the lymph that circulates slowly in the vessels of the lymphatic system and to the cerebrospinal fluid in the cavities of the central nervous system. It is also allied to the fluids contained within the eyeballs and the inner ear.

There are certain important differences in the chemical composition of the tissue fluids around the body's cells and those that are actually contained within the cells, the intracellular fluids. The main difference, however, between blood and tissue fluid, or lymph, is the presence of red blood corpuscles and the proteins called plasma proteins.

Shortly after birth, the total volume of blood contained in the body is only about half a pint, but, by adulthood, a man, depending upon his size, has about nine or ten pints in his body, and a woman six or seven pints. (During pregnancy, a woman has an extra pint or so to enable her to cope with the extra needs of the growing fetus.) The blood contained in the body is about one-third of the total amount of fluid found outside the body's cells, the interstitial fluid.

The plasma is a complex fluid, in a constantly changing equilibrium with the tissue fluid and the fluid within the cells. Every moment the plasma is gaining and losing chemical substances, but nevertheless manages to maintain a remarkably constant composition. Plasma is about 90 percent water, with 7 or 8 percent by weight of soluble proteins that give it a slightly syrupy texture and that have important roles to play in nutrition and defense. The remaining 1 or 2 percent is made up of a variety of small organic molecules—urea, glucose, amino acids and fats, and the all-important hormones and vitamins. There is an additional 1 percent of salts.

The main proteins in the plasma are albumins, globulins and fibrin, one of the components that helps blood to clot. Serum, thin fluid that oozes from a scrape of the skin, is simply plasma without the protein fibrin.

While the components of the plasma that have small molecules can pass in solution through the walls of the capillaries, which are only one cell thick, the albumins and globulins have rather larger molecules, and so tend to remain within the capillaries. They produce an osmotic pressure on the inside of the capillary wall that attracts water into the blood from the tissue fluids on the other side. Balancing this pressure is, of course, the blood pressure itself, which forces water, and the small molecules which are dissolved in it, outward into the tissues. Not all the water that leaves the bloodstream returns to it through the capillary walls.

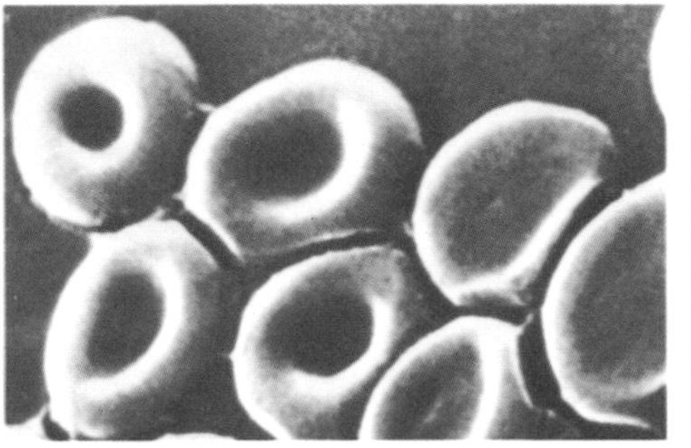

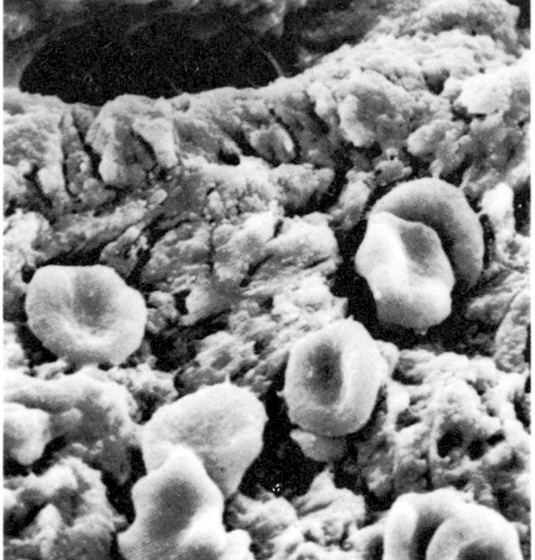

The biconcave shape of normal red blood cells is shown above, magnified four thousand times, in a scanning electron micrograph. With a diameter of .0003 of an inch and a rim thickness of only .00001 of an inch, these nonnucleate cells have elastic properties which allow them to squeeze through the smallest blood vessels. The red blood cells in a fractured femur, right, were viewed and photographed under an electron microscope at a magnification of 3,400 times. In the top left-hand corner is a lacuna, a space which was once occupied by a bone cell, or osteocyte.

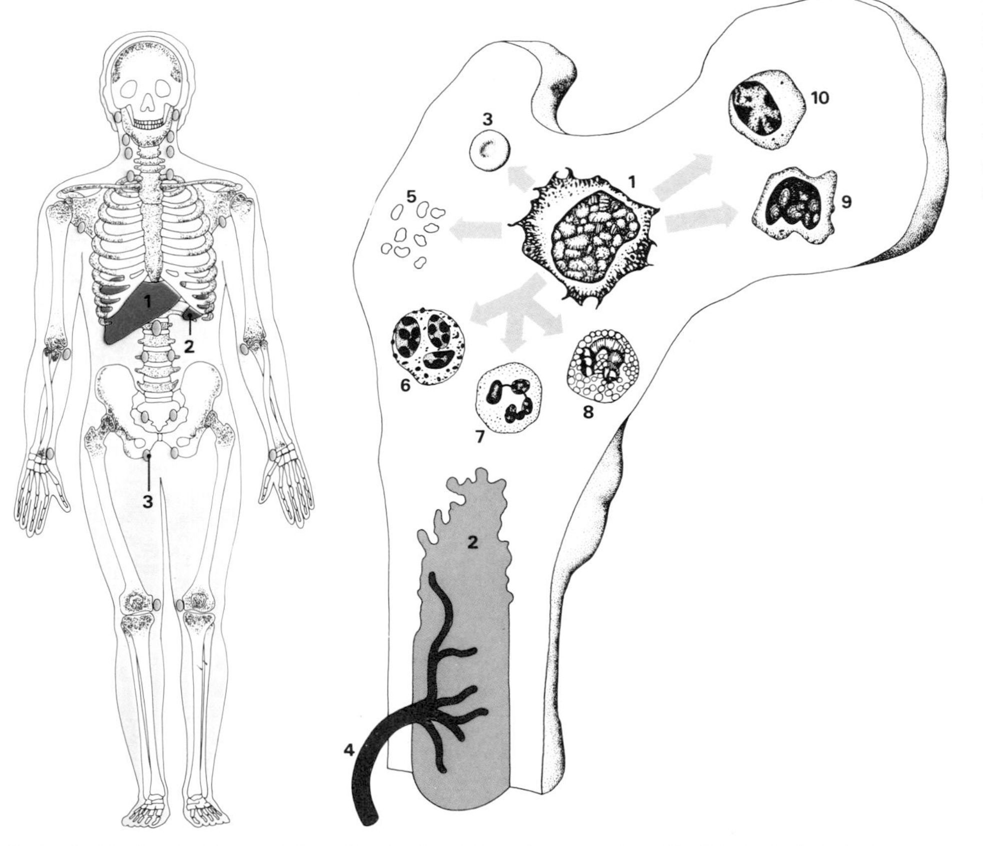

Red and white blood cells are made in bone marrow (the dotted areas shown above). White cells are also produced in the liver (**1**), in the spleen (**2**) and in the lymph nodes (**3**). After a relatively short life span expired blood cells are removed from the circulation by scavenger cells in the spleen, liver and lymph nodes.

Blood cells originate from stem cells (**1**) in bone marrow (**2**). In the diagram above, the blood cells are shown magnified one thousand times. Red cells (**3**) are produced by a series of changes in stem cells in which the cell nucleus is lost and hemoglobin accumulates. The red cells then enter the circulation (**4**).

The "blood-clotting" platelets, or thrombocytes (**5**), are fragments of cytoplasm from the stem cells. White blood cells—basophils (**6**), neutrophils (**7**) and eosinophils (**8**)—all have segmented or lobed nuclei, while monocytes (**9**) and lymphocytes (**10**) have single, more spherical nuclei.

lymph capillaries within the tissues, and returns in the lymph stream, which empties into the jugular vein in the neck, not far away from the heart.

Some enters the quite separate system of lymph capillaries within the tissues, and returns in the lymph stream, which empties into the jugular vein in the neck, not far away from the heart.

Together, the red and white corpuscles and the platelets are sometimes called the formed elements of the blood. Seen under the microscope, the red cells are orange-colored biconcave disks, rather like tiny balloons that have been tightly pinched in the middle. They have no nuclei, and each is only about three-ten-thousandths of an inch across. These are the cells responsible for the transport of gases in the circulation. There are normally about five million red blood cells in every cubic millimeter of blood. It is estimated that about two million of these cells are created and destroyed every second.

Blood is formed in the marrow of certain bones, in adults mainly in the skull, spine, ribs, breastbone and thighs. In infants, all the bones make red blood cells, and before birth the liver and spleen also produce them. Red cells appear first as normal cells, but they soon lose their nuclei. When they are released into the bloodstream they are simply "envelopes" to contain the purplish-red pigment hemoglobin, which makes up about one-third of their weight. Hemoglobin is a compound of a protein, globulin, with iron. The average life of a red cell is about one hundred and twenty days, during which it makes a journey of nearly a thousand miles through the circulatory system, finally to be destroyed in the liver, spleen or another organ of the body, so that the essential iron can be recycled.

Hemoglobin gives the blood its red color. When it combines with oxygen in the capillaries of the lungs, where the pressure of this gas is high, it forms a loose chemical bond to give the compound oxyhemoglobin. Oxyhemoglobin is a bright red color, that of arterial blood, by contrast with the more purplish color of venous blood returning from the body's tissues.

In the tissue capillaries, where the pressure of oxygen is low, because it is continuously being taken up into cells and entering into chemical reactions within them to release energy, the loose bond in the oxyhemoglobin molecules breaks up to release the oxygen into solution. The dissolved oxygen can then diffuse into the tissues. It is an extremely efficient method of oxygen transport. It has been estimated that we would need to have more than seventy-five times as much blood circulating in our bodies if there were no hemoglobin and the gas were to be carried simply dissolved in plasma, rather than in chemical combination.

The uptake and release of oxygen is linked very closely to the transport of the waste product carbon dioxide that is going on at the same time. In the tissues there is a relatively high pressure of carbon dioxide, and it diffuses into the bloodstream. Only a small proportion of carbon dioxide is transported in the plasma and the red cells as a simple solution. About one-fifth reacts with the oxyhemoglobin, to push out the oxygen and replace it to form the compound carbaminohemoglobin. In this form it is carried to the lungs within the red cells. The remainder of the carbon dioxide reacts with water to form the compound carbonic acid. This reaction takes place rapidly in the red cells because they contain more of the special enzyme carbonic anhydrase, which speeds up the reaction, than does the plasma. Carbonic anhydrase can speed up the rate of combination of carbon dioxide and water about fifteen hundred times.

The carbonic acid, in turn, splits up into hydrogen and bicarbonate ions. The hydrogen ions react with oxyhemoglobin to drive off the oxygen. So carbon dioxide is carried in the bloodstream as carbonic acid, as bicarbonate ions and as potassium and sodium bicarbonate. When the blood reaches the lungs, all of the chemical reactions are reversed. Because of the lower concentration of carbon dioxide in the lung capillaries, the complex exchange of gases results in the elimination of the waste carbon dioxide into the expired air.

Formation of lymph

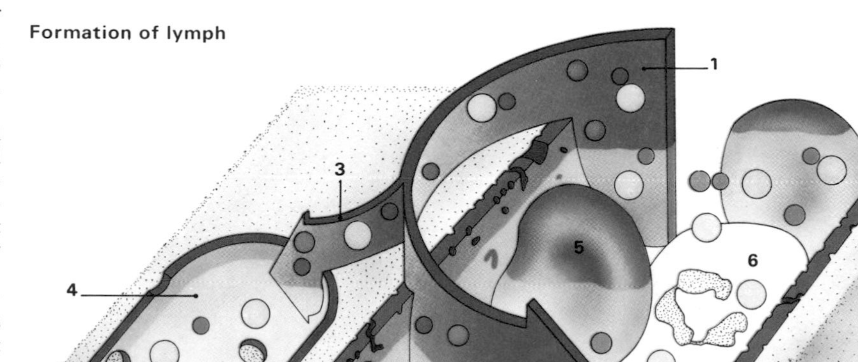

Blood pressure forces a watery fluid (**1**) from capillary walls (**2**) into tissue spaces. Most of this fluid reenters the capillaries; the remainder (**3**) passes into adjoining lymphatic vessels (**4**) as lymph. Red cells (**5**) remain in the blood capillaries and some white cells (**6**) pass out to act as scavengers. The inset shows the way some materials flow out of the capillaries by diffusion (**A**), and others through pores of varying sizes (**B, C**) at junctions between capillary cells. The lymphatic system eventually pours its contents back into the bloodstream at the junction of the jugular and subclavian veins.

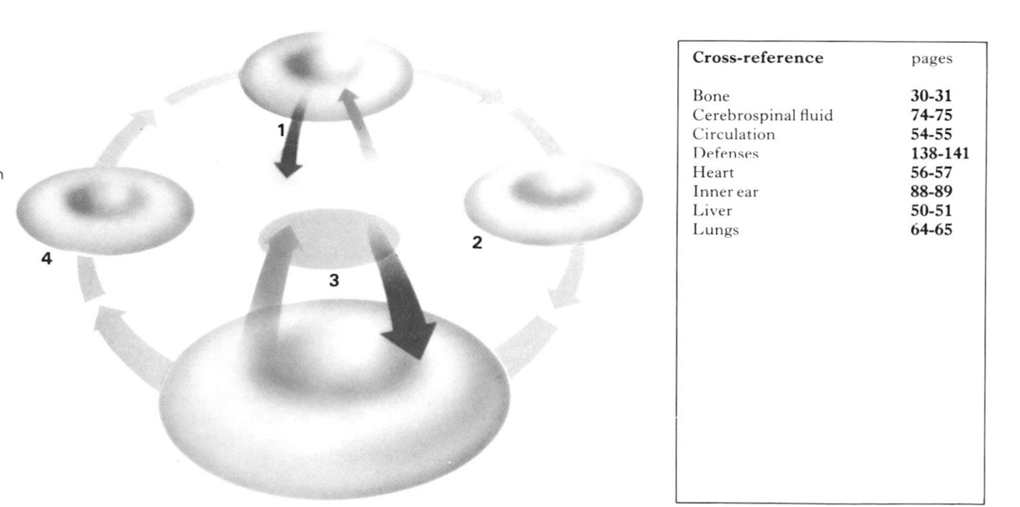

Transport of oxygen and carbon dioxide between the lungs and tissue cells is the prime function of red blood cells. Essentially, red cells are packets of hemoglobin, a molecule which readily combines with oxygen. As the red cells flow through the body tissues, oxygen diffuses from them to the tissues in exchange for carbon dioxide (**1**). The deoxygenated red cells carry a little carbon dioxide (**2**), then pass through the bloodstream to the lungs. Here the reverse exchange of oxygen and carbon dioxide takes place (**3**). The oxygenated red cells are then conducted back to the body tissues (**4**).

Cross-reference	pages
Bone	30-31
Cerebrospinal fluid	74-75
Circulation	54-55
Defenses	138-141
Heart	56-57
Inner ear	88-89
Liver	50-51
Lungs	64-65

Blood: function

The millions of microscopic red corpuscles in the bloodstream are the transporters of gas within the body, but this is by no means the only transport that must be arranged if the tissues of the body are to stay alive. Every cell needs food materials that can be "burned," using the oxygen, to provide energy, and the transport of these materials is one of the many functions of the plasma.

From the network of fine capillaries that surrounds the lining of the digestive tract, food materials are taken up and passed to the liver through the hepatic portal system. There they are processed and passed into the general circulation for the uses of all the body's cells. Proteins are broken down in the intestine into their building units, amino acids, and travel in this form in the blood. Carbohydrates in the diet are reduced to the small sugar unit glucose, and contribute to the sugar level in the blood. Fats are largely absorbed into the network of lymph vessels, the lacteals, as minute globules, called chylomicrons, direct from the intestinal wall, while a small part of digested fat, as glycerol and short-chain fatty acids, is taken up by the intestinal capillaries. The chylomicrons that have been taken into the lymphatic system rejoin the bloodstream via the neck veins, so initially bypassing the liver.

Although so many different substances are entering and leaving the blood every moment, its composition is regulated to stay reasonably constant, and there are upper and lower limits between which the concentrations, or "levels," of each of its components can normally be expected to lie.

After a meal rich in carbohydrates has been digested, for example, there is a period when the uptake of glucose from the intestine exceeds the amount that is being used up in the tissues. The excess glucose is quickly taken up by both the liver and the muscles, and converted to glycogen, in which form it is stored for possible later use. From the liver, glucose can be released if the blood level is falling too low. The uptake and release of blood sugar is regulated by hormones.

When blood passes through the kidneys it loses many of the waste products that, if allowed to accumulate, would gradually poison the body. And, of course, the volume of water that is contained in the blood is also controlled by this filtration system.

Constantly trickling into the circulating blood in minute amounts are the hormones that are released by the glands of the endocrine system. These "messengers" in the blood, produced by such glands as the thyroid, the pituitary, the adrenals and the sex glands, all circulate in the plasma to exert their influence on many different tissues, or "target organs," throughout the body.

The plasma in which the blood's cells are suspended contains the proteins albumin and globulin. One particular group of globulins, gamma globulins, is responsible for protecting the body against many infecting organisms. When foreign proteins, called antigens, enter the body's internal environment, defensive gamma globulins, called antibodies, are produced to destroy the antigens.

Antibodies combat antigens in several different ways. They may combine chemically with them and so neutralize them. They may cause the invaders to clump together, or agglutinate, so preventing their further spread and penetration of the tissues. Or they may actually attack and break down bacteria, or render invading cells more liable to attack by the phagocytic (cell-eating) white blood cells found in the blood, lymph nodes and other tissues.

White blood cells are two to three times larger than red cells, but almost a thousand times less numerous. Usually transparent, they are seen best in specially stained microscope preparations, when both their nuclei and cytoplasm take up color. White blood cells are subdivided, by appearance and by the role that they play in the body, into three main kinds, granulocytes, monocytes and lymphocytes. The granulocytes are further subdivided, according to their functions and staining characteristics, into neutrophils, which are by far the most numerous, basophils and eosinophils. The granulocytes and monocytes are formed in the bone marrow from parent cells that probably also give rise to red cells. The lymphocytes are produced mainly in lymph nodes and certain other lymphoid tissues.

The smallest particles found in the blood are the platelets, or thrombocytes. Platelets are produced by special cells in the bone marrow and live for only a few days. They contain a chemical that is released at the site of an injury and which converts the enzyme prothrombin, which circulates in the plasma, into an active form, thrombin. It is the thrombin which precipitates the protein fibrin in the blood so that it coagulates, or clots.

The many functions of blood

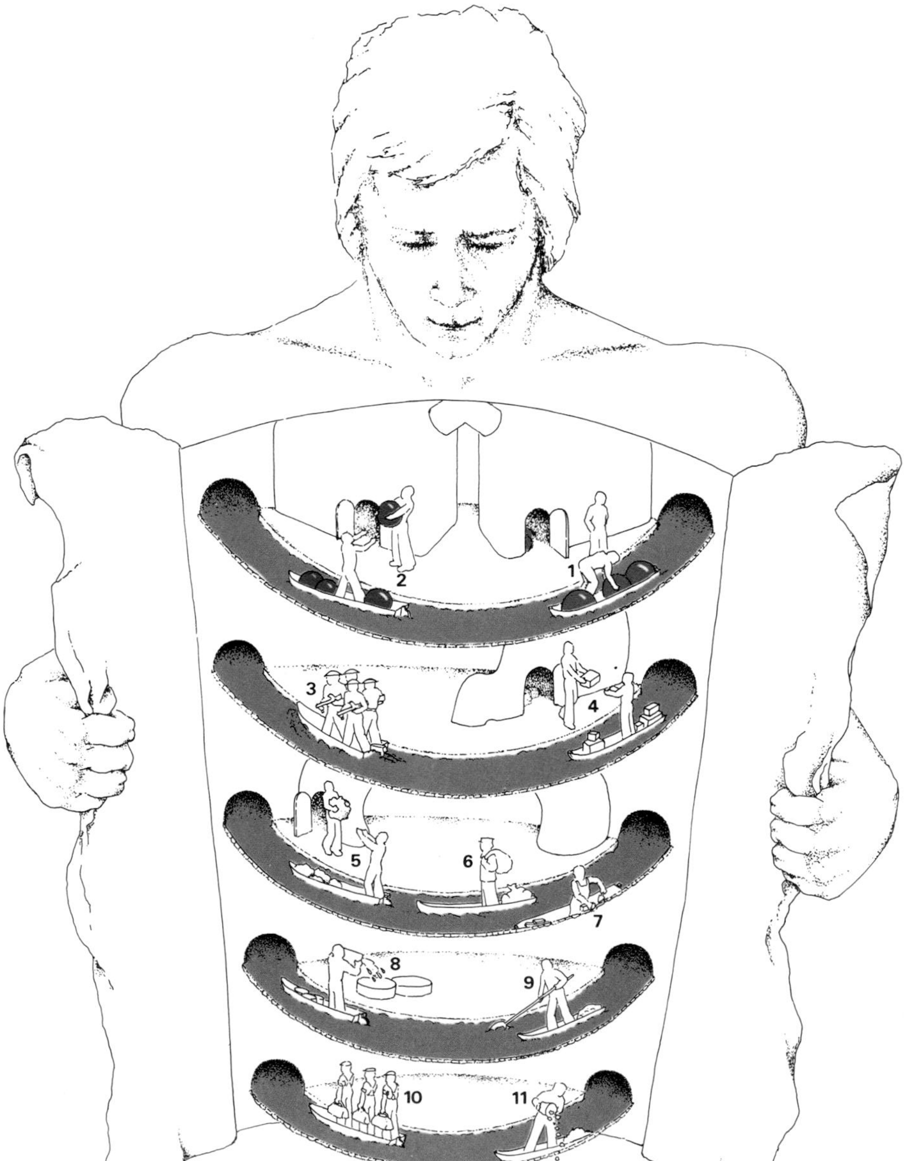

Blood transports many different materials around the body. Red blood cells carry oxygen from the lungs to the tissues (**1**) and carbon dioxide from the tissues to the lungs (**2**). Blood plasma transports antibodies, which protect the body from infection (**3**), carries food substances absorbed from the intestines (**4**) and takes waste products to the kidneys for excretion (**5**). It also transports hormones, secreted by endocrine glands, to their sites of action (**6**). Clotting factors circulating in the plasma prevent blood loss whenever blood vessels are damaged or broken (**7**). Blood regulates the water content of tissue cells (**8**). Blood buffers, chemicals within the blood, maintain correct acidity of body fluids (**9**). Enzymes may be transported by blood plasma (**10**), which also distributes the heat produced by metabolism (**11**).

Blood transfusion is only possible between people that have the same blood group. The first human blood groups were discovered in 1900. Before then it was not understood why some attempts at transfusion were successful and others disastrous. More often than not, blood cells transfused from one person to another would be caused by the recipient's blood to clump together, or agglutinate.

It is now known that present on the surface of blood cells are different antigens—factors that provoke in other people who do have them the production of proteins called antibodies, which are part of the body's defensive mechanism. Antibodies react to antigens by combining with them chemically. In the ABO blood group system, some individuals have in their cells antigens of type A, some have type B and some have both A and B. A fourth group of people, in O blood group, have neither A nor B.

Blood which has a certain antigen does not, of course, carry in the plasma antibodies against that particular antigen. If it did it would agglutinate itself. Consequently, blood of the same group can be safely mixed. Blood of groups AB and O have additional properties. Because a person with group AB can receive blood from a person with group A, B or O without agglutination occurring, he or she is called a "universal recipient." A group O person can give blood to a group A, B, AB or, of course, another group O person, and is, therefore, a "universal donor."

The important Rhesus blood group system was discovered in 1939, when antibodies were produced by injecting red cells from a Rhesus monkey into a rabbit. The antibodies that developed in the rabbit's blood were afterward found to agglutinate not just Rhesus monkey red cells but also red cells from a high proportion of human beings. More important, the serum from women who had given birth to babies suffering from a disease called hemolytic disease of the newborn was found to cause agglutination in the same cells, although the ABO grouping was compatible.

The Rhesus blood groups are called positive and negative. If a Rhesus negative person receives red blood cells from a Rhesus positive donor, antibodies are produced in the blood, so that if a similar transfusion takes place later, the donor's cells will be agglutinated. Tests are, therefore, always carried out to ensure that Rhesus grouping as well as ABO grouping is compatible before a transfusion is undertaken.

In the case of a pregnant woman who is Rhesus negative, trouble may arise if the father is Rhesus positive, because there is a possibility that the fetus may be Rhesus positive, too. If red cells from the fetus happen to escape through the placenta into the mother's blood, then she will develop antibodies to the Rhesus positive red cells, and leakage of her plasma back across the placenta will start to destroy the baby's cells within its own body. The baby could be born dead, or suffering from extreme jaundice. It is possible to detect if a pregnant woman is developing a high concentration of antibodies, which warns of this situation, and if necessary the baby's blood can be completely replaced, even before it is born. The greatest danger occurs not during the first pregnancy but during subsequent ones. It is possible to give injections that prevent antibodies from developing in women who may be at risk.

There are numerous other blood groups, such as the MNS, P, Lutheran, Kell Lewis, Duffy and Kidd systems, but antibodies to them are rare and they have little importance in ordinary transfusions.

The Rhesus factor

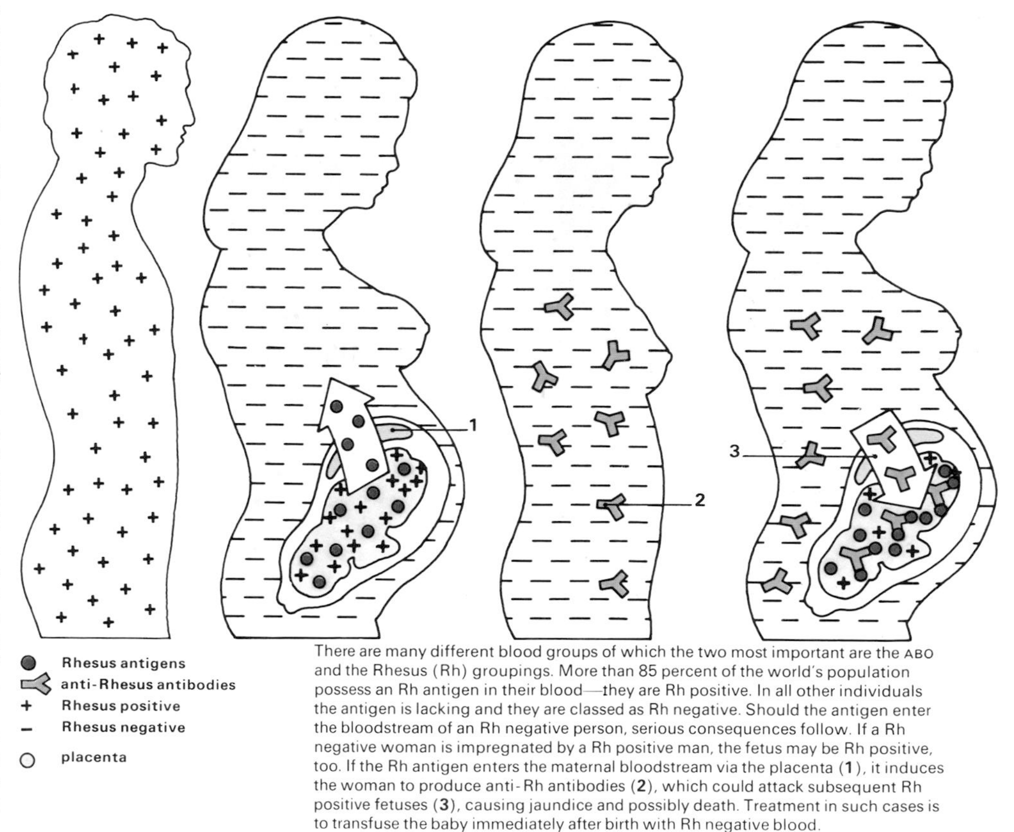

There are many different blood groups of which the two most important are the ABO and the Rhesus (Rh) groupings. More than 85 percent of the world's population possess an Rh antigen in their blood—they are Rh positive. In all other individuals the antigen is lacking and they are classed as Rh negative. Should the antigen enter the bloodstream of an Rh negative person, serious consequences follow. If a Rh negative woman is impregnated by a Rh positive man, the fetus may be Rh positive, too. If the Rh antigen enters the maternal bloodstream via the placenta (**1**), it induces the woman to produce anti-Rh antibodies (**2**), which could attack subsequent Rh positive fetuses (**3**), causing jaundice and possibly death. Treatment in such cases is to transfuse the baby immediately after birth with Rh negative blood.

Plugging a broken blood vessel

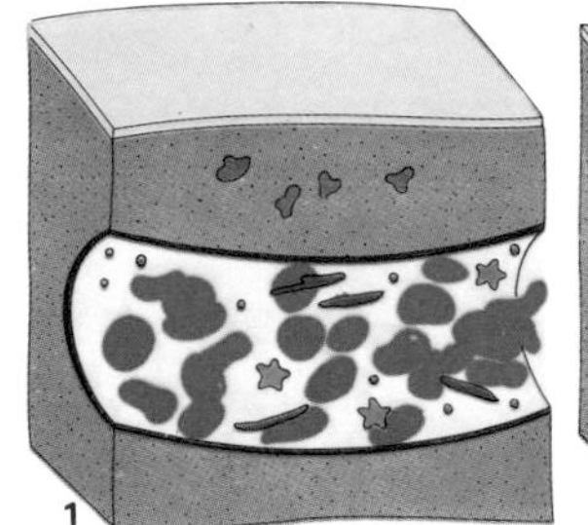

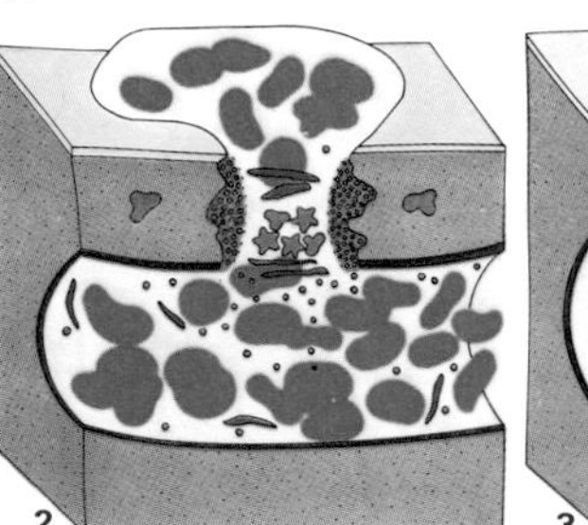

3

4

tissue factor
plasma factor
fibrinogen
platelets
red blood cells
fibrin

Blood clotting takes place in a sequential process to prevent excessive blood loss from a wound. Normally, circulating blood contains red cells, platelets, plasma-clotting factors and fibrinogen, a soluble protein. Tissue-clotting factors lie trapped within cells surrounding each blood vessel (**1**). When an injury occurs, blood escapes from the broken vessel. Platelets aggregate at the site and partially plug the break. Tissue-clotting factors are released (**2**). The interaction of the platelets with plasma and tissue-clotting factors results in the conversion of fibrinogen into insoluble fibrin threads, which form a mesh across the break in the vessel wall (**3**). Blood cells and platelets become trapped in the mesh. The semisolid mass shrinks and a yellow fluid, serum, is pushed out, leaving the familiar dark red clot (**4**). Soluble fibrinogen levels in the blood at the site of injury return to normal.

Cross-reference	pages
Defenses	**138-141**
Diet	**44-45**
Hormones	**70-73**
Kidneys	**66-67**
Lungs	**64-65**
Lymphatic system	**54-55**
Lymphatic system	**136-139**
Stomach and intestines	**48-49**

Respiration and speech

Every minute, usually without ever giving it a thought, we breathe in and out between ten and fifteen times. During an average day we move enough air to and fro to blow up several thousand party balloons—about five hundred cubic feet. The object of sucking in and blowing out this considerable volume of air is primarily to extract from it the oxygen that it contains and to pass into it the waste carbon dioxide from the blood.

Only about one-fifth of the atmosphere is oxygen, but that small fraction is vital. Just as a flame must be supplied with oxygen to stay alight, so every cell in our bodies must be supplied with oxygen to release the energy locked up in food materials. Brain cells are particularly sensitive to oxygen lack; starved of the gas for just a few minutes, they die—and cannot be replaced. We must breathe to stay alive.

Two systems in partnership are responsible for supplying all the body's cells with oxygen and, at the same time, for disposing of the waste carbon dioxide and water resulting from the cell's metabolism. These are the respiratory system, of airways and lungs, and the circulatory system, of heart, arteries, capillaries and veins.

Air enters and leaves the body through the nose and mouth. At the entrance to the nasal cavities, paired passageways separated by a bony septum, are the nostrils. Here coarse hairs act as a trap for large dust particles that may be inhaled. Within the nasal cavities the air that is breathed in passes over the thin, glistening membrane lining the walls, and is warmed and moistened. This mucous membrane is richly supplied with blood vessels and is capable of raising the temperature of the air from near freezing, on a cold day, to almost that of the body by the time it reaches the back of the throat. Projecting from the cells of the membrane are microscopic hairlike cilia in constant movement. The concerted action of the cilia propels the film of mucus, and any bacteria or dust particles which have been trapped in it, back toward the throat to be swallowed.

Warmed, moistened and filtered, the air now passes into the pharynx, the cavity behind the nose and the mouth. In the sides and roof of the pharynx are collections of lymph glands, the tonsils and adenoids, which pick up and eventually destroy the bacteria trapped in the film of mucus. The air is sucked down into the trachea, or windpipe, passing through the larynx, the boxlike structure that supports the vocal cords, and then goes into the lower respiratory tract.

During swallowing, the entrance to the trachea is momentarily closed by a flap called the epiglottis, to prevent food entering, but the tube itself is always kept open by C-shaped struts of cartilage. The lining of the trachea has cilia, like those in the nasal cavity, that waft particles caught in the mucus layer upward away from the lungs, for disposal by swallowing.

At its lower end, the trachea divides into the bronchi, one of which enters each lung, there to divide and subdivide and lead eventually to several hundred million air sacs, called alveoli. Here the exchange of gases between the inhaled air and the blood flowing in networks of capillaries takes place.

The lungs fill most of the thoracic, or chest, cavity, and are spongelike structures that are inflated and deflated by the movements of the chest wall and the rise and fall of the diaphragm—the sheet of muscle that divides the thorax from the abdomen below. Each lung is enclosed in a double layer of fine membrane called the pleura; the outer pleura is closely attached to the walls of the thoracic cavity and to the upper surface of the diaphragm, and the inner pleura is attached to the lungs.

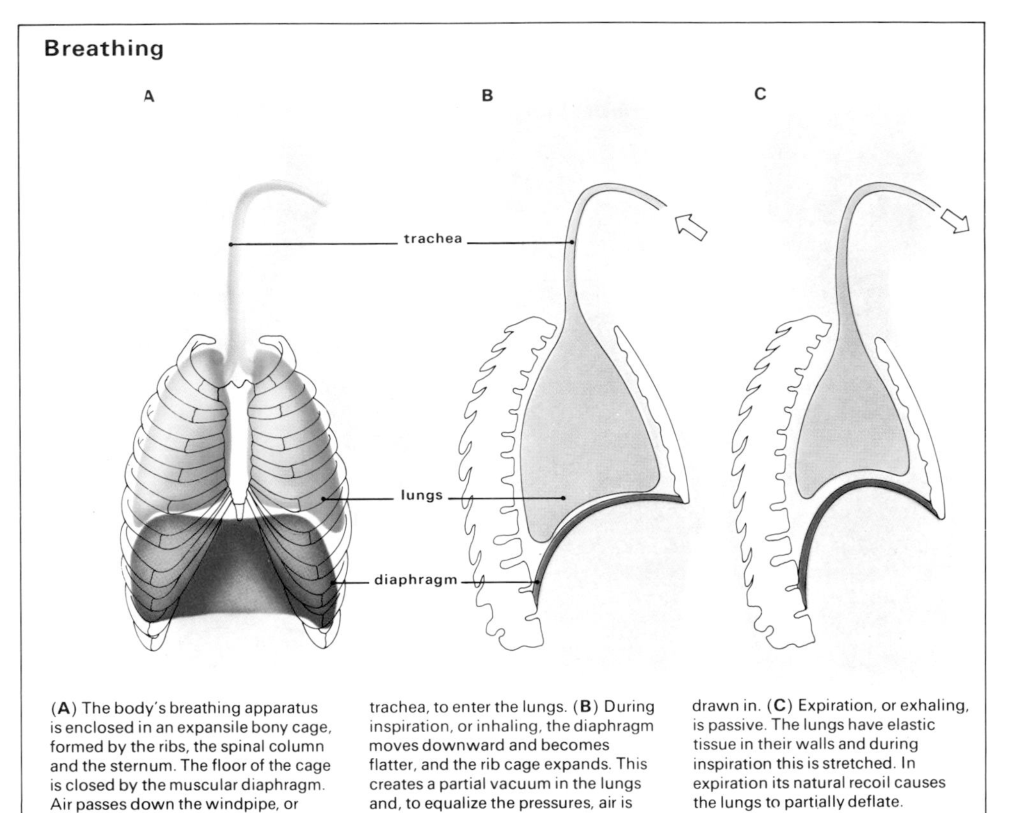

(**A**) The body's breathing apparatus is enclosed in an expansile bony cage, formed by the ribs, the spinal column and the sternum. The floor of the cage is closed by the muscular diaphragm. Air passes down the windpipe, or trachea, to enter the lungs. (**B**) During inspiration, or inhaling, the diaphragm moves downward and becomes flatter, and the rib cage expands. This creates a partial vacuum in the lungs and, to equalize the pressures, air is drawn in. (**C**) Expiration, or exhaling, is passive. The lungs have elastic tissue in their walls and during inspiration this is stretched. In expiration its natural recoil causes the lungs to partially deflate.

During normal breathing it is mainly the diaphragm that does the work. As the muscle contracts, stimulated by the slow, rhythmic impulses carried by nerves from the brain, the volume of the thoracic cavity is increased. At the same time, the ribs, hinged where they join the spine, and linked to the sternum, or breastbone, at the front, move upward and outward, pulled by the muscles between them, the intercostal muscles. As the volume of the chest increases, so the pressure within drops, and air is sucked into the lungs through the trachea, restoring the pressure. This is inspiration.

Expiration is normally an effortless process. The diaphragm and the intercostal muscles relax, and the elasticity of the spongy lung tissue forces air out as the volume of the thoracic cavity decreases. With muscular effort, considerably more air can be squeezed from the lungs.

The respiratory system is able to adapt to changes in the body's needs. As the rate at which energy is used up increases, so the rate and the depth of respiration are automatically adjusted. During quiet breathing, each breath lasts for some four to six seconds and involves the movement of about one pint of air. The quantity involved is called the tidal volume. During physical exertion—as when you are running for a bus—the amount of air needed may be some twenty times as much as during rest. The rate of respiration may be more than doubled, and, at the same time, the amount of air per breath is greatly increased. With effort, up to five pints more of air can be taken in during inspiration. This is called the inspiratory reserve.

Breathing is normally an involuntary activity, but it can, at least within certain limits, be controlled voluntarily. At the back of the brain, in the region called the medulla oblongata, is a group of cells known as the respiratory center, with nerve connections to the higher centers of the brain and spinal cord. We can override the actions of this center to some extent, and hold our breath or pant.

The respiratory center receives, through branches of the vagus nerves, information from stretch receptors in the lungs, and responds by sending messages to the diaphragm and intercostal muscles telling them to contract or to relax appropriately. The most important factor regulating respiration, however, is the level of carbon dioxide in the blood. As soon as the concentration exceeds a certain level, the respiratory center is stimulated, and sends messages to the muscles that bring about inspiration.

Messages also reach the respiratory center from receptors in the aorta and the carotid arteries, which similarly monitor the carbon dioxide level in the blood, as well as its acidity. As the lungs expand during inspiration, their stretch receptors are activated, and now the respiratory center responds by telling the diaphragm and intercostal muscles to relax again, and we breathe out. The respiratory center in the brain responds very quickly to slight changes in carbon dioxide concentration: an increase of only 0.3 percent in the carbon dioxide content doubles the volume of air breathed in and out.

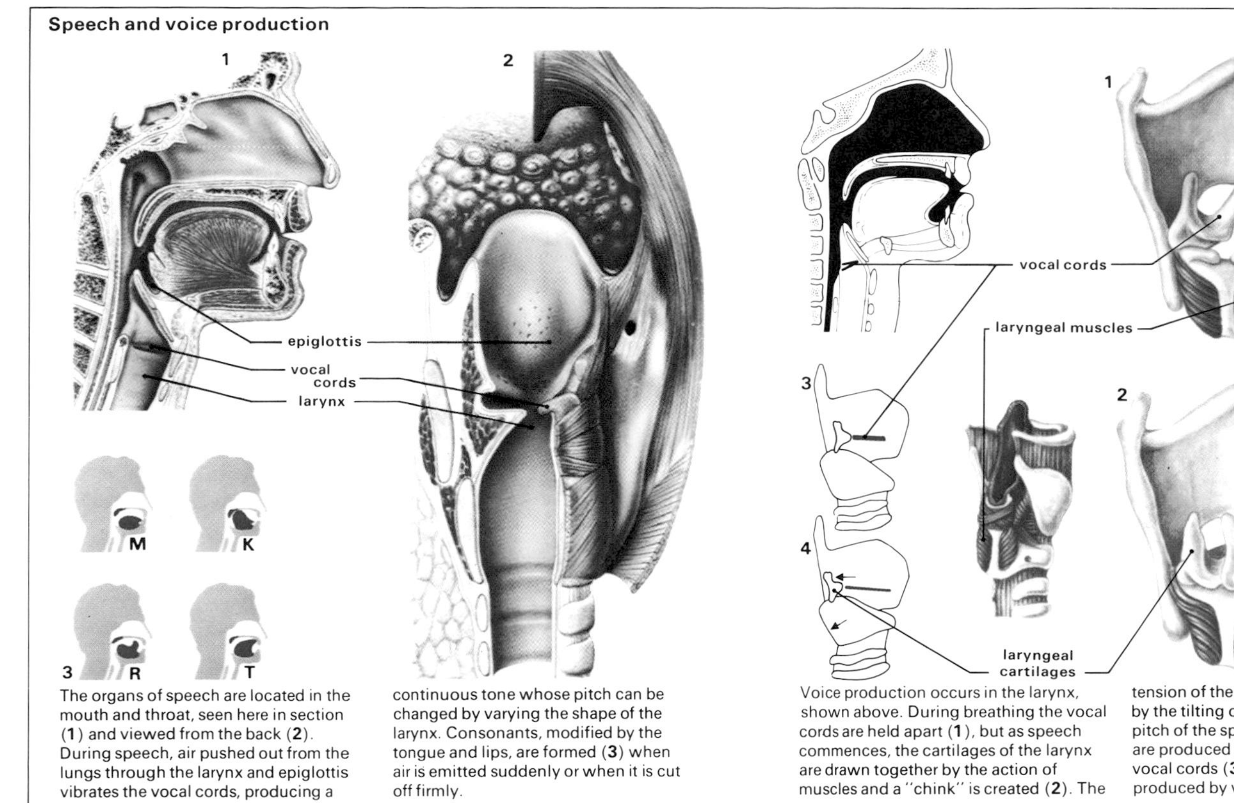

The organs of speech are located in the mouth and throat, seen here in section (**1**) and viewed from the back (**2**). During speech, air pushed out from the lungs through the larynx and epiglottis vibrates the vocal cords, producing a continuous tone whose pitch can be changed by varying the shape of the larynx. Consonants, modified by the tongue and lips, are formed (**3**) when air is emitted suddenly or when it is cut off firmly.

Voice production occurs in the larynx, shown above. During breathing the vocal cords are held apart (**1**), but as speech commences, the cartilages of the larynx are drawn together by the action of muscles and a "chink" is created (**2**). The tension of the vibrating cords, changed by the tilting of the cartilages, alters the pitch of the spoken sound. High notes are produced by the vibration of tight vocal cords (**3**) and low notes are produced by vibrating loose cords (**4**).

Coughing, sneezing and hiccuping, sighing, yawning, crying and laughing are all unusual forms of respiration and occur in response to physical or emotional triggers. Strong smells or dust particles, for example, can irritate the nasal lining and cause sneezing, while the cough is a response to the presence of irritant material, such as dust in the bronchi, or to excess mucus produced in response to infection or noxious fumes. Hiccups result from sudden, sharp contractions of the diaphragm, stimulated by the vagus, a nerve which is also involved in digestion. That is why hiccuping can be associated with eating certain foods. Crying and laughing—which are both long inhalations followed by a burst of short, staccato exhalations—are respiratory reactions to emotional stimuli.

The larynx, situated between the upper air passages and the bronchial tree, uses expired air to produce sound. Air passes between the vocal cords, membranes which fold across the larynx, causing them to vibrate. Movement of the cords varies the size of the gap through which the exhaled air passes, varying the frequency of the note produced. The most common tone of the human voice is 125 cycles per second, but much higher and lower tones are possible. During normal breathing the vocal cords do not vibrate.

The faster the air is forced past the vocal cords, the louder the sound that is produced. The quality and loudness of the sound is modified by the cavities of the mouth, nose, sinuses, throat and chest. Acting together they make the voice resonant and give our voices their individual intonations. Although the vocal cords produce only a series of different notes, a wide variety of different sounds that make up the consonants and vowels we use in speech can be produced.

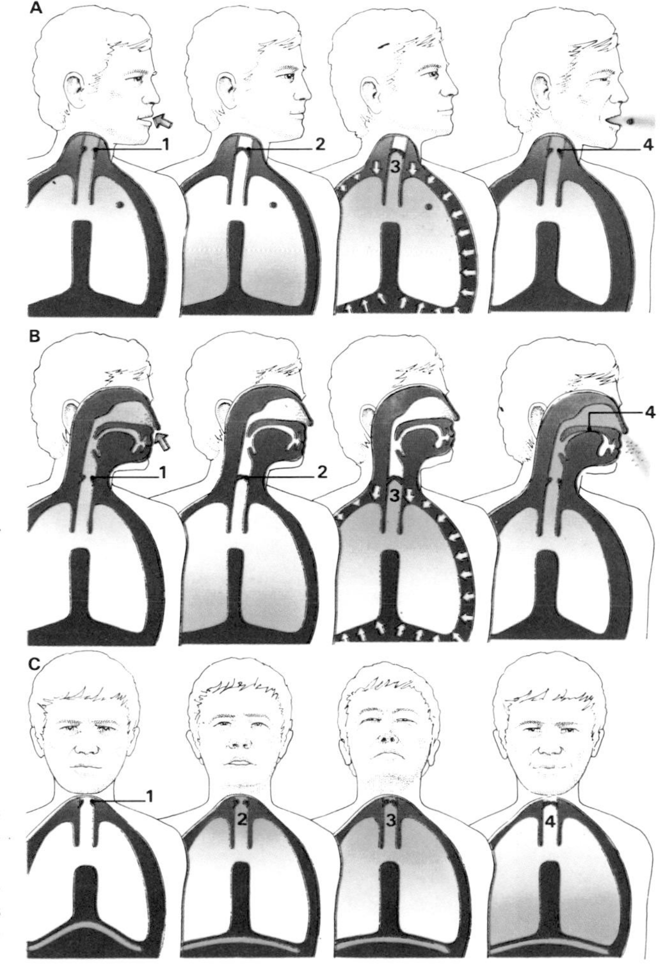

Coughing, sneezing and hiccups are all involuntary actions. The cough reflex (**A**) is stimulated by dust entering the air passages of the lungs. With the glottis (red) fully open, deep inspiration occurs (**1**). The glottis then closes (**2**). The lungs contract, compressing the air (**3**). As the glottis opens, the air is released (**4**), producing a cough. Sneezing (**B**) is triggered by irritation of the nasal lining. The actions of the lungs and the glottis (**1, 2, 3**) are the same as in coughing, but the tongue blocks off the back of the mouth (**4**) as air is violently ejected through the nose. In hiccups (**C**), the diaphragm (green) is initially relaxed and the glottis is open (**1**). The diaphragm contracts spasmodically, causing sudden inspiration (**2**). As air is drawn into the lungs, the glottis snaps shut, creating the characteristic sound aptly named hiccups (**3, 4**).

Cross-reference	pages
Brain and central nervous system	74-77
Defences	138-139
Lungs	64-65
Respiratory center	104-105

Lungs: the exchange of gases

Housed in the protective, flexible cage formed by our ribs arching around and forward from the spine at the back to meet the sternum at the front of the chest are the lungs. Forming the base of this cone-shaped cage is the sheet of muscle called the diaphragm, which is domed and pierced by apertures through which pass the esophagus and the major blood vessels. It is in the lungs that gases are exchanged between the circulating blood and the air that we breathe. The total area in these two vital spongelike organs is vast—they present an area for gaseous exchange that is something like forty times that of the outside surface area of the body.

The lungs occupy twin cavities, one on each side of the chest, separated by the heart, blood vessels and the esophagus. The left lung is slightly smaller than the right, and is indented on its inner surface by the heart. The tops, or apices, of the lungs bulge upward inside the C-shaped first ribs, and lie behind the clavicles. The bases of the lungs rest upon the diaphragm below.

The trachea, which conducts air into and out of the lungs, is about five inches long, and extends downward from the pharynx through the neck to enter the thorax. The trachea can readily be felt through the skin at the front of the neck, and the bulge of the cartilages that make up the larynx at its upper end forms the "Adam's apple," usually more prominent in men. Within the thorax the trachea forks into two branches, the left and right main bronchi, and a bronchus enters each lung. Because the angles at which the bronchi branch from the trachea are slightly different, anything which is accidentally "swallowed the wrong way" is more likely to enter the right bronchus.

Running close to the bronchi, and entering each lung, are the blood vessels which take blood to and from the lungs. The branches of the pulmonary artery bear stale, deoxygenated blood. The branches of the pulmonary veins carry fresh, oxygenated blood from the lungs. Branches of the bronchial arteries and the bronchial veins supply blood to the tissues of the lungs themselves. The region of each lung where all these structures enter is known as the root of the lung.

The left lung has two main divisions, or lobes, and the right has three. Each lobe of the lungs is further divided into a small number of sections, called segments. As soon as each bronchus has entered its lung it divides and redivides, and each segment receives its own branch. The branches themselves divide within the segments, and these smaller divisions are then known as bronchioles. The smallest of the bronchioles that form a meshwork in the sponge are the respiratory bronchioles, which are surrounded with smooth muscle and lined with paper-thin epithelium. Bulging like tiny bubbles from each respiratory bronchiole are the numerous sacs called alveoli. It is through the moist internal surface of the alveoli that all the exchange of gases takes place. The alveoli are the end of the journey for the oxygen of the air breathed in, before it is taken up by the red blood cells.

Lining the inside of the thoracic cavity is the fine membrane called the pleura, which extends in a continuous layer around the root of the lung on each side and then covers the outside of the lungs. There is, therefore, a double layer of pleura between the lungs and the inside of the rib cage and the upper surface of the diaphragm. The two surfaces of the pleura are lubricated with a little thin fluid. The space between them is known as the pleural cavity, which normally contains nothing but a film of this fluid.

The exchange of gases between the air breathed in and our circulating blood takes place through the microscopically thin lining of the alveoli and the walls of the tiny capillary blood vessels that form a network around each one. The capillaries are so fine that only one red blood corpuscle at a time can pass through, taking less than a second to do so.

Oxygen diffuses into the bloodstream and is taken up by the pigment called hemoglobin in

Microanatomy of the lungs

A group of the lung's tiny air sacs, the alveoli, are illustrated here at more than five hundred times life-size. An individual alveolus has been pulled out to show a network of blood capillaries surrounding the thin alveolar walls.

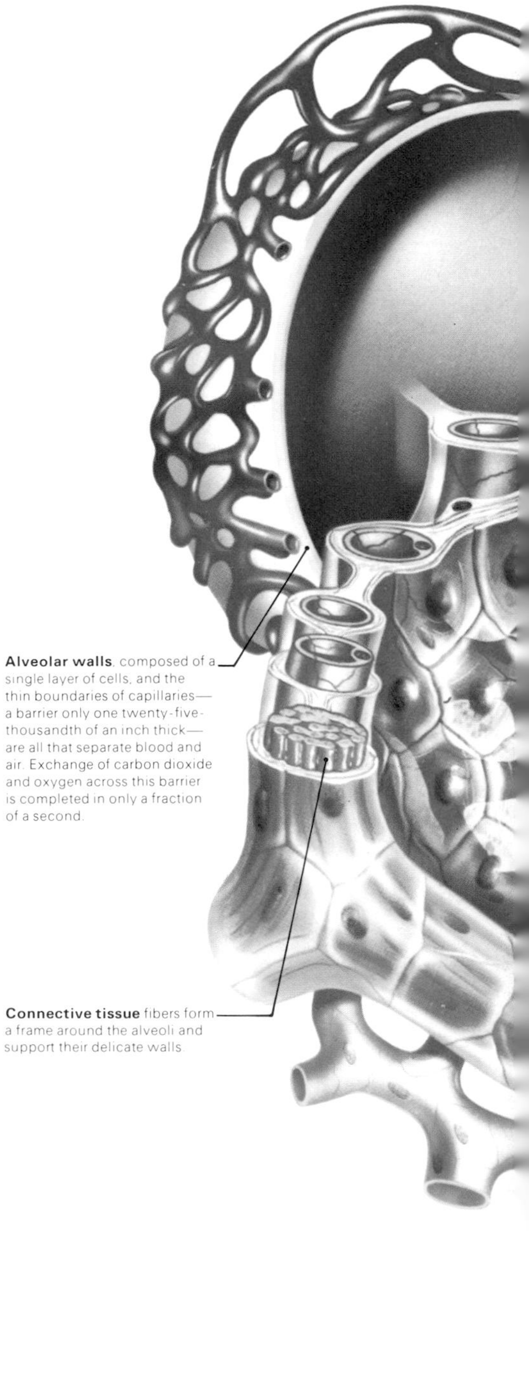

Alveolar walls, composed of a single layer of cells, and the thin boundaries of capillaries—a barrier only one twenty-five-thousandth of an inch thick—are all that separate blood and air. Exchange of carbon dioxide and oxygen across this barrier is completed in only a fraction of a second.

Connective tissue fibers form a frame around the alveoli and support their delicate walls.

The respiratory system

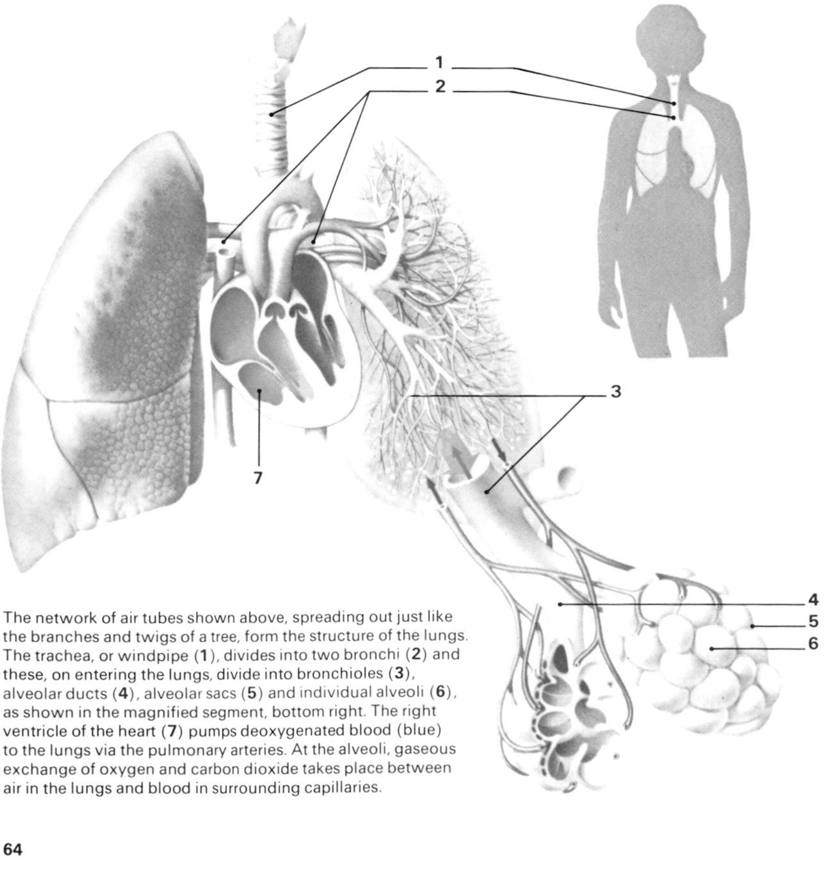

1
2
3
7
4
5
6

The network of air tubes shown above, spreading out just like the branches and twigs of a tree, form the structure of the lungs. The trachea, or windpipe (**1**), divides into two bronchi (**2**) and these, on entering the lungs, divide into bronchioles (**3**), alveolar ducts (**4**), alveolar sacs (**5**) and individual alveoli (**6**), as shown in the magnified segment, bottom right. The right ventricle of the heart (**7**) pumps deoxygenated blood (blue) to the lungs via the pulmonary arteries. At the alveoli, gaseous exchange of oxygen and carbon dioxide takes place between air in the lungs and blood in surrounding capillaries.

each red corpuscle, which then becomes oxy-hemoglobin and is brighter red in color. Simultaneously, the blood releases the carbon dioxide, mainly dissolved in the plasma, that it has transported from the tissues.

Because the alveoli must be lined with a thin film of water to enable gases to dissolve, expiratory air is saturated with water vapor. During a day the equivalent of four cupfuls of water are lost from the lungs. The cloud of water droplets that condenses on expiration in cold weather is this water vapor.

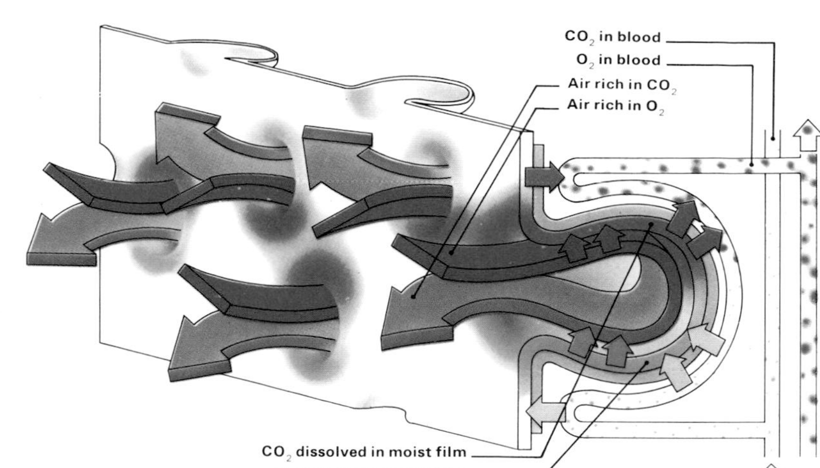

As air is breathed into the lungs, exchange of gases takes place at the highly folded, moist surface of air sacs, as shown above. These are richly supplied with blood capillaries, which take up oxygen and give up carbon dioxide.

Alveolar sacs are the terminal portions of the breathing tube system. Composed of up to thirty tiny air pouches, or alveoli, they provide a lung total surface-area of about 850 square feet for exchange of gases.

Pulmonary venules take oxygen-rich blood back to the heart.

Terminal bronchioles, measuring only .012 of an inch in diameter, carry air to and from alveolar sacs. Cells lining bronchiole walls bear cilia—minute hairs whose beating action prevents large dust particles entering the alveolar sacs.

Lymph vessels carry excess tissue fluid away from alveoli.

Pulmonary arterioles conduct carbon dioxide rich, oxygen-depleted blood to the alveoli.

Blood capillaries form a meshwork around the alveoli. As blood flows through the capillaries it loses carbon dioxide and takes up oxygen, and changes from dark to bright red in color.

Macrophages on the inner alveolar surface ingest and destroy dust, soot and other foreign particles. These cells are the body's main line of defense against airborne bacteria drawn into the lungs.

Cross-reference	pages
Blood	**58-61**
Circulation	**54-55**
Diaphragm	**62-63**
Metabolism	**52-53**

Kidneys: filtration units

The urinary system is concerned with the formation and elimination of urine. In an adult, more than 2,500 pints of blood pass through the kidneys (**1**) each day. Blood enters via the renal arteries (**2**) and is filtered to remove most of the waste products of metabolism. Seven pints of filtrate are produced every hour. Purified blood returns to the body circulation via the renal veins (**3**). The filtering process is carried out by more than two million tiny kidney units, or nephrons, which produce a highly concentrated solution of chemicals known as urine, which is harmful to the body if allowed to remain. Urine flows from the nephrons, first into the funnel-shaped renal pelvis (**4**) and then into the ureter (**5**). Waves of muscular contraction passing down the ureters push the urine into the bladder (**6**). With continuous filling, the bladder, a muscular bag, expands until it holds about one pint of fluid. A circular band of muscle around the neck of the bladder, the sphincter (**7**), controls the release of urine from the body.

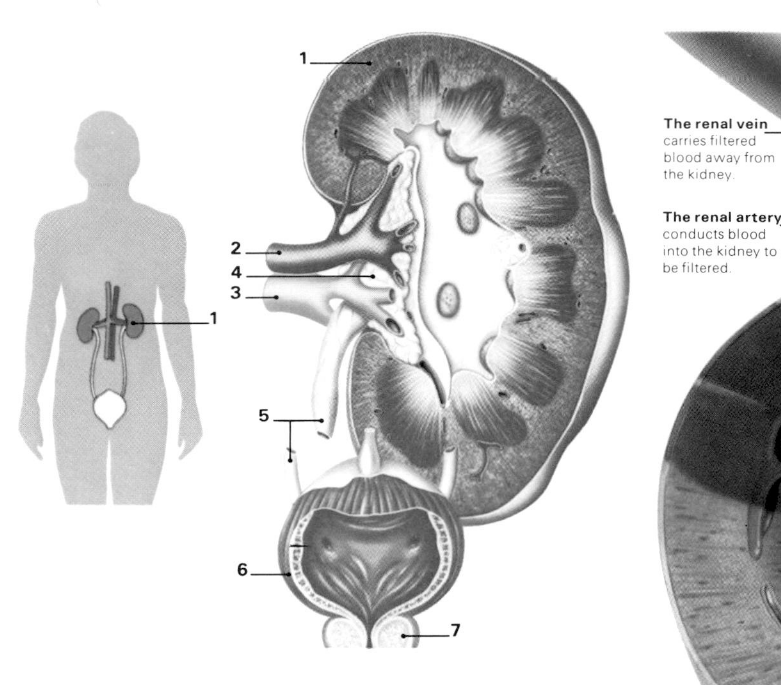

The renal vein carries filtered blood away from the kidney.

The renal artery conducts blood into the kidney to be filtered.

It would be remarkable, considering the vast number of processes that are involved in the day-to-day running of the human machine, if there were no waste products needing disposal. Many of the by-products of the chemical reactions that maintain life are "recycled" to obtain maximum value from their molecules, but there comes a stage when our bodies can find no way of utilizing certain materials. To avoid their accumulation, and the pollution and poisoning of our internal environment, this waste must be excreted. There are four main excretory routes.

The chief waste products of metabolism are carbon dioxide and water, resulting from the burning of carbohydrates and fats as fuel, urea, one end product of protein breakdown, bile salts and pigments, derived from the destruction of worn-out red blood corpuscles, and various other mineral salts from tissue breakdown and excessive intake in the diet. Carbon dioxide we breathe out through the lungs, in the company of water vapor. Water, salts and a little urea are lost through the pores of the skin as sweat. Bile salts and pigments, contributing as digestive agents, are partly reabsorbed and partly excreted in the feces.

The most important route of excretion, however, is through the kidneys—the paired filtration units lying at the back of the abdomen. These produce a constantly adjusted trickle of urine which is voided when the bladder becomes uncomfortably full. Urine, a solution in water of urea, salts and other soluble wastes, is produced by the kidneys from the blood that enters them, under high pressure, through the renal arteries. The renal arteries are large side branches of the aorta, and together receive about a quarter of the volume of blood pumped out by the heart at each beat. All of the blood in our bodies passes through the kidneys about twenty times every hour, but only about one-fifth of the plasma is filtered into the kidney tubules.

Each kidney is roughly the size of a child's fist. Seen cut open it has an outer pale layer called the cortex, and a darker core, densely packed with blood vessels that divide and subdivide, called the medulla. On the medial aspect of each kidney lies a urine-collecting cavity, the renal pelvis, leading down through a thin-walled tube, the ureter, into the distensible bladder.

Each kidney has about one million separate filter units, called nephrons. A nephron consists of a cup-shaped structure, Bowman's capsule, which encloses a knot of capillaries, known as a glomerulus, that carries blood from the renal artery, and a renal tubule. Each tubule is between one and two inches long, and loops, rather like a trombone, in and out from the cortex, where the capsule lies, into the medulla. The tubules unite into collecting ducts, and the ducts empty their contents into the renal pelvis.

As blood under high pressure passes through each glomerulus, water, dissolved salts, sugar, urea and many other small molecules are forced through the walls of the capillaries and into the capsule. Blood corpuscles and large protein molecules are kept back by the filter. The blood, now more concentrated because it has lost about one-fifth of its plasma, passes from the glomerulus into a network of fine vessels surrounding the loops of each tubule, and will eventually rejoin the general circulation through the renal vein. Before it does so, however, another important process must be carried out.

Although the kidney is described as excreting water, it has in fact a duty to conserve water. Our bodies need water as a medium in which every biochemical process can take place. Only if we were waterlogged could water be said to be waste material with a nuisance value. It is simply because a solvent is needed for true wastes, to remove them from the body, that large amounts of water pass into the kidney tubules. Nearly all the water is, therefore, recovered from the filtrate in its journey through the loops of the renal tubule, passing back into the bloodstream through the tubule walls into the network of blood vessels.

At the same time, by a process of chemical transfer through the tubule walls, sugar and other materials that the body needs, but which have passed into the filtrate because of the small size of their molecules, are also conserved by reabsorption into the blood. So efficient is the reabsorption of sugar, for example, that not a trace is normally detectable in the urine. After passing down the ureters, the urine is stored in the bladder, to be discharged through the urethra to the outside when convenient.

The activity of the kidneys in regulating the body's water content is controlled by hormones, of which the most important is antidiuretic hormone (ADH), secreted by the pituitary gland. Reaching the renal tubules in the blood, ADH stimulates reabsorption of water from the filtrate into the blood. If our intake of water is inadequate to make up for the water lost in perspiration and respiration, the slight increase in concentration of the blood is detected in the brain by special receptors, which stimulate the posterior pituitary gland to release more ADH than normal. Water loss in the urine is thus reduced. Conversely, blood that is too dilute reduces ADH secretion, producing a diuresis, or copious flow of dilute urine, to restore the water balance.

Anatomy of a kidney

This section through a kidney, almost three times life-size, shows its blood supply and one of the filter tubules, a nephron, magnified many times more. The kidneys filter unwanted materials from the blood and regulate the levels of water and chemicals in the body. As blood flows through a glomerular tuft, fluid is forced into the Bowman's capsule, here pulled out to show its cellular structure. The filtrate then flows along the coiled tubule. Further waste material is secreted directly into the nephron tubule. At the same time, reabsorption of essential chemicals into the capillaries takes place. Urine, the yellowish liquid remaining, passes from the nephron-collecting ducts into the renal pelvis, down the ureters and into the bladder.

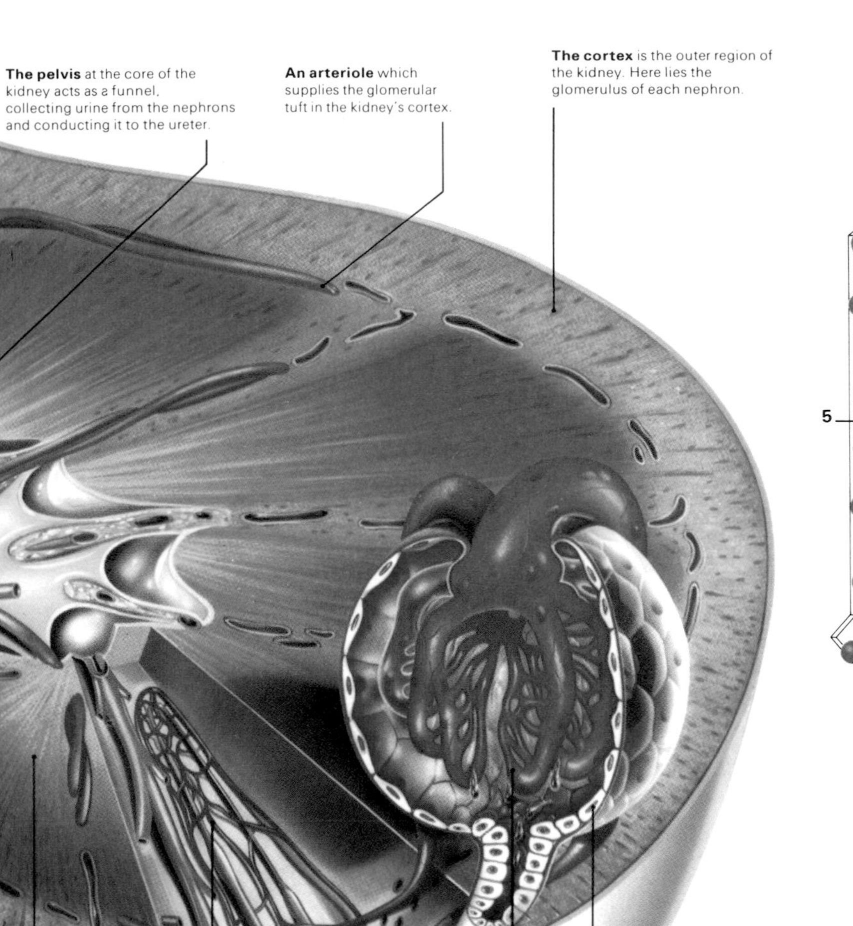

The pelvis at the core of the kidney acts as a funnel, collecting urine from the nephrons and conducting it to the ureter.

An arteriole which supplies the glomerular tuft in the kidney's cortex.

The cortex is the outer region of the kidney. Here lies the glomerulus of each nephron.

The ureter is a tube, about ten inches long, which conducts urine away from the kidney. Wavelike contractions moving along its muscular wall force urine down into the bladder.

The medulla, the inner region of the kidney, contains the loops of Henle and the collecting ducts surrounded by blood capillaries.

A capsule of tough, fibrous tissue surrounds and protects the kidney.

The loop of Henle is the central region of the nephron. Here reabsorption of essential materials from the filtrate into the bloodstream takes place.

The nephron is a highly specialized coiled tubule measuring up to one inch in length, but only about 0.00024 of an inch wide. There are more than one million nephrons in each kidney. They function as the kidney's filter units. With the Bowman's capsule at its beginning, each nephron extends through the medulla and terminates as a collecting duct opening into the renal pelvis.

The glomerular tuft is a tight knot of blood capillaries from which water and chemicals filter into the nephrons. Each tuft measures only 0.005 of an inch in diameter, but is often composed of more than fifty capillaries.

The Bowman's capsule, the expanded beginning of the tubule, is tightly clasped around the glomerular tuft. It is the nephron's collecting cup for fluid filtering from the blood.

The filtering mechanism, shown above, occurs at the Bowman's capsule of each nephron. As blood flows through the glomerular tuft (**1**), small molecules are forced through pores in the capillary walls (**2**) into the nephron (**3**). Red blood cells and large protein molecules remain. The filtrate flows down the tubule and is joined by organic wastes and excess salts secreted from surrounding kidney tissues (**4**). Most of the glucose, amino acids, vitamins and essential salts are reabsorbed into the blood (**5**), leaving a concentrated solution, urine (**6**).

Cross-reference	pages
Blood	**58-61**
Circulation	**54-55**
Hormones	**70-73**
Metabolism	**52-53**
Muscles of the bladder	**40-41**
Thirst	**112-113**

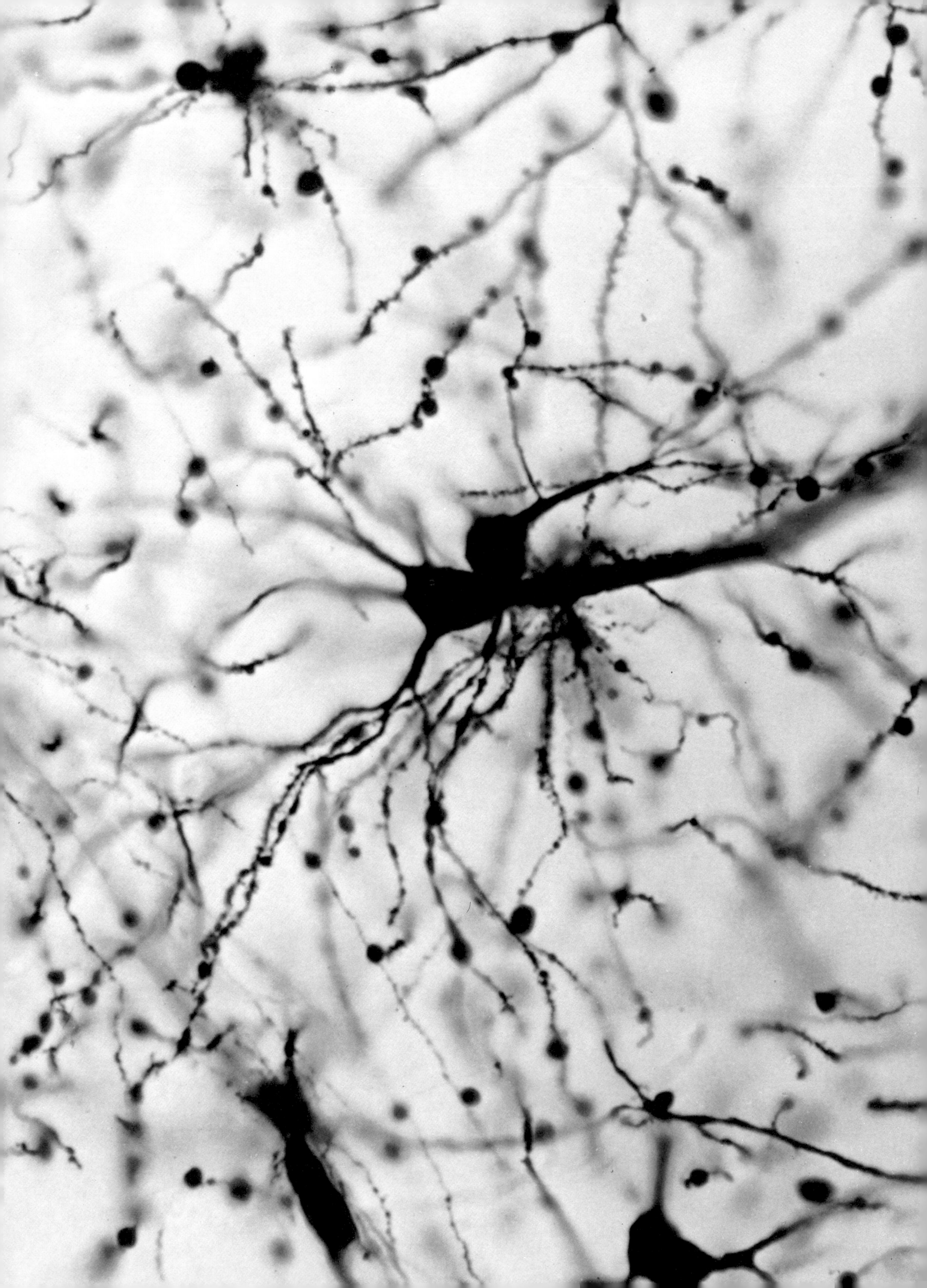

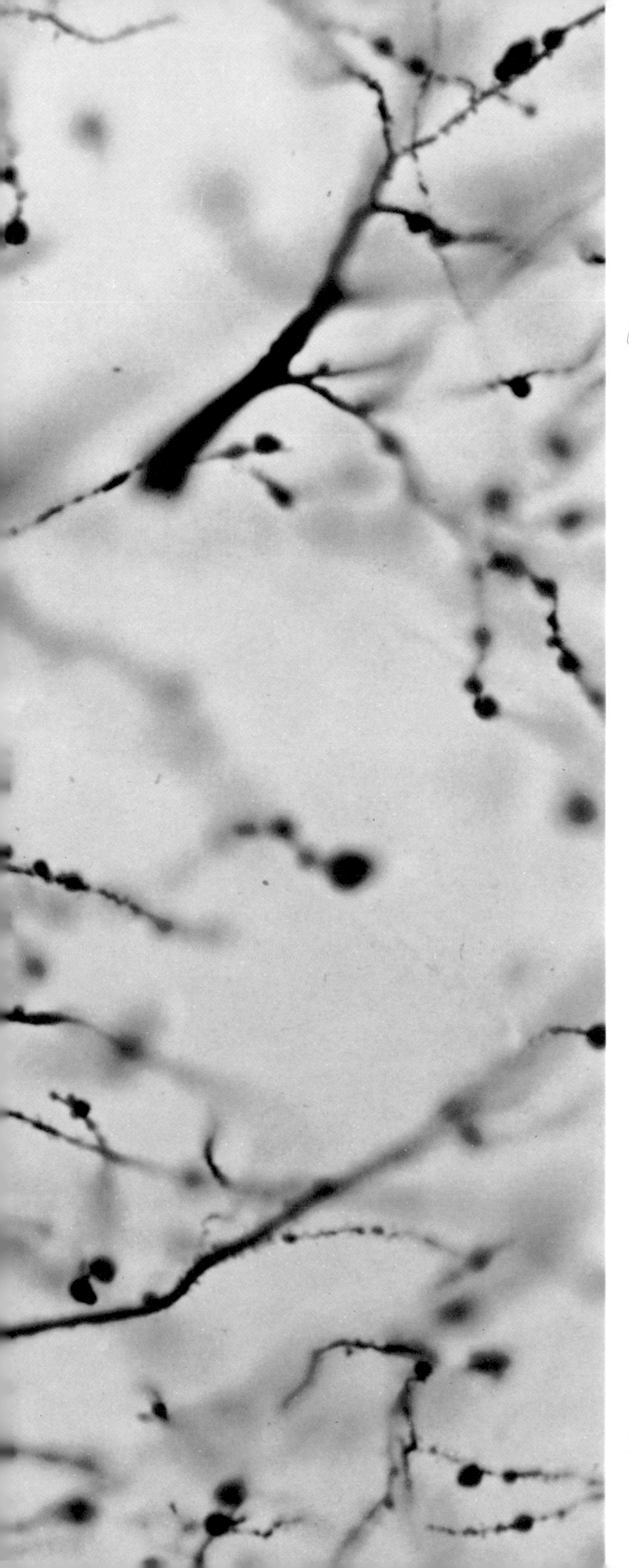

Control systems

It is perhaps in the complexities of the control systems, which direct and coordinate our activities, that mind meets body most closely. Consciousness is a property of the brain, and linking consciousness with bodily functions are the integrated, partner systems of sensory nerves, motor nerves and endocrine glands. We think and then act, imposing our will on the activities of muscles by means of nerves. If we are frightened, our pulses quicken as adrenaline goes to work. In both cases, mental activity has brought about physical activity.

But there are aspects of control that, fortunately, do not involve our conscious minds. We do not have to direct our fingers to stop touching something too hot—reflex nervous activity sees to that. Nor do we have to be conscious to digest food—the autonomic nervous system working together with the endocrine system organizes the secretion of digestive juices as they are required. There are control systems to maintain the constancy of the internal environment of which we need never be aware. Left to itself, the body controls itself, but pervading every activity is the influence of the brain, the master control unit, with its servants the endocrine and nervous systems.

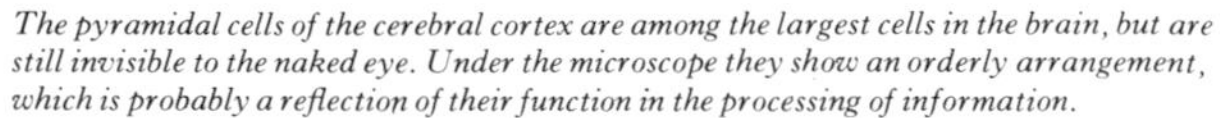

The pyramidal cells of the cerebral cortex are among the largest cells in the brain, but are still invisible to the naked eye. Under the microscope they show an orderly arrangement, which is probably a reflection of their function in the processing of information.

Hormones: chemical coordinators

Although a single-celled organism can regulate its own internal metabolism, in a highly complex organism like man, cells are adapted to specialized tasks, and whole body controls are required. The human body works efficiently only when the equilibrium within each organ, tissue and cell is closely monitored and controlled. Activity, growth and repair of tissues must be maintained, together with the supply of fuels and the removal of waste material.

These controls are achieved by two systems. The first is the network of nerve fibers which carries messages between the brain and the rest of the body. The second is the endocrine system. This system consists of a number of endocrine glands situated in different parts of the body whose functioning is not always linked directly to each other. Their name derives from the Greek words *endon*, which means within, and *krinen*, which means to separate.

The endocrine glands release their chemical products, or hormones, directly into the bloodstream, rather than into specialized channels, or ducts. The hormones are slower in their action than nerve impulses, with the exception of two hormones, adrenaline and noradrenaline, produced by the adrenal gland, which do act quickly, and for a short time, on many tissues. The hormones that are produced by the endocrine glands are, like all the constituents of the body, derived from food materials —protein, fat and carbohydrate. Hormones travel in the bloodstream to cells in all parts of the body. The membrane of every cell has receptors to one or more hormones, and the binding of a hormone to its specific cell receptor initiates particular changes in the internal metabolism of this "target" cell. One hormone may cause changes in more than one function of a particular cell, and the pathways altered vary according to the specialized tasks of the cell affected.

The way in which hormones cause changes inside cells has recently been the subject of major advances in our knowledge. A stimulating hormone fits to a specific receptor area on the cell membrane. This receptor has a special shape to which only its own hormone will fit, rather like a jigsaw puzzle piece will only fit to a correctly shaped second piece. By their combination with the receptor most, but not all, hormones lead to the production within the cell of a substance called cyclic adenosine monophosphate (cyclic AMP). This is a long name for a fairly simple chemical which, like hormones acting as chemical messengers within the blood, functions as a messenger inside the cell. The messenger, cyclic AMP, then alters such cellular functions as protein synthesis, or production, the storage or release of excess fuel of glycogen, or triglyceride, and the production of other hormones.

The endocrine glands consist of the hypothalamus and the pituitary at the base of the brain, the thyroid and parathyroid glands at the front of the neck, the adrenals in the abdomen above the kidneys, the pancreas, lying in the loop of the duodenum, and the ovaries in the pelvis or the testes in the scrotum. In addition, various other organs produce hormones, such as the several produced by the gastrointestinal tract, which are involved in controlling digestion.

In much the same way, for example, as the brain's control of muscle activity is constantly monitored and altered to suit the information received by the brain, so, too, is the activity of the endocrine glands constantly monitored and altered by nervous, hormonal and chemical information being fed to them. Hormone production is controlled in many cases by a negative feedback system, in which overproduction of a hormone leads to a compensatory decrease in the subsequent production of the hormone until balance is restored. The operation of these systems can be best understood when the various hormones of the anterior part of the pituitary gland are examined together with the other endocrine glands which are related to it.

The pituitary gland, which is about the size of a small pea, sits protected in a bony depression—the sella turcica, or "Turkish saddle"—at the base of the skull. It is in effect two separate glands, the anterior and posterior, each of which has a separate function. The anterior pituitary has been called the major, or controlling, gland because almost all its hormones direct the activity of its "target" glands elsewhere in the body. These hormones are called trophic hormones, and are thyroid-stimulating hormone (TSH or thyrotrophin), adrenocorticotrophic hormone (ACTH), follicle-stimulating hormone (FSH) and luteinizing hormone (LH).

TSH and ACTH, as their full names imply, increase the activity of the thyroid gland and the

How hormones work

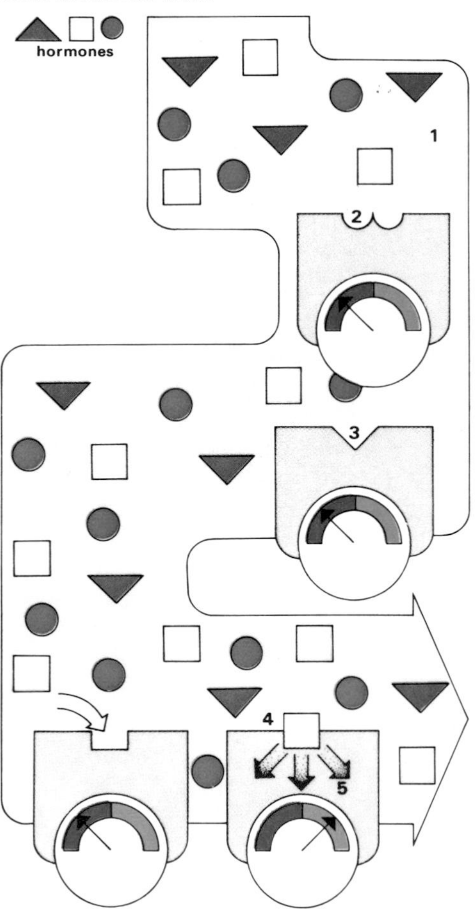

Hormones act as chemical messengers and are secreted into the bloodstream by the endocrine glands for distribution around the body. Each hormone acts only upon specific types of cells—those bearing the correct membrane receptor sites. These are the "target cells" for the hormone. This is illustrated above. Within the bloodstream (**1**) there are many different hormones. When specific receptor sites are not filled, cells remain "turned off" (**2, 3**). As hormones attach to receptor sites (**4**), they initiate a "switch on" reaction in the cell cytoplasm—this reaction is usually the production of the chemical cyclic adenosine monophosphate (cyclic AMP) from ATP. Cyclic AMP acts as a messenger—within the cell—to activate the cell's production lines (**5**). The activity in the bloodstream of some hormones, such as adrenaline, may last for only ten minutes. Others, including some sex hormones, have a longer life. Hormones may be inactivated by the target cells or by the liver, and the breakdown products, when their actions have been completed, are either excreted or reused in the further manufacture of hormones.

Major endocrine glands and their hormones

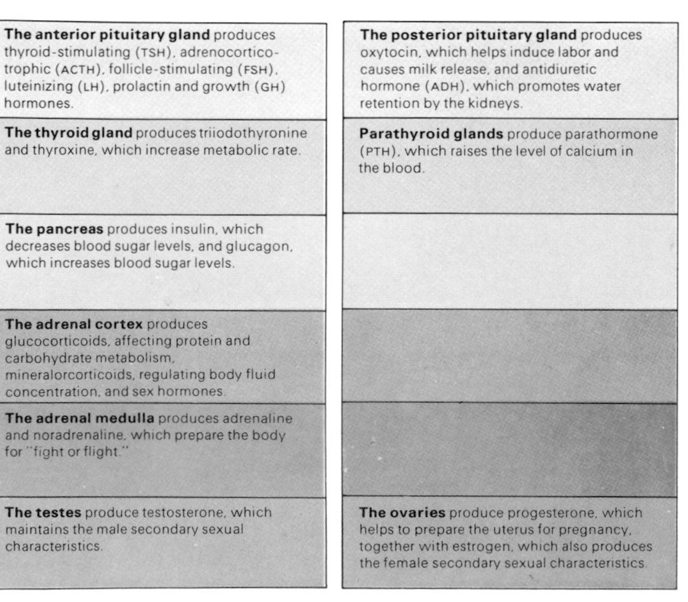

The anterior pituitary gland produces thyroid-stimulating (TSH), adrenocorticotrophic (ACTH), follicle-stimulating (FSH), luteinizing (LH), prolactin and growth (GH) hormones.	**The posterior pituitary gland** produces oxytocin, which helps induce labor and causes milk release, and antidiuretic hormone (ADH), which promotes water retention by the kidneys.
The thyroid gland produces triiodothyronine and thyroxine, which increase metabolic rate.	**Parathyroid glands** produce parathormone (PTH), which raises the level of calcium in the blood.
The pancreas produces insulin, which decreases blood sugar levels, and glucagon, which increases blood sugar levels.	
The adrenal cortex produces glucocorticoids, affecting protein and carbohydrate metabolism, mineralorcorticoids, regulating body fluid concentration, and sex hormones.	
The adrenal medulla produces adrenaline and noradrenaline, which prepare the body for "fight or flight."	
The testes produce testosterone, which maintains the male secondary sexual characteristics.	**The ovaries** produce progesterone, which helps to prepare the uterus for pregnancy, together with estrogen, which also produces the female secondary sexual characteristics.

Hormones maintain the equilibrium of the body's chemistry and regulate growth and development. They are produced by the system of endocrine glands, which secrete them directly into the bloodstream for distribution throughout the body.

adrenal cortex, while FSH and LH stimulate the ovaries or testes. The other major hormone of the anterior pituitary gland is growth hormone (GH), which acts on body tissues in general to produce growth in childhood and adolescence, and to influence the metabolism of protein, fat, carbohydrate and minerals. Some of its actions are thought to be due to production by GH in the liver of a small protein, called somatomedin, which encourages the growth of skeletal tissues. Another hormone from the anterior pituitary is prolactin. Its production increases during pregnancy and promotes milk production, lactation, by the breasts after childbirth. Its function in males and non-pregnant women is not clear.

The production and release of hormones from the anterior pituitary are affected by the amount of hormone circulating in the blood that has been produced by the target gland, and by the activity of the hypothalamus, a small area of the forebrain situated just above the pituitary. The hypothalamus is the main coordinating center between the endocrine and nervous systems. It produces a number of small peptides, chains of amino acids, which it secretes into a specialized local blood portal system.

The arterioles that serve the hypothalamus carry to the pituitary gland these peptides, which either prevent or promote the release of the anterior pituitary hormones. Thus there is a peptide called thyrotrophin-releasing hormone (TRH), which promotes the release of TSH, and similar peptides for the release of FSH and LH, and ACTH and GH. Growth hormone and prolactin are controlled by hormones from the hypothalamus, which inhibit their release.

The hypothalamus is influenced in its turn by nerve impulses from parts of the brain which control the circadian rhythms, our biological clock, and from higher centers in the brain, too. The levels of hormones from target glands, such as the thyroid and the adrenals, feed back not only on the pituitary itself, but also on the hypothalamus, and high levels result in inhibition of the release of the appropriate hormone.

The posterior pituitary is rather different from the anterior part to which it is so closely attached. It bears no secretory cells, but stores two hormones which are produced by specialized nerve cells in the hypothalamus and which are passed down the nerve fibers in the pituitary stalk. These hormones are oxytocin, which acts on smooth muscle, especially that of the uterus and breasts, and vasopressin, or antidiuretic hormone (ADH). ADH is secreted if the blood salt level rises due to excess salt in the body or to water lack, and causes the kidneys to reabsorb more water and to excrete a more concentrated urine. This is an important mechanism for conserving water which is necessary for the body to function properly.

The anterior pituitary gland, situated at the base of the brain, secretes several hormones which control the activity of other endocrine glands and influence general body growth. (**A**) Messages from the higher centers of the brain (**1**) stimulate hypothalamic cells (**2**) to secrete hormones into the pituitary portal blood system (**3**) Some of these hormones stimulate, and others inhibit, secretion of anterior pituitary hormones. (**B**) Hormones produce numerous effects —TSH stimulates thyroid hormone production (**1**) and GH affects bones and muscles (**2**). ACTH triggers hormone release from the adrenal cortex (**3**). Prolactin stimulates lactation (**4**). LH and FSH control hormone release by the testes (**5**) and ovaries (**6**); and in maturity, production of spermatozoa and ova in these organs.

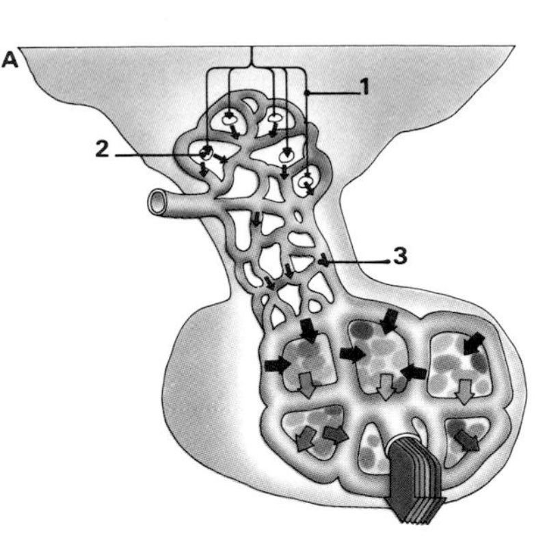

- thyroid-stimulating hormone (TSH)
- growth hormone (GH)
- prolactin
- follicle-stimulating hormone (FSH)
- luteinizing hormone (LH)
- adrenocorticotrophic hormone (ACTH)

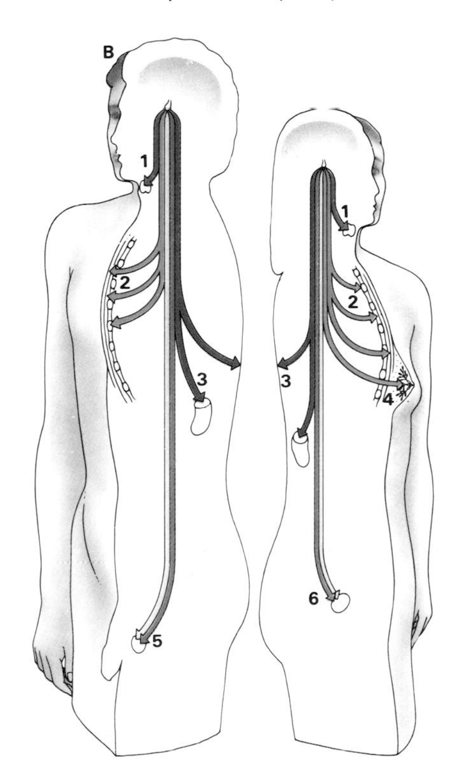

Control of anterior pituitary gland activity

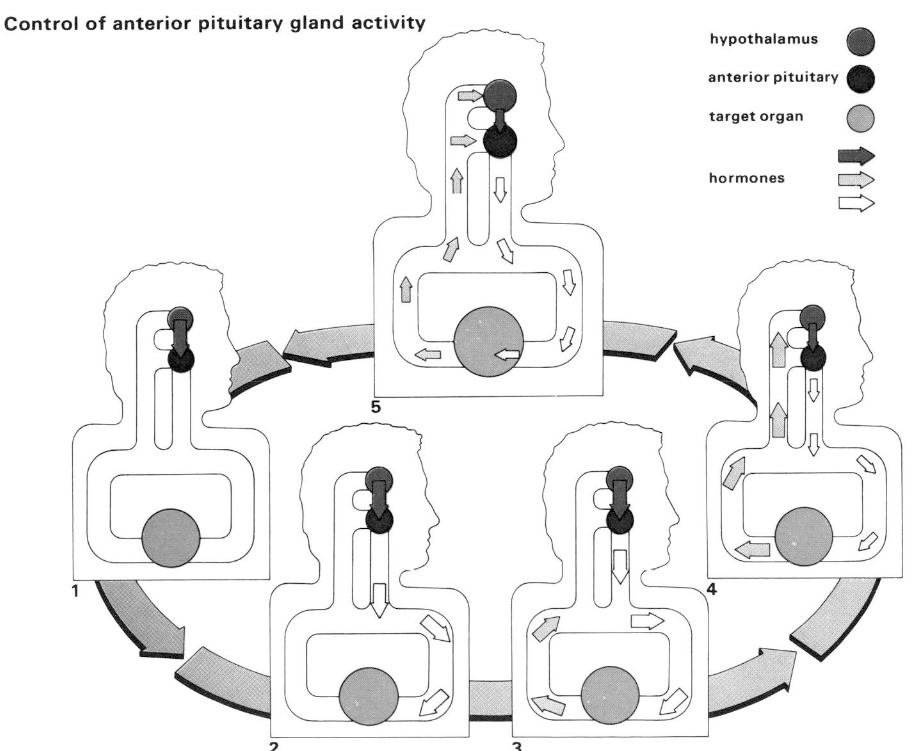

Hormone production is controlled by the levels of chemicals and other hormones circulating in the bloodstream. The secretion of hormones by the anterior pituitary gland is controlled by the hypothalamus of the brain and by the hormones secreted by the target glands on which the anterior pituitary hormones themselves act. The hypothalamus releases a hormone (**1**) which allows the anterior pituitary to secrete its own hormone messenger (**2**). This is carried in the blood to its target endocrine gland, which is stimulated to release its hormone (**3**) into the blood. The circulating target gland hormone, besides acting in the body, feeds back to the pituitary and hypothalamus (**4**) to inhibit their activity. This allows a balance between the activity of the hypothalamus, the anterior pituitary and the target gland (**5**).

Cross-reference	pages
Birth	**156-157**
Brain and central nervous system	**74-77**
Digestion	**46-47**
Female physiology	**146-147**
Hypothalamus	**112-113**
Kidneys	**66-67**
Lactation	**156-157**
Male physiology	**144-145**
Metabolism	**52-53**

Hormones: chemical messengers

The function of endocrine glands is to help control the internal environment and composition of each cell and organ, and of the whole body. Without this control and balance of body processes the highly sophisticated activities of man would not be possible.

The endocrine glands are anatomically separated, but some are connected functionally. The thyroid gland situated at the front of the neck is stimulated by TSH from the pituitary gland to produce the two thyroid hormones, thyroxine and triiodothyronine. They act throughout the body to maintain the activity of all the metabolic pathways in the cells at a steady rate. Energy-producing and energy-using processes are increased and fuel storage is reduced by excess thyroxine, and in its absence a person becomes slow and lethargic. The thyroid gland requires much of the dietary iodine intake and cannot function adequately if this element is lacking.

The other glands in the neck are four tiny parathyroid glands, which produce parathormone. When the level of calcium in the blood falls, parathormone levels increase and act at the kidney tubules to reduce the calcium loss in the urine, at the intestinal wall to increase absorption and at the bones to remove some of the element from its largest store in the body, so restoring a low calcium level to normal. A hormone, calcitonin, from the thyroid has actions which partly oppose those of parathormone. Very close control of calcium levels is necessary to maintain normal cell functions and bone structure. Low calcium levels may lead to tingling in the limbs and muscle cramps. High calcium levels may result in muscle weakness.

The adrenal glands sit above the kidneys in the abdomen, and are in two distinct parts—an outer cortex and an inner medulla. The cortex of the adrenal glands produces several similar hormones, the two most important of which are cortisol and aldosterone. These hormones are produced in response to ACTH from the pituitary, and have effects on the control of carbohydrates, proteins and fats, and on the handling of sodium, potassium and water in the body.

Cortisol primarily affects the glucose, amino acid and fat metabolism, and is one of several hormones that help maintain a steady supply of "building blocks" and "fuel" to maintain normal body repair and growth in all tissues. Aldosterone mainly promotes the reabsorption of sodium from the distal renal tubules, intestine and sweat glands, and increases the loss of potassium. Potassium is maintained at high levels within cells, and at much lower levels in blood plasma and in the fluid around every cell. Sodium ions, however, are maintained at high levels outside cells and at lower levels inside. Close balance of these ions is essential for the health of cells, and sodium or potassium is retained or lost as necessary in the urine to achieve this. Aldosterone release is effected by ACTH, and via renin produced in the kidneys. The enzyme renin leads to the production in the blood of a small peptide, angiotensin, which raises blood pressure and stimulates the formation of aldosterone.

The cortices of the adrenal glands also produce sex hormones—androgens, male hormones, and estrogens, female hormones. Both types are found in men and women, for it is the balance of these hormones, and those from the ovaries and testes, that is responsible for the secondary differences between the sexes.

The medulla, the other part of the adrenal gland, operates as a distinct endocrine organ, manufacturing two hormones, adrenaline and noradrenaline, which have widespread and rapid effects throughout the body. A quiet day in the country suddenly disturbed by an angry bull requires rapid evasive action, stimulated in part by these hormones. The emergency over, breakdown of the hormones occurs, but takes appreciable time, and sweating, anxiety and palpitations last a little while. These hormones achieve their effects by increasing blood flow to all the organs which will be needed to meet a potential threat to the person concerned. Flow to the heart and skeletal muscle increases with a compensatory reduction to skin and intestines. Blood pressure and pulse increase, and the body is thrown into a state of readiness. Accompanying these actions are dilatation of the lung bronchioles, and increased levels of fuels—glucose and fatty acids—in the blood. These alterations are preparations for what may be termed "fight or flight."

The hormone adrenaline also acts as a chemical transmitter at many nerve endings, for the adrenal medullae develop from the

Structure and functions of the pancreas

The pancreas is shown here with the foreground highly magnified to reveal its inner anatomy. Ninety-nine percent of pancreatic tissue is composed of acinar glands, which secrete an alkaline digestive juice into the duodenum via the pancreatic duct to help digest food. The endocrine areas of the pancreas, known as the islets of Langerhans, are composed of two major types of cell. The alpha cells secrete the hormone glucagon, and the beta cells secrete insulin, into the bloodstream. These hormones work in opposition to control the blood glucose level. Insulin promotes the uptake and usage of glucose in the tissues, particularly skeletal muscle and fat, and reduces glucose production in the liver. Glucagon has an anti-insulin effect in the liver, increasing glucose release.

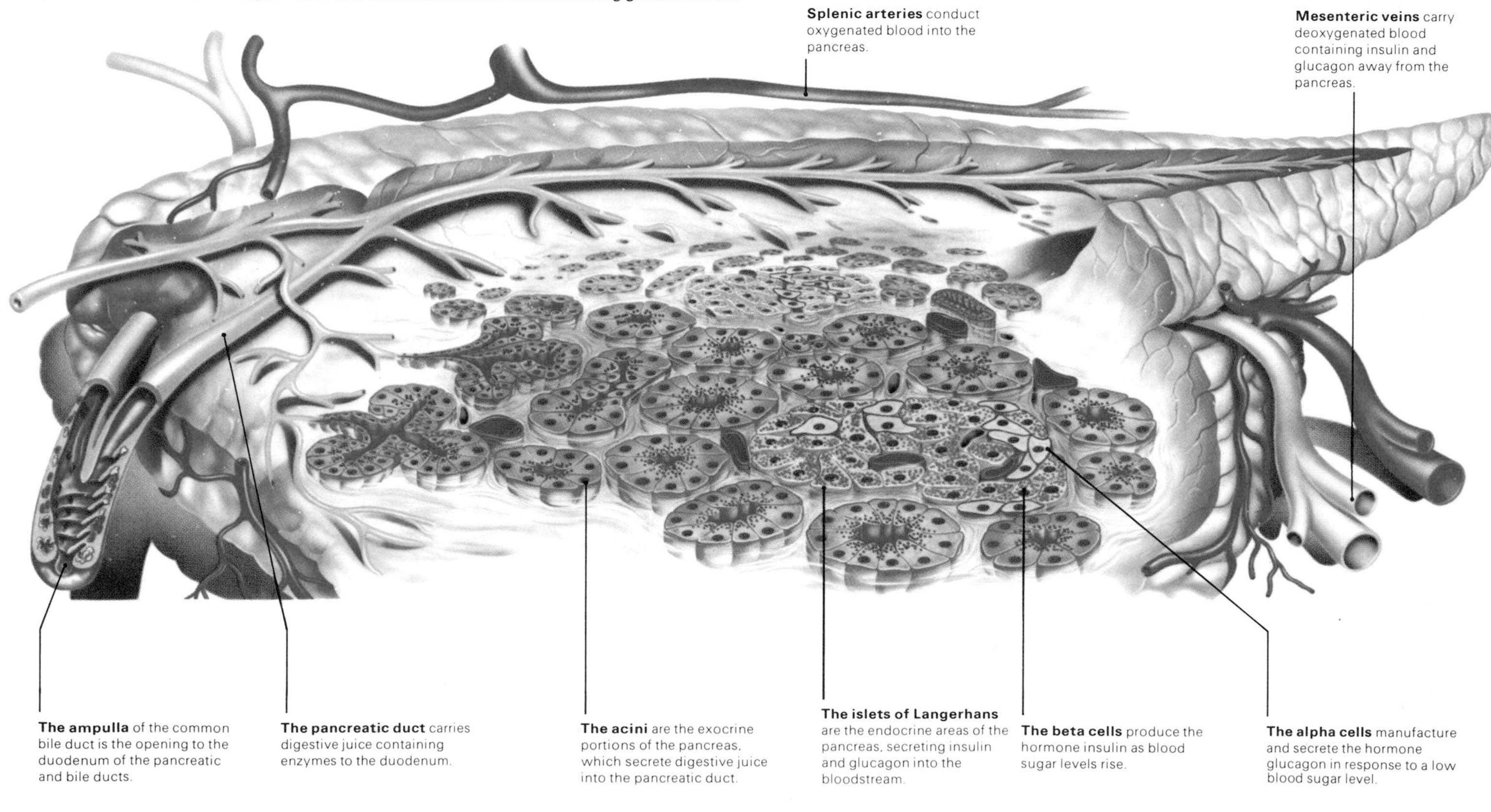

same tissue in the fetus that gives rise to the sympathetic nervous system.

In the abdomen there is a large gland, the pancreas, whose bulk is exocrine and digestive in function, but in which a very important one percent is composed of about a million clumps of cells, called the islets of Langerhans. These islets produce two major hormones that control the handling of the glucose and fatty acid fuels in the body. After a meal when blood glucose rises, it directly causes insulin release from the beta cells of the islets. Insulin reduces glucose production in the liver and encourages its uptake and usage by tissues. Similarly, it reduces the release of fatty acids and glycerol from adipose tissue. Inadequate production of insulin by the islets in response to glucose leads to the increase of blood glucose levels and to the appearance of glucose in the urine, for the kidney tubules are not able to completely reabsorb the increased quantities of glucose that are filtered. This is the disturbance of hormone production that occurs in diabetes mellitus. The other pancreatic hormone, glucagon, from the alpha cells of the islets, has opposite effects to insulin in the liver, and increases glucose production.

The control of fuels in the body is then a balance of the actions of these two hormones, and of the hormones adrenaline, noradrenaline, growth hormone, cortisol and thyroid hormones. This allows a controlled, regular supply of fuels to all cells for metabolism and growth under all circumstances.

The adrenal cortex and medulla

(**A**) The adrenal glands (**1**), sitting above the kidneys (**2**), are richly supplied with blood (**3**) and with sympathetic nerve endings (**4**). Block sections show the blood supply (**B**) and cellular arrangement (**C**) of the adrenals. Two different regions are distinguishable—the cortex (**5**), controlled by the pituitary hormone ACTH, produces hormones which maintain normal body chemistry, and the medulla (**6**), which secretes adrenaline and noradrenaline to increase body activity.

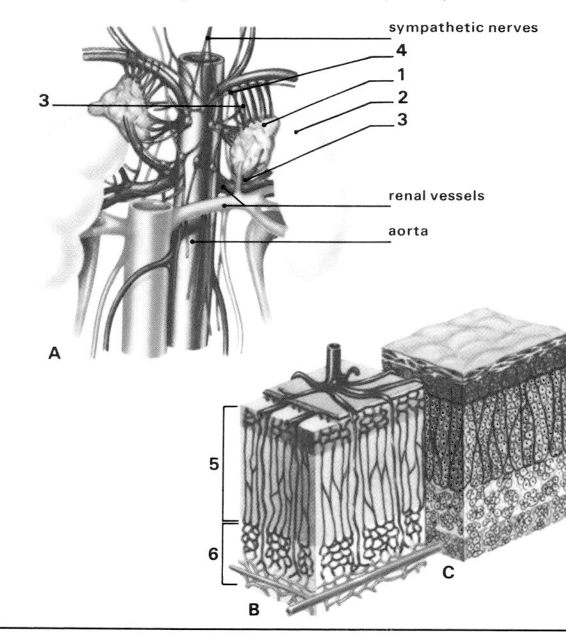

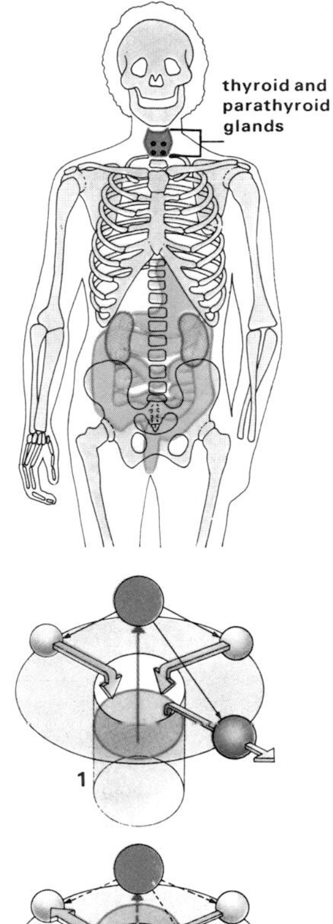

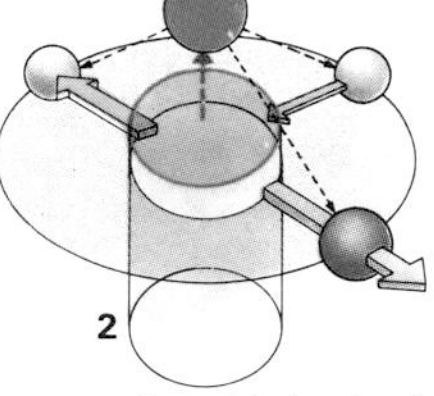

The four parathyroid glands, the black dots adjacent to the thyroid, secrete parathormone (PTH), which is involved in regulating blood calcium levels and, indirectly, in controlling blood phosphate levels. (**1**) Low blood calcium level (central orange column) stimulates output of PTH, which mobilizes calcium from bones (blue) and aids its absorption from intestine (green) and kidneys (red) to increase the circulating level. (**2**) PTH secretion is inhibited, calcium is deposited in the bones and less calcium is reabsorbed from the intestine and kidneys until the blood level returns to normal.

Fight or flight—the adrenal medulla

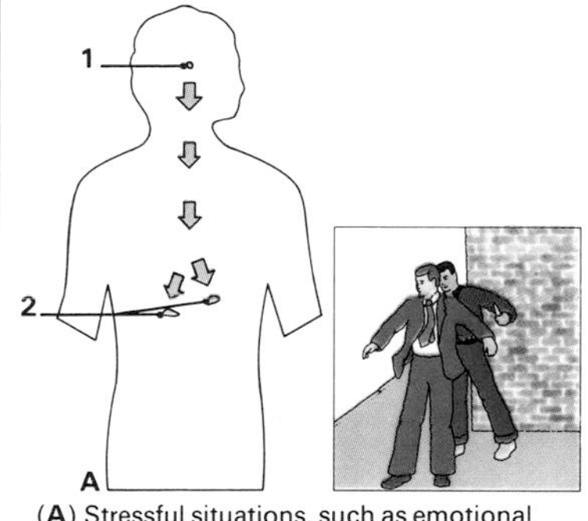

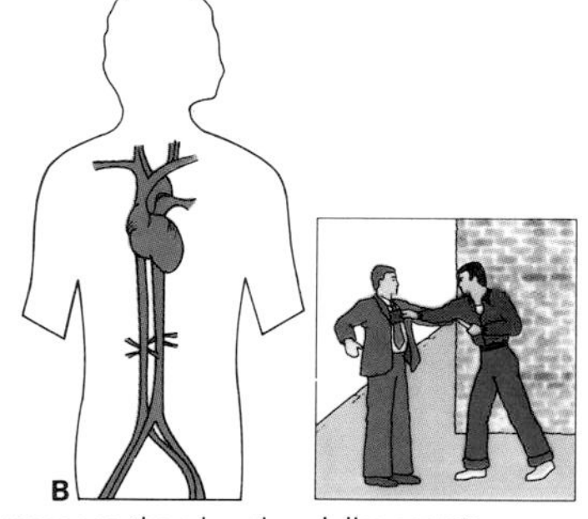

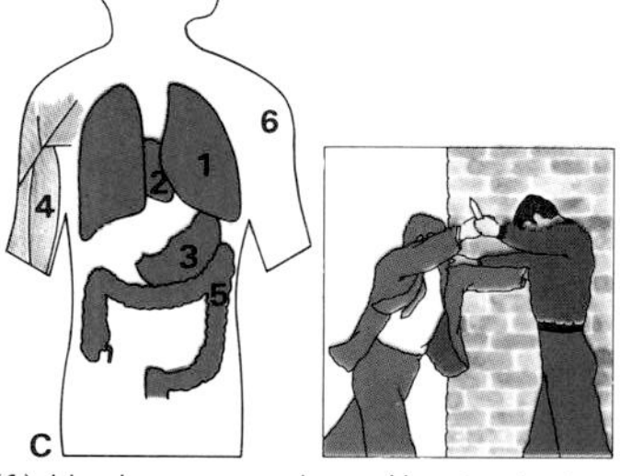

(**A**) Stressful situations, such as emotional disturbance, pain, exposure to extremes of temperature or low blood sugar levels, stimulate the hypothalamus (**1**) to transmit nerve impulses, via the sympathetic nervous system, to the adrenal medullae (**2**). (**B**) In response, the adrenal medullae secrete adrenaline and noradrenaline, in the ratio of about four to one, into the bloodstream. These hormones differ in their actions, but their combined effect is to prepare the body for "fight or flight." (**C**) Breathing is stimulated (**1**), blood pressure, pulse and heart output rise (**2**), blood levels of glucose and fatty acids for tissue fuel increase (**3**), muscle activity is enhanced (**4**) and the blood supply to the intestines (**5**) and the skin (**6**) decreases in volume.

Cross-reference	pages
Digestion	**46-49**
Emotions	**116-119**
Hypothalamus	**112-113**
Kidneys	**66-67**
Liver	**50-51**
Metabolism	**22-23**
	52-53
Ovaries	**146-147**
Pancreas	**46-49**
Sympathetic nervous system	**74-75**
Testes	**144-145**

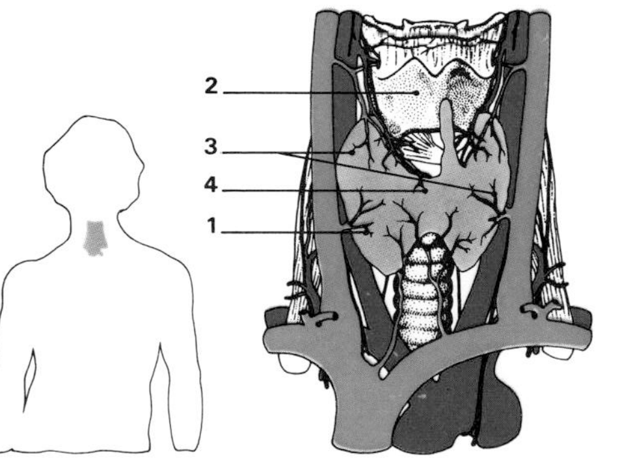

The thyroid gland (**1**), situated at the front of the neck and overlying the trachea (**2**), consists of two lateral lobes (**3**) and a connecting isthmus (**4**). Thyroid tissues trap iodine circulating in blood and use it to produce the hormones thyroxine and triiodothyronine. When secreted into the bloodstream, these hormones increase the metabolic rate of all body tissues. The thyroid gland also produces the hormone calcitonin, which helps regulate the calcium content of bones.

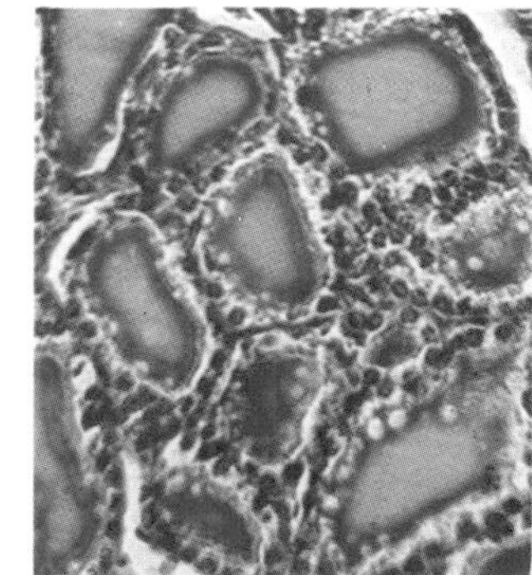

Thyroid hormones are stored in vesicles prior to secretion into the blood.

The brain and central nervous system

The anatomy of the nervous system

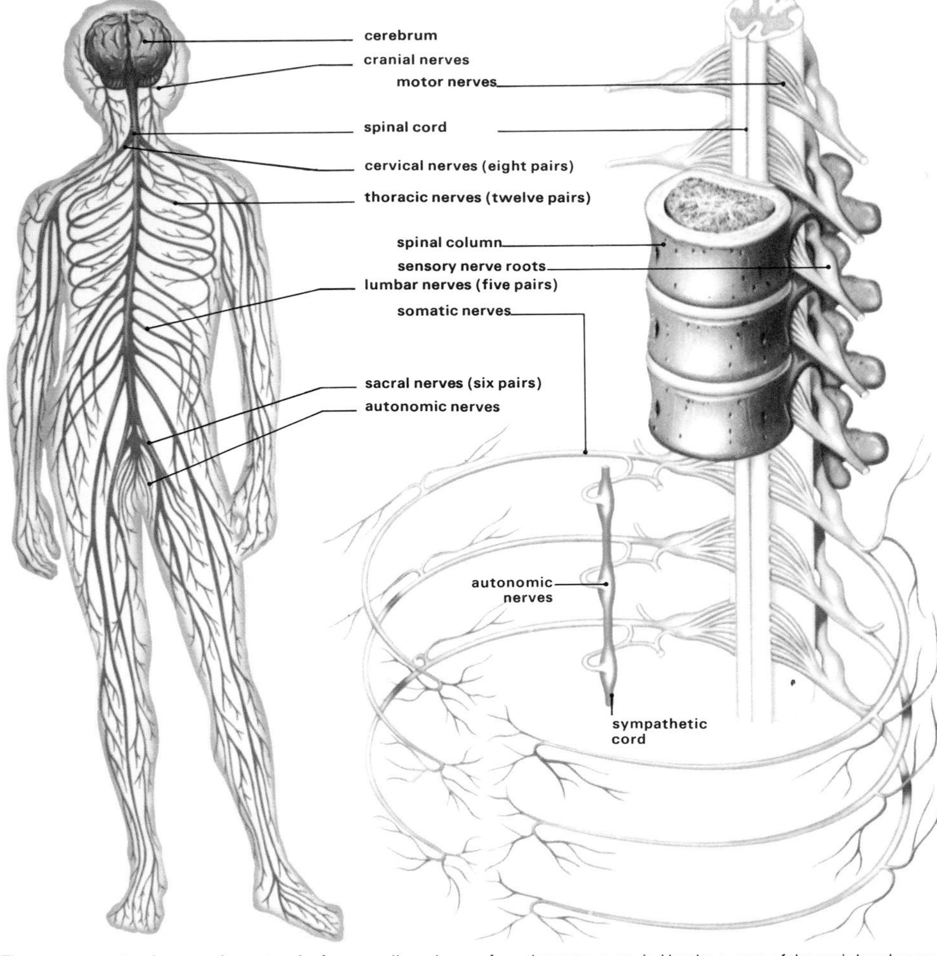

The nervous system is a complex network of nerve cells and nerve fibers spread throughout the body. Its function is to interpret, store and respond to information received from inside and outside. The central nervous system (CNS) consists of the brain and spinal cord and is responsible for processing information gathered from the rest of the nerves and transmitting instructions to the body. Messages passing to and from the CNS are carried by the nerves of the peripheral nervous system. This system includes twelve pairs of cranial nerves and thirty-one pairs of spinal nerves. The cranial nerves and the spinal nerves control voluntary movements and sensations. The autonomic nervous system, consisting of sympathetic and parasympathetic nerve fibers, controls such involuntary body functions as heartbeat.

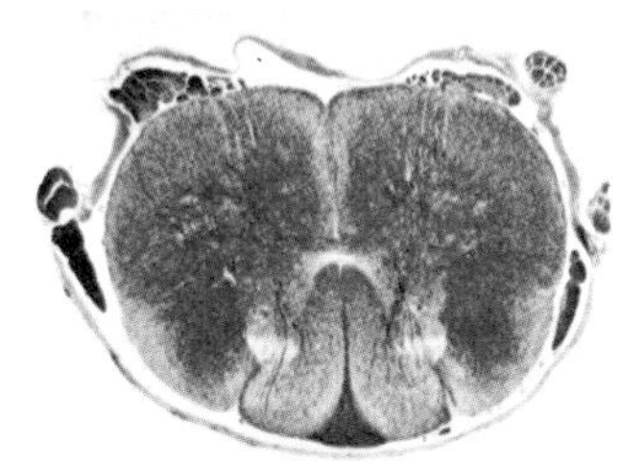

The spinal cord consists of a central H-shaped area of gray matter and a surrounding area of white matter.

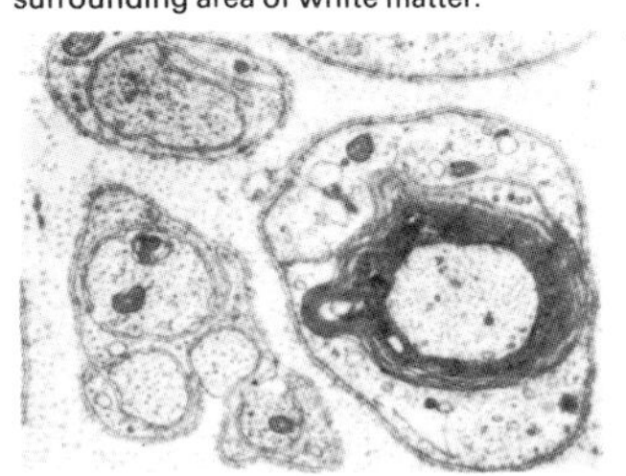

Nerve fibers seen in cross section through an electron microscope and magnified sixty thousand times.

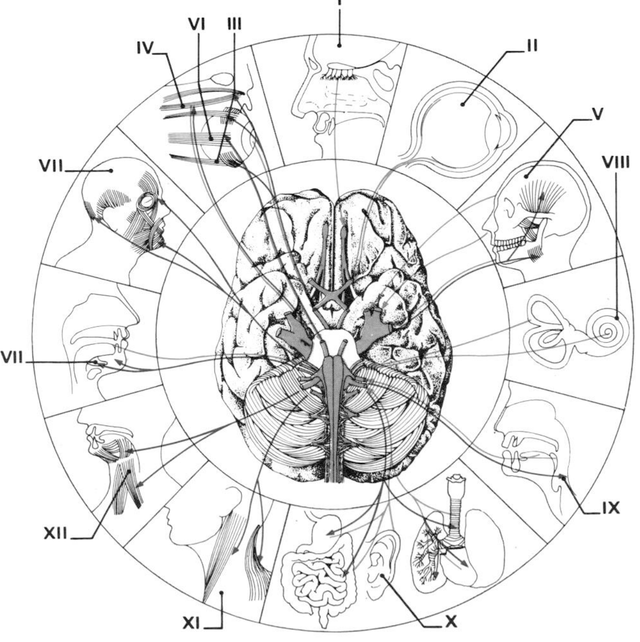

Monitoring and controlling every aspect of the body's activities, no matter whether we are awake or asleep, is the vast and complex communications network called the nervous system. At the very heart of the network is the central nervous system, consisting of the brain and spinal cord. And weaving through and penetrating to every near and outlying region of the body are the almost countless branches of the peripheral nervous system.

The central nervous system is better protected from damage than the peripheral system. The all-important brain, a densely packed mass of nervous tissue weighing nearly three pounds, is housed within the bony skull, and the spinal cord, continuous with the brain, runs down within the bony channel formed by the arches of the spinal vertebrae. Acting as channels of communication between the central nervous system and the rest of the body are the afferent, or sensory, nerves of the peripheral system, which bear incoming messages from sensory receptors, and the efferent, or motor, nerves, which carry outgoing messages to effector structures such as the voluntary muscles of the limbs.

Sensory and motor nerves of the peripheral system leave the spinal cord separately, between the vertebrae, but then unite to form thirty-one pairs of spinal nerves, in which there are both sensory and motor nerve fibers. Branches of the spinal nerves spread out to reach all parts of the body surface and all the skeletal muscles. Leaving the brain directly are twelve pairs of cranial nerves, each passing through a separate aperture in the skull.

There are more than ten thousand million nerve units, or neurons, in the brain, but even these account for only one-tenth of the brain cells. The remainder are surrounding and supporting cells called the neuroglia, or just glia. Together the neurons and glia make a soft, jellylike tissue; without the support of the surrounding skull it would distort and sag.

Closely enclosing the brain within the bony casing are three membranes, the meninges, with a protective and nutritive function. The dura mater is the outermost and toughest membrane, the pia mater is the innermost and thinnest, and between lies the arachnoid mater, weblike and pierced with numerous intercommunicating channels filled with cere-

The cranial nerves

The twelve cranial nerves, which arise from the undersurface of the brain, are usually denoted by Roman numerals. Their function is to send and receive information to and from the head, neck and most of the larger organs in the chest and abdomen. Three of the nerves are sensory (blue) and carry information only from the sense organs to the brain. The olfactory nerve (**I**) relays smell, the optic nerve (**II**) transmits visual information and the auditory nerve (**VIII**) is concerned with hearing and balance. Two of the cranial nerves only have motor function (red) and carry instructions from the brain. The accessory nerve (**XI**) supplies two neck muscles, the sternomastoid and the trapezius; the hypoglossal nerve (**XII**) supplies the muscles of the tongue and some of the small muscles of the neck. The sensory and motor fibers are mixed in the remaining seven pairs of cranial nerves. The trigeminal nerve (**V**) supplies muscles concerned with chewing and relays sensations from the face. The facial nerve (**VII**) controls the muscles of facial expression and serves the taste buds on the front two-thirds of the tongue. The glossopharyngeal nerve (**IX**) carries sensation and taste from the back of the tongue and throat, and helps to control swallowing. The vagus nerve (**X**) carries both sensory and motor connections to many organs in the chest and abdomen. The trochlear nerve (**IV**), the abducent nerve (**VI**) and the oculomotor nerve (**III**) supply all the external muscles of the eyeball with both motor and sensory fibers, making possible the very fine adjustments necessary to vision.

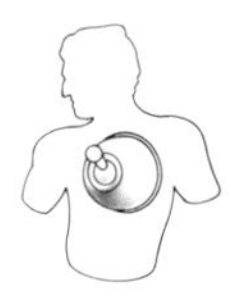

brospinal fluid. Cerebrospinal fluid is also found circulating in the cavities, or ventricles, of the brain, and in the tiny canal at the center of the spinal cord. The meninges are continuous with those covering the spinal cord. In their layers run blood vessels and nerves.

Seen with the skull removed, the brain is reminiscent of a walnut kernel in shape and in the nature of its surface, which is folded and wrinkled. The major mass of the brain is formed by the cerebrum, with left and right cerebral hemispheres separated in front, above and behind by a deep cleft, at the bottom of which is the corpus callosum. The corpus callosum is a thick bundle of nerve fibers linking the two hemispheres and allowing communication between them.

The cerebrum in man forms a higher proportion of the brain than in any other animal, and it is responsible for our unique intelligence and all our intellectual skills. Here are received impulses from all of our sensory organs, bearing in coded form information from both outside the body and within, for analysis and comparison. Here are stored all our memories of sensations recent and long past and here we carry on the processes of thought, make decisions and issue commands to be obeyed by the effector systems of the body.

The outer layer of the cerebrum, the cerebral cortex, contains a high proportion of cell bodies and is known as the gray matter, contrasting with the inner, white matter, which is composed largely of nerve fibers, with their sheaths of light-colored insulating material, myelin. The cortex is deeply folded, with grooves called sulci, and ridges between called gyri. This gives the cortex a surface area about thirty times greater than it would be if it did not have the folds.

Sensory information entering the brain from the eyes, ears, mouth and skin reaches special areas of the cortex after being relayed by the thalamus, a structure set centrally in the brain. A similar relay-station function for outgoing impulses from the cortex is performed by deeply buried areas of gray matter called the basal nuclei, or basal ganglia, in each cerebral hemisphere.

Situated beneath the thalamus is the hypothalamus, made up of nuclei that are active in regulating sleep, hunger and thirst, body temperature and sexual activity. The small pituitary gland is attached to the hypothalamus, which thus provides a close link between brain activity and the endocrine system. Structures surrounding the hypothalamus make up the limbic system, which is responsible for the control of our emotional responses.

At the base of the brain, and continuous with the spinal cord, is the brain stem. The brain stem extends up into the middle of the brain, and has at its center a collection of many millions of neurons arranged in a network called the reticular formation. This is responsible for monitoring all incoming sensory signals and filtering out those which are unimportant. Impulses from other regions of the brain are also scrutinized here. Out of perhaps a hundred million different impulses entering the system, a few hundred only may be allowed to reach other parts of the brain. The reticular system regulates the body's level of arousal, from a state of deep sleep at one extreme to a state of intense mental activity at the other.

Behind the brain stem and below the overhanging cerebral hemispheres is the lobed cerebellum, vaguely resembling the cerebrum but about one-eighth of its size. Here orders from the higher centers of the brain to the muscles are "processed." The cerebellum does not itself initiate movements, but is responsible for their smooth and balanced execution. Rather like a computer, the cerebellum stores the patterns that are necessary for coordinated muscular activity. If the cerebrum commands "write the figure eight," it is the cerebellum that integrates the necessary muscular activities. It also keeps the muscles in a slightly tense condition and prepared for action; this is called muscular tone.

The regions of the brain

The brain has three major functional and anatomical parts—the forebrain, midbrain and hindbrain. The forebrain consists of the cerebral hemispheres, thalami, hypothalamus and limbic system. The top inch of the brain stem is the midbrain. The hindbrain includes the cerebellum, the pons and the medulla.

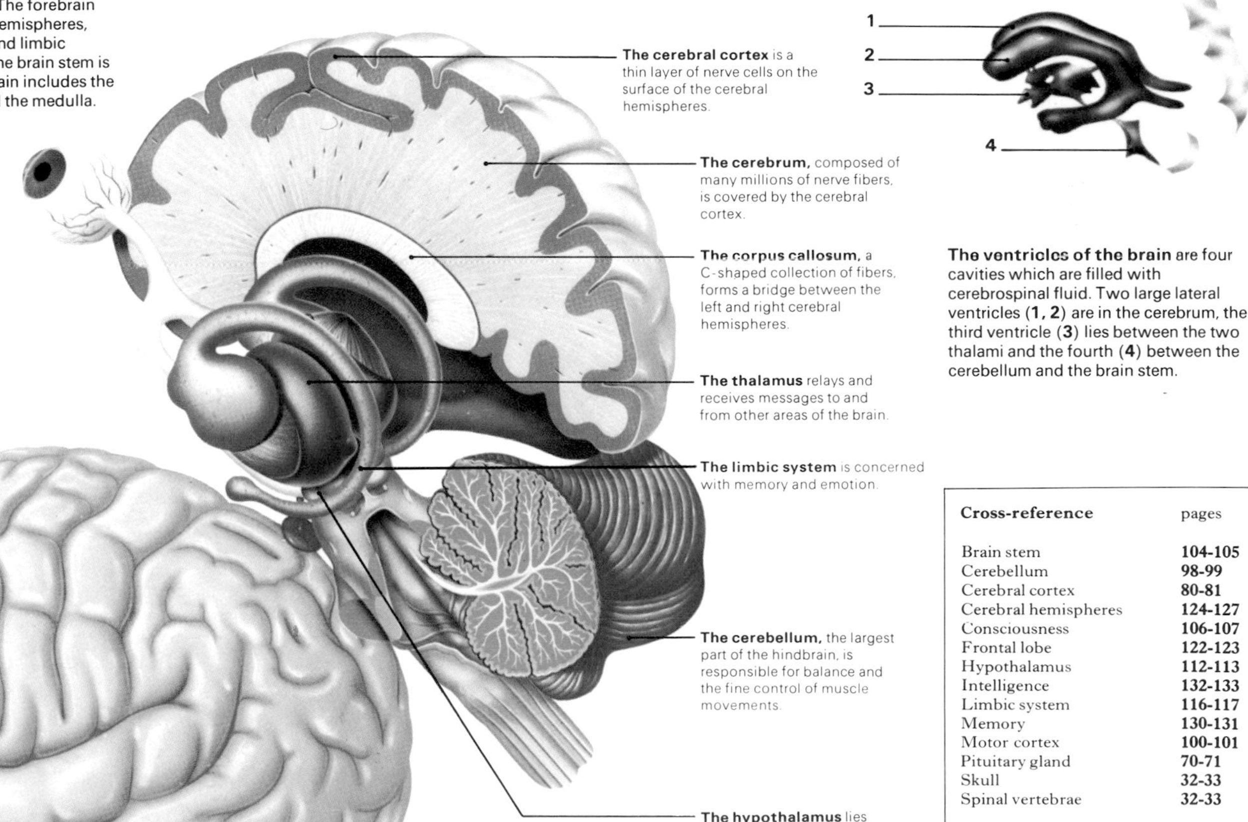

The ventricles of the brain are four cavities which are filled with cerebrospinal fluid. Two large lateral ventricles (**1, 2**) are in the cerebrum, the third ventricle (**3**) lies between the two thalami and the fourth (**4**) between the cerebellum and the brain stem.

Nerves: how they function

The anatomy of a nerve

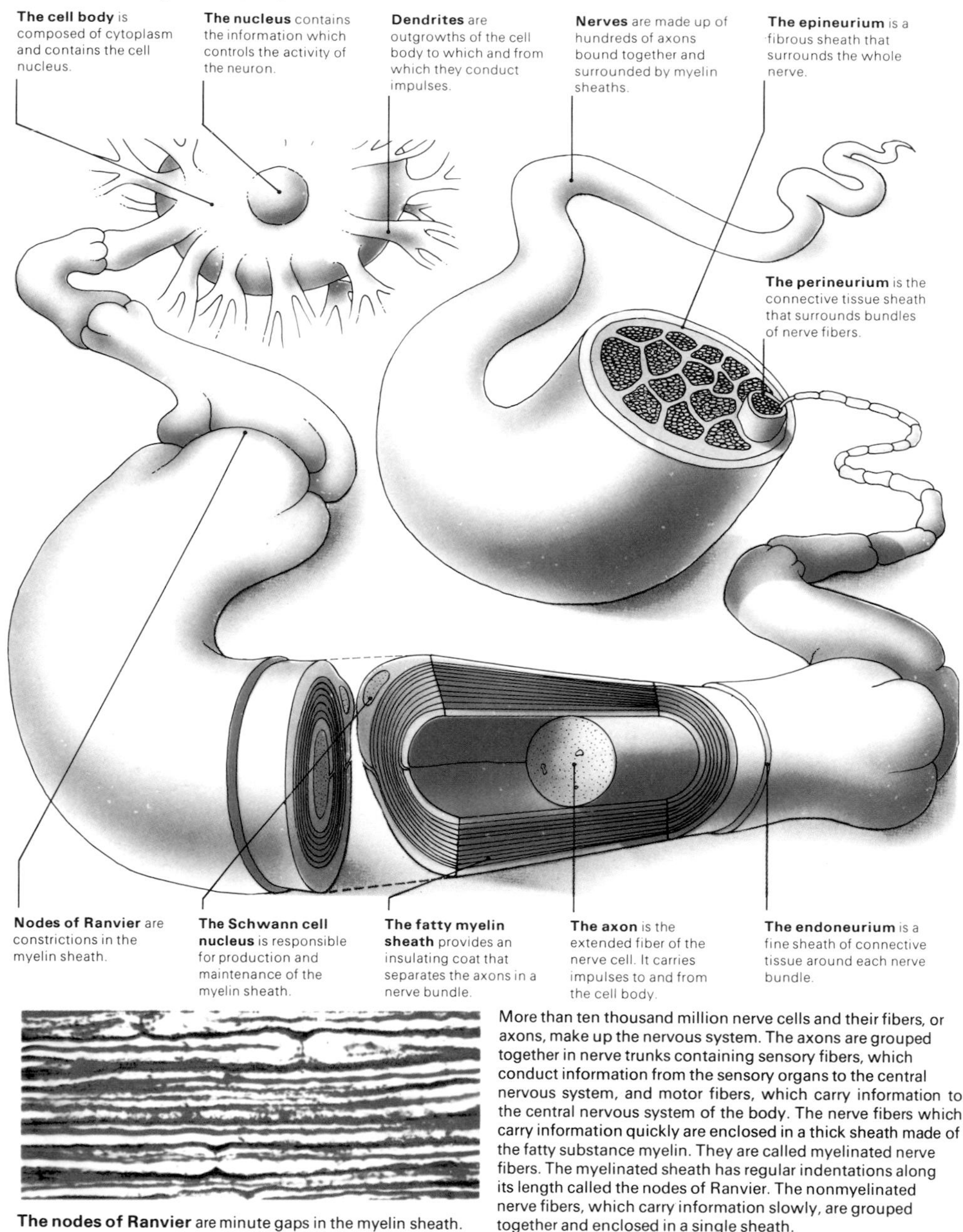

The nodes of Ranvier are minute gaps in the myelin sheath.

More than ten thousand million nerve cells and their fibers, or axons, make up the nervous system. The axons are grouped together in nerve trunks containing sensory fibers, which conduct information from the sensory organs to the central nervous system, and motor fibers, which carry information to the central nervous system of the body. The nerve fibers which carry information quickly are enclosed in a thick sheath made of the fatty substance myelin. They are called myelinated nerve fibers. The myelinated sheath has regular indentations along its length called the nodes of Ranvier. The nonmyelinated nerve fibers, which carry information slowly, are grouped together and enclosed in a single sheath.

The conduction of a nerve impulse relies upon the movement of electrically charged ions across the nerve cell membrane (**1**). When a nerve is resting, or polarized, there are more potassium ions (K^+) than sodium ions (Na^+) inside the cell, with an opposite ratio outside. Sodium ions are actively kept out of the cell by an energy-consuming pump mechanism. This maintains a negative charge on the inside of the cell and a positive charge on the outside (**2**). When an impulse travels along the nerve, sodium ions flood into the cell and make the inside of the cell positive with respect to the outside (**4**). This produces a rise in the electrical potential across the cell membrane. After the impulse has passed, potassium ions leave the cell, restoring the negative charge within the cell and the positive charge outside it (**3**). While this resting situation is being restored another impulse cannot be generated.

Although distinction is drawn between the central and peripheral divisions of the nervous system mainly for purposes of anatomical description, in fact they form an integrated whole. Both are built of the fundamentally similar nerve units called neurons, both have the function of receiving and transmitting messages in the form of nerve impulses from one part of the body to another—and neither would be of much use without the other.

Like most other cells, neurons have an enveloping cell membrane, cytoplasm containing mitochondria and ribosomes, and a controlling nucleus. However, they have the unique duty of carrying nerve impulses over long or short distances, and this specialization is seen in their microscopic structure.

From the cell body, or main part of the neuron, containing the nucleus, a number of short, branching outgrowths, or processes, fan out. These are called dendrites. Their function is to pick up messages from other cells with which they make contact, and to conduct them in the direction of the cell body. The cell body's longest process, which carries messages away from it, is called the axon. The axon of a neuron with its cell body in the spinal cord may extend all the way to a finger or to a toe to reach the muscle on which it acts. Axons give off branches along their length, and all branch profusely at their ends, to become axon terminals at the point where they make contact with other cells.

Neurons may be classified into three types, according to their functions. Afferent neurons carry messages inward from sensory receptors. Efferent neurons carry messages outward to effector structures, instructing muscles to contract and glands to secrete. Interneurons, as their name implies, link together other neurons. Interneurons start and finish in the central nervous system, and make up about 97 percent of all nerve cells. It is the vast complexity of their interconnections in the brain that gives us memory, thought and emotion and makes possible all our higher mental activities.

Neurons are able to carry messages because their membranes have a special property called excitability. In the resting, or excitable, state a nerve membrane is polarized. This means that there is an electrical charge across it, arising from a high concentration of potassium ions on the inside and a low concentration on the outside, and the reverse for sodium ions. The ions cannot simply diffuse through the membrane to equalize their concentrations because the membrane is not equally permeable to potassium and sodium ions.

However, if anything happens that alters the molecular structure of the membrane and suddenly increases its permeability, then ions can surge through, equalizing their concentrations and in so doing causing a burst of electrical activity as the charge across the membrane is altered. That is what happens when a nerve cell is excited, or becomes depolarized. In one region the molecular structure of the membrane is temporarily disrupted, potassium flows out and sodium flows in and there is a resulting change in the charge across the membrane—an action potential.

An action potential spreads rapidly from one region to the next because the change in electrical charge itself brings about a rearrangement of the molecules in the neighboring part of the membrane, so that the

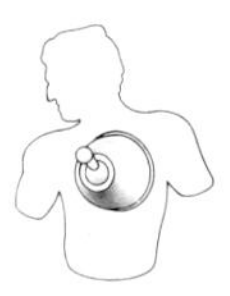

process is continued. The waves of alteration in molecular structure and inflow and outflow of ions constitute a nerve impulse. While we can conveniently think of a nerve impulse as being somewhat similar to an electric current passing in a wire, it can be seen that in fact the changes are very much more complex.

After it has been depolarized a membrane rapidly reverts to its resting state, but until the molecules and ions have been rearranged it is not excitable, it cannot conduct impulses, and it is said to be refractory.

The speed with which an impulse is conducted along a nerve fiber depends upon the size and type of nerve. There are large, rapidly conducting myelinated nerves, which have a sheath of the fatty material called myelin surrounding and insulating them, and more slowly conducting nonmyelinated fibers. In a myelinated nerve the impulse travels in a series of jumps between regions called the nodes of Ranvier, where there are gaps in the myelin sheath, and may reach a speed of three hundred feet a second. In smaller nerves, and nonmyelinated nerves, the speed may be only five or six feet a second.

Under normal circumstances, impulses travel through nerves in one direction only—from dendrite to cell body to axon and so to another cell. This is not because the neuron is unable to conduct in either direction, but because messages usually arrive from only one direction, across a one-way junction, or synapse.

At a synapse there is a gap between the slightly enlarged ending, or synaptic knob, of an axon terminal and the membrane of the next cell. When a nerve impulse reaches the synaptic knob, it causes the release of a chemical called a neurotransmitter (usually noradrenaline or acetylcholine) from minute vesicles, or bubbles, in which the substance is formed and stored. The neurotransmitter diffuses across the gap and causes depolarization of the receptor cell membrane, so starting off an impulse. The time taken for this to happen varies from about one-thousandth to three-hundredths of a second. Having passed on the message, the transmitter is rapidly destroyed by enzymes, and the synapse becomes ready for further action.

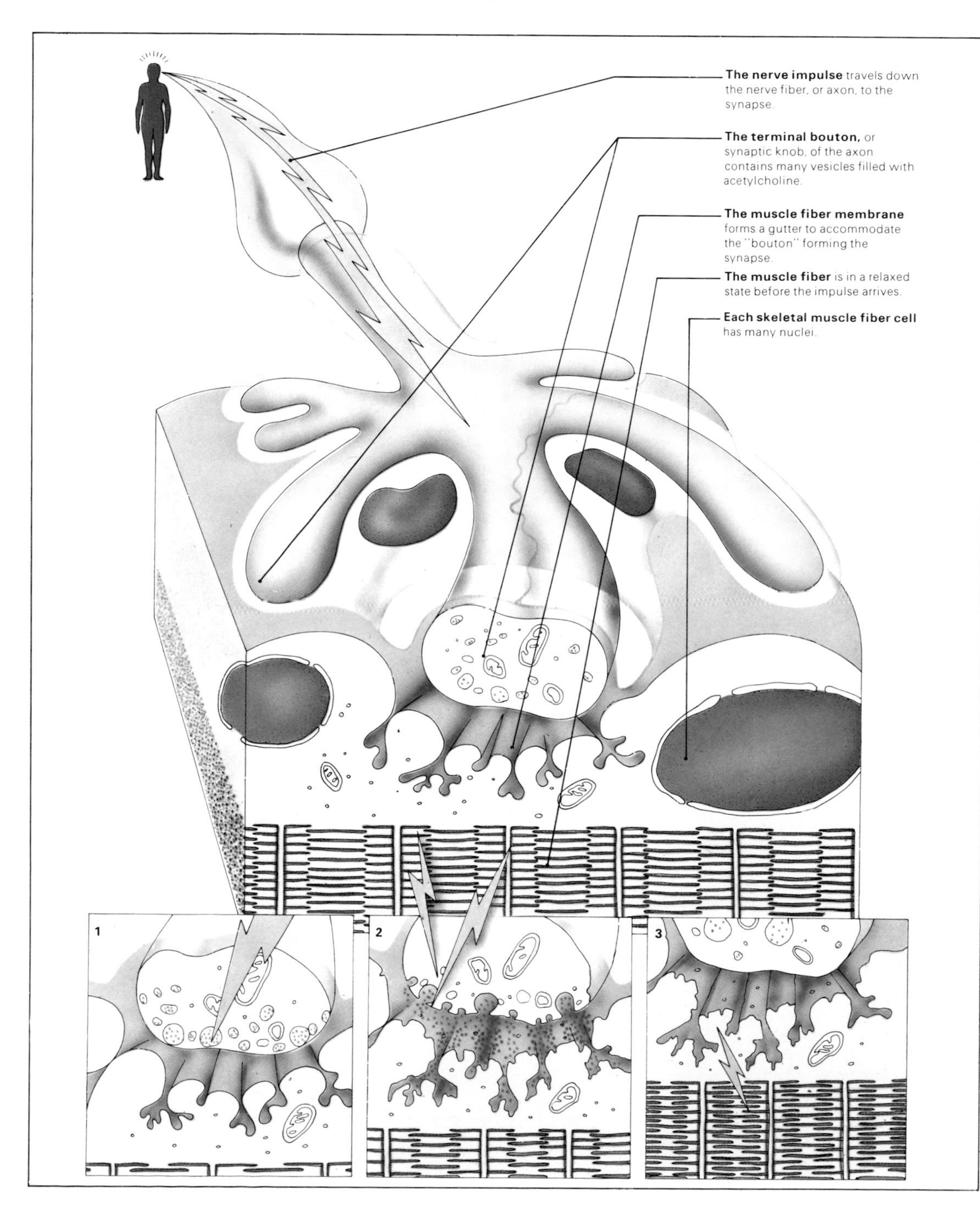

The motor end plate

The motor end plate is the end of a motor nerve fiber, or axon, which is attached to a muscle fiber. The motor end plate is covered by Schwann cells on its upper surface while on the lower surface the axon expands to form folds. Each expanded end contains many small bubbles, or vesicles, full of a chemical transmitter called acetylcholine (**1**), which is released when the nerve impulse arrives. When the transmitter is released (**2**) it passes into the space, or synapse, between the motor end plate and the muscle fiber and causes a change in the permeability of the muscle fiber membrane. This allows sodium ions to enter the muscle fiber and, in the same way as in conduction of an impulse in a nerve fiber, raise the electrical potential across the muscle membrane. This stimulation of the muscle lasts for only a few milliseconds because the transmitter substance is destroyed by enzymes. The change in the electrical potential across the membrane initiates the actin and myosin filaments in the muscle fiber to slide across each other, shortening the muscle and producing movement (**3**).

Cross-reference	pages
Cell structure	22-23
Central and peripheral nervous system	74-75
Hormones	70-73
Muscles, structure	36-37

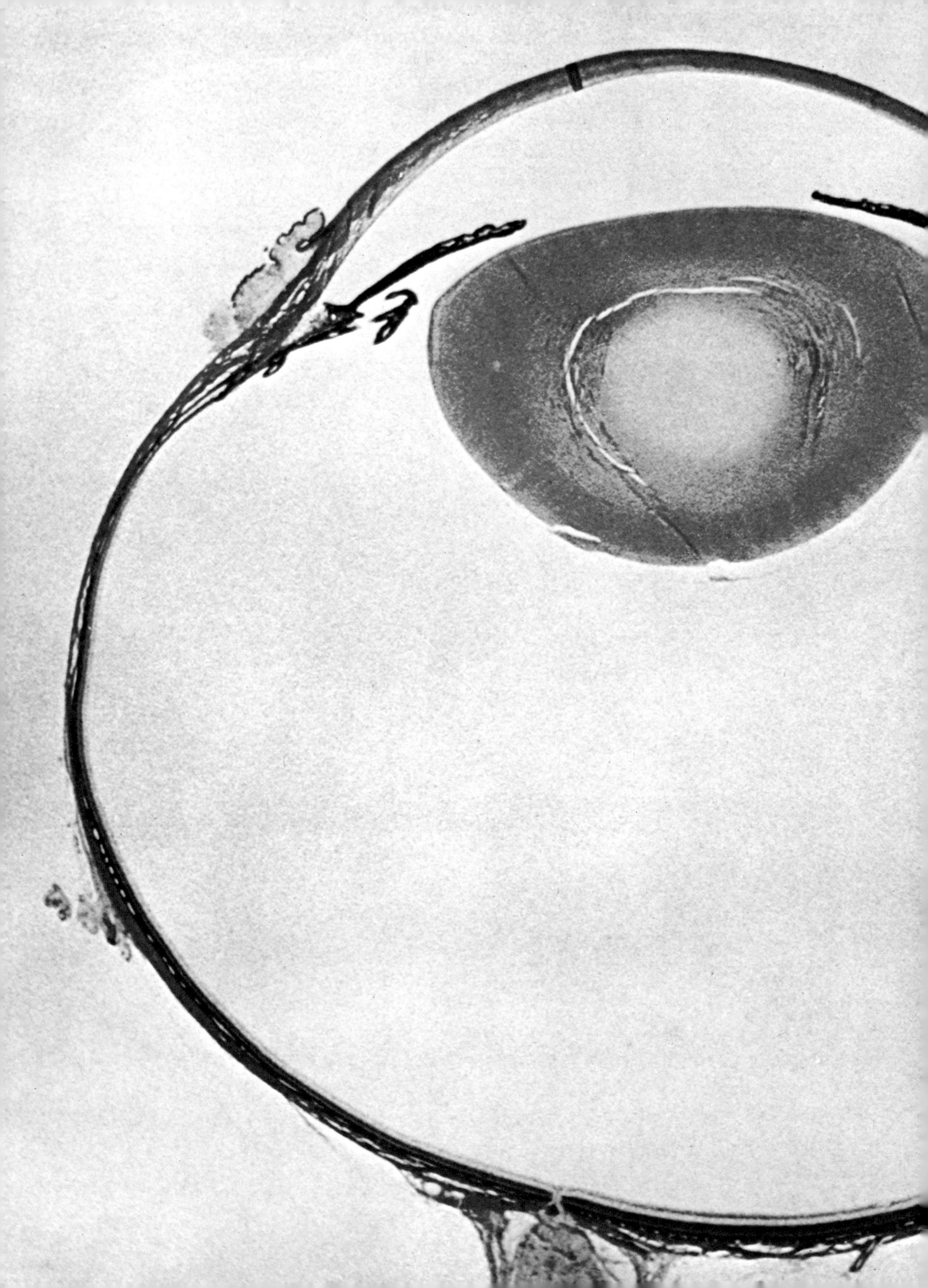

The senses

Through the five senses—sight, hearing, touch, taste and smell—we are acutely aware of what is going on in the world around us. The human body is equipped with sense organs, structures sensitive to the various kinds of energy in the environment. Without these organs, which collect this energy and translate it into nerve impulses which travel to the brain, where they are interpreted, all links with the environment would be lost. The eyes receive and convert energy in the form of light. The ears receive and convert energy in the form of sound. The skin is sensitive to energy arriving at the body as temperature, pressure and touch. Chemical changes at the tongue and nose induce electrical changes which are finally translated as taste and smell.

Besides receiving and transmitting stimuli to the brain, the sense organs provide us with built-in protection against excessive energy input. Loud noise and intense light, for example, may produce sensations of pain. The senses also reinforce the collection of information with aesthetic satisfaction and pleasure.

In addition to the five major senses, we have proprioception, the ability to receive information about our position in space and in relation to gravity. But despite their seeming efficiency, our sense organs only use a tiny proportion of the information which floods into them. And there are many other forms of energy in the environment, for example ultraviolet light and X rays, for which we possess no appropriate sensory receptors. The only way we can become aware of these energy forms is through the discomfort of sunburn or if they are presented in another guise—such as image patterns on a photographic screen—which the sense organs can pick up and convert to nerve impulses for the brain to process.

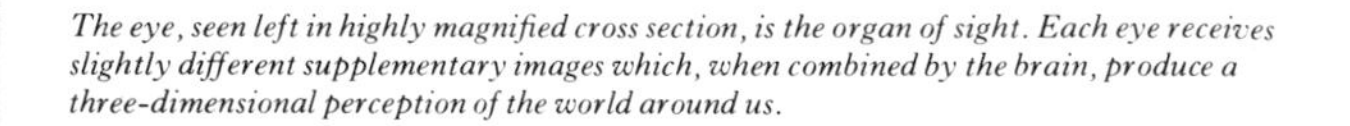

The eye, seen left in highly magnified cross section, is the organ of sight. Each eye receives slightly different supplementary images which, when combined by the brain, produce a three-dimensional perception of the world around us.

The cerebral cortex: information processing

Covering the surface of the two large hemispheres at the top of the brain is a thin, wrinkled blanket of gray nerve tissue—the cerebral cortex. It is called cerebral because it lies on the top of the white cerebrum, and cortex because it is the outer covering, or bark, of the brain. Any similarity it may have to the bark of a tree, however, ends with the definition. The cortex is very much a pulsating, live and thriving tissue which sifts, sorts, collates, organizes and presents information to us which we perceive as sight, sounds, thoughts, emotions and memories.

The cerebral cortex has about eight thousand million active nerve cells, enough to place one every two hundred yards from the earth to the planet Saturn and all contained in an area no thicker than the sole of a shoe and, if it were spread flat, no larger than a card table. These many millions of cells are, in turn, held together by a group of cells known as the glia (from the Greek *glia*, meaning glue), of which there are eight times as many as the cells they support. There are numerous explanations for the functions of the glia, but they seem to be mainly supportive, although it may be possible that the glia supply a limited quantity of nutrient for the cortical nerve cells.

The cerebral cortex, which if it were flat would cover an area of 0.5 square meters, fits into the skull because it is folded into a number of convolutions. The valleys of these convolutions are called sulci, the hills are the gyri and together, by fissures, or grooves, they divide the surface of the cortex into a number of areas. There are two main fissures which produce the principal landmarks on the surface of the cortex. The first is the central fissure which runs vertically downward from the crown of the brain to meet the second groove, the lateral fissure, which runs horizontally backward from the front of the brain. Within these boundaries arise the four lobes that mark out the territories of the cortex. Their frontiers are well defined and identical for both sides of each hemisphere. The four lobes take their names from the skull bones they underlie.

In man's evolution, the frontal lobe is the newest as well as the largest lobe; it is also the most complex in function. Primarily, it has

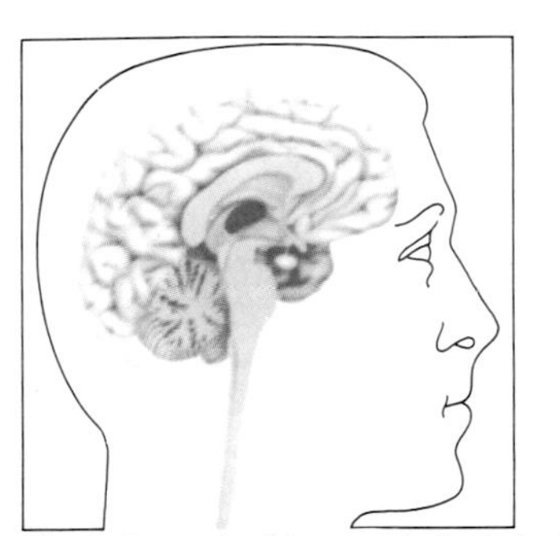

The thalamus (red) is part of the forebrain. It plays an integral part in relaying sensory information from the sense organs to the cortex.

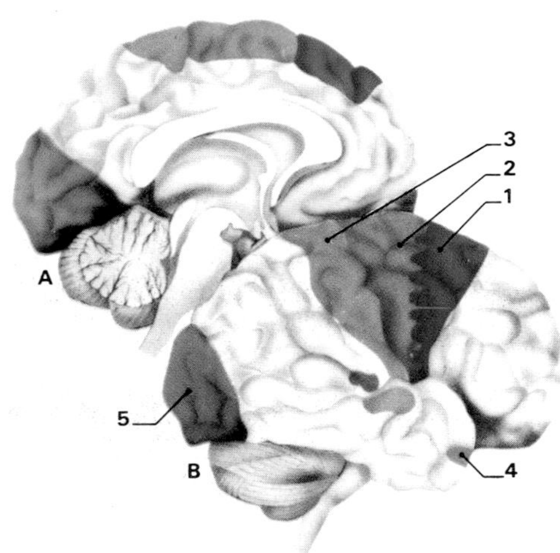

Sensory and motor areas are shown on the inside of the left cortex (**A**) and the outside of the right cortex (**B**). The premotor area (**1**) works with the motor area (**2**). Physical sensations pass to the sensory area (**3**) and smell and sight to the olfactory (**4**) and visual (**5**) cortices.

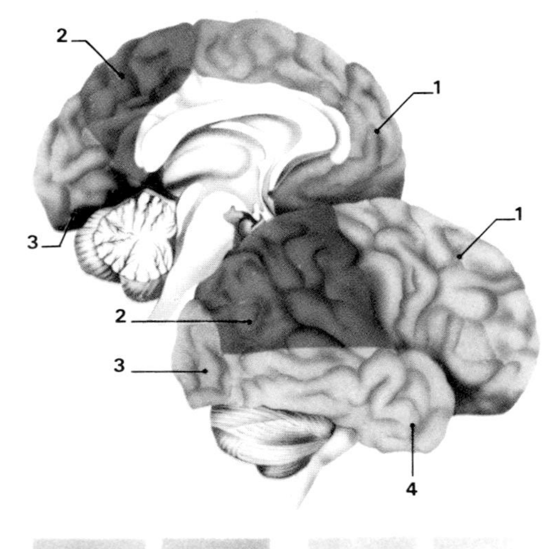

Brain waves are electrical patterns of activity recorded on an electroencephalograph (EEG). The diagram shows the normal patterns in the brain of a subject when resting, with his eyes closed but awake. (**1**) Frontal lobe. (**2**) Parietal lobe. (**3**) Occipital lobe. (**4**) Temporal lobe.

Human information processing

This illustration shows the organization of the nerve cells, the vertical and horizontal position of the fibers and the position of the blood vessels within the cortex. The convoluted covering gives it a large surface area within the restricted confines of the skull. The cerebral cortex is responsible for receiving, interpreting and storing information from the body and its external environment and organizing relevant responses to the incoming messages.

Convolutions are the folds on the cortex. Each "hill" is a gyrus and each "valley" is a sulcus. The hills and valleys give the brain its wrinkled appearance.

conscious control over all voluntary movement. It issues the orders for all patterns of movement from running and jumping to the subtle control required for speech. The prefrontal area of the frontal lobe is not concerned with movement. It is thought to be the seat of the highest forms of mental activity and also plays a role in determining our personalities and overall intelligence.

Lying behind the frontal lobes and divided by the central groove are the parietal lobes. Within them are the primary receiving areas for all sensations of touch. Body position is also sensed by the parietal lobes. Therefore a person with damaged parietal lobes has difficulty in orienting his body in a world that will seem spatially confused.

At the back of the cortex are the two smallest lobes, the occipital lobes. Their size, however, has no relationship to their importance. All sensations and images received by the eyes are sent on a well-laid-out route to the occipital lobes to produce what we know as sight. If this small area of the brain is damaged, some degree of blindness is inevitable.

Running laterally along both sides of the hemispheres and bound by the lateral fissure are the temporal lobes. Information picked up by the ears is sent for interpretation to a small area within the lobe, for it controls auditory perception. Both ears are totally represented on each temporal lobe in the auditory areas; therefore, if one lobe is damaged hearing is not affected, unlike the consequences of damaging one occipital lobe.

In each of the four lobes there are some nerve cells, or neurons, which are concerned with receiving information and others which are concerned with transmitting information. Since the information from one sense alone rarely gives a complete picture, all incoming sensory information is supplemented and modified by other pieces of information which are being processed at the same time in the cortex. In this way we have a clear picture of exactly what is happening around us.

The modification and integration of information is carried out by areas throughout the cortex which are known as association areas. It is through the activity of these areas that the selected use of combined information gives us our keen sense of awareness. Not all the available sensory information reaches the cortex, however, for much is lost through weak signaling and never achieves a level of consciousness necessary to activate the cortex.

With the exception of smell, all sensory information is sent to the cortex via the thalamus, a small, plum-shaped area in the center of the brain. From this strategic position the thalamus acts as a relay station to direct the flow of information between the body and the brain.

The areas of the thalamus are composed of closely packed clusters of neurons which form themselves into the thalamic areas, or bodies. Recent research has resulted in the thalamus being mapped to the areas of the cortex it serves. Some areas of the thalamus are particularly specific, but others deal with general transfer of information to and from different parts of the brain. The boundaries that have been imposed, however, are not necessarily definite, for it is probable that they will change as brain research progresses.

Fusiform, basket and stellate cells are all types of neurons present in the cortex. They communicate with each other and sift and sort out incoming information.

Horizontal cells are found only in the top layer of the cortex. They communicate with the vertical fibers emerging from the depths of the cortex.

Pyramidal cells have long axons which transmit information out of the cortex.

Afferent fibers bring information from the body and sense organs into the cortex, where they are connected with the neurons that process it.

Martinotti cells are found throughout the cortex. They are small cells with one long axon which runs vertically upward.

Glial cells, or neuroglia, support and bind all the nerve cells together. They provide nutrition for the cells that they surround. It is not known if they are concerned with processing information.

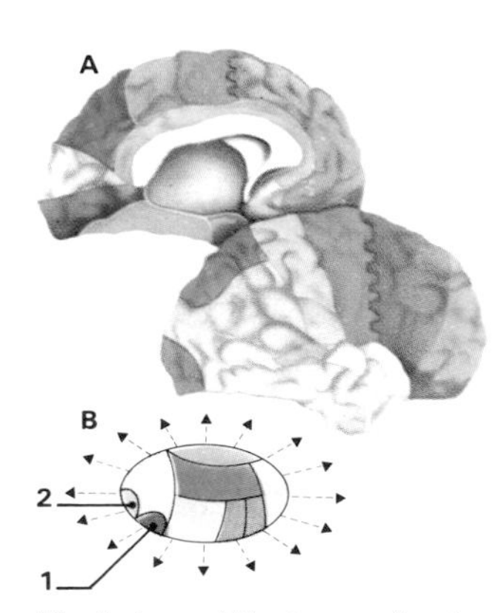

The thalamus (A) acts as a major relay center for incoming messages, passing them on to higher centers in the cortex. A thalamus with its nuclei color coded (**B**). Some nuclei deal with several different types of message, while other nuclei deal with only one type. The lateral geniculate body (**1**) relays information to the visual cortex, and the medial geniculate body (**2**) relays sound information to the auditory cortex.

Cross-reference	pages
Brain and central nervous system	**74-77**
Cerebral hemispheres	**124-127**
Evolution of the brain	**14-15**
Frontal lobe and personality	**122-123**
Hearing	**90-91**
Intelligence	**132-133**
Tongue and taste	**92-93**
Vision	**84-85**

The eye: structure

Cross section through an eyeball

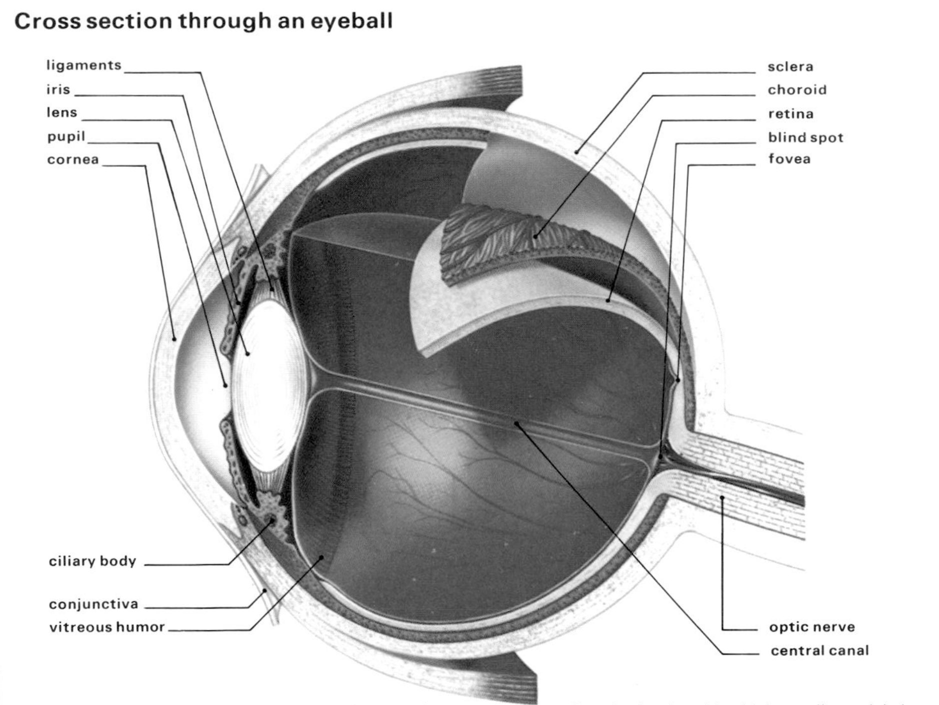

The eyeball has three major coats—the smooth, protective, outer sclera, the middle, pigmented choroid and the inner, light-sensitive retina. In this cross section, the retina and choroid have been peeled away to show the copious blood vessel network under the choroid, which supplies and drains the inner part of the retina as well as the choroid. The focusing mechanism of the eye involves the ciliary body, the ligaments and the lens.

Almost every animal is sensitive to light. Even the primitive amoeba contracts into a ball if subjected to illumination brighter than is good for it. And the lowly earthworm, which has no sign of eyes, rapidly squirms back into the safety of its burrow if the light intensity is suddenly increased. As human beings we are not merely sensitive to the particular spectrum of radiations that we call visible light; we are able to use light entering our eyes to find out what is happening around us. We can distinguish colors, focus for near and far vision and, because we have two eyes, with overlapping fields of vision, estimate relative sizes and distances and perceive things in three dimensions.

Our eyes are globular, jelly-filled organs, set in, and largely protected by, the deeply cupped cavities in the bones of the skull called the orbits. Each eye, which rotates in its cavity under the action of the orbital muscles arranged about it, is protected at the front by the eyelids. At the rear of each eye a thick optic nerve leaves the eyeball to pass backward to the visual cortex of the brain.

The outer layer of each eye is tough, fibrous

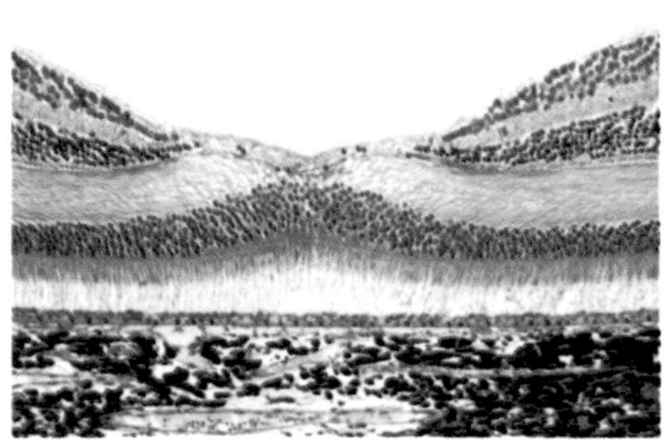

The macula, seen photographed in cross section, lies in the center of the retina and is the point of sharpest vision, where images are resolved and colors are most clearly seen. Sharpness of vision is enhanced by the one-to-one relationship between the cones and the nerve fibers. At the fovea, the central pit of the macula, the retina is very thin and contains only cones.

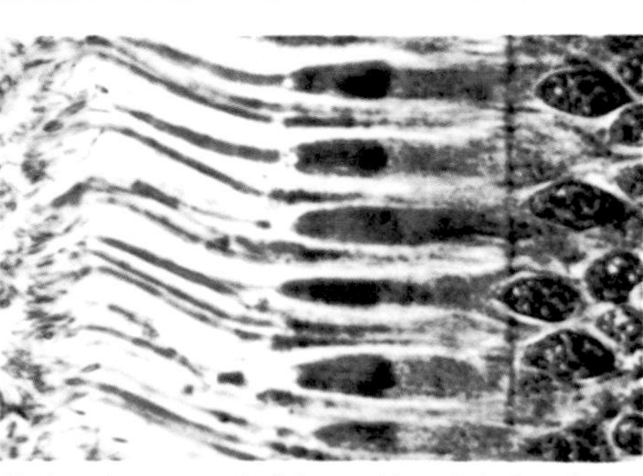

Rods and cones are the light-sensitive cells in the retina. There are about 125 million rods, the thin cylindrical cells, seen above, packed closely between some of the five million more bulbous cones. The rods are concerned with black and white vision, while the cones, which consist of two segments, effect color vision.

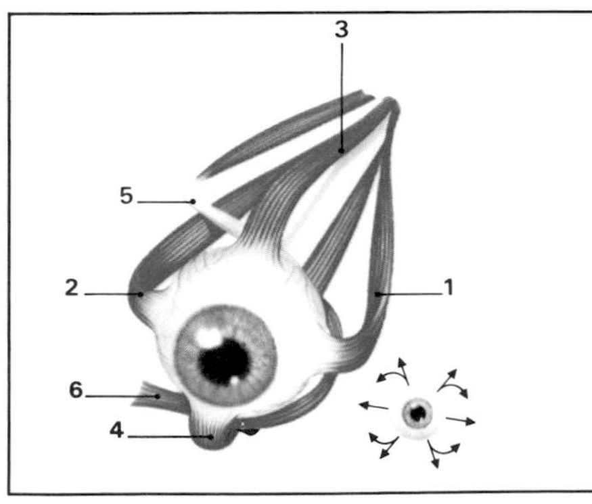

The six external ocular muscles rotate the eyeball. The lateral rectus muscle (**1**) moves the eyeball away from the midline of the body. The medial rectus (**2**) moves the eyeball toward the midline of the body. The superior rectus (**3**) moves the eyeball upward. The inferior rectus (**4**) moves the eyeball downward. The superior oblique muscle (**5**) moves the eye downward and outward. And the inferior oblique (**6**) moves it upward and outward.

Exploring the eye

The eyes, the organs of sight, are situated in the orbits, the sockets in the skull, the walls of which protect them from injury. The eyelashes, eyelids, muscles and lacrimal glands also protect these vital and delicate organs. The eye is divided into two segments by the lens and ciliary body. The front segment contains the fluid aqueous humor, and is in turn divided, by the iris, into anterior and posterior chambers, which are connected through the pupil's aperture in the iris. The back segment, called the vitreous body, contains a jellylike substance, known as the vitreous humor, and is lined by the light-sensitive retina. The cornea, aqueous humor, lens and vitreous humor are all transparent, thus allowing the unobstructed passage of light from the exterior, through the eyeball, to the retina. The cornea is the most important structure for refracting light, although the lens provides the fine control needed to converge the incoming rays onto the retina. The most striking external feature of the eye is the iris, the pigment-filled membrane that gives the eye its color, varying from light blue to dark brown.

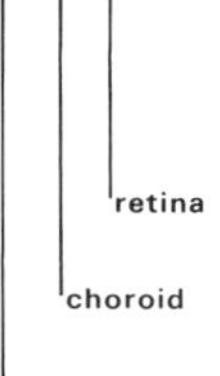

The hyaloid canal is the remains of a channel that carried an artery during the development of the eye in the fetus.

and, except at the front, opaque to light. This is the sclera, or "white" of the eye. At the front, and bulging forward, the transparent cornea replaces the sclera, and outside the cornea is the thin protective membrane called the conjunctiva. The conjunctiva is constantly cleaned and moistened by a salty, bactericidal fluid that is secreted by a single large lacrimal gland on the inside of each upper eyelid.

Behind the bulge of the cornea lies the anterior chamber of the eye, separated from the main, or posterior, chamber by the lens. A watery fluid, the aqueous humor, circulates in the anterior chamber. In the posterior chamber is a much firmer jelly, the vitreous humor. The aqueous and vitreous humors are under slight pressure and give the eyeball its firmness.

In front of the lens is a circular muscular diaphragm, the iris, which gives the eye its color, which may be blue, gray, green or brown or any combination, depending upon the amount of pigment present and the way the particles of pigment are arranged. The central hole in the iris, which looks like a black dot, is the pupil, through which light is admitted to the back of the eye.

The size of the pupil is regulated automatically by circular and radial muscle fibers in the iris which are under the control of the autonomic nervous system. In bright light the pupil constricts, reducing the amount of light entering the eye to the optimum. In dimmer light the pupil widens to admit as much light as possible.

Immediately behind the pupil is the soft, transparent lens of the eye, slung from a ring of suspensory ligaments that are attached to the circular and radial ciliary muscles. When the circular fibers contract, the tension in the suspensory ligaments is reduced and the lens is allowed to bulge. This shortens its focal length so that near objects can be seen clearly. When the radial fibers of the ciliary muscle contract, the lens is pulled into a flatter shape, with a longer focal length.

The main focusing of the light that enters the eye is not in fact carried out by the lens itself, but by the cornea and the aqueous humor at the front of the eye. The lens gives fine focus to all objects whether they are close, middistant or distant.

Within the sclera is the choroid layer of the eyeball, rich in blood vessels. Within the choroid is the light-sensitive, multilayered retina. The retina receives the image cast upon it by the lens at the front of the eye, and its nerve cells respond by translating the stimulus of light waves into nerve impulses.

The cells of the retina have a layer of the dark pigment melanin behind them. This helps to prevent the scattering of light within the eye. The actual light-sensitive cells are called rods and cones because of their distinctive shapes when seen under a microscope. Together they number some 130 millions. Both rods and cones contain special pigments that decompose as soon as light falls on them, and re-form at once with the aid of enzymes and vitamin A.

Cones are responsible for the perception of color, and rods, which are more plentiful at the edges of the retina, for vision in dim light. At the very back of the eye, in an area called the fovea, cones are especially plentiful, and this is the area of clearest vision. In the "blind spot" there are neither rods nor cones. Here the nerve fibers from the retina converge to form the optic nerve.

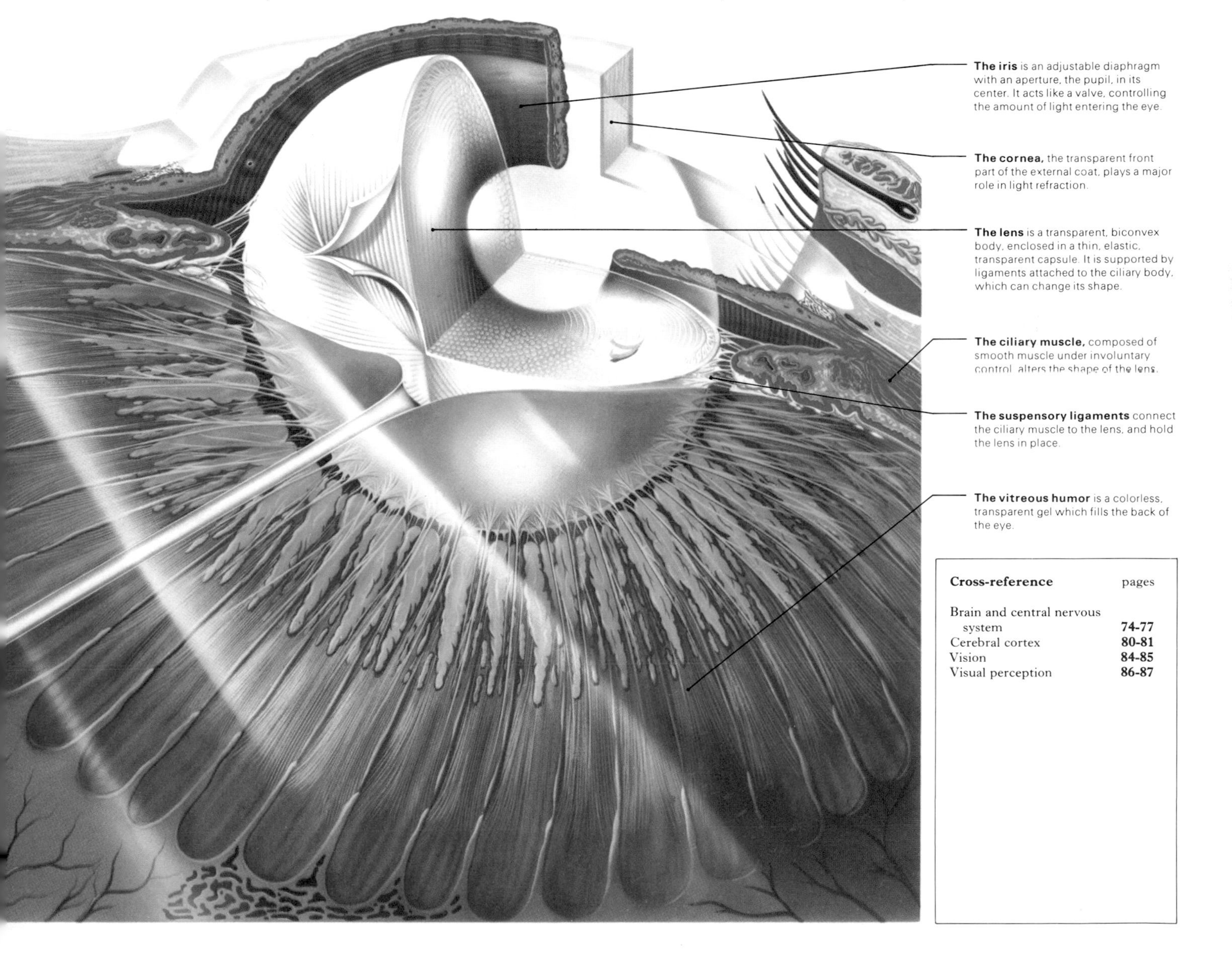

Cross-reference	pages
Brain and central nervous system	**74-77**
Cerebral cortex	**80-81**
Vision	**84-85**
Visual perception	**86-87**

Eyes: how they function

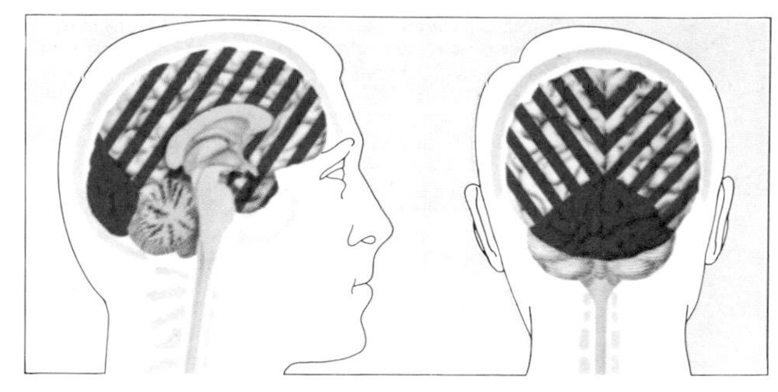

The primary visual cortex lies in the occipital lobes of the cerebrum, but other visual association areas are located throughout the cerebral hemispheres. The visual cortex together with these areas interprets images.

The visual mechanism

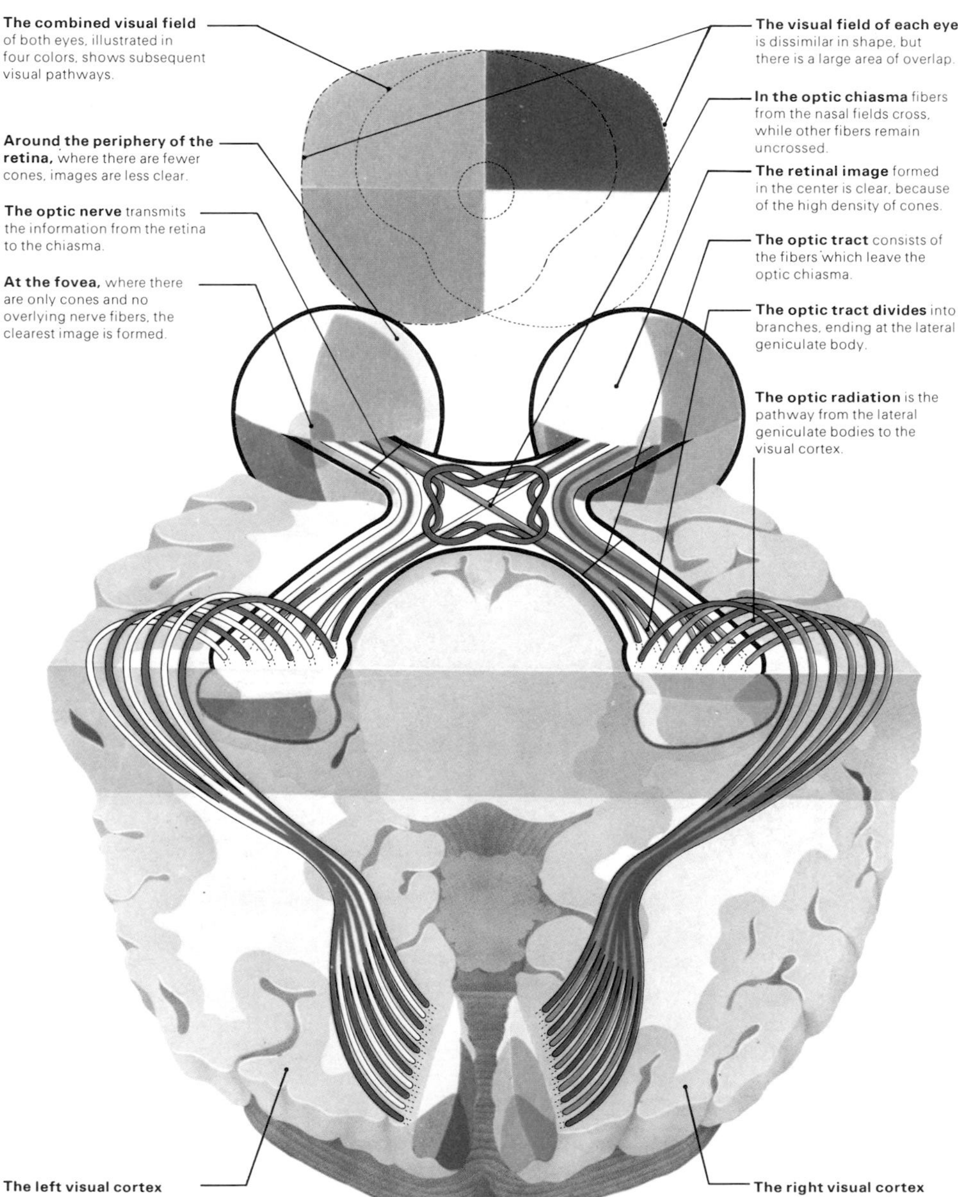

This cross section of a brain, viewed from above, shows how the image which reaches the retina is coded and relayed to the visual cortex. Light falling on the retina stimulates the fibers of the optic nerve. These fibers join to form a cross, called the optic chiasma, where fibers from the inner side of each eye pass to the visual cortex on the opposite side. Fibers from the outer field of each eye are uncrossed, passing to the visual cortex on the same side. From the optic chiasma, the information passes, via the optic tracts, to the lateral geniculate bodies, where perception of depth occurs, and then on to the optic radiation, which transmits the information to the primary visual cortex, situated in the occipital lobes. The primary visual cortex is responsible for the perception of the position of objects in space, and their relationship to each other, as well as the perception of light and shade. In this way, an overall composite picture of any object is formed.

Many more nerve cells are devoted to serving our sense of sight than are devoted to any of our other senses, emphasizing the extreme importance of vision in our lives. With this extraordinary faculty we have the ability to discern objects and events, not just in the immediate vicinity, but at great distances—everything from the ends of our noses to stars twinkling millions of miles away in space.

To a certain extent, there is an analogy between the eye and a camera. Working together, the lens and the cornea form images on the light-sensitive photographic plate of the retina. The iris, besides controlling the amount of light that enters, aids the lens in producing a sharp and well-defined picture of whatever we wish to bring into focus, exactly like the diaphragm of a camera. Like a photographic plate, too, the retina contains chemical substances that are changed by light of different wavelengths. Here, however, the analogy ends.

Light falling on the rods and cones triggers off nerve impulses by a photochemical process, and coded messages travel in the optic nerves back to the brain. Exactly how these nerve impulses produce living, moving, colorful and meaningful "pictures" within our minds is not known, any more than it is known how any of our other senses produce mental images that can be stored in the memory, later to be evoked, or recalled, at will. Nevertheless, more is being learned of the mechanisms involved.

Passing backward and slightly inward from the eyes, the two optic nerves meet at the optic chiasma, or crossing. Here about half of the fibers from each eye cross over the midline, to run with the remaining fibers on the opposite side. Beyond the chiasma, the two tracts so formed each contain fibers that started in the retinas of both eyes. The nerve fibers that cross over are not just a random selection from each optic nerve, however, but fibers that started at the inner, or nasal, side of each retina. Thus the right optic tract carries only messages corresponding to objects seen to the left of the field of vision, and the left tract carries messages corresponding to objects seen to the right.

From the optic chiasma the tracts pass to each side of the thalamus, entering the brain and connecting with relay stations—the lateral geniculate bodies in each cerebral hemisphere. Pathways known as the optic radiations then conduct impulses to the areas of cortex at the back of the occipital lobes—the visual cortex.

Experiments have shown that different parts of the retina are represented point for point in the central area of the visual cortex. An electrode used to stimulate a small area of the cortex causes the subject to see a flash of light, apparently before his eyes. If the electrode is moved, the light moves, too, to another part of the visual field. Farther from the center of the visual cortex, an electrode causes the subject to see objects or definite shapes. Farther still and the stimulus may vividly call up whole scenes from the person's past.

Other experiments have shown that in cats there are cells in the outer layers of the visual cortex that respond only to certain characteristics of objects in the visual field. Thus some cells respond to vertical lines and no others, or only to lines at a particular angle, or only to objects moving in a particular direction. As this is probably true in human brains, too, it means that there are mechanisms at work picking out distinctive features in everything that is presented to our eyes.

We see things not only in terms of shape and movement, but in terms of color as well. This ability is conferred by the cones of the retina, which contain light-sensitive pigments that trigger impulses when light of certain wavelengths falls upon them. Some cones contain pigment that responds to red light, some contain green-sensitive pigment and others blue-sensitive pigment. White light is, of course, a mixture of light of all the different wavelengths, but it is possible to reproduce it by using just red, blue and green light, and indeed to produce any color in the spectrum by adjusting the proportions of these three colors.

Depending upon the mixture of wavelengths in the light entering the eye, the red, green and blue cones will be stimulated in different proportions and to different degrees, and the coded signals that they transmit to the brain are interpreted by the cortex as a sensation of color. It may be difficult to believe that, for example, the color yellow could ever be produced by mixing red and green light—all three colors seem so fundamentally different from one another—but this is in fact the case.

It must be remembered that color is not a physical property of the objects that we see, but a sensation in the brain resulting from the arrival in the cortex of different sequences and combinations of nerve impulses. For the sake of communication, we agree to call the sky blue and grass green, attributing to them certain properties, but we are really giving names to our own sensations and can never know whether other people's sensations are the same as those we personally experience.

Color blindness, which is better called color-defective vision, may result from deficiency in one or more of the pigment cone types in the retina, or from faults in the nerve circuits responsible for relaying appropriate impulses to the cortex. The most common defect is an inability to distinguish red from green. For an unknown reason about 10 percent of men are affected, but less than 1 percent of women. The nature and degree of the defect can be ascertained with specially designed test cards, such as those of the Ishihara series.

In dim light we all become color blind, because the cones receive insufficient stimulation to make them work, and only the black- or white-discriminating rods respond, giving sensations of light and dark. Hence the saying, "At night all cats are gray."

The focusing mechanism of the eye consists of the lens, which is completely encircled by the ciliary muscles and attached to them by the suspensory ligaments. (**A**) In distant vision the ciliary muscles are relaxed and the lens is pulled flat. It does little focusing because the almost parallel light rays from the distant object need to be only slightly refracted to bring them to a point on the retina. (**B**) In near vision, the ciliary muscles contract, the suspensory ligaments loosen and the lens becomes more convex. The curved surface of the lens reinforces the cornea's focusing of the more divergent rays from the near object.

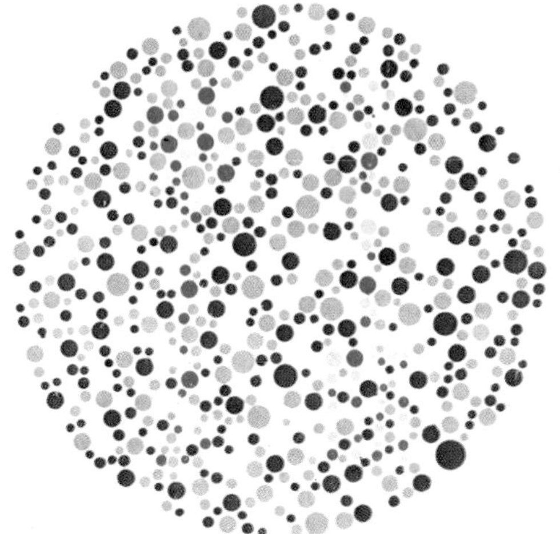

Color blindness occurs when a single type of color-receptor cone is missing from the retina, due to a genetic defect, rendering the person unable to distinguish some colors from others. There are three main types of color blindness—green, red and, rarely, blue. In the test card above, both the fork and spade can be seen by people with normal color vision. The spade is purple, a color which is received by the receptors for blue-green in the retina. Since green-blind people have no green-receptor cones they cannot distinguish the spade from the background. Similarly, red-blind people have no red-receptor cones and so fail to see the fork, which is red and should, therefore, stimulate the cones which receive red. Using a series of such cards, the nature of color blindness can be determined.

Color vision is a synthesis of red, green and blue light (**1**) which is brought about by changes in pigments within the cone cells of the retina. The human eye can recognize hundreds of different colors, which are combinations of these three colors. It is believed that three specialized types of cones exist—one responding to red light, one to green light and one to blue light. Some of these cones, however, can be stimulated by the whole visual spectrum. The variable stimulation of all the cones can register all of the colors known to man. Absence of, or defects in, the cones that respond to red, blue or green light, causes various types of color blindness. Color-blind people may not be able to see green and red (**2**), blue and yellow (**3**) (although blindness to blue is rare), or any color at all (**4**), with everything appearing gray.

Animated films are composed of many still pictures of a subject, each of which represents a split second in time. In this cartoon, each separate picture shows a minute change in position of the character, so that when put in sequence a complete pattern of movement is shown. When the film is run, the eye is presented with twenty-four static images per second, but the retina can only register two separate signals per second, so the images run into each other, and continuous movement is seen.

Cross-reference pages

Visual perception: the mind's eye

Visual perception, the way in which the brain sorts out a recognizable picture from the battery of impulses sent to it from the retinas of the eyes, is one of the great miracles of human functioning. This is the mechanism by which tiny, upside-down retinal images—formed in two dimensions—are transferred into neural codes, the language of the brain, and perceived as three-dimensional objects. It is the mechanisms of visual perception which shape and maintain our picture of the outside world.

Perception is ruled by certain basic laws, the constancies—size, shape, color and brightness. Together these enable us to see objects as they really are, despite the angle, distance or other factors which might distort perceived shape, size or color. In the mind, a ship maintains its apparent size whether it is seen on the horizon or close to, berthed in port. Similarly, people, cars and houses seen from the top of a high building are perceived as real, although they look like toys.

Shadow is one of the most reliable visual cues, and changes in brightness or color intensity are critical in the judgment of depth or distance. Terrain empty of shadow cues can mislead the inexperienced observer. Notorious in this respect are desert or virgin snowscapes, which lack both contrast and landmarks, and judgment of distances can be hopelessly difficult in both these settings. There are similar problems, too, with the lunar landscape, according to American astronauts, who found that they had difficulty estimating distances on the moon.

Eye movements, necessary to provide adequate visual input, are vital to accurate perception. The eye moves through a series of rapid flicks and tremors, called saccades, which subtly change the position of the image on the retina, resulting in the continual stimulation of the retinal cells. With this changing input, the code of impulses transmitted to the brain is always changing, even when we stare fixedly at a static object.

To show what happens if the eye movements are stopped, psychologists use a device that stabilizes the retinal image. With this device the image cast on the retina remains unchanged, but there is a difference in the way it is perceived due to various modifications in the coded information which reaches the brain. Gradually, colors fade, corners and planes disappear and lines begin to merge as the perceived image is lost.

Somewhat surprising is the perception of movement in situations where in fact there is none. Known as the phi phenomenon, this experience is sometimes brought on by street signs: the lights are switched on in sequence, giving an impression of movement, although nothing actually moves. The stimuli registering on the retina are merely those of lights flicking on and off, but we perceive the lights as moving units.

The threshold at which the mind becomes consciously aware of visual stimuli is not necessarily the same as that at which the image is detected. Consequently, if an object is seen only briefly, information as to its nature may fail to reach the brain at all.

Perception involves far more of a digest of information than can be fed in from the bare, two-dimensional image of any particular object. It requires, too, knowledge of the object derived from a fund of past experience which, besides vision, may include the other senses, together with such associations as temperature or pain. There is some controversy as to whether perception is learned or innate, with the learning faction gaining some support from evidence that some people born blind but later acquiring sight are at a disadvantage, profiting less from their new visual environment than people normally sighted from birth.

Sometimes, however, the eye and the brain, working together to convert an image into something recognizable and easily understood, between them yield up the wrong conclusion. This is the time when perception goes awry, that we are misled and succumb to the effects of distortion or illusion. Some geometric figures play tricks on the mind because they do not conform to the perceptual constancies; others are visually disturbing.

One of the most famous illusions concerns the moon, perceived as larger when it is just above the horizon than when it is overhead, although the size of the retinal image is nearly the same in each case. The illusion occurs because distance cues near the horizon—silhouettes of buildings and trees and the skyline itself—are far more positive than anything to be seen directly overhead. Looking overhead, therefore, we underestimate distance and perceive the moon as being smaller.

For thousands of years, art was influenced by a lack of understanding of how perception works. The Egyptians painted figures in profile without perspective, while the Chinese developed their own stylized perspective based on cultural convention. In Chinese painting, lines diverge with distance rather than converge. It was not until the Italian Renaissance that the rules of linear perspective were widely introduced, giving depth to works of art.

Visual perception

This eye-level view of a landscape architect's model appears three-dimensional. However, the image it forms on the retina of the eye is only two-dimensional. Distance and depth are perceived not by the eye, but by the brain, which recognizes certain visual features as indications of the relative positioning of each object. Observe the following cues in this picture: objects on the horizon, or high in the field of view, appear far away. Colors fade and detail is lost in the most distant objects. Note the relative size of the main objects. The open space between objects decreases in the distance and lines seem to converge. Shadows indicate the positions of objects in relation to one another, to the ground and to the light source. If the outline of one object breaks the contour of another, the object with the broken contour appears farther away.

Here the same model as the one above is viewed from a more elevated position. There are now differences in the positioning of the objects. The trees and the arch are the same distance away from the wall in the foreground. The rows of bushes and the flower bowls are all behind the arch, not in front of it. The lawns extend only a few feet in front of the arch. The model has been made in such a way that all these cues are present in the first picture, giving the brain the impression of immense depth. From this elevated view, it is evident that the depth does not exist and the ambiguities of the model are revealed. The brain has deceived itself because the scene contravened normal experience.

Visual illusions

The eyes and the brain work together to convert an image into something recognizable and easily understood. However, the brain is often deceived by what the eyes see and perception goes awry. Thus we are misled and succumb to the effects of visual illusions. If the cues provided by an illusion are ambiguous, the brain may perceive images in two quite different ways. Some objects, for example, appear distorted, yet when measured with a compass or a ruler they are found to be geometrically perfect. Other illusions provide normal distance and depth cues and appear three-dimensional, yet they are, on analysis, impossible objects. Some illusions are simply visually disturbing. The mechanisms within the brain which produce these effects are not fully understood.

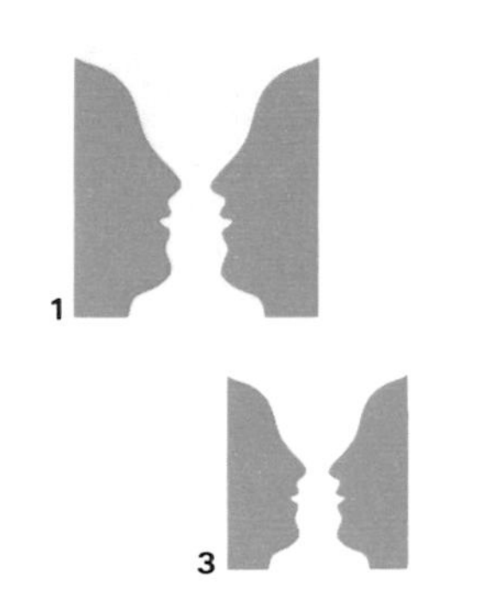

Figure and ground

When, in an illustration, the differentiation between the object and the background is unclear, a spontaneous reversal of object and background occurs. Here the drawing is perceived as a vase or as two facing profiles (**1**). Attempts to isolate the vase (**2**) or the profiles (**3**) with color fail, because sufficient cues are lacking.

An illusion of imperfection

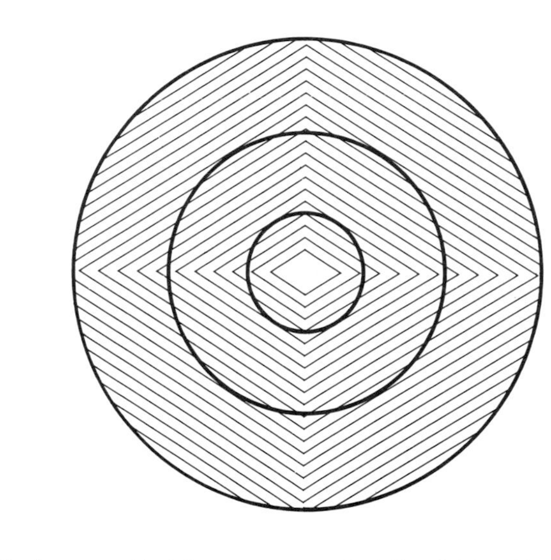

The three concentric circles are geometrically perfect. This can be checked with a compass. The distorted appearance of the circles is caused by the angles the straight lines make with the circles. This gives the perceived effect of circles with flattened sides. In a similar way, the straight lines seem to be broken where they intersect with the circles. Just why the mind creates this illusion is uncertain.

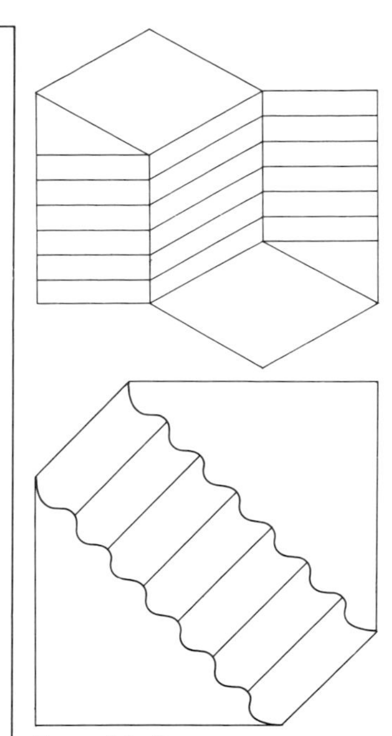

Reversible figures

In both illusions above, depth cues are given, but the drawings are ambiguous—the stairs appear to be going up as well as down. In confusion, the brain continually alternates between the two possible interpretations.

Perceptual motion

When we look at an object, our eyes do not remain still, but move continuously. This can produce a strong feeling of motion, as in the static picture above.

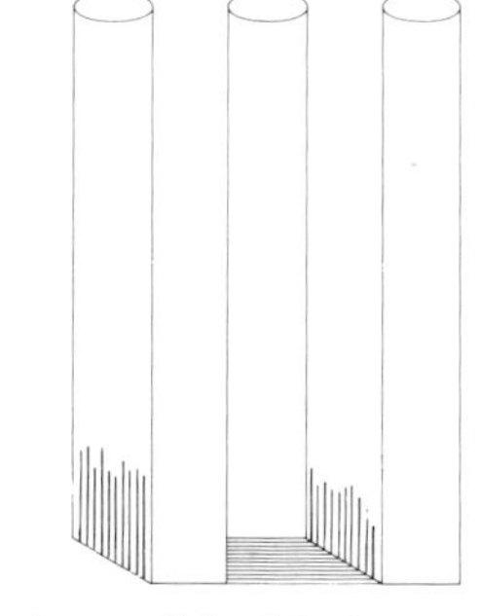

An impossible object

This object is drawn in a manner that deceives the brain. Although ordinary depth cues are provided, the mind cannot decide what it should perceive.

The distorted room

Here is a picture of a room which the brain perceives as rectangular in shape. But can the two people in the room be so different in size? In fact, the room is distorted to produce the same image on the retina as a rectangular room (see inset). The man on the left is actually standing much farther away.

Gray where there is no gray

This is simply a pattern of black squares on a white background, yet we see gray areas between the squares. The mechanism of the mind which produces this effect is still unknown.

Afterimages

Stare intently at the center of the flag for about forty seconds. Then look immediately at a plain white surface—a sheet of paper or a wall. The flag should now be seen in its true colors. The black becomes white, the yellow is blue and the green is red. It is believed that this phenomenon occurs because retinal cells are sensitive to complementary colors. The "black," "yellow" and "green" cells situated in the retina are overstimulated and respond by producing white, blue and red afterimages against a white background.

Cross-reference	pages
Brain and central nervous system	**74-77**
Eye	**82-83**
Vision	**84-85**

The ears: structure

Most people think of the ear only as an organ of hearing. In reality it is an organ with three functions. Besides enabling us to perceive sound waves, it sends impulses to the brain that tell us which way up we are, and also messages that tell us of movements of the body in the three dimensions of space.

Like the eyes, the ears are protected within cavities in the skull, but the delicate parts of the ears are set far more deeply in the bone. The outer visible ear flap and the curving tube leading inward on each side of the head are together known as the outer ear. At the end of the tube, which is called the auditory canal, lies the eardrum, or tympanic membrane. Inside the eardrum is the cavity of the middle ear, which communicates with the back of the throat through the Eustachian tube, leading downward and inward.

Set in a chain leading across from the inside of the eardrum are the ear ossicles, three tiny bones (the smallest in the body)—the malleus, or hammer, the incus, or anvil, and the stapes, or stirrup. These are responsible for transmitting the vibrations that sound waves cause in the eardrum across the cavity of the middle ear to the inner ear. The inner ear is almost completely encased in bone. It contains coiled and looped tubes which are filled with fluid and are provided with special sensory cells. From the sensory cells lead fine nerve fibers that join into bundles and transmit messages to different regions of the brain.

The outer ear flap, or pinna, is fairly small, but serves to direct sound waves into the auditory canal and gives some protection to the opening. In humans the pinna is immobile, but many animals, including dogs and cats, are able to prick up their ears and turn them to face the direction from which sound is coming.

The auditory canal has hairs and modified sweat glands that secrete earwax, or cerumen, that gradually makes its way to the outside with particles of dust and bacteria that have found their way into the interior and have become trapped.

The eardrum vibrates according to the frequency of the sound waves that strike it, and the vibrations are passed onto the ear ossicles. The ear ossicles are suspended from the roof of the middle ear cavity by tiny ligaments and muscles, and act as a series of levers to magnify the inward and outward movements of the eardrum into shorter but more powerful vibrations. Thus the stapes, the last of the chain, vibrates at the same frequency as the eardrum, but with a force more than twenty times greater. Acting as a footplate to this stirrup-shaped bone is a tiny oval disk, which is in close contact with a membrane stretched across an oval "window" in the bony casing of the inner ear. Within the membrane is the fluid that fills the cochlea of the inner ear, where the vibrations are translated into nerve messages.

The cochlea is a hollow chamber in the solid bone of the skull, spiraling rather like a snail's shell and narrowing as it does so. Within the cochlea, and extending across the width of the narrowing canal, is the basilar membrane, and extending for the length of the membrane is the organ of Corti, a specially sensitive strip of complex structure. Pressure waves traveling in the fluid contained in the cochlear canal activate different regions of the organ of Corti, and fine nerve fibers carry appropriate messages to the brain through the auditory nerve bundle.

The inner ear also contains the mechanisms that enable us to maintain our posture and balance, functions really quite separate from the hearing functions of the cochlea. Three fluid-filled canals, the semicircular canals, set at right angles to one another, are adapted to detect even the slightest movement of the head in any particular plane.

Swirling of the fluid in the canals stimulates tufts of fine hairs in the ampullae swellings at the ends of the chambers. In a meshwork around the roots of the hairs are nerve fibers that translate movements of the hairs into nerve impulses, and the impulses travel to the brain along the vestibular nerve. Movement of the head in the horizontal plane will produce messages from the horizontally set canal, while movement in the vertical plane will produce messages from the vertically set canals, appropriate to the direction of movement.

The swirling of fluid in the semicircular canals does not stop for an appreciable length of time after the head has been moved vigorously. After spinning around and then stopping, the feeling of dizziness that we experience is due to the continued stimulation of the fine hairs which, by movements of the fluid, indicates to the brain that the head is moving, although the eyes are telling the brain that movement has stopped. The result is temporary confusion and disorientation, as well as loss of balance.

Two more fluid-filled chambers in the inner ear, the utricle and the saccule, are one means by which the brain detects changes in the posture of the head. Like the ampullae, they have fine hairs attached, at their roots, to nerve fibers. The hairs are surrounded by a jellylike material, and among them are otoliths, tiny crystals of calcium carbonate.

Under the influence of gravity, the otoliths press upon the hairs and, according to the posture of the head, stimulate different nerve fibers. Thus we can tell which way is "up" and which is "down," even when our eyes are closed. The nerve fibers from the utricle and the saccule join those from the ampullae in the vestibular nerve to make bundles, and pass to the regions of the cerebellum called vestibular nuclei, where the body's movements are coordinated.

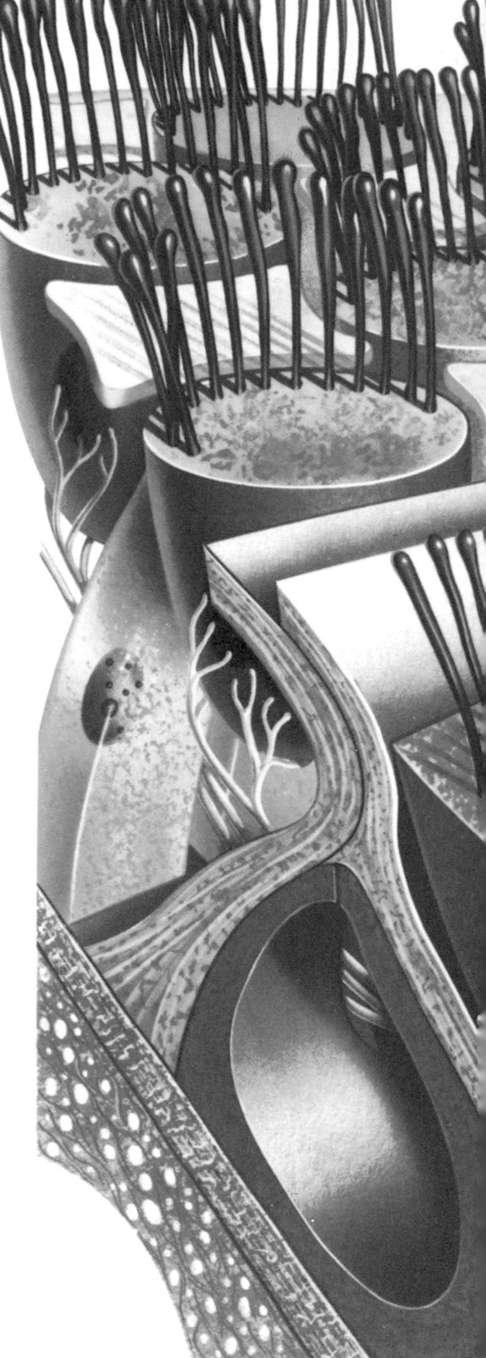

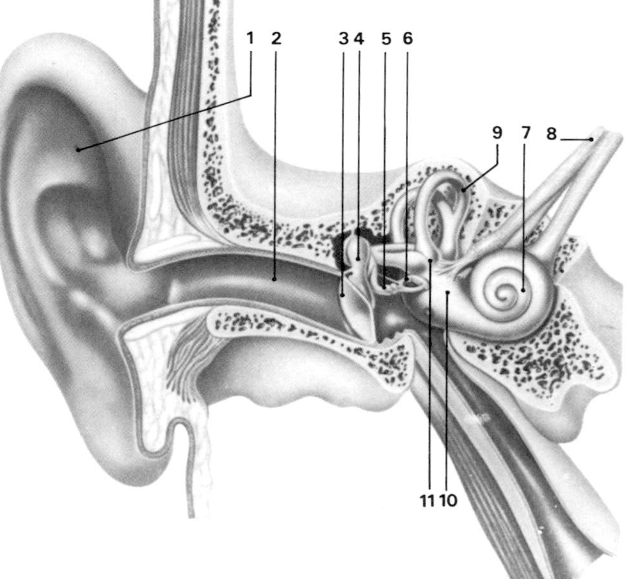

The ear is the organ of hearing. The pinna of the outer ear (**1**) collects sound waves and channels them along the external auditory meatus (**2**) to the eardrum (**3**). The eardrum vibrates and mechanical waves are transmitted by the ossicles, three small bones of the middle ear—the malleus (**4**), incus (**5**) and stapes (**6**)—to the cochlea (**7**) of the inner ear. The cochlea, a coiled tube filled with fluid, contains the organ of Corti. Fluid waves, created within the cochlea, excite nerve cells in the organ of Corti, which send impulses along the auditory nerve (**8**) to the brain. The semicircular canals (**9**), the saccule (**10**) and the utricle (**11**) are concerned with balance and the posture of the body.

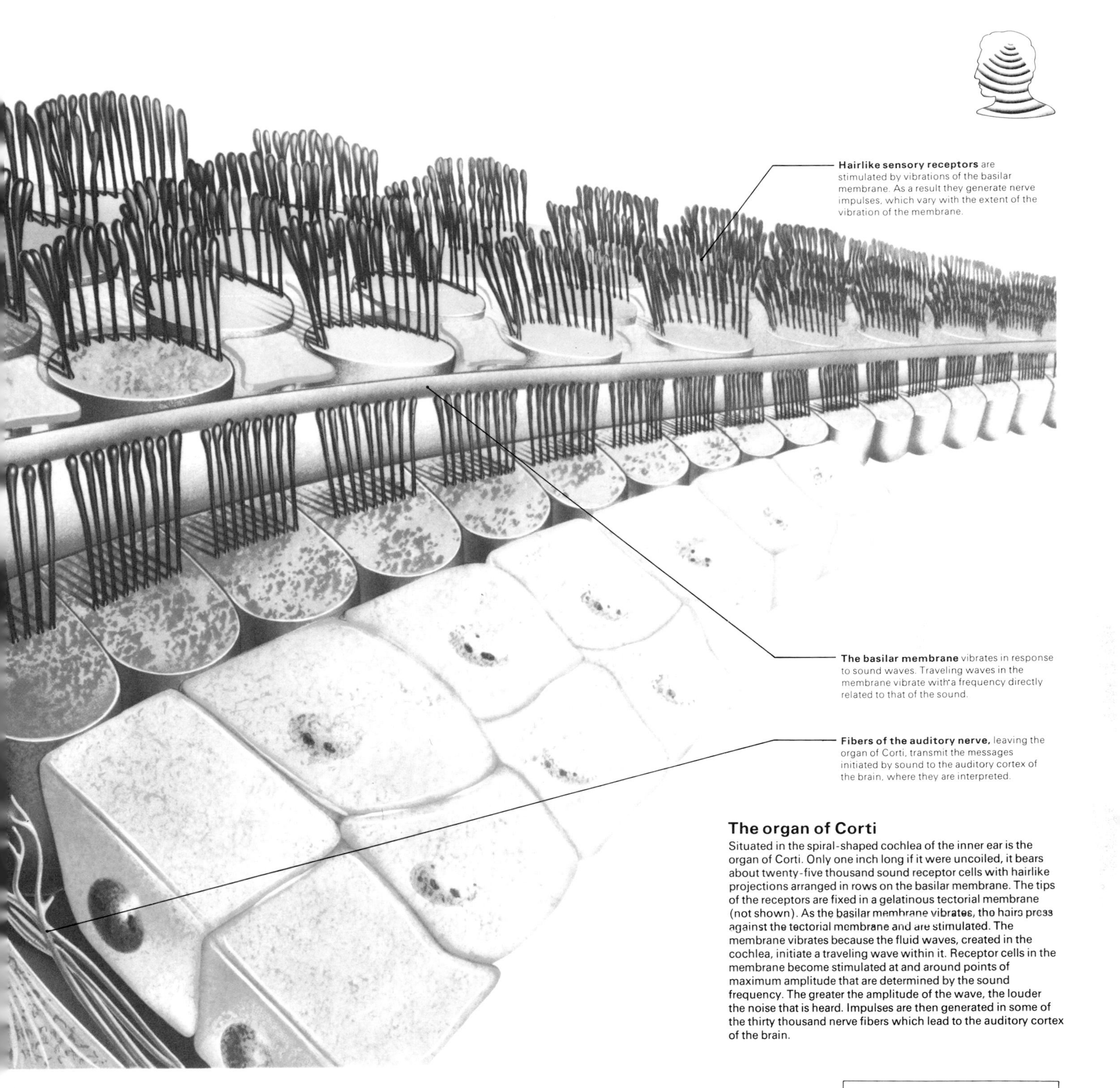

Hairlike sensory receptors are stimulated by vibrations of the basilar membrane. As a result they generate nerve impulses, which vary with the extent of the vibration of the membrane.

The basilar membrane vibrates in response to sound waves. Traveling waves in the membrane vibrate with a frequency directly related to that of the sound.

Fibers of the auditory nerve, leaving the organ of Corti, transmit the messages initiated by sound to the auditory cortex of the brain, where they are interpreted.

The organ of Corti

Situated in the spiral-shaped cochlea of the inner ear is the organ of Corti. Only one inch long if it were uncoiled, it bears about twenty-five thousand sound receptor cells with hairlike projections arranged in rows on the basilar membrane. The tips of the receptors are fixed in a gelatinous tectorial membrane (not shown). As the basilar membrane vibrates, the hairs press against the tectorial membrane and are stimulated. The membrane vibrates because the fluid waves, created in the cochlea, initiate a traveling wave within it. Receptor cells in the membrane become stimulated at and around points of maximum amplitude that are determined by the sound frequency. The greater the amplitude of the wave, the louder the noise that is heard. Impulses are then generated in some of the thirty thousand nerve fibers which lead to the auditory cortex of the brain.

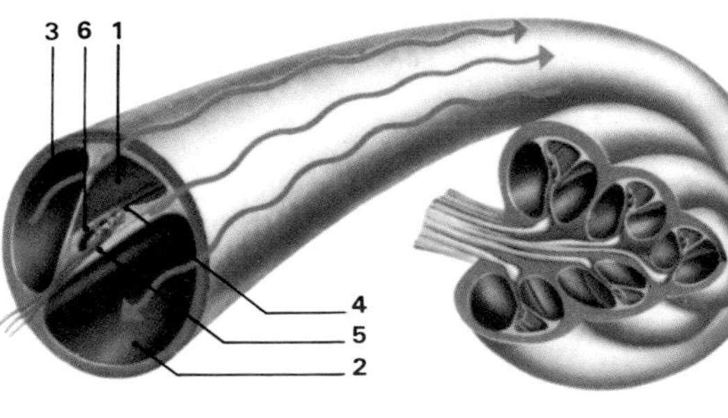

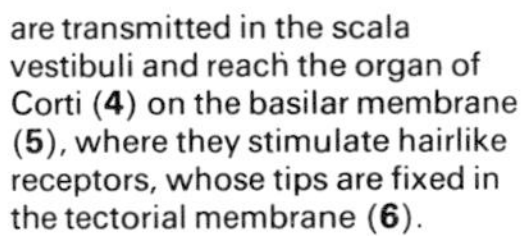

Were the cochlea uncoiled it would be a continuous tube divided into three separate compartments—the scala media (**1**), the scala tympani (**2**) and the scala vestibuli (**3**). Sound waves are transmitted in the scala vestibuli and reach the organ of Corti (**4**) on the basilar membrane (**5**), where they stimulate hairlike receptors, whose tips are fixed in the tectorial membrane (**6**).

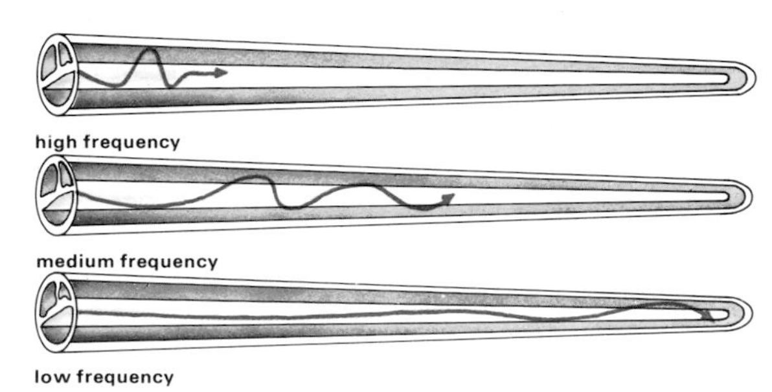

Hearing is dependent upon sound waves entering the inner ear. Sound waves of different frequencies create different wave forms in the basilar membrane within the cochlea. High-frequency sound vibrates the basilar membrane at the base of the cochlea, medium-frequency sound vibrates it in the middle and low-frequency sound vibrates it at the apex.

Cross-reference	pages
Brain and central nervous system	74-77
Cerebellum and balance	98-99
Cerebral cortex	80-81
Hearing	90-91

Ears: function

Many people who are totally deaf claim that if they had the choice they would rather be blind. This may seem surprising, since sight is always regarded as our most precious sense. But a deaf person has lost a very basic means of communication with other people for which there can be only poor substitutes. Without hearing, speech, music, telephone and door bells, car horns and other warning sounds are all irrelevant, and life loses a unique part of its fabric. Although excessive noise may be a menace to well-being, ordered sounds provide an invaluable source of information as well as of pleasure.

The vibrations in the air that constitute sound waves have the qualities of amplitude, the size and energy of the waves, and frequency, the number of vibrations or cycles per second. The louder a sound, the greater the amplitude; the higher the pitch of a sound, the greater is its frequency. Adults can detect sounds that have a frequency anywhere in the range of sixteen to about twenty thousand cycles per second, but our ears are best adapted to register frequencies in the one thousand to two thousand range. Young children are able to hear higher-pitched notes than adults. After puberty, however, the sensitivity to higher notes begins to decline, although the perception of low notes remains reasonably constant throughout life.

It is sound waves, traveling into the auditory canal, that set up vibrations in the eardrum which are passed through the ear ossicles to the oval window of the inner ear. From here the waves of alternately high and low pressure travel in the fluid contained in the spiraling cochlea. Because, unlike air, this fluid, which is called perilymph, is not compressible, the pressure waves need to be dispersed. This function is carried out by the round window that lies below the oval window, and, like it, has a membrane stretched across an opening in the bone surrounding the inner ear. Sound waves that have traveled through the perilymph make this membrane bulge into the middle ear, dispersing the energy.

The cochlea, which has two and a half turns, could be seen, if it were uncoiled, to contain three canals, separated from each other by two thin membranes, the basilar and the vestibular membranes, and coming together almost to a point at the apex. The oval window is connected to the base of one of these tapering canals, the vestibular canal, and the round window is at the base of the tympanic canal. These two canals are continuous at the apex of the cochlea. The third, narrower canal is called the cochlear canal. This contains not perilymph but endolymph, which has a slightly different chemical composition. It is within this canal that the actual organ of hearing, the organ of Corti, lies.

The organ of Corti has five rows of cells extending for the whole length of the cochlea. Each cell rests upon the basilar membrane and has tiny hairlike projections in orderly rows. The hairs have overhanging them a third membrane, the tectorial membrane, in which their tips are embedded.

Sound waves traveling in the perilymph produce a "ripple" in the basilar membrane, and the resulting shearing action activates the hair cells, which are connected to the branches of the auditory nerve. The upper end of the basilar membrane responds to sounds of low frequency, and the base of the membrane responds to high-frequency sounds. Between the apex and the base, different regions of the membrane, and so the organ of Corti, respond to different frequencies. A sound with components of more than one frequency will cause a response in more than one region of the membrane. More than thirty thousand different nerve fibers make up the auditory nerve, which transmits complex messages to the brain.

The nerve messages carried by the auditory nerve are not simply the equivalent in living tissue of the electrical impulses that travel along a cable connected to a microphone. While it is true that it is possible to insert electrodes into the auditory nerve and pick up signals that can be reconverted into sounds recognizable as those fed into the ear, thus using the cochlea as a mere microphone, it is probably not these signals that the brain interprets. Other streams of impulses can be picked up, apparently carrying information about the frequency, amplitude and other qualities of the sounds in a coded form, for decoding by the cortex. It is likely that the code employs variations in signal frequency, differences in the fibers that are used—both in number and position within the cochlea—and possibly the straightforward microphone effect as well. But this is not yet known.

The auditory nerve enters the brain stem as the eighth cranial nerve. Here, in the cochlear nucleus, nerve impulses are relayed by complex links, some to another collection of cells called the inferior colliculus, some to relay stations lower in the brain stem and some to the opposite side of the brain. Finally, how-

The auditory pathways

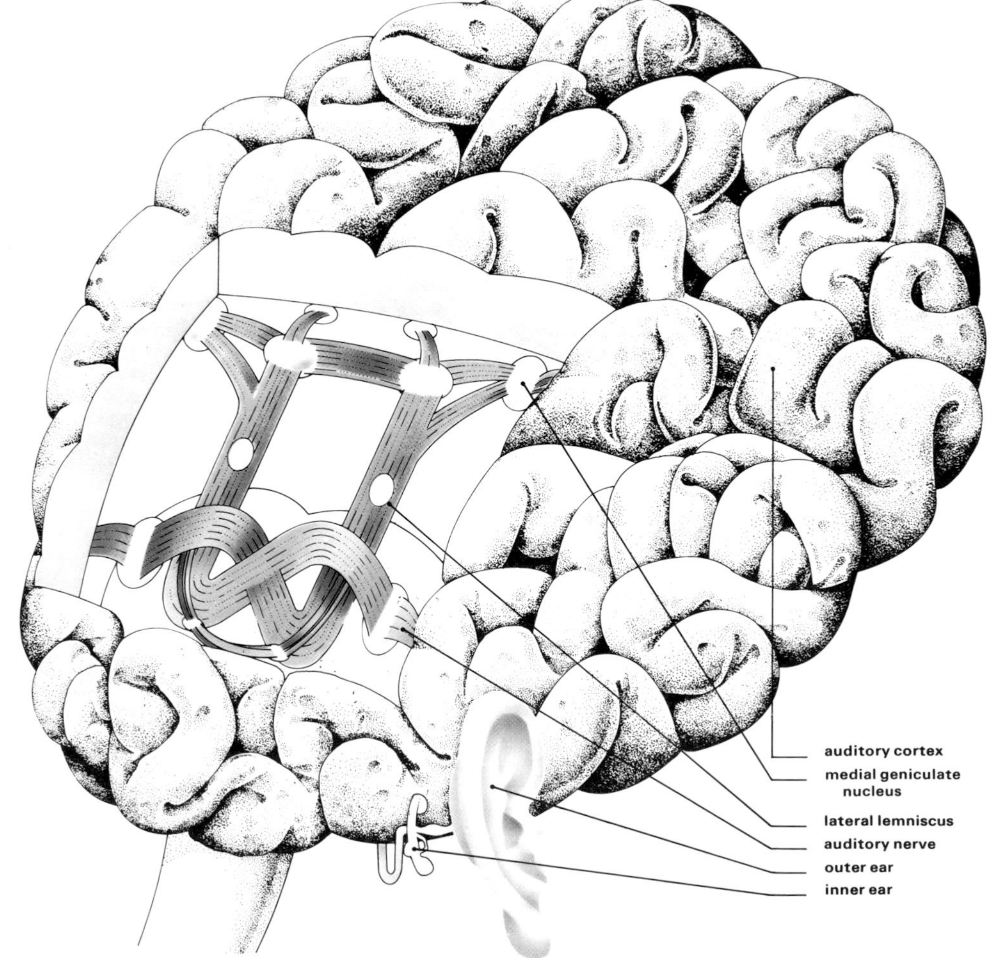

The auditory pathways begin in the nerve fibers of the organ of Corti in the inner ear, where sound waves are converted to nerve impulses. These impulses travel in the auditory nerve to the auditory cortex of the brain. During their passage to the auditory center, some of the auditory nerve fibers cross in the brain stem. This results in the sound which enters one ear passing to both cortices. The impulses are relayed in the lateral lemnisci and from there, via the medial geniculate nuclei, to the auditory cortices in the temporal lobes of the brain, where sound is perceived.

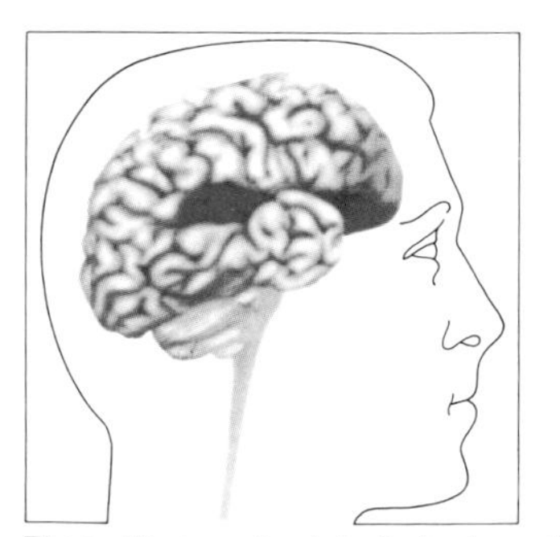

The auditory center, in the brain, shown above in red, is concerned with hearing. Although shown here on the outer surface of the brain, most of this area lies within the temporal lobes and is found within the surface of the cerebral cortex on both sides ol the brain.

ever, all the impulses reach the hearing center of the brain, the auditory cortex. Sounds of low frequency are relayed to the outer surface of the cortex and those of higher frequency to deeper areas.

To make the impulses reaching it meaningful, the brain relies upon all the stored memories within its millions of cell networks —not just sound memories, but visual memories, memories of taste and smell and touch. By comparing and contrasting and rejecting information, the brain forms an integrated mental picture. Exactly what sequences and combinations of sounds will mean to an individual depends entirely upon his personal past experiences.

The brain begins to amass a store of sound memories almost from the moment of birth, and by adulthood we are probably able to distinguish nearly half a million signals that are meaningful. Should the sound memory center be accidentally destroyed, a condition called auditory agnosia occurs, in which, although nerve impulses corresponding to sounds reach the brain, they no longer have any meaning.

Our ears are extremely effective listening devices, and have evolved to a point where any further increase in sensitivity would prove a nuisance rather than an advantage. We should even be able to detect the random movement of the molecules in the air with a resulting constant hissing in our ears. The amplitude, or the loudness, of sound is measured in decibels. The range of sound heard in the world around us varies enormously. A barely audible whisper registers about ten decibels while a jet taking off can generate as much as one hundred and thirty decibels. Sounds above one hundred decibels can cause pain to the listener and recurrent exposure to sound at this level may damage the structure of the ear, producing temporary or even permanent deafness. Deafness can, in fact, be due to damage to any part of the auditory apparatus, but loud, sharp noises are usually responsible for perforating the tympanic membrane, or eardrum. The degree of deafness depends upon the size and position of the perforation.

In many industrial countries throughout the world legislation has been enacted to protect workers from excessive noise. Wherever the noise cannot be kept below the accepted level, workers must be supplied with earplugs or helmets.

Selective attention

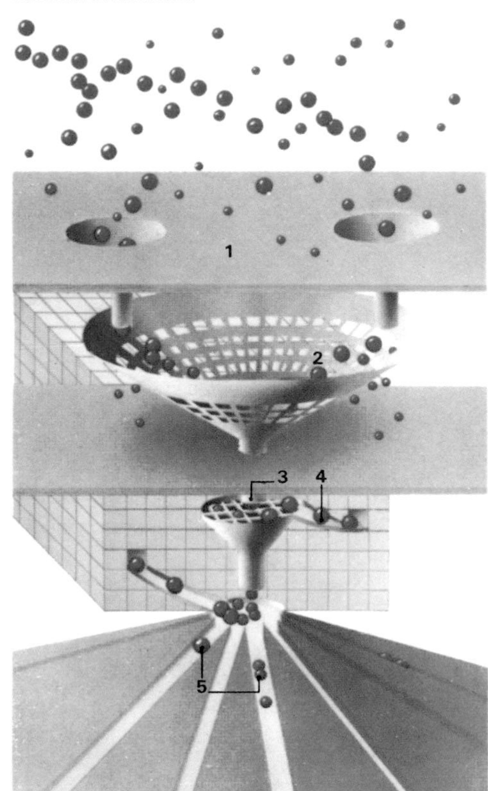

Selective attention is the ability to concentrate on a specific flow of sound amid a babble of background noise, for example listening to a conversation in the middle of a crowd. In the diagram, the red particles represent items of sound information. All sound enters the ears (**1**) and passes through a selective filter (**2**), which discards irrelevant information, then goes into a limited-capacity decision channel (**3**), which decides on the specific source of the sound. Both mechanisms probably operate below the level of consciousness. Some items pass into the memory store (**4**). Information from the decision channel and memory store determines which actions are taken by the brain (**5**).

The range of sound heard in the world around us varies enormously. A space rocket may, for example, emit one hundred and sixty decibels (db) of noise, a jet taking off emits one hundred and thirty db, an express train one hundred db, whereas a bee buzzing emits twenty-five db and rustling leaves only twenty db. Sounds above a level of one hundred decibels cause pain to the listener and repeated excessive noise can cause permanent damage to the ear, resulting in deafness. People whose occupation brings them into contact with such noise levels are advised to wear ear protectors such as earplugs or helmets.

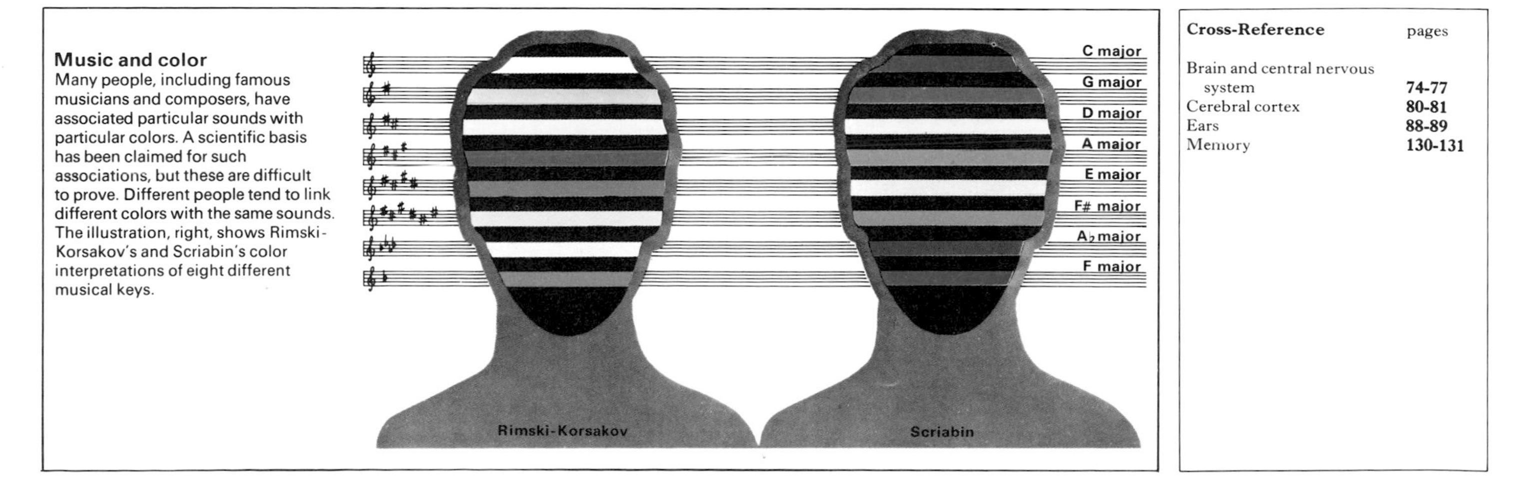

Music and color

Many people, including famous musicians and composers, have associated particular sounds with particular colors. A scientific basis has been claimed for such associations, but these are difficult to prove. Different people tend to link different colors with the same sounds. The illustration, right, shows Rimski-Korsakov's and Scriabin's color interpretations of eight different musical keys.

Cross-Reference	pages
Brain and central nervous system	**74-77**
Cerebral cortex	**80-81**
Ears	**88-89**
Memory	**130-131**

The tongue and taste

Quicker than an eye can blink, a frog's tongue shoots out and captures a tasty fly. We eat our own meals with less haste and more taste, but nevertheless our tongues perform the same basic function. The tongue is a mobile, muscular structure that aids the intake of food and bears organs to receive the sensation of taste. An additional and very important function in human beings is that of assisting articulation, the production of speech.

The tongue consists of groups of interlacing muscle fibers, arranged in different planes and covered with an epithelium containing numerous cells capable of responding to chemical substances. The tip and sides of the tongue are free and the back bulges into the throat, overhanging the epiglottis and the entrance to the trachea. The root of the tongue is attached to the floor of the mouth and at the back to the small hyoid bone embedded in muscle. A fold of tissue in the midline on the underside, the frenum, limits the tongue's movements.

We taste things by means of taste buds, groups of chemoreceptor cells that detect substances in solution around them and send nerve impulses to the brain accordingly. The taste buds are arranged in tiny projections, called papillae, from the surface of the tongue. Each papilla has one or two hundred taste buds. There are four kinds of papillae—filiform, which are pointed, foliate, which are leaflike, fungiform, which are like the tops of mushrooms, and vallate, which are rather like flattened doughnut rings. The ten or eleven larger vallate papillae can easily be seen in a V-formation at the back of the tongue. The other forms, less discernible to the eye, give the tongue its rough texture.

An adult has about nine thousand taste buds, mainly on the upper surface of the tongue, with a few sprinkled in the palate, at the back of the throat and on the tonsils. A baby has a great many more, and they are more widespread. In babies there are even papillae inside the cheeks. With advancing age the number of taste buds decreases, and the ability to distinguish flavors is diminished. Taste buds are normally replaced very rapidly, new ones replacing the old every ten to thirty hours.

A taste bud consists of clusters of chemoreceptor cells, each with minute hairlike projections, called microvilli, projecting from its free upper surface, and linked below with a network of nerve fibers. Each cell is linked to more than one nerve fiber, and one nerve fiber may connect with just one taste cell or with many, thus making precise study of the taste mechanism rather difficult.

From the front two-thirds of the tongue, two different kinds of nerve bundles carry impulses back to the brain. One kind relays sensations of temperature, pain and touch, and the other relays pure taste. These sensations reach the brain separately, but the brain mingles the information to give the flavor that is finally perceived. Thus warm milk has a different flavor from cold milk, and dry cornflakes taste different from soggy cornflakes. If the nerve bundle carrying sensations of temperature, pain and touch is destroyed, these sensations are, of course, lost. But so, too, is the sensation of taste, strangely enough, only to return after a period of hours or perhaps months. If the pure taste bundle is destroyed, the sensation is lost permanently. All the sensory nerves from the back third of the

The organ of taste

The surface of the tongue is covered with thousands of tiny fronds, or papillae, which give it a velvety sheen. The taste buds, the primary organs of taste, are found within these papillae. There are four types of papillae—filiform, fungiform, foliate and vallate. Filiform and fungiform papillae are found on the front of the tongue and foliate papillae at the back. The vallate papillae form a V across the back of the tongue.

Fungiform papillae are mushroom-shaped structures situated on the front half of the tongue.

Filiform papillae, found on the front half of the tongue, are threadlike in shape and more numerous than the fungiform type.

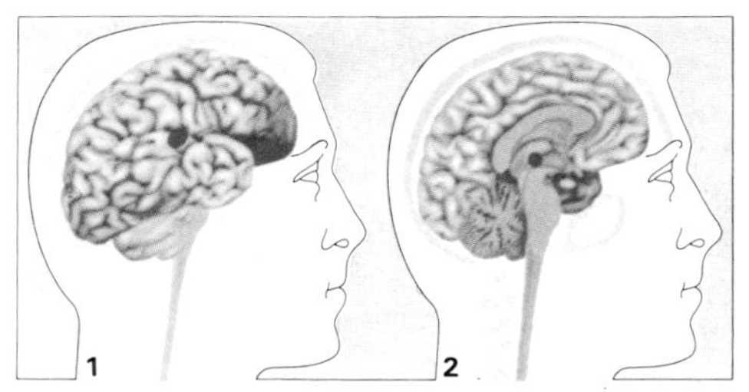

The taste centers are in the cortex (**1**) and in the thalamus (**2**) of the brain.

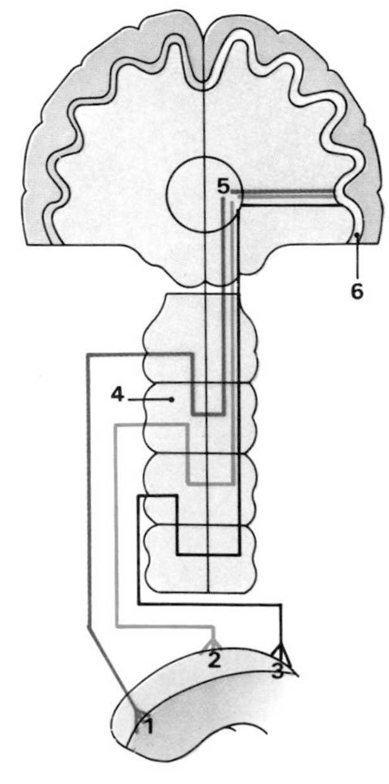

bitter
sour
sour
sweet
salt

The tongue recognizes four qualities of taste, which can blend to produce subtle variations.

The taste pathways begin in the tip (**1**), middle (**2**) and back (**3**) of the tongue and travel to the brain. Sensations of sweet and salt are carried mainly by nerves from the tip, sour tastes are carried by nerves from the middle and bitter tastes by nerves from the back of the tongue. The nerves cross in the medulla (**4**) and pass to the thalamus (**5**) and the cortex (**6**). Taste from one side of the tongue is interpreted in the opposite side of the brain.

tongue reach the brain through a different, single nerve.

Traditionally, there are said to be four basic tastes—sweet, salt, sour and bitter. By simple testing with solutions of various chemical substances, it has been found that different areas of our tongues have different sensitivities. Salt and sweet substances are best tasted at the front of the tongue, sour substances at the sides and bitter substances at the very back. There is virtually no taste sensation in the middle of the tongue. The different tastes do not correspond with the different kinds of papillae. Each papilla seems to be sensitive to a mixture of tastes, although the vallate papillae register only bitter.

The reason for different substances having particular tastes is still not clearly understood. The chemical structure of many gives a guide, as in the case of sour-tasting substances, which all contain acid groups. Lemon contains citric acid, for example, and vinegar contains acetic acid, but saccharin bears no relation in structure to sugar and yet is many times sweeter. And many compounds unrelated to common salt are salty. Bitterness is also a property of many totally unrelated chemical compounds, such as quinine and magnesium sulfate.

The chemical substances found in food and drink may possibly have an effect on various enzymes in the fluid content of the taste cells, and so trigger off electrical impulses in the taste nerves, because of the altered pattern of the cell's metabolism, but this is not certain. Taste discrimination is extremely complex and probably depends upon a code in which different patterns of nerve messages combine in the brain to give the final sensation.

The sensations of taste and smell are almost inextricably linked in our ability to appreciate food and drink. The sense of smell is by far the most sensitive. It has been estimated that a person needs twenty-five thousand times as much of a compound in the mouth to taste it as is needed by the smell receptors in the nose to smell it.

What is interpreted by the brain as taste may in many cases in fact be smell. The odors arising from food and drink in the mouth and passing up to the nose, there to stimulate the olfactory receptors, give the main impression of "palatability." Taste sensations, entering the brain through the lingual and glossopharyngeal nerves, in completely different areas, are more a reinforcement of this first impression than a true guide. The basic "tastelessness" of most of the food we eat becomes apparent when we have blocked noses; a wine taster with a cold is put completely out of action.

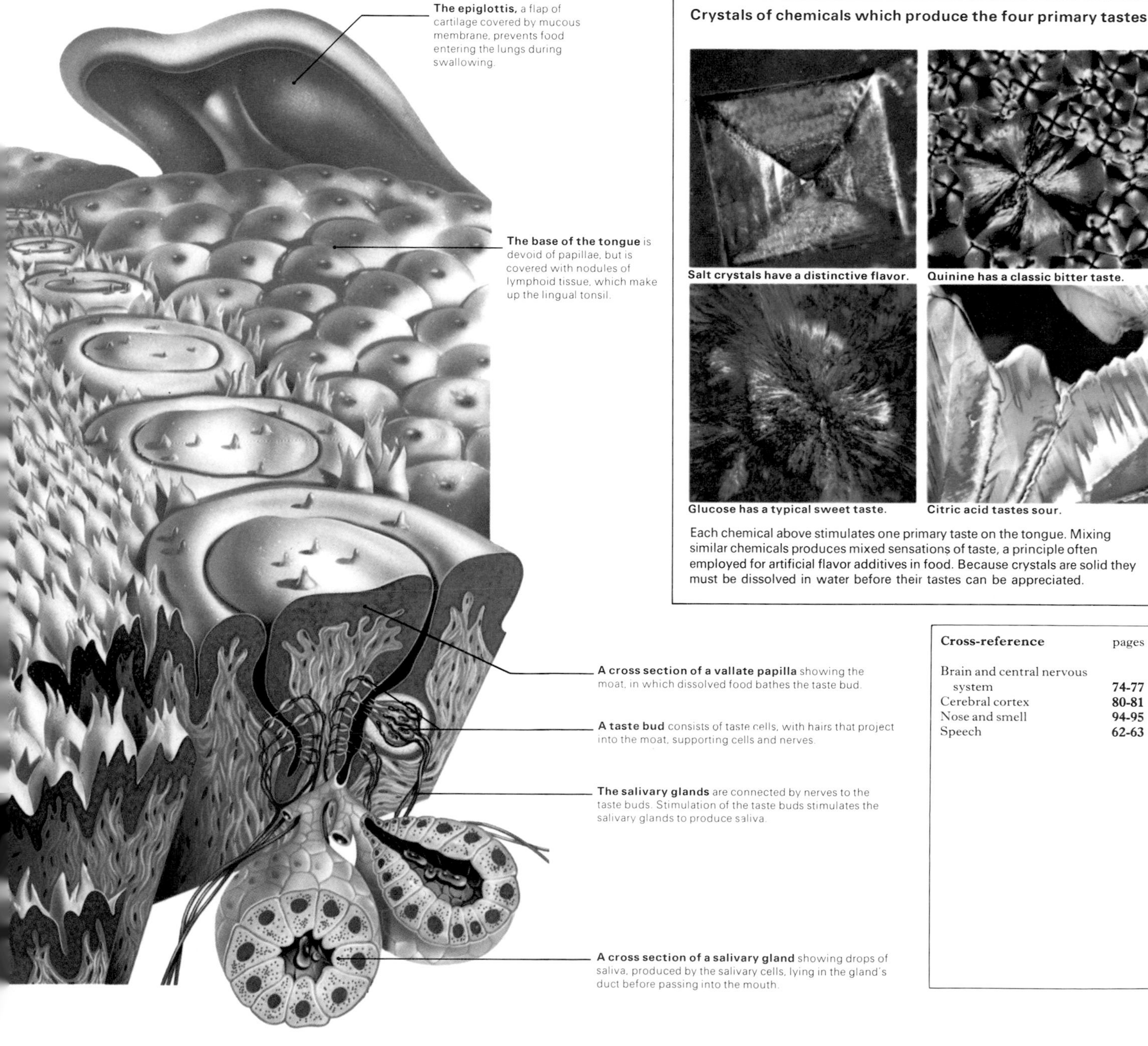

Crystals of chemicals which produce the four primary tastes

Salt crystals have a distinctive flavor.

Quinine has a classic bitter taste.

Glucose has a typical sweet taste.

Citric acid tastes sour.

Each chemical above stimulates one primary taste on the tongue. Mixing similar chemicals produces mixed sensations of taste, a principle often employed for artificial flavor additives in food. Because crystals are solid they must be dissolved in water before their tastes can be appreciated.

Cross-reference	pages
Brain and central nervous system	**74-77**
Cerebral cortex	**80-81**
Nose and smell	**94-95**
Speech	**62-63**

The nose and smell

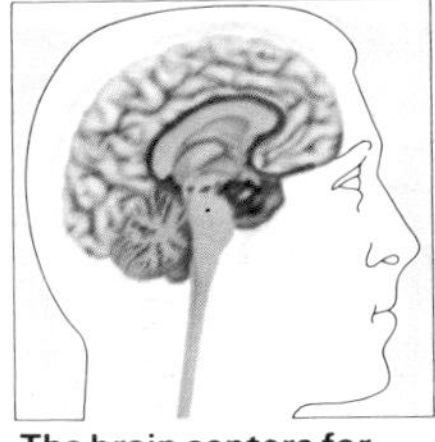

The brain centers for smell are shown (red) in this cross section of the brain.

The smell pathway

Air inhaled through the nose passes over the olfactory membranes, where chemicals that are in the air stimulate the numerous olfactory receptor cells. The smell information passes from the receptor-cell neurons to the bulbs and tracts of the first cranial nerve, which pass into the frontal lobes of the brain. Each tract divides into medial and lateral striae, which pass the information to the olfactory cortex, where smell is perceived. The exact route taken by the striae is uncertain, but two possible pathways are shown here.

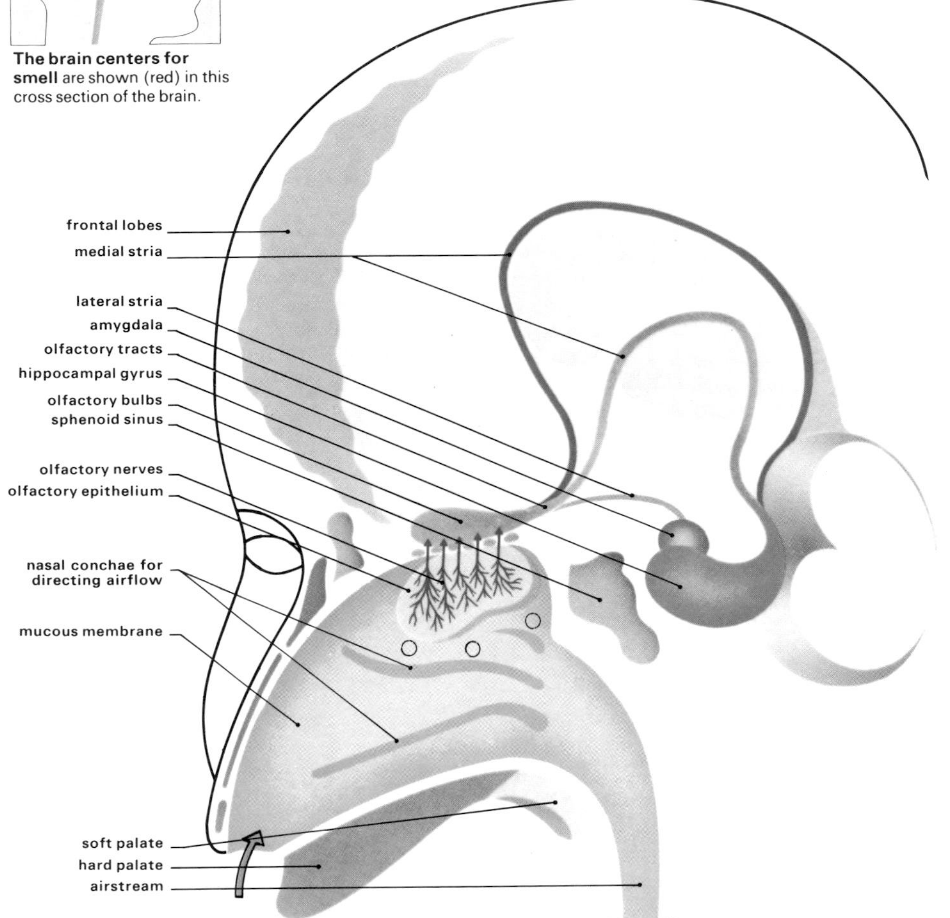

The smell membrane

The olfactory membrane is a thick yellow-brown structure, about one inch square, located in the upper part of each nasal cavity. It consists of about one hundred million smell receptor cells, which are surrounded by supporting cells. The smell receptor cells have an olfactory vesicle (**1**) bearing cilia, which project into the mucus that covers the smell membrane. Chemicals in the air react with the cilia and stimulate the receptor cells. The smell information is passed by the receptor cell axons (**2**), which leave the membrane as the first cranial nerve and which relay with mitral cell axons (**3**) to the olfactory cortex. The supporting sustentacular cells (**4**) contain a pigment that colors the membrane yellow. The membrane also contains Bowman's glands (**5**), which secrete mucus. Their function is to keep the membrane moist so that chemicals can dissolve and stimulate the cilia.

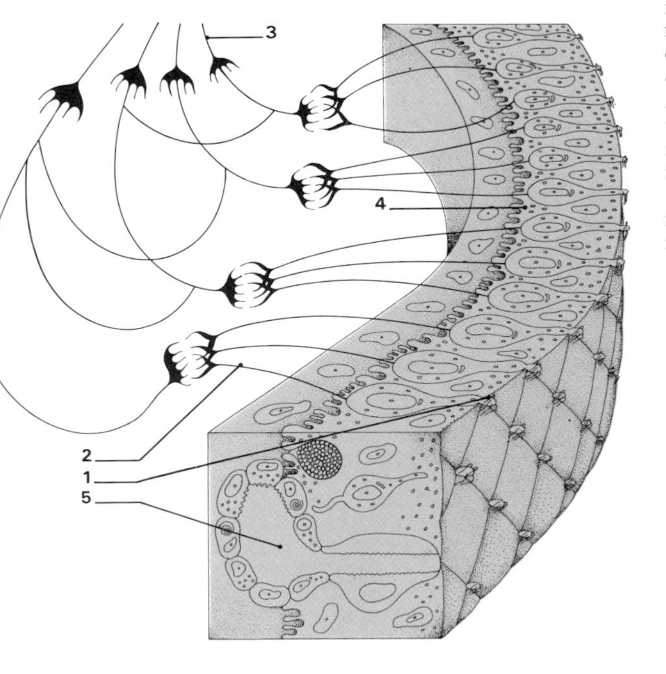

Of all our senses, that of smell is for many people the most immediately evocative. While we may struggle to "place" a face that somehow seems familiar, or to attach significance to a fragment of music heard only once before, sometimes the merest hint of an odor can trigger instant and vivid recall of a scene from the past, and revive emotions associated with it. Smell almost seems to be a more basic sense than our others, and to have a more direct link with our subconscious minds and memories.

Smell is indeed one of man's most primitive senses. In early, sea-living creatures, smell and taste were probably but a single sense, with the important, life-preserving function of monitoring the watery environment for the presence of food material and noxious substances. Many fish, with which man shares common distant ancestors, still sample the water in which they live with combined smell-taste chemical receptors scattered over their body surfaces. We, however, like most other air-breathing animals, have evolved two separate sets of chemoreceptors.

Both smell and taste depend upon the detection of molecules in solution. We can taste only substances that will dissolve in the secretions of the mouth, and can smell only volatile, airborne molecules trapped in the sticky surface moisture of the lining of the nose.

The receptor sites for smell are two small patches of olfactory epithelium, with a total area of less than a square inch, in the roof of the nasal cavities, behind the bridge of the nose and slightly above its level. This is the olfactory mucosa. It consists of thousands of closely packed cells, columnar in shape, with the highly specialized chemoreceptors for smell scattered between them. Each chemoreceptor cell has a body more or less buried among the supporting cells, but has two extensions. One reaches to the surface of the mucosa, where it terminates in minute, hairlike cilia that protrude into the thin covering film of mucus, and the other makes a connection with nerve fibers at the base of the epithelium.

Somehow—and it is still not understood precisely how—molecules of odorous substances that reach the cilia trigger off nerve impulses in the receptor cells. Chemical transmitter substances, released from the endings of nerve fibers to pass on messages to adjoining cells, are well known in the nervous system. Acetylcholine and noradrenaline are two such transmitters. Their molecules, diffusing across nerve–nerve junctions, or synapses, and nerve–muscle junctions, are thought to attach themselves temporarily to receptor sites on the membranes that they affect, distorting the molecular structure of the membrane.

Each membrane normally has an electrical charge across it—it is polarized—but now, for an instant, ions can flow through in each direction, and the electrical charge is altered. But altering the electrical charge disturbs the arrangement of neighboring membrane molecules, and so the charge spreads rapidly over the membrane. The "wave" of depolarization is a nerve impulse. It is thought that odorous molecules may cause depolarization in a similar way at special receptor sites on the surface of olfactory cells.

Even if the molecules that we smell do start off nerve impulses in much the same way as chemical transmitters do, it is still necessary to explain how we manage to distinguish between so many different odors. A person

with an acute sense of smell can discern at least ten thousand different odors, and one possibility, therefore, is that there are at least this number of different receptor sites, each triggered by only one particular kind of molecule. It is, however, extremely unlikely.

Drawing an analogy with vision, we know that there are only three different kinds of color receptor cell in the retina of the eye and yet we are able to distinguish all of the colors in the spectrum. It is likely that there are just a few different smell receptors, each of which responds to molecules of a basically similar geometric shape. Different combinations of molecules would stimulate a certain "pattern" of response in the receptor sites, and the brain might then interpret the pattern of incoming nerve impulses as being a particular smell.

One recent theory supposes that there are seven different basic types of scent—camphoraceous, musky, floral, pepperminty, ethereal, pungent and putrid. Different proportions and concentrations of these scents in the inhaled air could then give rise to thousands of quite different smells to be perceived by the brain. A strongly camphoraceous smell is given off by mothballs, for example, a pungent smell by lemons and a floral smell by roses. But a combination of these three scents would not give rise to a mental picture of the unlikely mixture of rosebuds and mothballs ground up in lemon juice, but of something completely different and unique.

Nerve impulses from the olfactory mucosa travel in fibers that pass through minute holes in the ethmoid bone of the roof of the nose and enter the cranial cavity. Here the fibers join to form the paired olfactory bulbs, and the pathways lead backward to enter the brain. Unlike the pathways for vision and hearing, which lead through the thalamus, the pathway for smell makes direct connection with the limbic system of the brain, which is known to be concerned with motivated behavior, memory and emotion. This might provide an anatomical basis for the impression that the sense of smell is somehow different from our other senses in its effect on our consciousness.

Recent evidence suggests that some odors are in fact recognized only subliminally—that is without our being consciously aware of their effect on our behavior. Certain chemicals known as pheromones are secreted by lower animals as sexual attractants. Thus some male moths are able to detect the scent of females of their own species more than five miles away and fly to them. It may well be that there are human pheromones at work, too, in our body odors, subconsciously influencing our own sexual instincts and behavior.

Compared with many other kinds of animals man has a relatively poorly developed sense of smell. We cannot imagine such a sensitivity of olfaction that would enable us, like any trained dog, to follow the trail of one particular person over miles of rough ground, or to detect the presence of drugs or explosives in luggage. Our smell receptors are not even directly in the main pathway of the air that we inhale, and unless a smell is specially obtrusive, it may take a deliberate sniffing action to intensify the stimulus before we can say that a certain substance is present or not. Nevertheless, even in human beings, the sense of smell is a remarkable phenomenon, relying on the presence of just a few molecules of a certain shape to conjure up meaningful images in our minds.

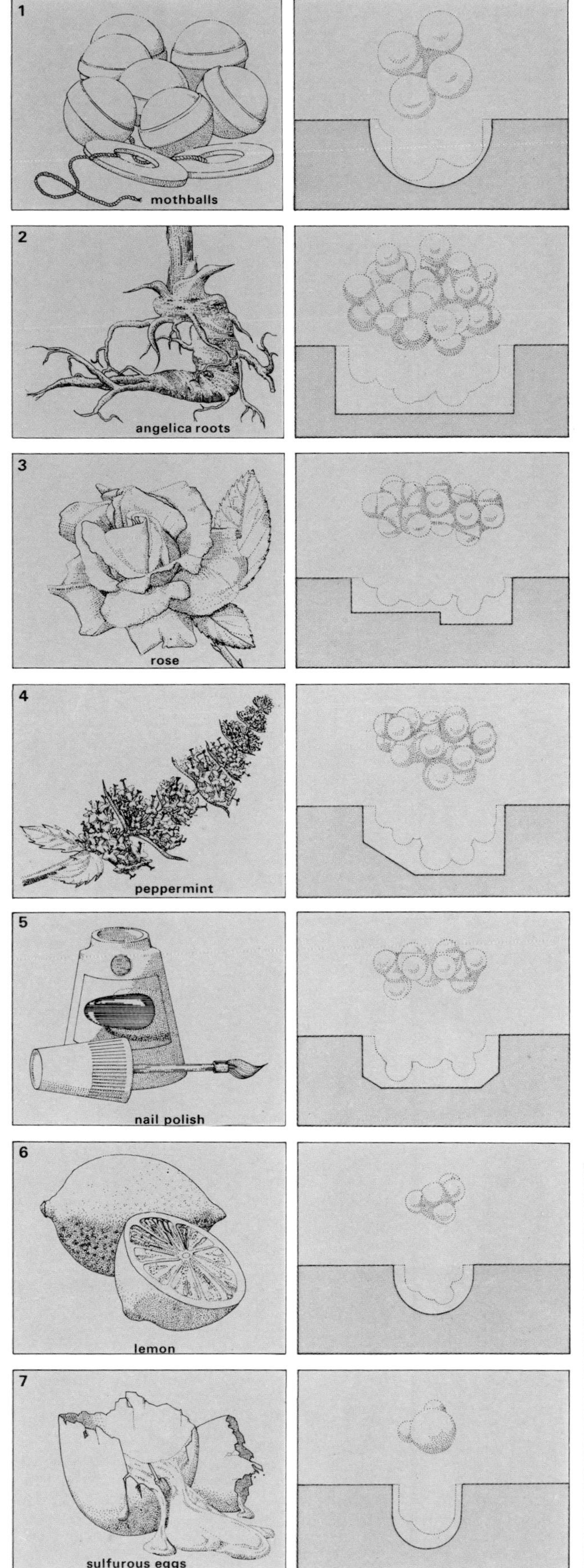

Receptors for different smells
Seven primary odors may exist from which all other smells are derived. Classification of these primary odors is difficult since it is extremely hard to achieve objective testing. One classification suggests that the seven primary odors are: (**1**) camphoraceous, (**2**) musky, (**3**) floral, (**4**) pepperminty, (**5**) ethereal, (**6**) pungent and (**7**) putrid. Substances which produce these types of odor are illustrated, in order, on the left side of the diagram. Before a chemical can be smelled, it has to be volatile to diffuse into the air. It also has to be slightly water soluble to dissolve in the mucus that coats the olfactory epithelium, and it must be fat soluble to enter the cilia of the receptor cells. The right column of the diagram illustrates the theory that the chemicals of the seven primary odors have different molecular shapes, and that there are special receptors in the olfactory membrane, each of which is able to accommodate only one of these shapes using a lock-and-key principle. When combinations of these receptors are stimulated, all of the other secondary smells are produced.

Cross-reference	pages
Brain and central nervous system	**74-77**
Limbic system	**116-117**
Sex drive	**114-115**

Touch, temperature, pressure and pain

When we put on clothes in the morning, for a while we are aware of their presence, as the sensory nerves of touch signal information to the brain about the extra weight and warmth, the texture of the material, and whether the garments fit loosely or are, perhaps, painfully tight. After a few minutes, these sensations no longer reach our consciousness, and throughout the day we are, for the most part, unaware of the presence of clothes, except for occasional reminders—or more frequent ones if the clothes are uncomfortable.

The sense of touch brings from the surface of the body a wide variety of information about the immediate vicinity. Tactile, or touch, receptors, in the skin covering the whole body, are triggered by different stimuli. Their activation gives rise to the sensations of fine touch, pressure, heat, cold and pain. Together with the receptors that are in tendons and joints, which give information about the position and movements of parts of the body, they form the somatic receptors. ("Somatic" means to do with the body wall or the framework of the body.) Until recently, it was not certain whether there were specific receptors for the different kinds of stimuli, but there do appear to exist distinct receptors for heat, cold, fine touch, pressure and joint position. The existence of specific receptors for the sensation of pain is, however, in doubt.

Unlike the receptors for our other senses, those for touch have to transmit signals a fairly long distance to the brain. While our eyes and ears, for example, feed in information through nerves that are only inches long, the information from our toes must pass up long nerve fibers to the spinal cord, and from there be relayed to the brain stem and the thalamus, before reaching the outer layer of the brain. This is a journey of several feet.

The sense of touch serves to provide a warning of danger for the body. Anything which causes a sensation of pain is also likely to cause damage to the body's tissues, and immediate action is called for to avoid this. Although the signals in nerve fibers move rapidly, it is obviously better if any waiting time between receiving a painful stimulus from, say, a burned toe and withdrawal of the foot can be reduced. Painful stimuli, therefore, tend to bring about reflex responses.

In such a reflex response, a message from the toe reaches the spinal cord, and from here messages are passed directly to the muscles of the lower leg, which bring about withdrawal of the foot before in fact the brain becomes aware of the pain. Such a "short circuiting" action is typical in the case of pain.

The region of the brain where sensations of temperature, pain and touch are perceived is the sensory cortex of the parietal lobe. There is a crossover in the pathways of sensation from the body before the brain is reached. The sensory nerves cross either in the spinal cord or in the brain stem, so that sensations from the left side of the body reach the sensory cortex of the right cerebral hemisphere, and those from the right side of the body are registered in the left cortex.

Each region of the body sends signals that reach a fairly well-defined area of the sensory cortex, and as a result of experiments it has been possible to map out the cortex to show the area of the body represented. The toes are represented by an area on the top of the brain, nearest the fissure between the left and right hemispheres. Passing to the side and downward are the terminations of the pathways from the leg, trunk, arm, hand, face, tongue, throat and the internal organs. The parts of the body with the greatest sensitivity are represented by the largest areas of sensory cortex, so that the fingers, thumbs and lips are each represented by almost as large an area as the whole of the rest of the body put together. The strip of sensory cortex in one hemisphere is duplicated in the other hemisphere.

The fact that the sensations we get from wearing clothes seem to recede into the background after a short while is in part due to adaptation of the brain, as the monotonous influx of information about roughness of trousers, tightness of waistband and so on continues. Such information is interesting, but does not warrant much attention. It remains simply as background information, much like background music. In part, too, the diminished awareness is due to adaptation of

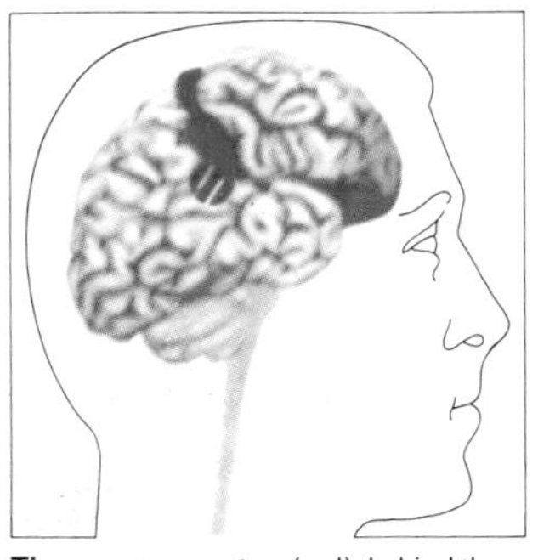

The sensory cortex (red), behind the central groove of the brain, is the primary area for the interpretation of incoming sensory information. The association area (striped) coordinates this information.

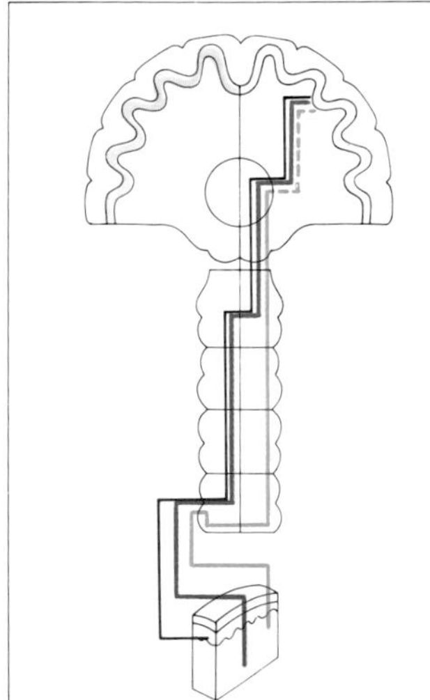

Some sensory pathways begin at the skin and pass upward, in tracts, or bundles of nerve fibers, within the spinal cord. The pathways cross in the brain stem and pass to the thalamus and sensory cortex, where sensations are perceived. Pain and temperature fibers (blue) run together, but touch and pressure (black and red) run separately, in different tracts.

The sensory homunculus

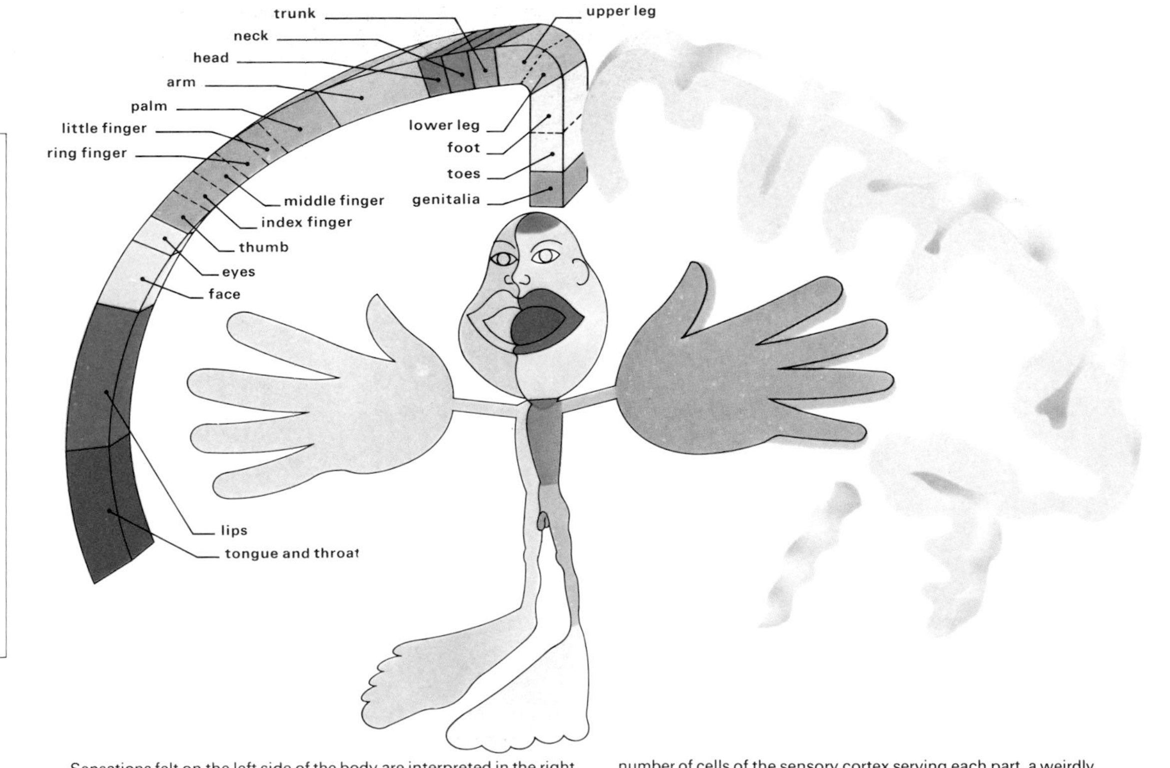

Sensations felt on the left side of the body are interpreted in the right sensory cortex and vice versa. Each part of the body is represented by a specific area on the surface of the cortex. The more sensitive the part of the body, the greater the area of sensory cortex needed to interpret its messages. If the parts of the body are drawn in proportion to the number of cells of the sensory cortex serving each part, a weirdly disproportionate man results, as shown in this sensory homunculus. The colored areas on the diagrammatic representation of the right sensory cortex correspond in size to the colored areas which they serve on the left side of the body.

the sensory receptors themselves, which stop sending so much information unless something different happens to them.

Microscopic study of the skin has revealed several different types of nerve ending. Some sensory fibers have free nerve endings in the epidermis and dermis of the skin, and have no specialized structure surrounding them. Free nerve endings are widespread over the body, and are believed to be responsible for sensations of pressure and pain. They are found in both hairy and nonhairy skin. In hairy skin the endings are close to the shafts and to the roots of hairs, which act like levers so that small movements of the hairs trigger off nerve impulses.

The other types of nerve ending all have receptors of a recognizable structure, called end organs. The ones about which most is known are the Pacinian corpuscles, found in nonhairy skin and particularly in the fingers. One finger may have as many as a hundred of these corpuscles, which consist of alternating layers of cells and fluid, structured rather like the skins of an onion. These are sensitive to pressure. Because pressure must be transmitted through the layers of cells and fluid, a rapid stimulus is more likely to trigger a nerve impulse—with only slowly increasing pressure, the effect is dissipated in the fluid before it can affect the nerve ending. Pacinian corpuscles are especially sensitive to vibration.

The oval-shaped Meissner corpuscles, found in the ridges of skin, are believed to be mainly concerned with touch, too, and the leaflike Merkel's disks are possibly variations of them, with the same function. The end bulbs of Krause contain loops of fibers like a loosely wound ball of wool, and the end organs of Ruffini have branching fibers with flattened ends. The end bulbs of Krause are believed to detect cold, and the end organs of Ruffini are thought to detect heat. The heat receptors may possibly be triggered by distortion of protein molecules in their nerve endings.

Pain is detected by any end organ that is stimulated excessively, and also by free nerve endings. The essential difference between pain and the other sensations is that it always produces a reaction, whether physical or emotional or both. The surface of the body varies greatly in the distribution of receptors. These have been mapped out experimentally on the skin using a needle and a hot or cold stylus to plot "spots" corresponding to the different sensations. In the fingers it is possible to distinguish points only one-tenth of an inch apart, while on the thighs the points must be separated by about three inches before they give rise to separate sensations of touch.

Reading by touch

A blind person using the Braille system relies upon the sense of touch to read. To distinguish sixty-three different combinations of small raised dots, of which the Braille alphabet is composed, a highly developed sense of touch is required. Sighted people can rarely achieve the same level of touch discrimination in their fingertips because this sense is not as important.

Sensory receptors in the skin

The sense of touch involves the stimulation of receptors in the skin. The most superficial receptors are found in the epidermis. They are the free nerve endings, which respond to touch and pain; Merkel's disks, which respond to continuous touch; and the end bulbs of Krause, the receptors that probably register cold. Below the epidermis is the dermis. In the dermis are Meissner's corpuscles, which send information to the brain about the texture of the object that is being touched; nerve endings around the bases of the hair follicles, which are stimulated by movement of the hair; Pacinian corpuscles, which respond to pressure; Ruffini corpuscles, which respond to changes in temperature.

End bulbs of Krause are believed to respond to cold. Each consists of a thin capsule which contains a nerve-fiber terminal.

Free nerve endings are in the epidermis and also, to a lesser extent, in the dermis. They are believed to be responsible for conducting pain and touch information.

Pacinian corpuscles, stimulated by pressure, are concentric rings of capsule cells surrounding central cores which contain nerve terminals.

Ruffini corpuscles are a flattened network of nerve fibers enclosed in sheaths of special capsule cells. They are thought to respond to heat.

Merkel's disks consist of a sheath enclosing a biconvex disk which is connected to a nerve fiber. They are stimulated by continuous touch.

Meissner's corpuscles are in the surface areas of the dermis. They are made up of many whorls of nerve-fiber endings in a sheath.

Hair-follicle nerve endings respond to movements of the hair when it is displaced by anything coming into contact with the skin.

Cross-reference	pages
Brain and central nervous system	74-77
Brain stem	104-105
Cerebral cortex	80-81
Cerebral hemispheres	124-127
Consciousness	106-107
Skin	28-29

The cerebellum: monitor of movement

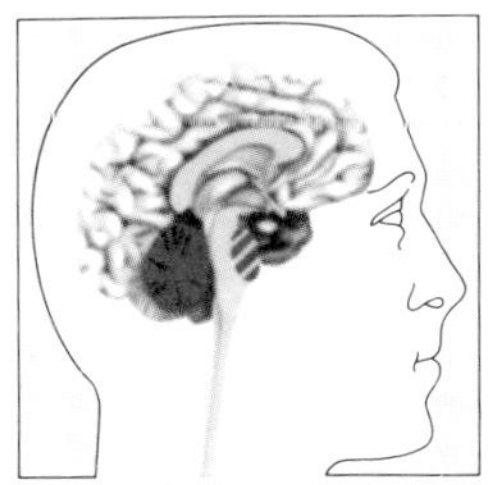

The cerebellum, shown in this cross section in red, lies below the occipital lobes of the brain's cerebral hemispheres. The striped area indicates where the cerebellum joins the pons on the brain stem.

The anatomy of the cerebellum

The cerebellum is part of the hindbrain and is concerned with the regulation of coordination and balance. It is an oval-shaped structure with a central furrow, which is flanked on either side by a bulbous lobe. The furrow is called the vermis and the expanded lobes the hemispheres. The surface is covered with folds, or folia. In the illustration below, a section through a folium has been drawn out and cut through to show its internal structure. The outer region of the cerebellum is known as the cortex, or gray matter. It contains two distinct layers of nerve cells, the molecular layer and the granular layer, which are separated from each other by a layer of Purkinje cells. Beneath the cortex lies the white matter, which is composed of nerve fibers, mossy fibers and climbing fibers that pass in and out of the cerebellum.

Eristratus, who lived in Ancient Greece, is said to have been the first man to claim that fast-running animals, such as the hare or deer, had a better-developed cerebellum than their less-active brethren. In fact, this is not true. The cerebella of the deer, ox and sloth are all similarly developed. What is true, however, and what Eristratus was really saying, is that the cerebellum is intimately connected with the control and organization of movement and balance in all animals.

The cerebellum is situated at the back of the brain, above the brain stem and below the cerebral cortex. In man it is most highly developed, and accounts for about 11 percent of the whole brain's weight. From birth until the age of two, the cerebellum grows much faster than does the cerebral cortex, so by this age it has almost reached its adult size. It is during this period of the child's life that the basic movements are being learned. These movement patterns are filed away in the cerebellum and are then, throughout life, called upon by the brain whenever a movement is to be made.

The cellular structure of the cerebellum is quite different from any other part of the brain. Here the cells are arranged in a precise and ordered fashion, in a rectangular matrix unlike, for example, the cerebral cortex, where their pattern is less precise. Not only is the structure of the cerebellum different but its function is also unique. It is the only major group of cells within the central nervous system whose function is to inhibit rather than to excite, to restrain rather than to act. And it is this restraining action of the cerebellum that organizes the flow of movements initiated by the cerebral cortex and ensures that they proceed in an orderly and effective way.

The cerebellum acts like a computer. Because of its precise structure and its inhibitory character, information coming into the cerebellum is dealt with and cleared within one-tenth of a second. There is no chattering between the cells as in the cerebral cortex. The cerebellum receives information, computes the answer and sends out the result as a command to stop some part of the action which is proceeding wrongly. It continuously monitors and adjusts the progress of an action.

The motor cortex of the cerebral hemisphere initiates an action. It says, for example, lift the glass. Impulses then flow from the pyramidal, or output, cells of the left motor cortex, down their output fibers, which cross over in the brain stem, where they relay and send branches to excite the cells in the cerebellum in the opposite side of the head. Thus, the cerebellum, within one-fiftieth of a second of a movement being initiated, knows all about it.

By the use of memory of how the movement is to proceed, it quickly computes the changes which are required in the cortical output and alters the way these cells discharge. At the same time as this is proceeding, the muscles themselves are beginning to move the arm, and information from sense receptors in the arm about the degree and extent of the movement are relayed back via the spinal cord to a group of cells in the brain stem which fire into the cerebellum. The cerebellum again compares this input with the blueprint pattern already extracted from memory, computes the necessary changes and quickly modifies the ongoing activity in the muscles by sending messages to change the action of the nerve cells which control them.

In this way, the cerebellum continually monitors the intentions of the motor cortex and alters its pattern of discharge so that the right movement is achieved, measures the actual progress of the movement and finely adjusts the range of the movement so that the arm swings smoothly out from the side of the body, the fingers grip the glass, neither too tightly nor too loosely, and the glass is then swung back and up to the mouth without any liquid being spilled. This is a fantastically complex maneuver and its execution requires intricate feedback of all stages as they occur.

Although it is easy to understand the function of the cerebellum when a movement is being made, its action is just as important in sitting or standing. Information is continually streaming into the cerebellum, not only from the motor cortex, spinal cord and muscles but also from the organs of balance, which are situated in the labyrinth of the inner ear. These organs provide information about position and changes in position of the head.

The labyrinth is made up of these semicircular canals set at right angles to each other in three different planes. The canals are filled with endolymph, a watery fluid which, whenever the head is moved, swirls against tiny hairs, which are the sense organs in the canal, so that they are stimulated to send information about rotation and change of position of the head to the cerebellum. Similar receptors in the jelly-filled utricle and saccule of the labyrinth are stimulated by the movements of the otoliths, minute crystals of calcium carbonate that move under the influence of gravity, giving information about the tilt of the head.

All this information flows into the cerebellum and allows it to continually modify the activity in the main motor pathway so that the body is held balanced in an upright position. Faulty information from these organs results in the feelings of vertigo, when the room appears to spin although the body is quite still.

It may seem surprising that the cerebellum acts by stopping or by inhibiting rather than by exciting. However, even when the body is apparently still, constant motor activity continues to keep it upright and maintain muscle tone. It is on this ceaseless barrage that the cerebellum acts by stopping activity here, allowing more there, changing the overall pattern, so that it can rapidly and efficiently control the ongoing pattern of movement.

The cerebellum can be compared to a woodcarver who sits down to carve a model from a block of wood. He chips wood away from the top, shaves it from the middle, files and scrapes away until, by removal of wood, the model is created. In the same way the cerebellum stops and inhibits and so alters the ongoing activity until the desired movement is produced.

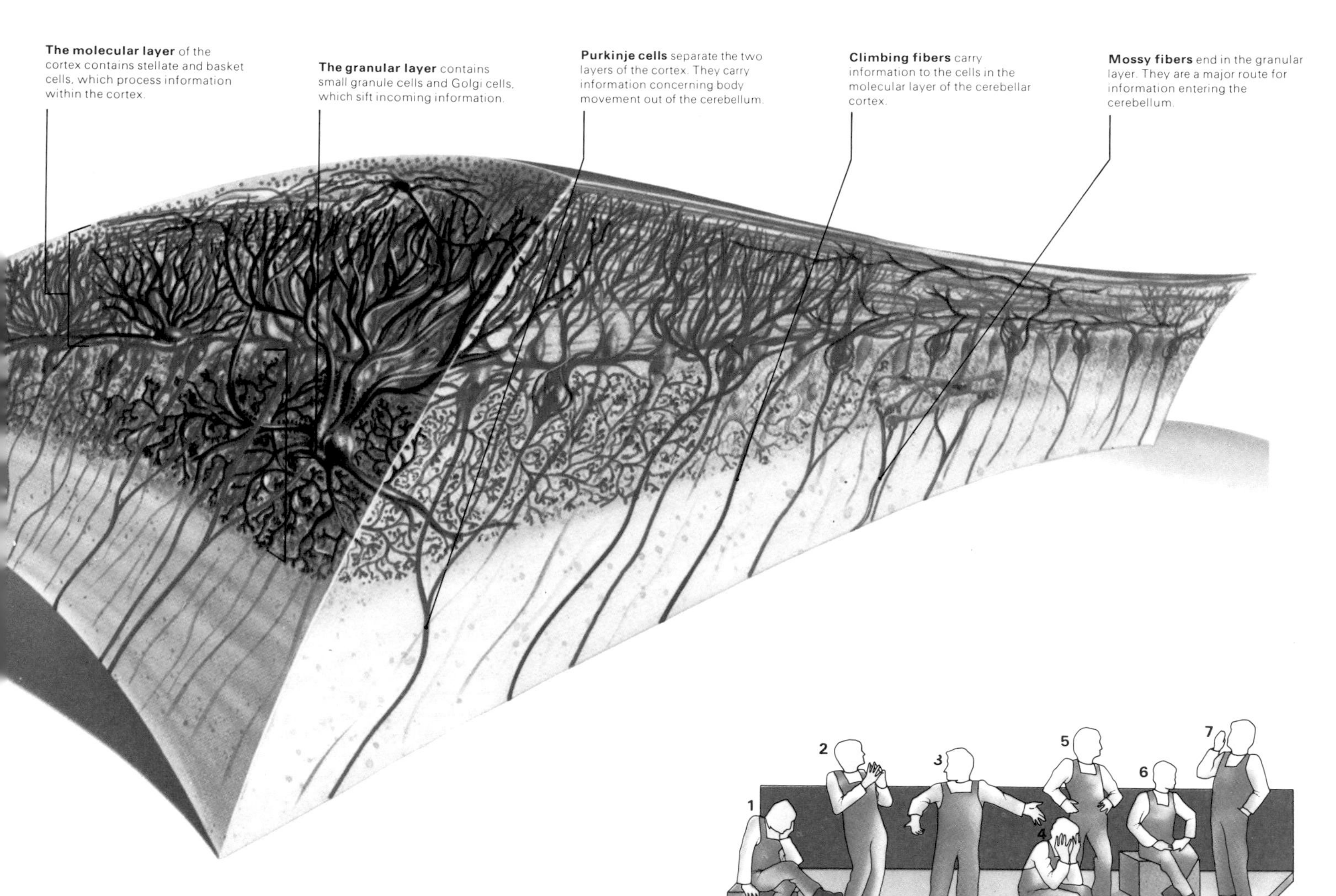

The molecular layer of the cortex contains stellate and basket cells, which process information within the cortex.

The granular layer contains small granule cells and Golgi cells, which sift incoming information.

Purkinje cells separate the two layers of the cortex. They carry information concerning body movement out of the cerebellum.

Climbing fibers carry information to the cells in the molecular layer of the cerebellar cortex.

Mossy fibers end in the granular layer. They are a major route for information entering the cerebellum.

Body postures, or gestures, both conscious and unconscious, communicate human feelings. The illustration shows seven of these postures. The head-in-hand posture (**1**) may communicate boredom, while the arched back (**2**) can indicate self-confidence. An open-handed gesture (**3**) reinforces openness and sincerity. Guilt or shame are demonstrated by the face in the hands (**4**). The hands-on-the-hips posture (**5**) is another gesture of self-confidence. Sitting straightbacked with head erect (**6**) gives an impression of interest and attention. Rubbing the nose (**7**) is a gesture of doubt. These postures, whether made consciously or unconsciously, involve the nerve pathways concerned with posture, coordination and equilibrium, regulated in the cerebellum.

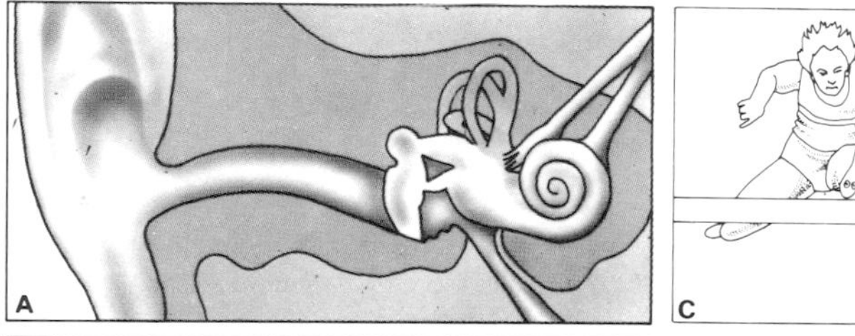

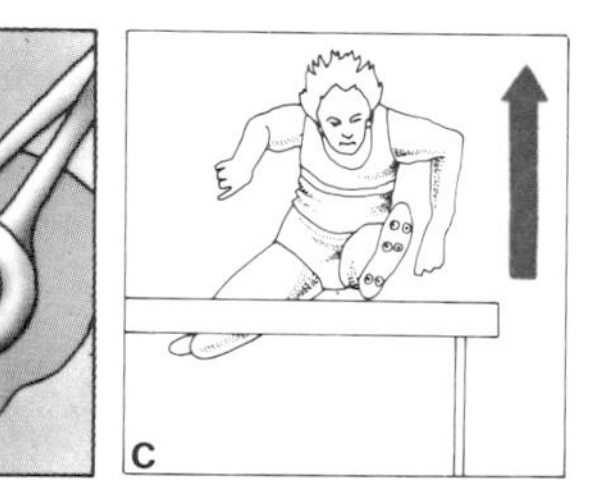

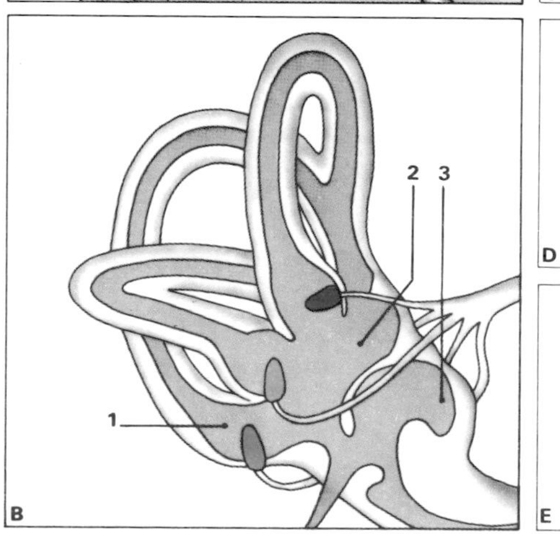

The organs of balance, which influence the action of the cerebellum, are located in the inner ear adjacent to the organs of hearing. (**A**) The position of the three semicircular canals and the saccule, the utricle and the endolymphatic duct are shown in relation to other structures in the inner ear. (**B**) A cross section of the semicircular canals, which consist of three tubes, positioned at right angles to each other in different planes. They are filled with a fluid known as endolymph. In the base of each semicircular canal (**1**) are hairlike receptors (colored red, green and blue). Any movement of the head causes the fluid within the canals to move and so stimulate the receptors to send information to the brain about the direction of the particular movement that has occurred. The utricle (**2**) and saccule (**3**) contain receptor cells with tiny stones of calcium carbonate—otoliths. The otoliths are relatively mobile and movement of the head causes them to move. This changes the pressure that they exert on the receptors, which then send information about the direction of the head's movement to the brain. (**C**), (**D**) and (**E**) illustrate the three planes of movement in which the head and body can move to stimulate the receptors in the semicircular canals.

Cross-reference	pages
Brain and central nervous system	**74-77**
Cerebral cortex	**80-81**
Ear and balance	**88-89**
Motor cortex and movement	**100-101**

The motor cortex: coordination of movement

Man has been liberally endowed by nature, not only with the ability to perform a wide variety of such movements as running, jumping, climbing and skipping, but also with the skill to perform delicate and intricate movements such as those which are needed to assemble a watch or thread a needle. The main center for the planning, initiation and execution of all the body's movements is the motor cortex.

There are, in fact, two motor cortices, one in each hemisphere. They are situated behind the frontal lobes of the brain, below the crown of the head. Each one controls the movements of the opposite side of the body, because the nerve fibers which leave it, carrying the messages for the control of movement, cross over in the brain stem before they enter the spinal cord.

Each motor cortex is in the form of a strip that runs down from the middle of the head to end in the sulcus, or groove, which separates the frontal lobe from the temporal lobe. This strip acts like a large terminal block, collecting information about the planning and co-ordination of movements from the areas of the cortex in front of it, and then assembling this information and sending it out down the main motor pathways to initiate movement.

The number of motor fibers originating in this area depends on the parts of the body which they supply. The greater the body's need for fine control of movement, in the hands for example, and the lips, the greater the number of fibers supplying this area, and so the larger the number of nerve cells in the motor cortex. Areas like the thighs and calves, needing little control and few fibers, are represented by a small number of cells.

The fibers that control movements originate in giant, or Betz, cells, the largest cells in the brain. These cells are organized in a definite pattern, so that movements of the foot and ankle are controlled from the top of the strip and then, proceeding down toward the sulcus, the leg, thigh, stomach, chest, then shoulder, arm, forearm, hands and fingers, then the neck, eyelids, lips and tongue are represented.

It is important that each side of the body is able to move independently of the other. Consequently, very few fibers cross the mid-line to carry information to the opposite motor cortex. The movements of lips, tongue and face used in speech are controlled by a special motor speech center situated behind the motor cortex which has a rich connection with other parts of the brain which deal with the generation of speech.

More important than the initiation and execution of movement, however, is the planning of complex actions. In front of the motor cortex lie two more strips, the premotor area, the first of which is involved in the construction and integration of complex patterns of movement. It is here that the movements needed to carry through a sequence of actions are put together and then transferred stage by stage to the motor cortex for execution. It is very important that during purposeful action both sides of the body should be coordinated and so large numbers of fibers cross over from one hemisphere to the other, carrying information about the synthesis and control of both sides of the body. Although we think of movements and action as being important, equally important is the ability to stop and be still. The inhibition of movement occurs in the second premotor strip.

In front of the premotor area lies the main body of the frontal lobe. Here judgment and planning take on a quite different dimension. These areas do not deal simply with the elaboration of schemes of movement, but with higher order systems whose function it is to produce schemes of schemes. The higher functions of man—planning, judgment and control of personality—reside in this area. Here there is control of behavior, rather than just motor control, integration of the organism, rather than just integration of movement.

The motor cortex, although it initiates and controls the sequence of voluntary movements, can only act if the body has been prepared to move. This preparation is the very basis from which movement can take place, and it is organized in different levels of complexity. At the simplest level is the spinal reflex, controlled only by nerve cells in the spinal cord, which ensures that as one group of muscles contracts, the opposing set relaxes.

At the next level, cell groups within the brain stem and midbrain act to keep the muscles in a state of readiness. This system maintains muscle tone, so that the muscles are neither too floppy nor too stiff to move easily. A further and more complex system exists deep within the forebrain, where the basal ganglia in the thalamus coordinates and ensures that as each movement takes place it progresses smoothly. It is the harmonious interaction of these three systems and the cerebellum, the organ which monitors and controls the progress of each movement, which makes our movements smooth and controlled and enables us to be still, neither jerking, trembling nor twitching when at rest.

All except the simplest movements are learned. We spend the first few years of our lives setting up complex memory banks of movements which we can later call upon and

Voluntary movements and the brain

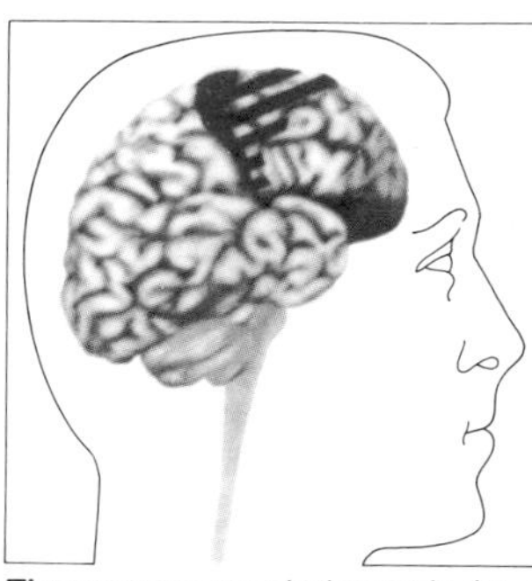

The motor cortex is the area in the brain which issues orders for body movement. The primary area is colored red, the premotor area is striped.

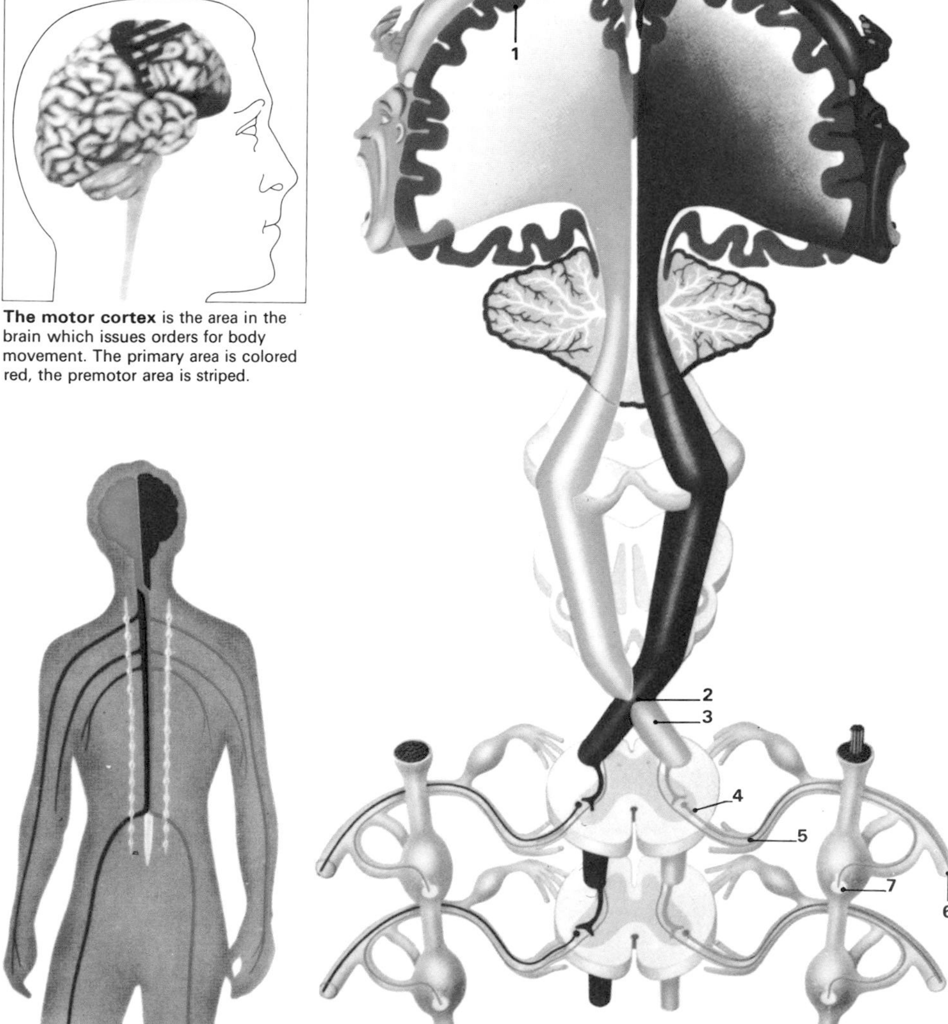

The motor cortex controls the voluntary muscle system of the body. In the diagram above, the left motor cortex is red and the right motor cortex is blue. The nerve impulses pass from the cortex (**1**) and travel into the brain stem, along bundles of nerve fibers which cross in the medulla oblongata (**2**). Impulses from the right side of the brain thus control the left side of the body, and vice versa. The nerve impulses then pass from the medulla oblongata to fibers in the spinal cord (**3**) and leave at each section (**4**) as the motor roots (**5**) to form the spinal nerves (**6**), which transmit the impulses to the muscles. Involuntary muscle movements are controlled by the autonomic nerves (green), which leave the motor root and then pass to nearby nerve chains (**7**).

use. Many of these movements are transferred straight from memory to the motor cortex and never come into consciousness. These movements, such as walking, become habitual and the programs which produce them are readily available from memory.

When a new skill is learned, however, conscious control is required. Each aspect of the skill has to be learned and practiced, and it is not until it has become habitual that it falls below the level of consciousness. In driving a car, for example, the movements of gear changing, turning the steering wheel, depressing the clutch, have all to be thought about, leaving little attention for looking at the road. Once these skills have been learned they occur subconsciously, without any conscious participation, leaving the attention free to concentrate on planning and judgment of the traffic conditions.

The motor cortex carries out the movements ordered by "will," but it is powerless to act unless the body has been prepared to move by the subconscious action of many different systems deep within the brain.

Before movement is carried out, the cells in the motor cortex which control the muscles involved come into action. The complex code required for the movement is elaborated in the motor cortex and is then transferred down the spinal cord, where it is further modified, and then out to the muscles themselves.

Recording electrodes placed over the motor cortex are able to pick up this initial activity before the movement takes place. If there is damage within the spinal cord, so that the pathway from the motor cortex to the muscles is interrupted, these early motor potentials still occur although no movement takes place. Scientists are investigating the possibility of initiating movement in such cases of spinal damage by picking up the electrical activity which occurs in the cortex at the beginning of a willed movement, amplifying it and then transmitting it through wires to the effective muscles. This work is still in its early phases, but the chief problem lies in making an adequate connection between the wires carrying the messages and the muscles. There is still a long way to go before this can be achieved, but this early work holds out promise for some groups of people who now spend their lives paralyzed because of a damaged motor pathway in the spinal cord.

The motor homunculus

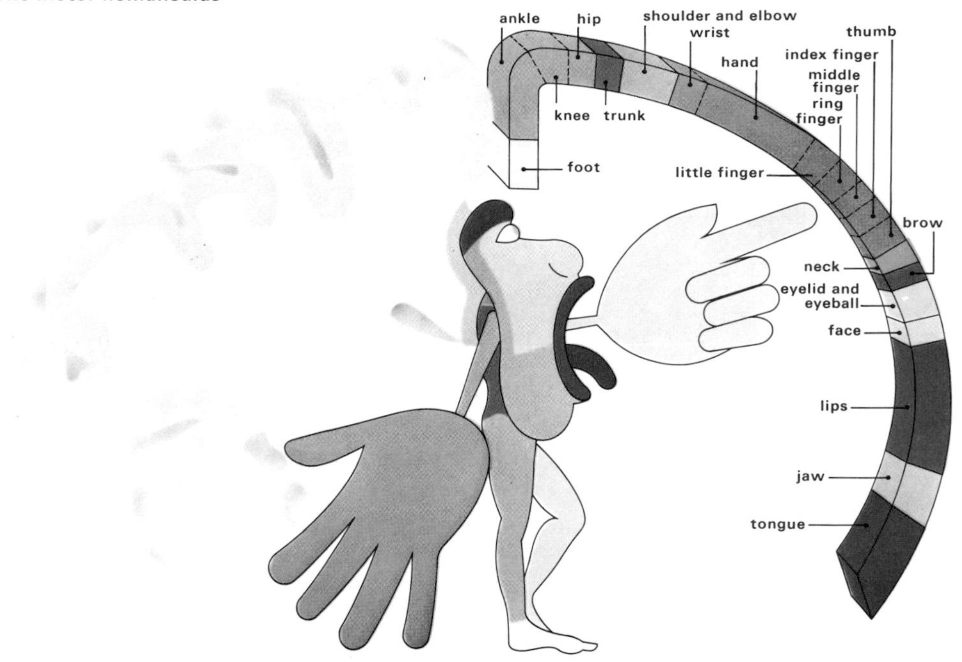

The diagram shows the left motor cortex divided into different-colored segments which control the corresponding colored parts of the right side of the body. The parts of the body of this man are drawn in proportion to the number of cells in the area of the motor cortex serving each part. Parts of the body which make very fine movements, such as the hands and fingers, are controlled by large areas of cortex, while parts like the trunk need only small areas. The right motor cortex controls the movements of the left side of the body in exactly the same way.

Unconsciously used skills are developed by learning processes. Most people can pour water into a glass, but if the water jug is held a few feet above the glass only someone who has practiced will succeed, at the first attempt, without any spillage. The expert's brain issues a general message from the motor cortex, which triggers the memory and the cerebellum, where movement patterns are stored. The novice does not have this previous experience and so his brain has no program for this learned skill. He, therefore, has to consciously guide his movements. He will usually miss the glass several times, since his actions are being controlled by several separate orders from the motor cortex and not, as in the case of the expert, by one coordinated instruction.

The final program for action

Many body actions learned during childhood are instinctive and require little conscious effort to be achieved. Other body actions require a particular skill which can be attained only through practice. The power of the hurdler, the poise of the ballet dancer and the coordinated strength of the pole vaulter all demonstrate the special blend of skills, combining power, agility, balance and purposeful muscular control, which are necessary to achieve these heights of skillful performance.

Cross-Reference	pages
Brain and central nervous system	74-77
Cerebellum	98-99
Cerebral cortex	80-81
Cerebral hemispheres	124-127
Consciousness	106-107
Frontal lobe	122-123
Learning	128-129
Memory	130-131

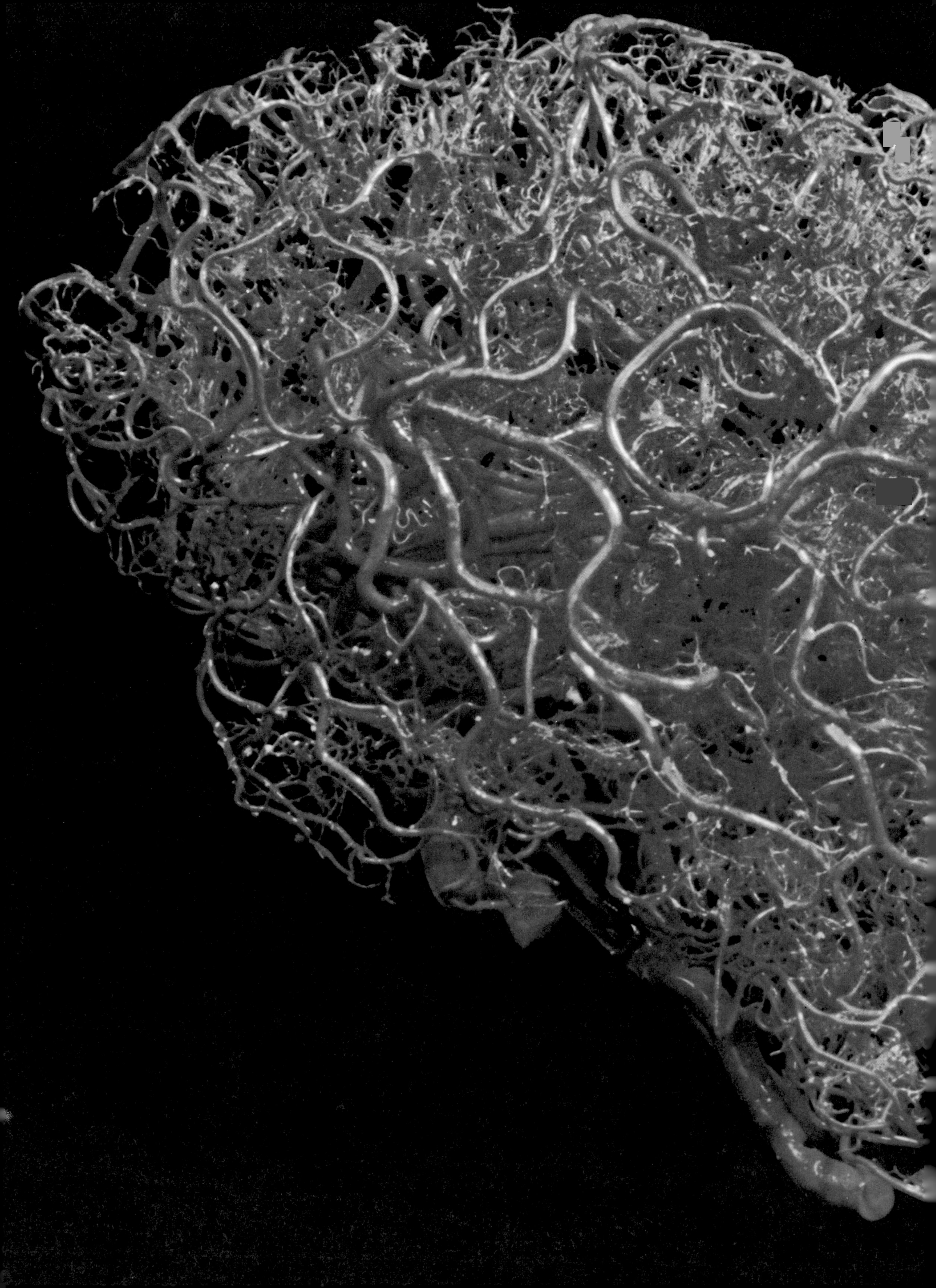

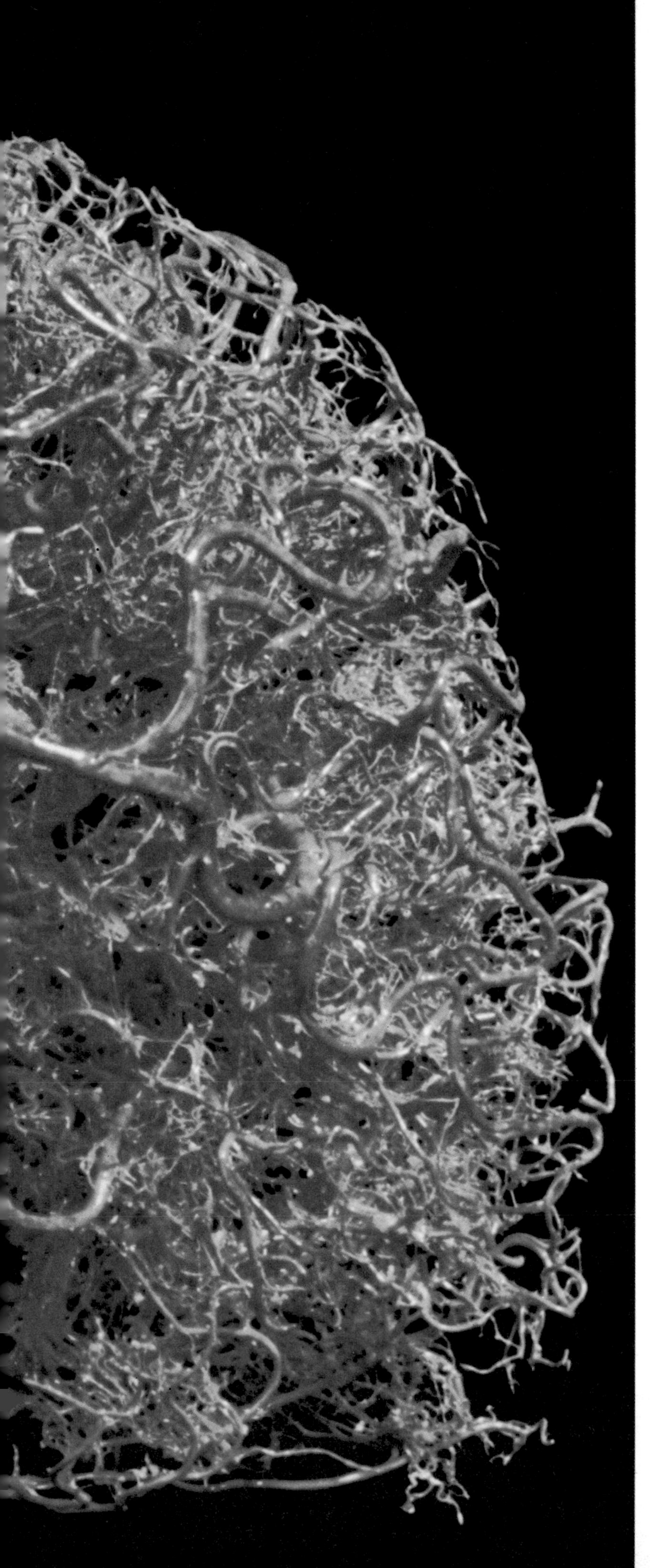

The brain

When asked to describe how they thought the brain worked, most people would inevitably reply or imply "like a computer." This, of course, is getting things rather the wrong way around. After all, which came first—brain or computer? If an analogy is to be drawn, then we can really only say that to a very limited extent man has managed to imitate just a few of the functions of his own brain by constructing computers.

Brains have been in existence for many millions of years, but it is only within the last few decades, by experimenting with increasingly complex arrays of disks, tapes, punched cards, electronic gadgets and electrical circuits, that we have been able to make machines with memories akin to our own. Now we can build computers with powers of calculation that far outstrip our own when it comes to speed, and computers that can even play chess. But we cannot build computers that display mixtures of emotions, can delight in their own abilities or fall in love.

The human brain, like that of all vertebrates, can be divided into three regions—the forebrain, the midbrain and the hindbrain. The forebrain is made up of the two cerebral hemispheres and such interior structures as the thalamus and the hypothalamus, and the midbrain is the intermediate zone between the forebrain and the hindbrain. The hindbrain, in turn, leads into the spinal cord, and so connects ultimately with the network of nerves that makes up the peripheral nervous system.

The cerebrum forms a larger proportion of the brain in man than in any other animal. Its thin outer layer, the cerebral cortex, intricately folded, is the gray matter. Here lies the secret of man's intellect, and the intelligence and mental skills that elevate us from being mere animals into being a superior and dominant species. But the brain is as much concerned with directing the "lower" functions of the body—those, like breathing and eating and excreting, that merely preserve life—as it is with the "higher" functions of remembering, learning and reasoning. There are whole sections of the brain that look after basic bodily functions with the minimum of direction from the thinking brain, although emotion and willpower modify their activities in numerous subtle ways.

As well as monitoring and directing the body's activities by means of electrical nerve impulses, the brain also exerts control through the endocrine system and the circulatory network. The pituitary gland is sometimes described as having half nervous and half hormonal functions. But this is rather simplifying the view of the brain as a whole, because the entire working of the brain is a complex mixture of electrical and chemical changes anyway, and they cannot be separated. Chemical compounds are involved in the transmission of every single nerve impulse in the nervous system. If we are to talk of the brain as working "like a computer," we should not forget that it also works "like a laboratory."

Possibly the cells in the body most sensitive to oxygen deprivation, brain cells cannot survive for more than a few minutes without freshly oxygenated blood. Seen left, twice life-size, is a resin cast of the blood vessels that supply the brain with vital nutrients.

The brain stem: life support

All the functions of the brain do not have equal importance. Breathing, and the control of heart rate and blood pressure, for example, are vital for the survival of the body, and assume a greater degree of importance than do other, less-essential mechanisms, such as speech, sight or hearing. In man's evolution it was these basic, vital functions which developed first and whose control is located in the earliest part of the brain—the brain stem.

The brain stem is the link that joins the newer parts of the brain, the cerebral hemispheres, which create man's intellect and will, and the thalamus and limbic system, which organize his moods, memory and appetites, with the most primitive part of all, the segmental nervous system within the spinal cord. Through the brain stem run the major sensory and motor pathways conducting information to and from the brain. It is here that these pathways cross over so that each half of the brain controls the opposite side of the body.

Deep within the cerebral hemispheres of the brain lies a vascular plexus, the choroid plexus, which secretes the clear cerebrospinal fluid. This fluid collects in the cavities of the cerebral hemispheres, the lateral ventricles, and then passes out into a midline cavity, the third ventricle, from where it flows down through a channel, the aqueduct of the midbrain, to enter the fourth ventricle, situated over the brain stem. From here it flows out to bathe the surface of the brain and down to surround the spinal cord. Sometimes, after an infection such as meningitis, or due to an abnormality at birth, the holes in the brain stem, through which the cerebro-spinal fluid flows out from the ventricular system, become blocked. The pressure inside the ventricles of the brain increases, causing the brain to swell.

Deep within the brain stem is the reticular formation, a latticework of short-fibered nerve cells, within which lies the basic life-support systems. Here are the centers which control breathing, heart rate and blood pressure and, equally important, the level of consciousness. The cells controlling consciousness form a chain which extends through the center of the brain stem to the midbrain, from where their influence spreads out to affect the activity of both the right and left cerebral hemispheres.

Nervous pathways spread out from the thalamus, the deep core situated in the middle of the cerebral hemispheres, to all parts of the cerebral cortex. These fibers, when stimulated, are able to excite large groups of cells so that the cortex can start processing information in a detailed way. It is this information processing which is the very basis of consciousness. Within the cortex, information from the senses and from different areas of the brain are collated, altered, transformed and finally transmitted to different areas of the brain. Here, together with the information produced from memory and impulses conveying information about emotional and feeling states from the limbic system, they synthesize the private world of our consciousness. Changes in the level of excitation of the reticular formation determine whether we are asleep, awake or unconscious. For even in unconsciousness

The brain stem (red) consists of the pons and the medulla and forms part of the hindbrain. It lies beneath the cerebral hemispheres and in front of the cerebellum.

The anatomy of the brain stem

The brain stem, which is only two and a half inches long, is seen here cut from top to bottom and laid open, in a similar way to which a book is opened. The central dark area is the reticular formation. The left side shows the motor nuclei of the cranial nerves and the right side illustrates their sensory nuclei.

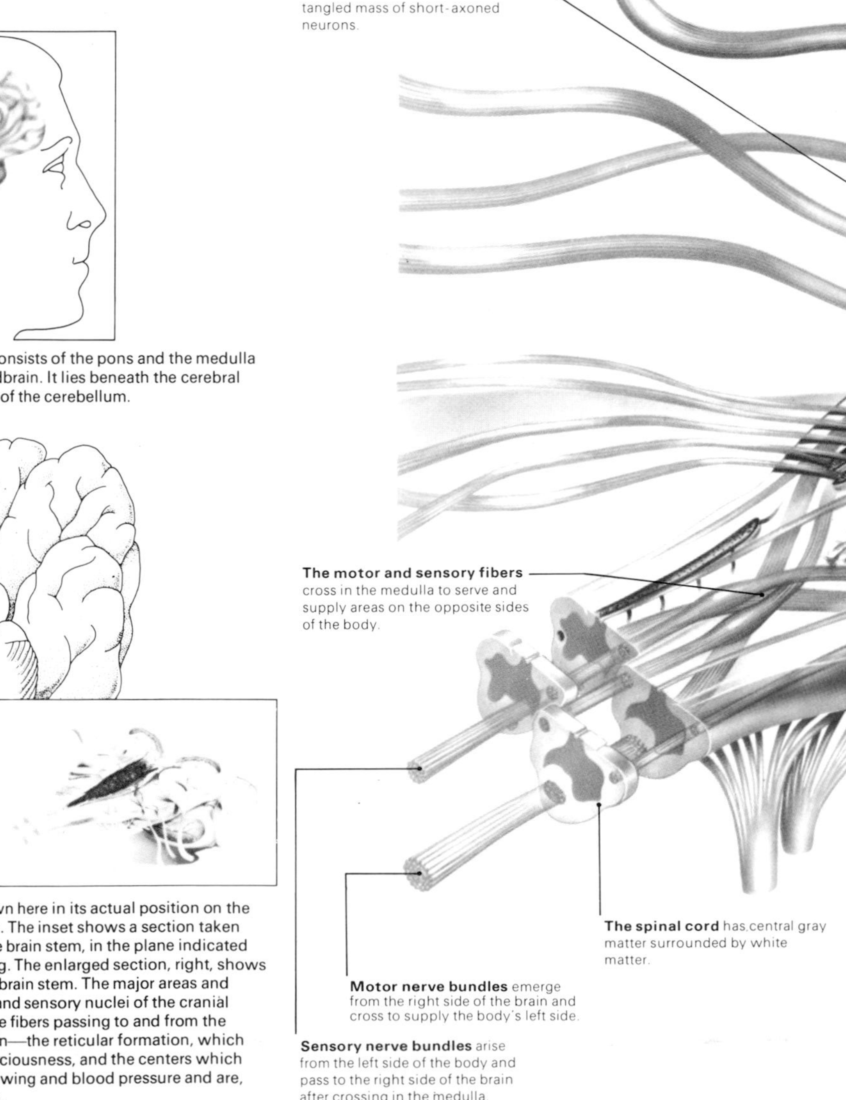

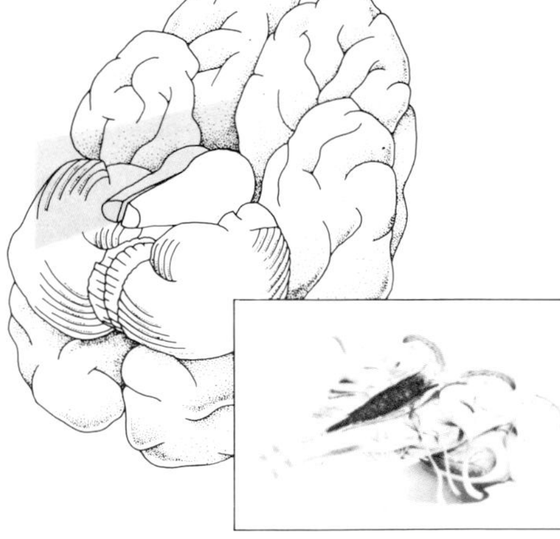

The brain stem is shown here in its actual position on the undersurface of the brain. The inset shows a section taken through the middle of the brain stem, in the plane indicated by the light brown oblong. The enlarged section, right, shows the structures within the brain stem. The major areas and structures are the motor and sensory nuclei of the cranial nerves—bundles of nerve fibers passing to and from the higher centers of the brain—the reticular formation, which controls the level of consciousness, and the centers which control breathing, swallowing and blood pressure and are, therefore, essential to life.

the activity of the brain stem continues to keep the life-support systems functioning.

To work effectively, the life-support systems need to receive a constant barrage of information about the state of the body. The reticular formation is ideally sited to do this. Nerves carrying information branch off from the main sensory and motor tracts into the reticular formation and keep it continually informed about the activity of other parts of the nervous system. Every time your body moves, some adjustment of heart rate, blood pressure or breathing rate is needed to compensate for the changes which have occurred. When you stand, for example, the diameter of the arterioles in the skin and muscles are narrowed to keep the blood pressure at a steady level. Heart rate is increased to provide the extra blood required and breathing rate goes up to provide the extra oxygen that is needed to maintain your body in an upright position and to remove the increased levels of carbon dioxide in the muscles. All these adjustments must be made continually, every moment of the twenty-four-hour day, to ensure that your body functions efficiently.

Besides controlling these vital life-support systems, the reticular formation contains other clusters of nerve cells which control the stomach, mouth, face, ears and eyes. Those nerve cells which supply the stomach regulate, among other things, the amount of hydrochloric acid that is secreted. In addition to these cells, another important group controls vomiting. This vomiting center can be stimulated by impulses from many different parts of the body—from the stomach, for example, or by a nasty taste in the mouth, from the organs of balance, as in motion sickness, and even, in some people, by stroking a small patch of skin behind the ear, known as the Alderman's Patch (because aldermen were said to use it to make themselves vomit in order to continue eating at municipal banquets).

Messages controlling swallowing, movements of the tongue, face and eyes all pass through a group of cells in the reticular formation that controls their output from the brain stem. Sensations of taste from the mouth and touch from the face pass back through columns of nuclei close beside those controlling the motor outflow. If the nerves which take the motor output from the brain stem to the muscles of facial expression are cut, that side of the face becomes paralyzed. However, if the pathways higher up within the brain are interfered with, all voluntary movement on that side of the face is lost, but movements concerned with emotion, including laughing, smiling or crying, can still take place.

At the upper end of the brain stem, but below the thalamus, lies the midbrain, one of whose functions is to control eye movement and pupil size. It is here that the motor messages controlling the movements of the two eyes are generated, to ensure that when the eyes move they do so together and to the same extent. Linked with eye movement is the diameter of the pupil. When the eyes look from a distant object to a nearer one, the pupil is seen to dilate. This occurs possibly to allow more light to enter the eye, as objects seen close are less bright than objects which are at a distance. These centers also control the flicking movements of the eye which occur, for example, when watching the roadside from a moving car.

Not only does the reticular formation receive sensory information from all over the body, but it is also concerned with the fine control of all the body's movement. Special fibers leave the reticular formation, travel down in the spinal cord to relay and send their fibers to the muscle spindles, special end organs deep within the muscles. This system of control is of enormous importance as it allows the fine adjustment which makes coordinated and smooth movement possible. Without it, we would jerk like automatons.

Whether asleep or awake, the reticular formation guards and controls the basic and life-supporting systems of the body. Although the intellect is created by the large cerebral hemisphere, it is this small area of the brain stem, the reticular formation, which holds the key to life itself.

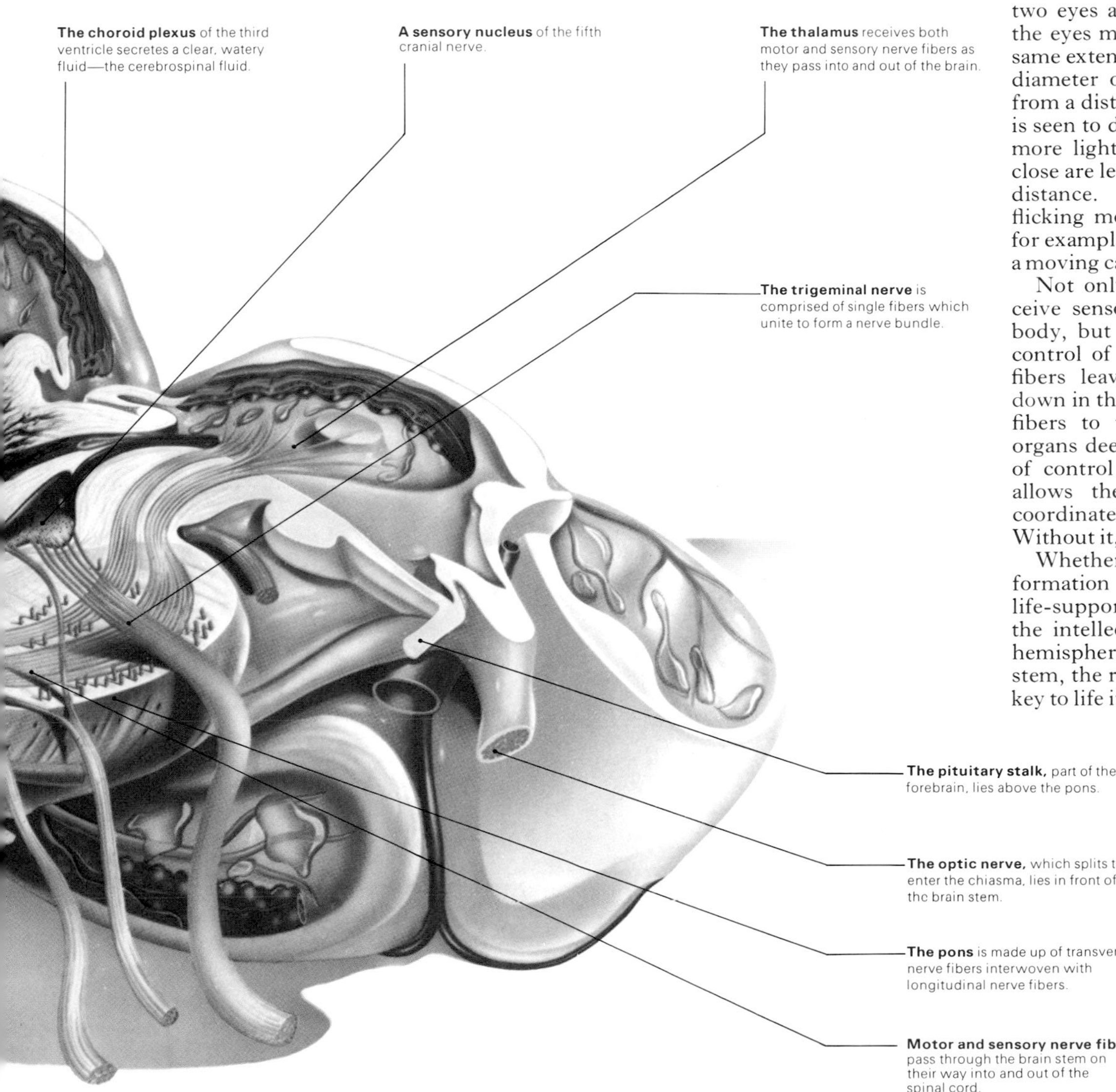

The choroid plexus of the third ventricle secretes a clear, watery fluid—the cerebrospinal fluid.

A sensory nucleus of the fifth cranial nerve.

The thalamus receives both motor and sensory nerve fibers as they pass into and out of the brain.

The trigeminal nerve is comprised of single fibers which unite to form a nerve bundle.

The pituitary stalk, part of the forebrain, lies above the pons.

The optic nerve, which splits to enter the chiasma, lies in front of the brain stem.

The pons is made up of transverse nerve fibers interwoven with longitudinal nerve fibers.

Motor and sensory nerve fibers pass through the brain stem on their way into and out of the spinal cord.

Cross-reference	pages
Brain and central nervous system	**74-77**
Cerebral cortex	**80-81**
Eye	**82-83**
Limbic system	**116-117**
Movement	**100-101**
Reticular activating system	**106-107**

Consciousness: the inner world

About one hundred thousand million cells make up the central nervous system. Most of these cells are in the brain's cerebral cortex, which is made up of two hemispheres. The surface of each of these hemispheres contains the cells which elaborate the nervous codes coming into the brain from the senses. These codes carry information about events in the outside world to receiving areas in the cortex. Here the information is processed by columns of cells and is then transferred to the association areas which lie adjacent, so that information from other senses, from memory and from other areas of the brain, can be combined.

It is the combination of information from these different sources and their elaboration by the cortex which produces the foundation of consciousness. Sherrington, the early twentieth-century physiologist, with his rare gift for poetical descriptions of the functioning of the higher nervous system, described it as "a dissolving pattern, always a meaningful pattern, though never an abiding one, a shifting harmony of subpatterns, the functioning of the enchanted loom."

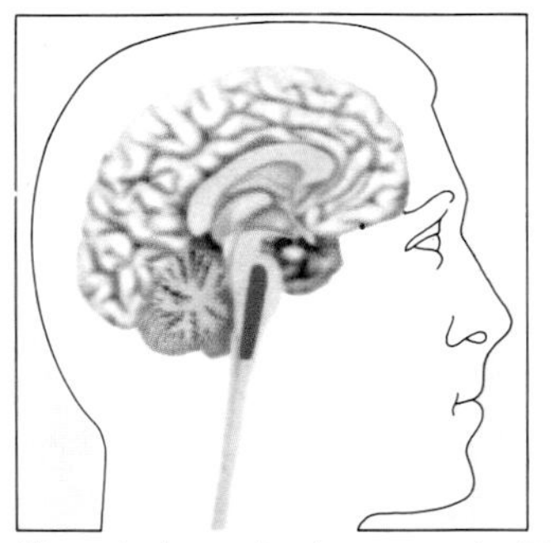

The reticular-activating system (red) is located within the brain stem. It is composed of hundreds of short-axon neurons, which produce a dense and intricate nerve complex that is devoid of any precise pathways.

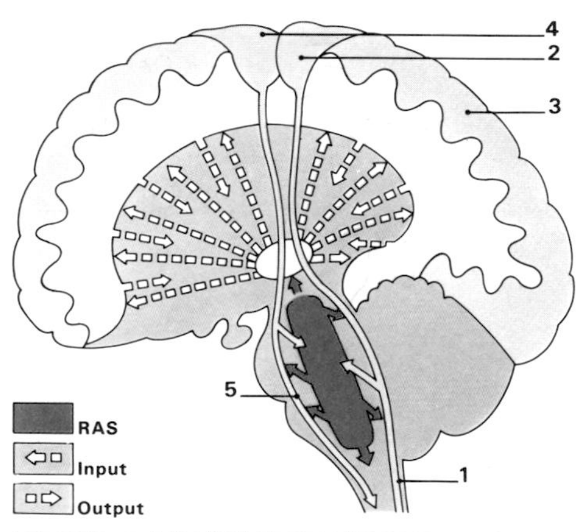

Information entering the brain along the sensory nerve pathway (**1**) passes to the sensory cortex (**2**). However, nerve branches from the pathway first send impulses to the ascending reticular-activating system (RAS), which stimulates activity and attentiveness throughout the entire cortex (**3**). The resultant outgoing information leaves the brain from the motor cortex (**4**) through the motor pathway (**5**) and then into the spinal cord.

For consciousness to be maintained, the cells within the cerebral cortex are kept in a state of continual excitation by the reticular formation, a group of cells in the brain stem. The reticular cells are so connected that they have a wide influence over the cortex, and they themselves are stimulated by nervous impulses from many different parts of the body. If the stimulation of the reticular formation is decreased, by cutting down sensory input, for example, by putting an individual in a darkened, quiet room, his hands and feet in soft woolly gloves, then the excitation of the cortex falls and the quality of consciousness is changed. He either becomes unconscious in sleep, or he hallucinates, seeing, hearing and feeling sensations which are not there.

The electrical activity of the cerebral cortex can be measured on the surface of the scalp as rapidly changing small voltages shown on the electroencephalogram, or EEG. In deep sleep or coma, large, slow waves of electrical activity sweep throughout the cortex at between one and three waves a second, keeping the level of excitation below that needed for consciousness. As the sleeper awakens, the reticular formation increases the level of cortical excitation until in light sleep the waves are sweeping far more quickly through the brain, at six to seven waves per second. When consciousness appears, the waves have accelerated into the alpha range, eight to thirteen waves a second, and when fully aroused, a diffuse and broken pattern of electrical activity is seen, the pattern of full alertness.

This broken pattern in the electrical activity of the cortex shows that the cells are deeply involved in processing the information which comes to them from the senses and from other parts of the brain. Here are constructed the feelings, sensations, ideas, thoughts and images which come into consciousness and which are the private world of each individual. It is not yet known quite how the change occurs from the very complex electrical patterns within groups of cells to the sensations and feelings of consciousness. What is known, however, is that if these cells are damaged or the patterns altered, the quality of consciousness changes.

The brain works by extracting patterns of information from the senses and setting up models of what it thinks the world is like. These models are continually updated with new information which comes into the brain from the outside world, so that an ongoing picture of events can be created. Sometimes it is possible to catch the brain out and then the inside picture and the outside world do not agree, as, for example, in visual illusions. There is no real world outside for each individual—only the world the brain constructs.

The brain can be made to function in many different ways, and each way produces a different private world. Drugs alter experience by their effect on brain function. Barbiturates, sleeping pills, depress the action of the reticular formation and slow down the exchange of information between cells in the cortex, thus encouraging sleep and producing slowed reaction times. Amphetamines and caffeine stimulate reticular-formation activity, allowing faster thinking and a clearer, brighter world. Tranquilizers dull the action of the limbic system, that part of the brain where the emotions are created, while LSD and other hallucinogenic drugs affect the way information is transferred between cells, thus allowing new patterns of brain activity to arise and creating new experiences, many of which do not occur in the normally functioning brain. Antidepressants affect the midbrain, the area which also controls mood. In all these cases, a different world is experienced because of the alteration of brain functioning by the drugs.

Other, gentler, techniques, such as hypnosis, can alter the perceived world by making a subject dissociate from different aspects of consciousness. He will not feel a pin pushed deep into the skin, for example, and can be instructed to "see" only selected pictures in a magazine. But the paradox here is that he has to recognize and reject in order not to "see," thus showing the separation of the different elements of consciousness.

The unconscious, or subconscious, part of the mind is considered by psychiatrists to have an important influence on behavior. Sigmund Freud was the first to recognize that the experiences from infancy and childhood long since forgotten, but stored in the unconscious, are responsible for many of the thoughts and actions of the adult.

There seems to be little doubt that the private, or inner, world of the adult mind depends on the complexity of information the brain received during the period of infancy and childhood. The information received by the brain helps to form the conscious and unconscious elements of the mind. Upsets or irregularities which are experienced in this early developing period may influence the content of the conscious and unconscious. These upsets may manifest themselves as emotions that may be beneficial and expressed, for example, through art, fashion or music. They may, on the other hand, be expressed through antisocial behavior. Consciousness seems to draw a thin line between feelings that can result in good or bad actions.

Art forms can, therefore, be seen as expressions of consciousness and they may in turn affect the development of the culture from which they emerge. Similarly, every generation will be a product of thousands of minds and will have an influence on the generation that follows it. Thus the activity of the consciousness, or inner world, may help to explain why we are not the same people as our ancestors and why we can certainly all be said to have unique minds, differing opinions and variable emotions.

Brain waves recorded from the scalp prove that human beings never "switch off" their minds. Here three levels of arousal—sleeping, relaxing and action—all have a characteristic brain wave pattern. The machine used to monitor and record this activity is an electroencephalograph (EEG). In deep sleep, the EEG pattern produces large, slow waves. When relaxing the waves become faster. In an active situation the EEG is very dense and is described as having a low voltage and a high frequency, which produce a fast wave.

Unconsciousness and consciousness

The two heads contain visual metaphors for the mind's unconscious and overlapping conscious. The unconscious is stylized as night with a single large wave of gentle activity flowing through it. The conscious is represented by a vibrant, lively community where the double bank of fast waves symbolizes the vigor of the conscious mind.

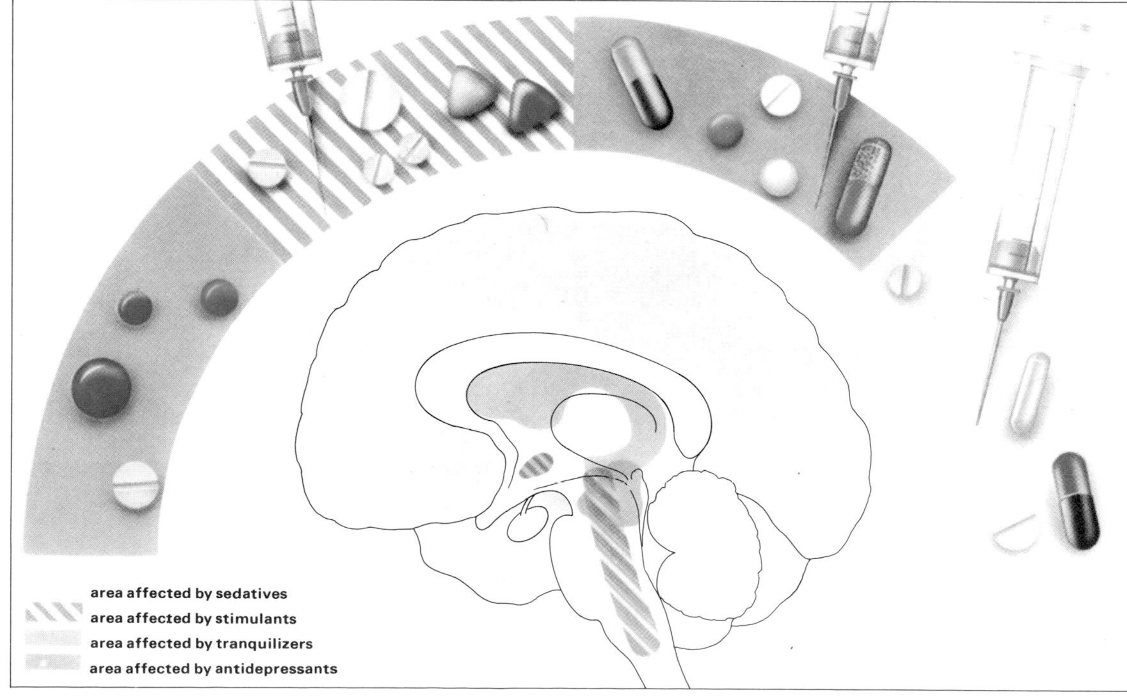

States of consciousness can be altered by the use of drugs. Antidepressants (blue) affect the midbrain, the area involved with the formation of mood. Thus they can prevent and help to cure depression. Stimulants (striped), such as amphetamines, act on the reticular-activating system (RAS) and the hypothalamus. Their effect is to decrease fatigue, increase alertness and elevate mood. If the mind is overactive and, therefore, anxious, tranquilizers (green) may be used to reduce the level of anxiety. They act principally on the limbic system and the RAS. Sedatives (light brown), for example barbiturates, are a group of drugs often administered as sleeping pills. They act on the RAS and on the cortex to reduce the conscious experiences that would normally induce wakefulness.

Cross-reference pages

Brain and central nervous system	**74-77**
Brain stem	**104-105**
Cerebral cortex	**80-81**
Limbic system	**116-117**
Sleep	**108-109**

Sleep: its nature and function

In a dream laboratory a subject sleeps in a soundproof room while changes in the electrical activity of her brain cells are monitored by electrodes attached to different areas of her scalp (below). These are recorded and printed as wavelike patterns, called an electroencephalogram (EEG), on a moving sheet of paper. Each of the four stages of sleep produces a characteristic EEG, and periods of dreaming are revealed by rapid movements of the eyes. After each dream period the investigator wakes the subject, who then recounts the dream.

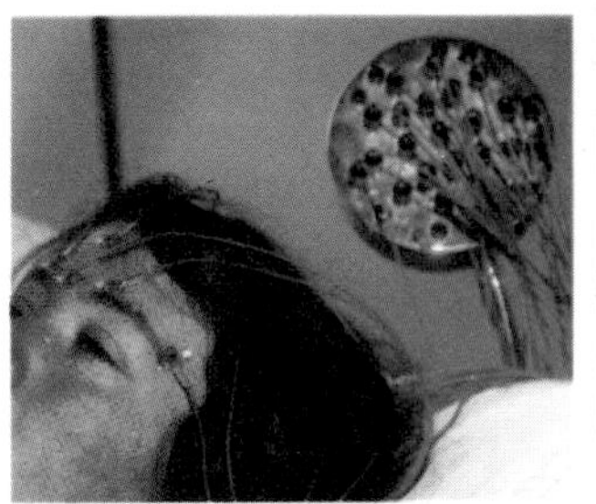

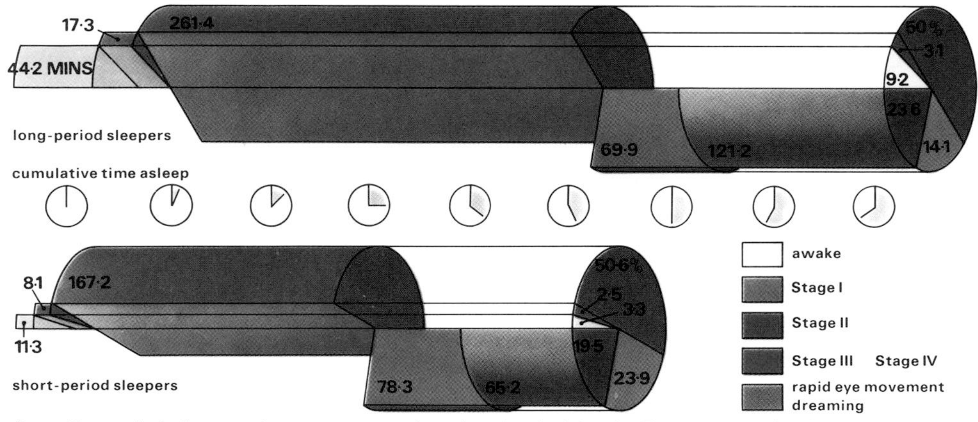

According to their sleep requirements, a group of people, between the ages of twenty and thirty-four, were classified as short-period and long-period sleepers. The average time each group spent asleep and the type of sleep it was is shown by the length of the tube. The percentage of total time per night spent in each type of sleep is indicated in the segmented circles. Both types had the same amount of slow-wave sleep, but the long sleepers had almost twice as much dreaming sleep.

Brain patterns and sleep

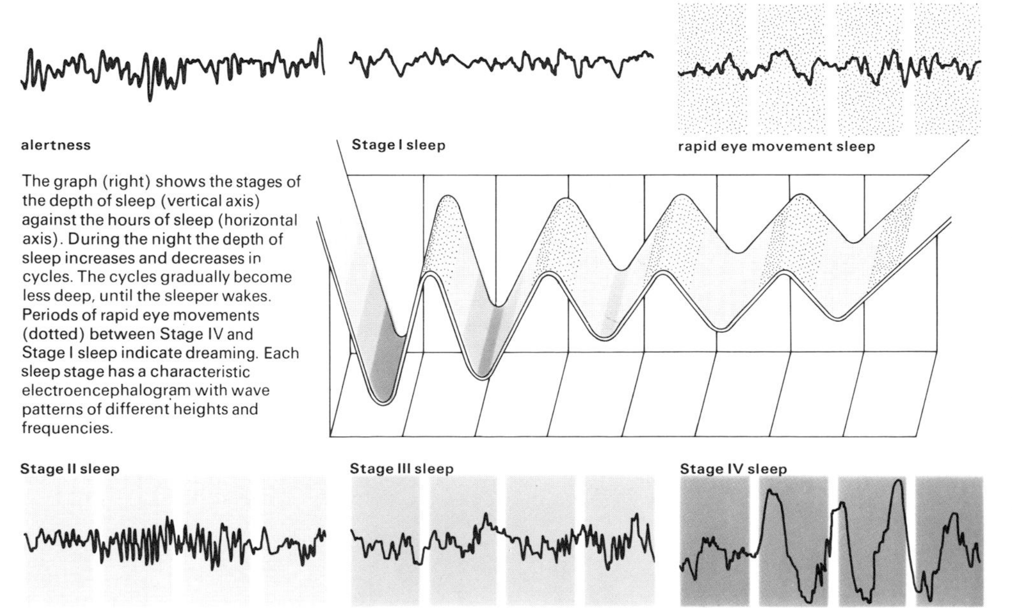

The graph (right) shows the stages of the depth of sleep (vertical axis) against the hours of sleep (horizontal axis). During the night the depth of sleep increases and decreases in cycles. The cycles gradually become less deep, until the sleeper wakes. Periods of rapid eye movements (dotted) between Stage IV and Stage I sleep indicate dreaming. Each sleep stage has a characteristic electroencephalogram with wave patterns of different heights and frequencies.

Sleep is the most obvious of our many internal biological rhythms, but only recently has its nature and function begun to be understood. It is now known that our nightly eight hours are spent not in one state but in several, and that the division between sleep and waking is far from absolute.

The scientific study of sleep has been made possible by the measurement of brain waves, tiny fluctuations of electrical potential between different parts of the scalp. The record of these fluctuations, the electroencephalogram, or EEG, shows that sleep consists of a series of stages occurring in a regular order in a cycle of about ninety minutes, a cycle which occurs four or five times during the sleep period and continues to influence our alertness during waking hours.

Many thousands of volunteers have slept "wired up" to recording devices in laboratories to give researchers a picture of a typical night's sleep. It was observed that as drowsiness sets in, the rapid, relatively low amplitude brain waves of waking become increasingly interrupted by trains of slow, high amplitude waves. At this point the subject's thoughts begin to wander. He may hear voices speaking his own name or declaiming nonsense. Vivid hypnagogic imagery—faces, landscapes, abstract patterns—may seem to move before his closed eyelids. Sometimes there are violent sensations of light, noise or falling, accompanied by sudden jerks of the body. Slow brain waves now dominate the EEG record, and the subject is in Stage I, or light sleep.

He now sinks through the second and third stages of orthodox, or nondream, sleep, each stage marked by progressively greater muscular relaxation and slower brain waves, into the deep sleep of Stage IV. In this state, the subject is difficult to rouse and is confused if awakened. It is in deep sleep that sleeptalking, sleepwalking and bedwetting tend to occur.

After about seventy minutes of sleep, spent mostly in Stages III and IV, the sleeper moves restlessly and drifts up through the sleep stages toward wakefulness. But instead of reentering Stage I, he passes instead into a quite different type of sleep. His EEG shows fast, low amplitude waves similar to those of wakefulness. His muscles are limp—including the neck muscles, which retain some tension throughout the other sleep stages—and some of the normal reflexes are lost (thus, nightmares of being unable to move are based on a genuine paralysis). Breathing and heartbeat become irregular. The sleeper's inertness is interrupted by sudden jerks, large body movements and facial grimaces. The eyes move rapidly under the closed lids, and the sleeper, if male, almost always has penile erection. This is "paradoxical," or rapid eye movement (REM), sleep.

This initial REM period lasts about ten minutes and marks the end of the sleep cycle which, perhaps after a few moments of semiwakefulness, begins again. With each repetition, the REM periods lengthen while the time spent in Stages III and IV becomes less. Stage IV may be entirely skipped after the first or second cycle. At the end of the fourth or fifth cycle the sleeper wakens, but according to recent research the same rhythm as that of sleep goes on throughout the day, causing a regular waxing and waning of alertness.

Mental activity seems to continue throughout sleep. People awakened from REM sleep report that they were dreaming, while the

various stages of orthodox, nonREM sleep seem to be occupied by thinking of a fairly everyday kind interspersed with muted and fragmentary dreams.

Sleep deprivation experiments show that sleep is a necessity. After two or three nights without sleep, subjects can still cope with novel and challenging tasks, but their performance on routine tasks is increasingly impaired by ever more frequent microsleeps—short periods of slow-wave EEG—which prevent sustained attention. As sleep loss continues, waking dreams, or hallucinations, may occur together with, in some people, delusions of persecution and other such symptoms.

A night's sleep restores normality. The first night is spent mainly in Stage IV sleep and the second in REM sleep, which shows a need to catch up on time missed in these states. Deprivation of Stage IV sleep alone leads to depression (clinically depressed patients show below normal amounts of Stage IV), malaise and apathy. The effect of REM deprivation in humans is unclear, but short periods seem to act as a mild stress, causing irritability, anxiety and difficulty in concentrating.

Stage IV and REM sleep each have a biochemical trademark. During deep sleep, high levels of growth hormone are produced; this regulates body growth, stimulates tissue growth, speeds healing and lowers blood cholesterol. Children need ample Stage IV sleep for proper growth, while in adults its function is probably mainly restorative. During REM sleep, bursts of stimulating adrenal hormones are released. This mechanism may be disrupted if the sleep-waking pattern is reversed. This is one reason that nightshift workers and sufferers from jet lag may have difficulty sleeping and poor concentration while awake. The body needs about three weeks to adapt to such a reversal.

Orthodox and REM sleep also appear to arise from different anatomical and biochemical systems. Their mechanisms are not yet clear, but orthodox sleep seems to be mediated by brain-stem cells rich in the brain transmitter serotonin, while REM sleep may be triggered by noradrenaline-producing cells in the pons.

The need for sleep varies among individuals and with age. Babies sleep fourteen to sixteen hours daily, most of this REM or Stage IV sleep. The average eight hours of the young adult drops to five or six in old age, little or none of this spent in Stage IV and with the proportion of REM sleep much reduced.

The internal biological clock may determine the timing of an individual's behavior patterns. The clock functions independently of the environment, but is synchronized by external stimuli to changes in the external world. The biological clock affecting times of sleeping and wakefulness is synchronized to daytime wakefulness and nighttime sleep. Clocks representing the same day over six weeks show little variation in activity (**1**). In constant darkness (**2**) the clock becomes "free running." It functions on a length of day greater than the normal twenty-four hours and results in the sleep period becoming later and later each day.

Transatlantic flights upset the body's internal clock. Prior to the departure of the flight, the internal clock is synchronized to local New York time, top left. On the first day after the flight, the biological clock is still on New York time although local Rome time is six hours ahead, top right. As a result, the person sleeps during the daytime. One week later, bottom right, the internal rhythm has adapted to the new local time. On the day after the return flight, bottom left, the internal rhythm is again desynchronized.

Cross-reference	pages
Brain and central nervous system	**74-77**
Brain stem	**104-105**
Cerebral cortex	**80-81**
Dreams	**110-111**

Dreams: pictures in the mind

Dreams have been regarded as windows on the depths of the personality, as predictors of the future or messages from gods or demons. The determinants and meaning of dream content are still far from understood, but in recent years scientists have made significant discoveries about the conditions in which dreams always occur.

Dreaming mostly takes place during the rapid eye movement (REM), or "paradoxical," stages of sleep. This state, which occupies one and a half to two hours of the average night's sleep, is a puzzling combination of mental alertness, as indicated by electroencephalographic (EEG) records of brain waves, and muscular near-paralysis. Detailed study reveals that REM sleep is not a unitary state, but consists of two alternating substages, REM-M and REM-Q. If the dream experience is regarded as a theatre play, then the REM-M (eyes moving) state may be likened to the acts of the play and the REM-Q (eyes quiescent) state may be compared to its intermissions.

During REM-M, the sleeper's eyes follow the dream action. Eye movements are usually horizontal, as they are during waking life when we scan a scene, but if the dreamer is watching events involving an up-and-down motion his eye movements will be vertical. If there are few eye movements, he is usually observing some distant scene. The dreamer tends to be totally immersed in his experience, participating in it rather than reflecting on it. He makes no large body movements, but there is usually subtle muscular activity which reflects, in a very attenuated fashion, the motions he is making in the dream.

As each dream, or each scene of a dream, comes to an end, the sleeper enters REM-Q. His eyes are still, fine muscle movements cease and are replaced by gross body movements. If awakened now, the subject will probably report that a dream has just finished and that he was trying to interpret it or was thinking about it. He may still be experiencing a dream, but if this is so he is probably aware that he is dreaming. When there are many REM-Q episodes during a REM period, dreaming is, accordingly, fragmented. When there are none, subjects will recall a single "one-act play."

Full-blown visual dreams can occur during ordinary nonREM sleep, but are less common and usually less vivid. In these nonREM

Rapid eye movements

Rapid eye movements can be measured by placing an electrode on each side of a sleeping person's eye. Eye movements between the front and back of the eye cause changes in the electrical potential, which can be recorded on a moving band of paper. Below left is a recording of a person undergoing a dreaming phase of sleep, the only stage at which rapid eye movements occur. The recording, below center, is of Stage III sleep, which is dreamless. Below right is a recording of a person whose eyes are closed and who is in a relaxed state. All readings are measured over a four-second time period.

Interpretation of dreams

Dream analysis involves interpreting elements of a dream as symbols of repressed wishes or other unconscious processes. The dream "I was going upstairs with my mother and my sister. When I reached the top, I was told that my sister was going to have a child," was recounted to the Viennese neurologist Sigmund Freud (1856–1939) and the Swiss psychologist Carl Jung (1875–1961) for their interpretations. According to Freud, this dream was strongly rooted in the family situation (**1**) and was the unconscious representation of repressed sexual desires. The stairs represented a censored version of coitus and the presence of the mother represented the boy's sexual desire for her (**2**). The expected child could indicate an incestuous relationship between brother and sister (**3**). Jung believed dreams were expressions of experiences in the form of universal symbols. A visualization of his interpretation is shown in terms of these symbols. He believed that the boy had homosexual inclinations and felt guilty about neglecting his mother, which Jung interpreted as neglect (shown as Dionysus in a relaxed pose). He interpreted the sister as "love of womanhood" (symbolized here as Aphrodite), climbing the stairs as his growing up (shown as the Ages of Man) and the child as rebirth (symbolized as spring).

stages, mental activity is more like ordinary thinking. As the night goes on, however, both dreams and nonvisual "sleep thinking" become less everyday and more bizarre and emotional. This may be because, as time asleep increases, REM periods become longer, as does the time spent in the lighter stages of non REM sleep, allowing more time for dreams or ideas to develop.

Everybody spends from one to two hours nightly in the REM state. But the resulting dreams are forgotten unless the sleeper wakes during a REM period or within ten minutes of its end. It seems that a memory cannot be formed if sleep is continued unbroken. "Dreamers"—those who often remember their dreams—may differ from the "nondreamers"—those who seldom recall them—in sleep pattern and in visual memory. Dreamers tend to wake briefly after each REM period, allowing the dream to be fixed in memory so that it may be recalled in the morning, while nondreamers pass into the next sleep cycle without waking. Awakened from the REM state, subjects with good visual memory almost always remember their dream, but those with poor visual imagery can do so only about half the time.

Most dreams are visual and in color, but nonvisual dreams can occur and people blind from birth dream of sound, smell and touch (although they usually have eye movements during REM). Outside stimuli may be woven into the dream experience, but dream content is less affected by internal sensations. Thirsty sleepers seldom dream of thirst. Similarly, although males from infancy to old age have penile erection during REM periods, only a minority of their dreams have any overt sexual content.

The typical dream has a "cast" of two to four characters, over half of them known to the dreamer in waking life. Dreams involving only the dreamer, and those in which five or more characters appear, are relatively rare. Dream mood is most often negative. Apprehension, anger and sadness are the dream moods most commonly reported.

People deprived of REM sleep become anxious or irritable, according to personality type. When they are allowed to sleep normally, they tend to fall straight into REM sleep, without the hour of nonREM sleep which usually precedes it, and pass much of the night in this state, as if to catch up on their dreaming. This rebound effect also occurs after stopping a course of barbiturates, antidepressants or amphetamines, all of which reduce REM sleep. Nightmares are likely at such times, when the need to dream is unusually great. Children, who have a higher proportion of REM sleep than adults, suffer many more terrifying dream experiences. Cats deprived of REM sleep begin to produce bizarre behavior, but this has not been experimentally observed in humans, although it has been suggested that the confusional or schizophrenic psychoses which can follow long-term barbiturate or amphetamine use may be linked to the REM sleep loss these drugs cause.

Theories of dreaming are numerous, but since serious adverse effects have not yet been definitely proved to result from dream deprivation, scientists cannot be absolutely certain that dreaming serves a crucial function. Some psychologists have suggested that dreams are merely random afterimages of the day's events. Others believe that while the REM state probably has a necessary metabolic function, the dreams that occur in it are meaningless by-products. The difficulty of proving that a dream is "random" or "meaningless" mitigates against acceptance of such theories, and most modern researchers continue to believe that dream content is significant, a belief that man has always held and which is integral to psychoanalytic views of personality.

Sigmund Freud, the father of psychoanalysis, believed that dreams express irrational, usually sexual, infantile wishes, of a sort necessarily repressed in everyday life and, indeed, so abhorrent even to our dreaming personalities that they can appear only in symbolic disguises. Carl Jung, a follower of Freud who later began his own school of psychoanalysis, held that dreams may also reflect the higher aspects of the mind and often provide the dreamer with wise guidance and sound advice.

The most popular modern theories concur with psychoanalytic beliefs in attributing to dreams a central function in regulating mental processes and bringing the individual's inner world into harmony with the external world. One theory, the "rehearsal" theory, suggests that dreams afford an opportunity for rehearsing such genetic behavior patterns as fighting, hunting, fleeing and copulating. Another theory, based on a computer analogy, claims that dreams embody the process of mental sorting and filing which is needed to assimilate the events of the day.

But, as the sheer multiplicity of current dream theories suggests, while the dream state is now yielding its secrets to science, dreams themselves continue to retain their mystery and their fascination.

A dream

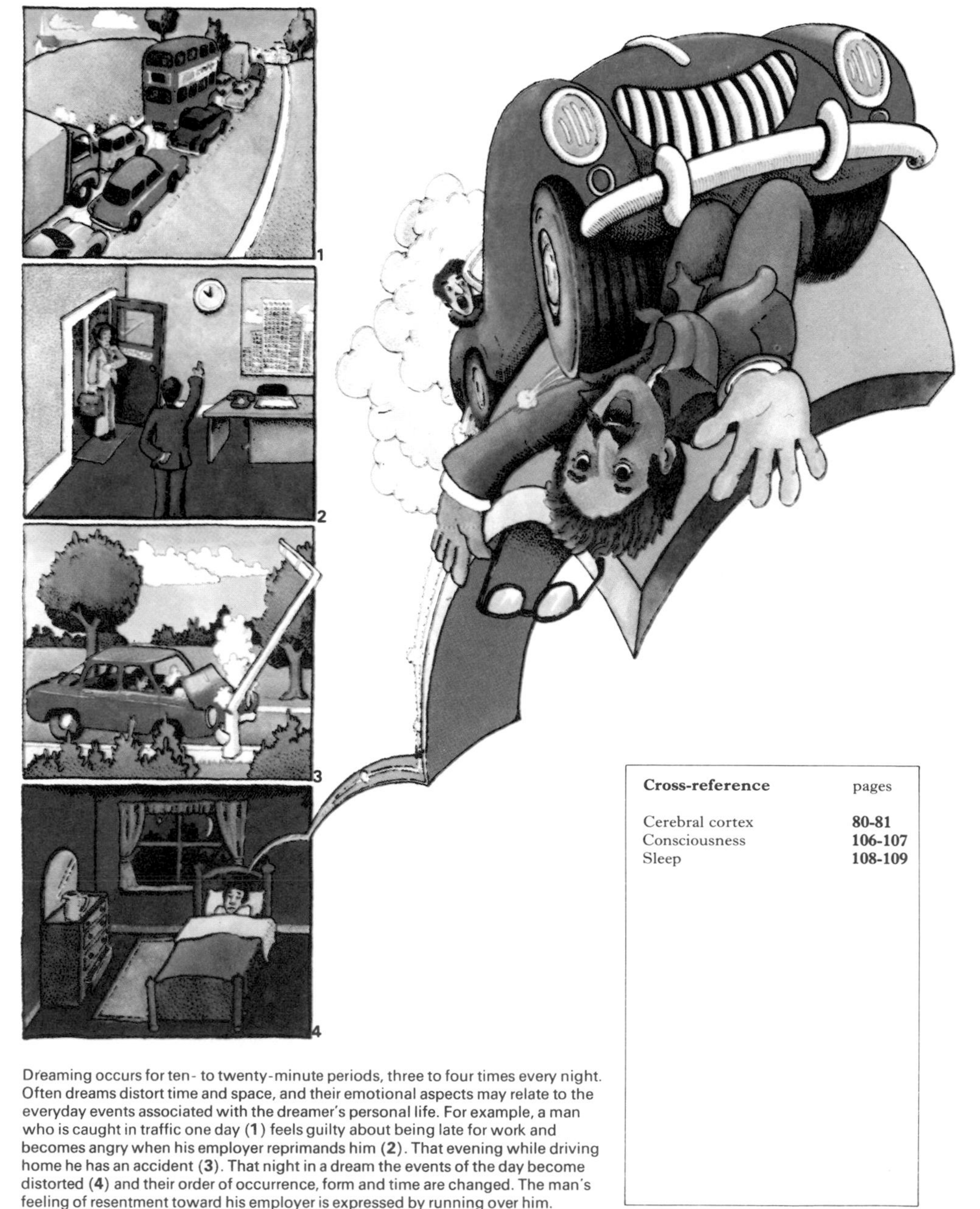

Dreaming occurs for ten- to twenty-minute periods, three to four times every night. Often dreams distort time and space, and their emotional aspects may relate to the everyday events associated with the dreamer's personal life. For example, a man who is caught in traffic one day (**1**) feels guilty about being late for work and becomes angry when his employer reprimands him (**2**). That evening while driving home he has an accident (**3**). That night in a dream the events of the day become distorted (**4**) and their order of occurrence, form and time are changed. The man's feeling of resentment toward his employer is expressed by running over him.

Cross-reference	pages
Cerebral cortex	**80-81**
Consciousness	**106-107**
Sleep	**108-109**

Hunger and thirst

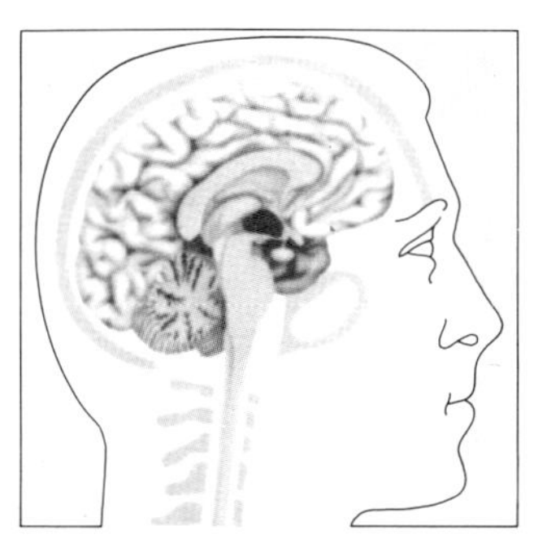

The hypothalamus is situated at the center of the underside of the brain, below the thalamus and above the pituitary.

The human machine, like any other, needs fuel and constructional material to keep it going and in good repair. Without an intake of food and water we would sooner or later die. In the case of water, it would be sooner. The body keeps reserves of fuel that can be used in emergency, stores of fat and glycogen that can be burned to provide energy, but although we are 70 percent water, we have no built-in water tank that we can conveniently tap when running dry.

Water is essential for every process in the body, and it is hardly surprising that its lack should lead to mental as well as physical changes. The physical aspects include dryness of the throat, muscular weakness, loss of elasticity in the skin and the paucity or absence of urine. The mental aspects include disorientation and the occurrence of hallucinations—the visions of bars and oases supposedly common to all travelers in cartoon deserts. But the most noticeable phenomenon is, of course, thirst.

Dehydration of the body produces a need for water, much as the deprivation of anything necessary to sustain life in an organism produces a need for that substance. It is not necessary to know what is missing to have a need; we may have a need, for example, for vitamin C without being aware that there is such a substance. In the case of a need for water, however, we know that we must drink,

The hypothalamus

Connected by many nerve tracts with the brain and spinal cord, the hypothalamus acts as the link between the endocrine system and the nervous system. It functions automatically, monitoring and regulating the autonomic nervous system as well as the state of the body's metabolism through eating, drinking, temperature control, sexual drives and "fight or flight" reactions. It also controls the menstrual cycle and is thought to contain a pleasure center. Finally, it directs hormones released by the anterior pituitary gland.

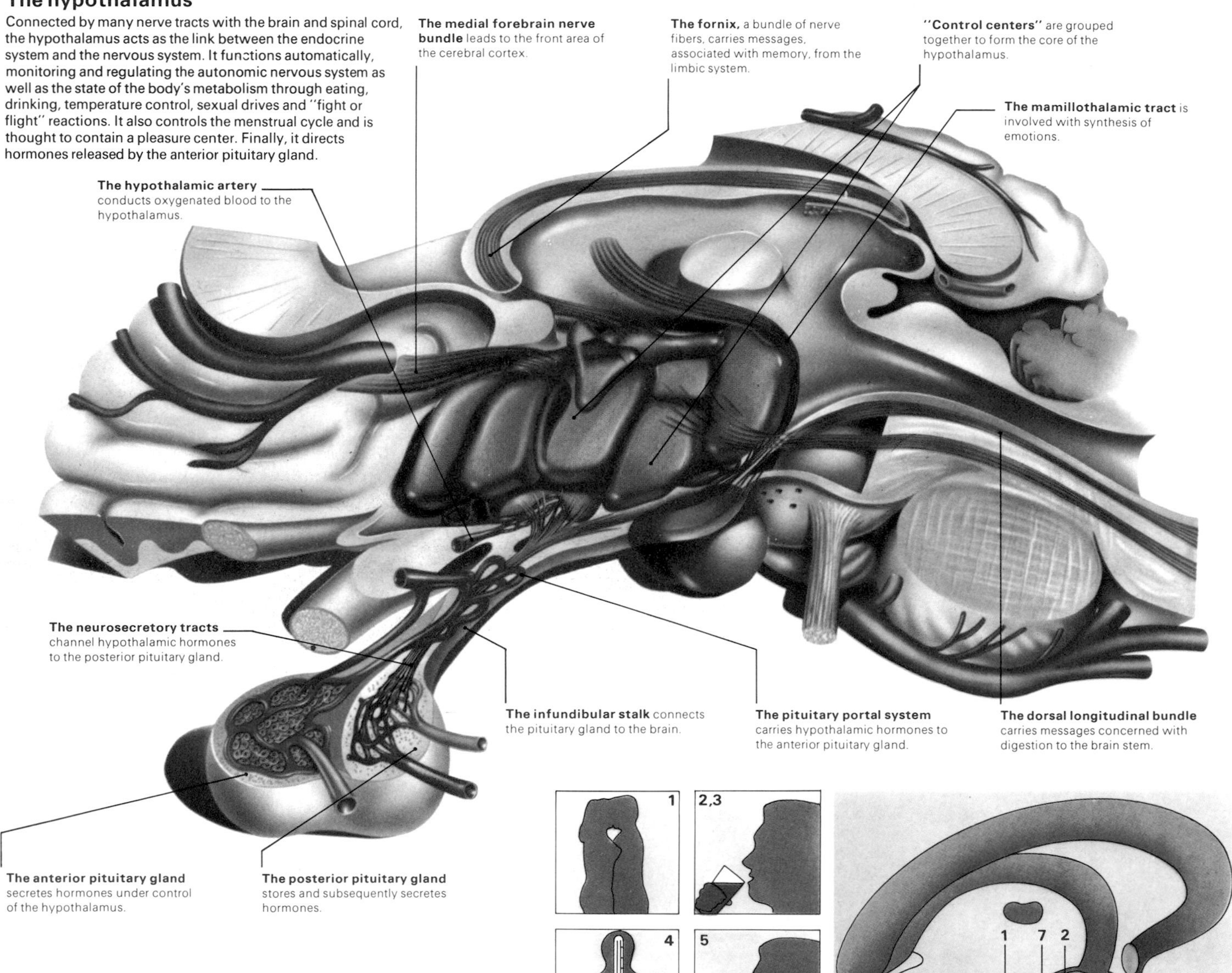

Functions of the hypothalamus

The hypothalamus consists of a number of areas that control the basic drives—hunger, thirst and sex—as well as the internal environment of the body. It is also concerned with emotions and with the sensations of pleasure and, possibly, with the sensations of pain and "displeasure." **(1)** The posterior area controls sexual drives and, therefore, the ability to reproduce the species. **(2)** The anterior areas control thirst and the drive to find water. **(3)** The supraoptic nuclei are also concerned with the thirst drive. **(4)** The preoptic nucleus is the body's thermostat and functions to control internal body temperature. **(5)** The ventromedial nucleus, or "appestat," controls the hunger drive. **(6)** The dorsomedial nucleus controls aggressive behavior. **(7)** The dorsal area is thought to be the human "pleasure center."

and we are driven to seek fluid. The drive behind our actions is thirst.

If we are thirsty and drink, then the need for water is satisfied in the body, and so the drive is reduced. We stop drinking and are no longer preoccupied with seeking fluid. But how does the body know we need water?

One of the ways in which the presence of too much or too little water in the body's tissues is detected is by means of special groups of detector cells in the brain, located in the region called the supraoptic area of the hypothalamus. These cells, called osmoceptors, monitor the concentration of the blood, which is normally kept constant between quite narrow limits. If the body takes in too little fluid, or too much, then these limits are exceeded.

When there is a water lack, the blood becomes more concentrated, so that there is more than the normal 0.9 percent of salt in it; this is known as hemoconcentration. Blood that is too concentrated stimulates the osmoceptors, and these send messages to the lower part of the hypothalamus, where the pituitary gland is situated.

The main route by which water is lost from the body is through the kidneys, where urine is formed. The amount of water that is taken back into the bloodstream through the renal tubules is controlled by antidiuretic hormone (ADH), a hormone secreted by the pituitary gland. This hormone promotes the reabsorption of water. A certain amount of ADH is always being secreted by the pituitary to adjust the concentration of the urine. When the blood is too concentrated, the osmoceptors stimulate the pituitary to secrete more ADH, more water is reabsorbed by the bloodstream from the urine, and the urine becomes more concentrated. Water is conserved, so that further concentration of the blood is hindered.

The system works also when there is too much water in the bloodstream. The osmoceptors detect the abnormality, less ADH is produced by the pituitary, the urine becomes more copious and dilute, and the excess water is excreted to reduce the risk of waterlogging in the tissues.

The water-balancing mechanism is also aided by hormones secreted, under the influence of the pituitary, by the adrenal cortex. The group of adrenal hormones called the mineralocorticoids regulates the amounts of sodium and potassium that are allowed to leave the blood in the urine, so giving a further adjustment of the concentration of the blood.

Hunger is equivalent to thirst in that it is a drive that can be reduced by appropriate behavior, in this case the seeking for and eating of food. We need a certain food intake to meet our daily energy requirements and to provide materials for growth and repair of our tissues. Too great an intake and we become overweight; too little and we become wasted and exhausted. There is obviously a mechanism to determine to some degree how much we need to eat.

From experiments in animals, it has been found that areas of the hypothalamus are again involved. If an area called the ventromedial nucleus is destroyed in experimental animals, they will overeat, swallowing any palatable food that is placed before them. But, at the same time, they show none of the other signs of a hunger drive. They do not search for food, nor do they eat any more after being starved than they did before. This area seems to be a control center for eating related simply to the availability of food, rather than to the body's caloric needs.

A nearby area, called the lateral hypothalamus, when stimulated produces all the signs of hunger, even in satiated animals. They overeat as though they were starving, even gulping down normally unpalatable food and drink until they become obese. It seems, therefore, that hunger is controlled by a balance of messages from the ventromedial nucleus and the lateral hypothalamus.

To prevent overeating, there are signals from the mouth and stomach which quickly reduce hunger. Simple distension of the stomach with a balloon reduces hunger, but not as much as distending it with food. And food taken by mouth is more effective than food introduced directly into the stomach.

The amount of glucose circulating in the bloodstream is certainly the main factor controlling hunger. Low blood sugar can produce hunger pangs—contractions in the stomach and intestines that give a painful reminder of the need for food. But there must also be other factors, because we need not just carbohydrate, but protein and fat as well as minerals and vitamins in our diet.

The hunger machine, shown here gathering food, is a mechanical representation of man's psychological and biological drives for food. Controlled by a system of weights and a balance in the cab of the vehicle, when the "hunger" side of the balance (green) is down, a gear engages, driving the machine forward to find food. As food is shoveled onto the receiving tray, it gradually tips, releasing the weights in the cab and altering the balance pans so that the "satiety" side (blue) descends. This disengages the machine from its food-gathering activity. Similarly, man's hunger is controlled by a balance between the hunger center and the satiety center in the hypothalamus of the brain. When man is hungry, cues from the hunger center stimulate searching and feeding behavior. When satisfied, the body cues the satiety center, there is a feeling of fullness and feeding stops.

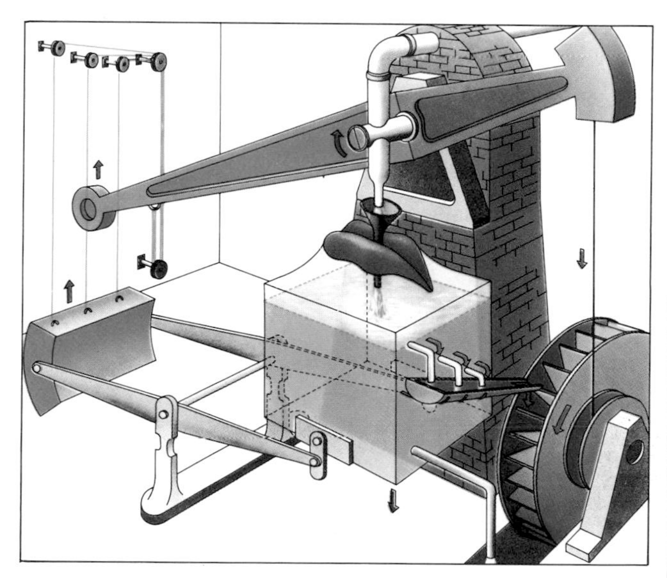

The thirst machine is a mechanical representation of man's psychological urge to drink. When the machine is thirsty, as shown here, the primary side of the balance (green) is down, and the water tap opens. The tank fills and, via pulleys, begins to raise the primary arm. Simultaneously, water overflowing from the full tank turns the water wheel, pulling down the secondary arm (gray). The tap closes and drinking ceases. On a similar balance principle, the thirst mechanism in man is controlled by the hypothalamic thirst center (primary arm) and the cerebral cortex (secondary arm), which regulates the body's intake of fluids. According to the quantity of fluid in the body, body mechanisms cue these controlling areas, to stop or to start drinking.

Cross-reference	pages
Blood	**58-61**
Diet	**44-45**
Hormones	**70-73**
Kidneys	**66-67**
Metabolism	**52-53**

Sex drive and response

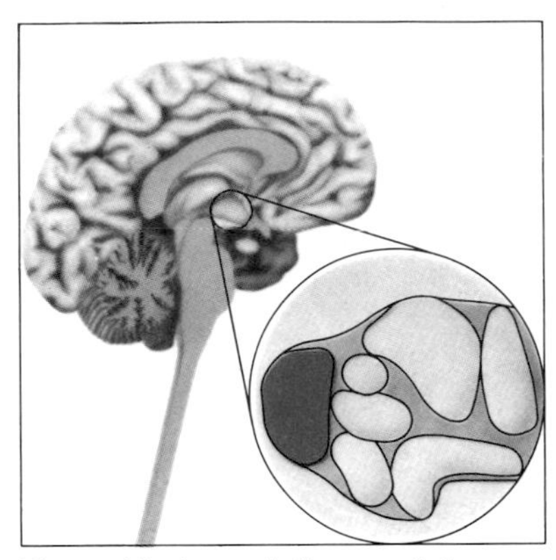

The sex drive is controlled by an area in the hypothalamus of the brain. This site, shown here in red, is called the posterior hypothalamic area.

Human sexuality has moved far from its biological origins. Both for the individual and for society it has come to serve many purposes besides that of reproduction, and its expression is influenced by cultural and psychological factors quite as much as by instinctual drives. Sigmund Freud believed that sex drive, or libido, is the fundamental source of the energy which man has drawn upon to create his civilizations and his cultures. While modern psychologists recognize that other deep-lying motivations have also contributed to human achievement, they agree that sex is the basic force that makes man a social animal. Although we are largely unconscious of it, all human relationships are to some extent eroticized, or powered, by the sex drive.

In man, as in all animals, the sex drive has three components—instinctive behavior patterns, which are genetically "programmed" into various sites in the limbic system and hypothalamus, sex hormones, which in part activate these patterns, and an innate tendency to respond sexually to certain environmental stimuli, or releasing factors.

The hormone responsible for sexual response in both men and women appears to be largely the male hormone testosterone. The female hormones, estrogen and progesterone, which determine sexual receptivity in the female of most animal species, seem not to be important factors in the human female's sex drive, although they are, of course, essential in ovulation and pregnancy and may be important for the development of the so-called maternal instinct.

Testosterone is produced in the testes of the male and in the adrenal glands of both sexes. Women, therefore, have about one-twentieth of the male amount. Since hormone levels alone do not determine the extent of sex drive in humans, female sexual capacity is comparable with that of the male, although women are more strongly influenced by convention and culture. Similarly, there is no very direct relationship between the amount of testosterone a man produces and his sex drive. Research has shown that once sexual activity is established as a source of enjoyment, its intrinsic rewards make it largely independent of hormonal status.

In many animals, especially the higher mammals, exposure to other members of the same species during infancy is crucial to the later development of sexual behavior. Chimpanzees kept in isolation until puberty, for example, never learn to mate, and studies of feral children, brought up in the wild by animals, suggest that the same may be true of humans. Although the development of adult sexual behavior depends on the surge of sex hormones at puberty, many components of the adult pattern appear in prepubertal mammals and in human children and are expressed in the form of play and experiment.

The environmental releasers for sexual behavior tend to be very specific for animals other than man, and sexual behavior is usually dependent on their presence. Thus a male chimpanzee will not mate unless stimulated by the smell of a female in heat and will ignore her sexual invitations if she is deodorized. Body smells, the sight of the secondary sexual characteristics of the opposite sex, the sight of the prospective mate's genitals in a sexually-ready condition, courtship movements and postures and specific types of physical contact, are all releasers in various animal species, and there is little doubt that they all have some importance in releasing the human sexual response.

In the case of man, however, learned and cultural factors supplement these and indeed can entirely overset or displace them. Excluding only the genitals, for example, there is no part of the female body which has not in some cultures been regarded as highly erotic and in others as sexually neutral. Although today the sexual body odors, resulting from genital and apocrine gland secretions, tend to be regarded as undesirable, modern research indicates that these odors may contain pheromones, chemicals which influence hormone release and which have an unconscious aphrodisiac effect.

In man, as the result of his evolution, the influence of the cerebral cortex on the limbic

Stimulus response pathways in sex

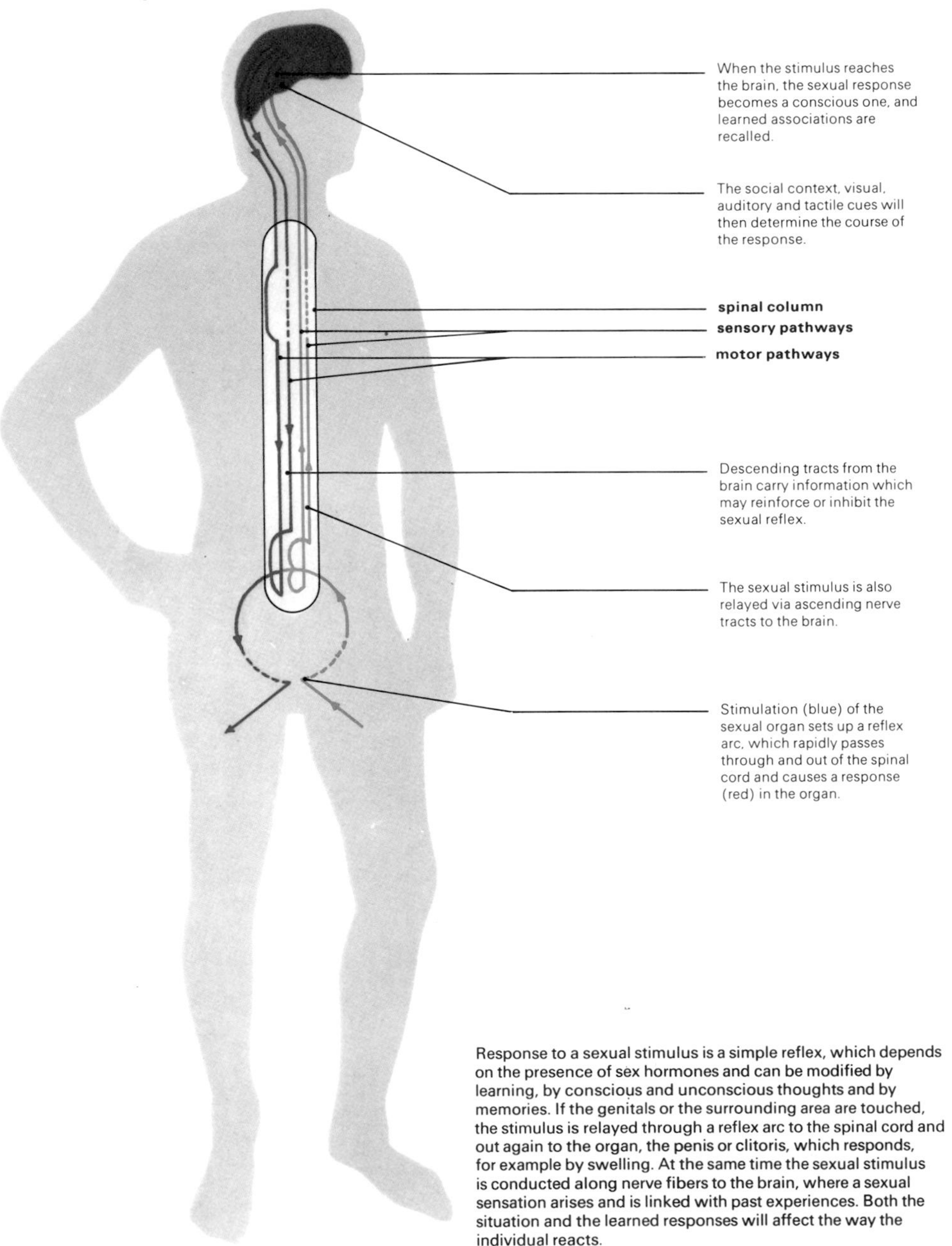

Response to a sexual stimulus is a simple reflex, which depends on the presence of sex hormones and can be modified by learning, by conscious and unconscious thoughts and by memories. If the genitals or the surrounding area are touched, the stimulus is relayed through a reflex arc to the spinal cord and out again to the organ, the penis or clitoris, which responds, for example by swelling. At the same time the sexual stimulus is conducted along nerve fibers to the brain, where a sexual sensation arises and is linked with past experiences. Both the situation and the learned responses will affect the way the individual reacts.

system and hypothalamus has largely freed human sexual response from its physiological roots. Reason, learning and experience, mediated by the cerebral cortex, are powerful influences on our sex drive. Once sexual response is initiated, however, a relatively determined and unvarying sequence of events occurs which is much the same in both sexes, and which may be divided into four stages—the excitement phase, the plateau phase, the orgasmic phase and the resolution phase.

Sexual capacity in humans can continue throughout life into old age, although it seems that, like many other activities, its continued performance depends on practice. Although celibacy carries no dire penalties, ill consequences may arise for either sex when there is frequent sexual arousal without the relief of orgasm. This is particularly likely to occur in women and, after several years, can result in a chronic congestion of the pelvic organs with consequent gynecological disturbances.

Recent evidence suggests that coitus may have previously unlooked-for effects on the hormonal status of both sexes. In men, testosterone production may increase as a result of regular intercourse, leading not only to increased sexual capacity but having general beneficial effects on energy and nonsexual drive. In some women, intercourse may trigger ovulation.

The whole cycle of sexual response is, like the sex drive itself, capable of being affected by conscious and unconscious states of mind. Arousal and orgasm can even occur as a result of erotic thoughts, without any physical stimulation. And at any point in the cycle adverse mental factors can intervene to prevent its completion. Certain drugs and illnesses can also cause loss of sexual capacity.

A number of studies agree that three or four sexual outlets per week is average in Western society, although there is very wide individual variation in both sexes. The outlets include orgasm produced outside intercourse, and by far the commonest source of these is masturbation, practiced at some time by virtually all males and most females. Orgasm can also occur during sleep, most commonly in males as a result of nocturnal emission.

The fact that, in humans, sex can and does occur regardless of whether impregnation is possible—a woman is only fertile for three or four days a month, for example—indicates that intercourse has nonreproductive functions. Most importantly, frequent sexual activity serves to strengthen the male–female relationship and thus helps to ensure a stable family background for the growing child during its many years of dependence.

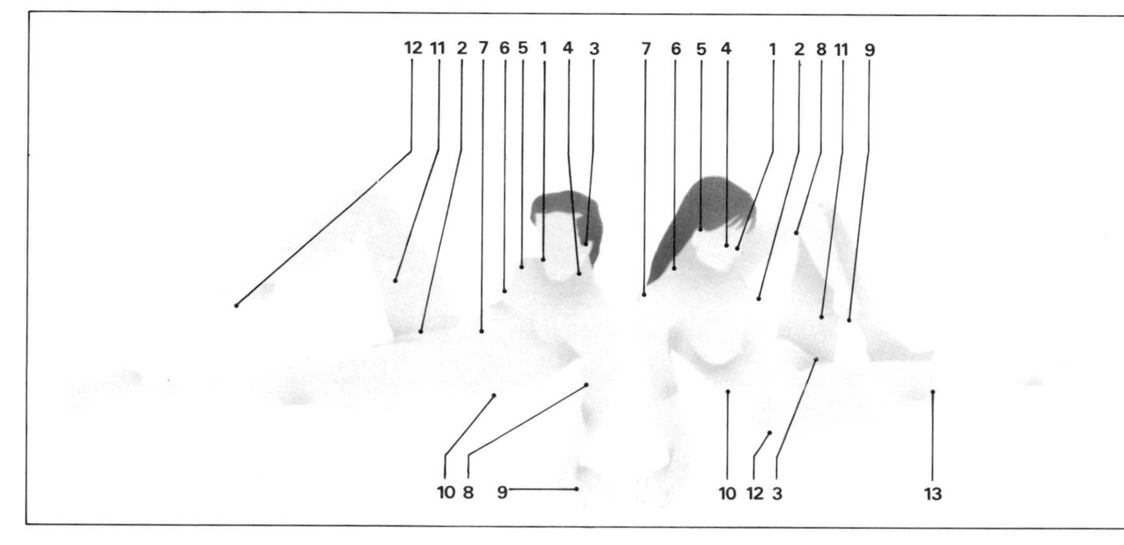

The areas of the body where contact is particularly effective in arousing sexual feelings are known as erogenous zones. Although there are variations between men and women and in individuals, the main female erogenous zones are the lips, the inside of the mouth and the tongue (**1**), the breasts and nipples (**2**) and the genitals (**3**). Also sensitive are the cheeks (**4**), the ears (**5**), the neck (**6**), the shoulders (**7**), the insides of the arms (**8**), the hands (**9**), the waist (**10**), the insides of the thighs (**11**), the buttocks (**12**) and the backs of the knees (**13**). The male erogenous zones include the lips, the inside of the mouth and the tongue (**1**), the genitals (**2**), the ears (**3**), the neck (**4**), the shoulders (**5**), the nipples (**6**), the navel (**7**), the insides of the arms (**8**), the hands (**9**), the base of the spine (**10**), the insides of the thighs (**11**) and the soles of the feet (**12**).

The sexual attraction that two people feel for each other is a basic drive, in the same way that the desire for food is a drive. The behavioral response to the sex drive is, however, more complex. Both men and women have an instinctive desire for sex, accentuated by the sex hormones, but the way they respond will depend on previous learning. Physical attraction is also a learned response and involves many factors, including body contour and smell.

Cross-reference	pages
Cerebral cortex	**80-81**
Coitus	**148-149**
Female anatomy and physiology	**146-147**
Hormones	**70-73**
Hypothalamus	**112-113**
Limbic system	**116-117**
Male anatomy and physiology	**144-145**
Puberty	**170-171**
Smell	**94-95**

The limbic system: mood and memory

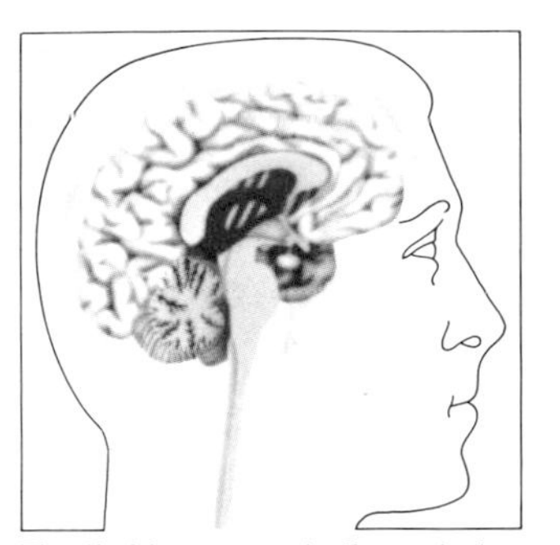

The limbic system (red), a paired structure, is concerned with memory and emotions. The striped area shows the position of those parts of the limbic system which are hidden by other parts of the brain.

The units of the limbic system

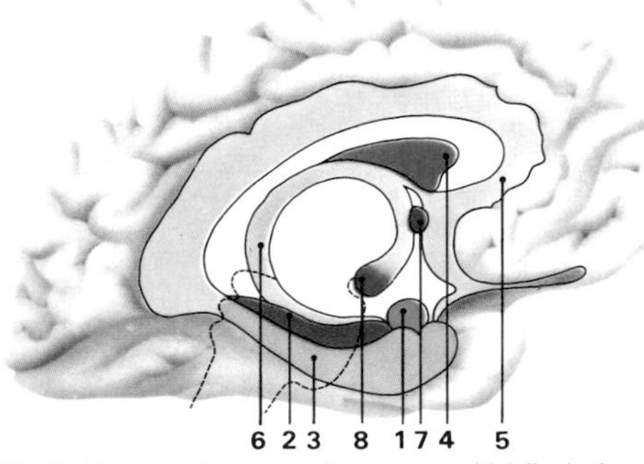

The limbic system is a composite structure which lies in the temporal lobes of the brain and in the region of the thalamus. It is concerned with emotions and memory. The amygdaloid bodies (**1**) are believed to be concerned with aggression. The hippocampus (**2**), lying above the parahippocampal gyrus (**3**), is concerned with memory. The septum pelucidum (**4**) is thought to be associated with pleasure reactions. The cingulate gyrus (**5**), the fornix (**6**) and the anterior commissure (**7**) carry nerve fibers to and from other structures in the limbic system, as do the mammillary bodies (**8**), which are also vitally concerned with memory. These eight areas make up the complete limbic system.

Man may be said to have not one brain but three, which work together in reasonable harmony but differ from each other in structure, biochemistry and function. Each of these brains marks a separate step on the evolutionary ladder.

The limbic system was the second of these three structures to develop, an addition in primitive mammals to the brain inherited from their reptile forebears. This reptilian brain forms, in man, the upper brain stem—much of the reticular formation, the midbrain and the basal ganglia—and includes the hypothalamus. Its task is largely the regulation of stereotyped, or instinctive, behavior and of vital biological functions and rhythms. Anatomically, the limbic system, or "old mammalian brain," surrounds these earlier structures and is intimately connected to them. In modern mammals it is in its turn surrounded by a later acquisition, the neocortex, or the "new mammalian brain," the source of reasoning and, in man, of language. The higher up the evolutionary scale a mammal is, the larger its neocortex is in relation to its limbic system.

Since the limbic system is structurally so primitive, and is found in much the same form in all mammals, it was once thought to be purely animalistic in function. Indeed, its links with the olfactory system led to its being referred to, even in man, as the rhinencephalon, or "smell" brain. It is now clear, however, that in higher mammals the limbic system has become adapted to wider purposes, until in humans it is involved in functions as diverse as emotion, attention, memory and learning.

The limbic system (as with all higher brain structures, there are two mirror-image systems one on each side of the brain) is nested deep in the temporal lobes. Each forms a ring around the brain stem and the basal ganglia. The hippocampus, fornix and hippocampal gyrus comprise the lower arc of this ring, the cingulate gyrus the upper. The front end of the ring is formed by the septum, amygdala and mammillary body. Within the ring is the anterior thalamic nuclei. These structures have intricate neural and biochemical connections among themselves and with both higher and lower areas of the brain. The details of these pathways are not yet precisely known, but it is clear that the interconnections within the limbic system and with lower brain stem areas at least are partially concerned with maintaining a balance, or equilibrium, of emotional state and of alertness.

This emotional function of the limbic brain has been the object of much research in animals. The lower part of the limbic ring, fed by the amygdala, seems to be largely concerned with feeding, fighting, fleeing and copulation. Electrical stimulation of these regions can cause emotion-based behavior indistinguishable from that produced in the normal way—except, for example, that the foe provoking rage or flight is nonexistent. Similarly, removal of portions of the ring can lead to lasting alterations in behavior. An amygdalectomy, for example, will render dominant and aggressive animals docile. To some extent the upper arc of the limbic system overlaps in function with the lower, but it seems to be especially concerned with expressive and feeling states which are conducive to sociability. Stimulation may cause grooming, courtship or sexual interest.

In humans, experimental stimulation of these deep-lying brain regions is only performed in the search for sites of abnormal brain activity, as a guide to future surgery. But rage, agitation, anxiety, elation, excitement, colorful visions, deep thoughtfulness, sexual interest and relaxation have all been produced as a result of electrical stimulation of the amygdala, hippocampus and septum.

Owing to the connections between the limbic system and the cortex, emotion is open to the influence of reason and reason may similarly be affected by emotion. Indeed, the balance between these brain areas is relatively easily disturbed, and there is some evidence to suggest that disordered limbic functioning may be at the root of some mental illness.

The direct links between the thinking cortex and the feeling limbic system are, however, only part of the story. Input from the sense organs is influenced by the current state of the limbic system before it gets to the areas of the brain in which it is "deciphered." This is how, as one psychologist has put it, the brain "transforms the cold light with which we see into the warm light with which we feel"—the reason the same scene may be rose-colored to the happy and blue to the sad. In extreme conditions, emotions can distort perception out of all relation to reality; in a milder form, this is the everyday experience of us all. A common example is that of looking for a friend in a crowd; the more anxious one becomes to find him, the more prone one is to "project" his features onto the faces of strangers.

Emotional response is an important part of the function of the limbic system. But it also has a crucial role to play in deciding which aspects of the environment we pay attention to and which we ignore. The hippocampus is especially implicated here. It seems to be continually comparing sense input with a learned pattern of expectation. So long as the environment remains unchanging, or at least reliably predictable, the hippocampus is active, and by being so inhibits or damps down the lower brain structures—the reticular forma-

The center of emotion and memory

The limbic system, which is concerned with memory and learning, is also believed to be primarily involved in emotional responses. Emotions are the conscious result of an interaction between the activities of the cerebral cortex, the limbic system and the visceral organs of the body which produce specific physical changes. A number of theories have been postulated to explain this relationship between the body and mind. The emotional responses of any individual are, however, also a product of his knowledge and experience.

tion—responsible for alertness. Once the scene changes or becomes unpredictable, the hippocampus ceases to be active and so ceases its inhibitory influence; released from this, the reticular formation becomes excited and alerts the organism. The change is noticed and attention paid to it. This is why one stops hearing a long-continued noise, for example, but instantly notices when it changes in quality or ceases. This has been nicknamed the "Bowery-El" phenomenon, after trains that ran on an elevated track in New York City. The trains were discontinued, and for several nights the police regularly received calls at the times when the trains used to pass, from people living along the line. The callers complained that something odd had awakened them. The odd phenomenon was, of course, the absence of a noise which the brain had come to expect.

The hippocampus, then, is a vital cog in the mechanism that allows us to select for notice only what is important in the mass of irrelevant stimuli that constantly bombards us. It seems also to play the same role where recall of stored experience is concerned. An individual who cannot separate his memory of what happened five minutes ago from what is happening now and what happened last year is, for practical purposes, an individual without memory, although his experiences may leave physical traces in the brain in the usual way. Patients with slight hippocampal damage exhibit disturbances in concentration and attention as well as in emotional response, and their thoughts, perceptions and recall tend to be interfered with by irrelevancies.

Being able to selectively attend to the present environment and relate this efficiently to the stored knowledge of the past is obviously an essential component of learning. But the limbic system is also involved with learning in another and equally important way. It contains most of the "reward" and "punishment" centers, which enable us to assess the results of our actions and to learn whether or not it is desirable to repeat them. These centers were first discovered when it was observed that rats would press a lever to obtain stimulation in certain brain areas, sometimes starving to death rather than leaving this pursuit to eat. With electrodes in the pain centers, the rat would work hard to avoid stimulation. Pleasure-center stimulation has been tried on depressed patients. Depending on the precise group of limbic cells involved, they experience sensations ranging from quasi-orgasmic bliss to deeply pleasurable relaxation.

One current suggestion is that recall of the pleasurable or unpleasant consequences of experiences and actions is mediated through the amygdala, since removal of this structure produces a state in which animals and people have great difficulty in learning by reward and punishment. The amygdala is thus believed to work in harness with the hippocampus; the hippocampus enabling us to choose the relevant aspects of a situation to respond to and the range of relevant possible responses to make, the amygdala reminding us which of these responses is likely to lead to pleasant results and which to unpleasant ones.

The limbic system still holds many mysteries, but overall its function may be likened to that of the unconscious mind as defined by Sigmund Freud. These deep brain structures transform the objective world of sense input and reason into the subjective world of human experience. They link feeling to thought and perception and impose emotional prejudices and learned expectations on reality, guiding our actions according to the imperative principles of pleasure and displeasure. With its decisive effect on behavior and perception and its essential irrationality, the limbic system may well hold the key to a fuller understanding of neurosis and of mental illness.

The amygdaloid bodies, which contain both incoming and outgoing nerve fibers, may help to regulate emotions, particularly aggression.

The fornix is a thick bundle of nerve fibers which forms the main outgoing pathway from the hippocampus.

Afferent fibers bring impulses into the dentate gyrus and the area of Ammon's horn.

The dentate gyrus lies within the hippocampal sulcus. It may relay impulses in the hippocampus, but its function is not fully understood.

The subiculum is a region of cortex composed of up to six different cell layers and a multitude of nerve pathways.

Ammon's horn, which merges with the subiculum, contains ovoid and pyramidal cells, the function of which are thought to be associated with memory.

Efferent fibers carry information out from Ammon's horn to other regions of the brain.

The hippocampus is composed of folded layers of cells and fibers and is associated with emotions, learning and short-term memory.

The parahippocampal gyrus is a highly specialized area of nerve cells concerned with memory pathways.

Cross-reference	pages
Brainstem	104-105
Cerebral cortex	80-81
Emotions	118-119
Evolution of brain	14-15
Hypothalamus	112-113
Memory	130-131
Reticular formation	106-107

Emotions: the range of human feelings

A man is driving along a highway in thick fog. Suddenly he sees a group of cars piled up in front of him. The man is frightened, his mind is temporarily paralyzed, his muscles tense up, his stomach is queasy and he feels as if his heart is trying to break out of his chest. His behavior is instantaneous and dramatic as he seeks evasive action. He jams on the brakes and tries to turn into a safe area. But he fails. There is a screeching of tires, the breaking of glass and the sound of smashing metal. Then all is quiet. He realizes he is alive. The fear turns to joy and then the joy becomes anger directed at an anonymous figure he blames for causing the accident. He gets out of the car, his body tense with anger. He approaches the devastated cars, sees the injured, feels compassion and turns to seek help.

This is a description of a range of emotions—reactions, some pleasant, some unpleasant—that have a motivational effect on behavior. Emotions are consciously experienced and cause both psychological stress and physical upheaval to effect changes in behavior. Consequently, any theory of emotion has to directly involve the body as well as the mind.

Two theories appeared at approximately the same time in the late nineteenth century. In 1884, William James, a Harvard psychologist, suggested that certain environmental stimuli

Physical responses to emotion

Fear is one of the most extreme emotional feelings and, like all extreme emotions, is accompanied by physiological changes in the body. This illustration shows how the body prepares itself for "fight or flight" when in a highly emotional state. The initial signal comes from the brain, which spurs the body to release adrenaline into the bloodstream. This then triggers off a series of interrelated responses in the body.

1. The mere thought of fear activates the frontal lobe of the cerebral cortex, which stimulates the hypothalamus into action.

2. The hypothalamus, positioned in the brain, activates the suprarenal medulla.

3. The suprarenal medulla releases adrenaline into the bloodstream and numerous responses in the body ensue.

4. The pupils of the eyes dilate.

5. Hair stands on end.

6. If the skin is broken, blood will readily coagulate to prevent severe loss.

7. The chest expands to increase the volume of inhaled air.

8. The bronchioles relax, allowing a greater volume of oxygen to enter the lungs.

9. The heart dilates, increasing the blood output.

10. Blood pressure rises.

11. Muscles contract.

12. Blood vessels near the surface of the skin contract, causing the skin to pale.

13. Other blood vessels dilate, and the liver releases glucose, which provides fuel for the muscles.

14. The bladder empties stored urine in cases of extreme fear.

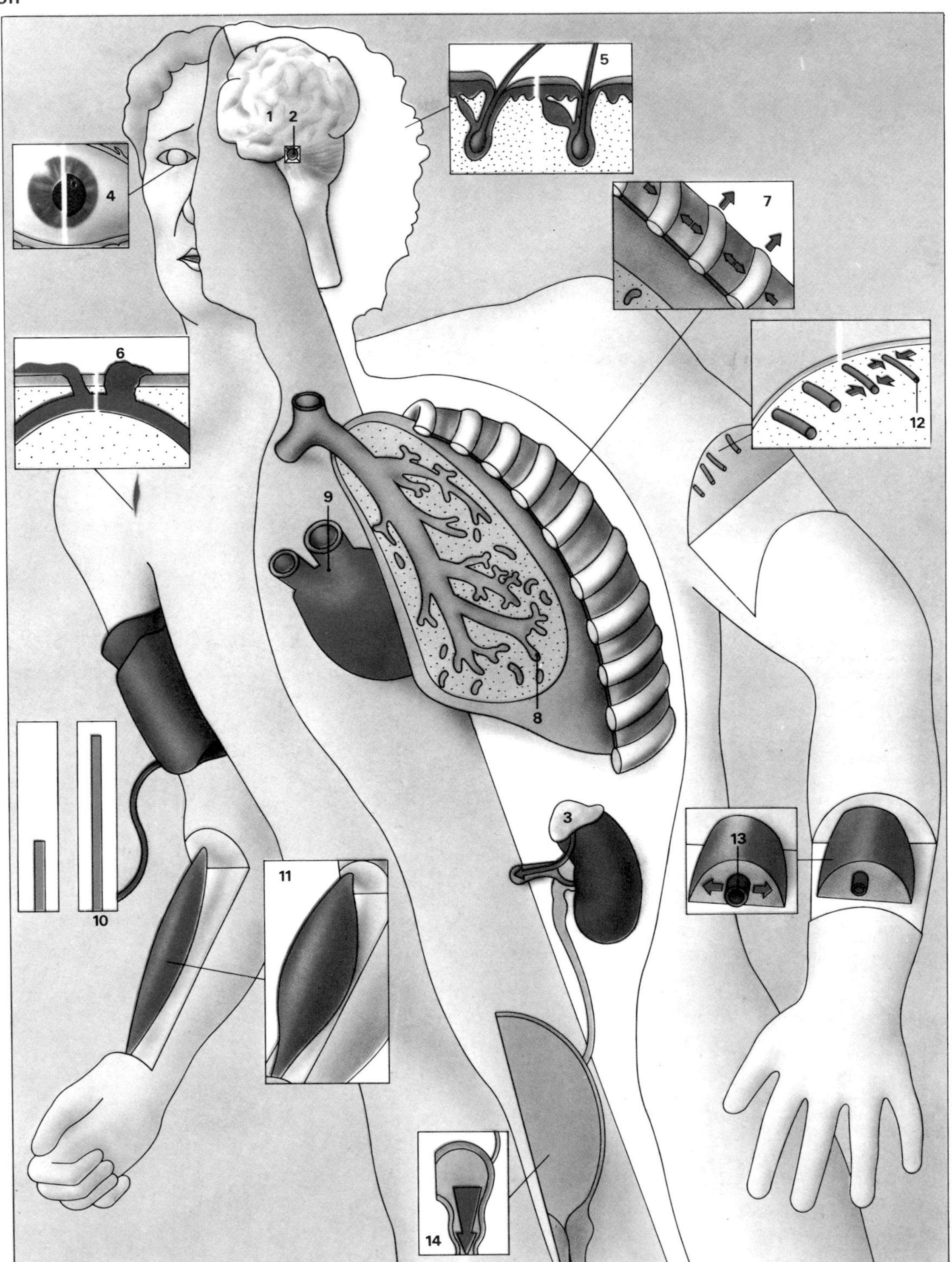

directly evoke responses from the body and that these visceral responses, from the internal organs including the heart, stomach, lungs and intestines, affect the mind to produce an emotional feeling. James summarized his entire theory in the statement, "We are afraid because we run, we do not run because we are afraid."

At about the same time, a Danish scientist, Carl Lange, produced a similar idea about the development of emotion. The two theories, bound together and known as the James–Lange theory, proved to be a landmark in understanding emotion. Almost immediately, however, the theory became the object of deep and strong criticism. The chief critic was Walter Cannon, who held that emotion is a state of heightened arousal and is part of an activation system. He believed that the activation continuum extends from sleep to panic. Together with P. Bard, Cannon produced an activation theory which contended that emotions serve as part of an emergency system for activating the body. The Cannon–Bard theory was, therefore, in direct opposition to the James–Lange theory. According to the Cannon–Bard theory, the seat of emotion was in the brain itself, specifically the thalamus. Outside stimulation was received by the thalamus, which in turn sent impulses to both the cerebral cortex and the viscera, in this way preparing the body for action. Emotion, they believed, resulted from the new heightened state of arousal.

Modern research has now shown that both these early theories are oversimplifications, but there still appears to be no definite explanation for the relationship in emotion between the mechanical analogies of information processing and the physiological changes of those in the body. Building on the work of James–Lange and Cannon–Bard, most modern theories emphasize the cognitive, or thinking, element in the formation of emotion. In other words, the brain seems to work out an appropriate emotion from what it has perceived around it or from information it has received from a third party. One study has shown, for example, that people, when told about a particular situation, tend to exhibit the correct emotion for the situation, whether or not their knowledge of the situation was accurate or firsthand. This would indicate that all emotions tend to be "in the eye of the beholder."

To study emotions in depth they must be analyzed, ordered and described and this no one has yet been able to achieve. The almost limitless vocabulary applicable to emotions gives some indication of the wealth and subtlety of human feelings. However much scientists and writers try to draw comparisons between the mechanical analogies of information processing, they will always fall short when it comes to describing emotion. One of the major difficulties in doing a descriptive analysis of emotion is that emotions are usually only relevant to the immediate context in which they occur, and even then they will vary with the intensity of feeling.

Feelings of pleasantness and unpleasantness, known as hedonic feelings, are regarded as the most fundamental emotional experience. Included in the range of pleasantness are joy, love, laughter and contentment, while fear, grief, hate and resentment are unpleasant. Every emotion, whether pleasant or unpleasant, is affected by the intensity of feeling and the amount of tension that the emotion generates.

Emotions, however, may not always easily fall into the hedonic categories as they can often be paradoxical. Pleasant feelings can sometimes become so overpowering that they then become unpleasant, as when a pleasurable stroking of part of the body becomes unpleasant and irritating when it is overdone.

To measure emotion, psychologists look at physiological changes in the body that are readily measurable. These normally include heart rate, blood pressure, breathing rate and skin conductance. The electrical conductance of the skin, commonly called the galvanic skin response (GSR), offers an indication of emotional feeling, particularly of anxiety. Any feeling of emotion is normally accompanied by a change in GSR. This is the principle upon which lie detectors work.

Facial expression communicates emotion. Different emotions are conveyed in the photographs above. The expression on the left shows surprise or horror, while the other expression, right, could be either joy or pleasure.

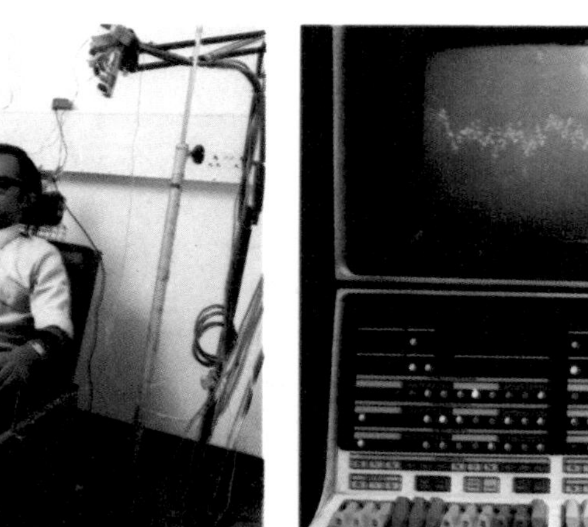

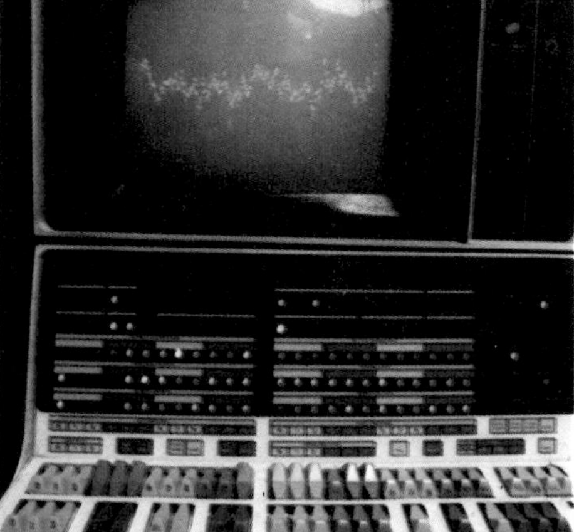

A polygraph measures physiological changes that occur during emotion. Attached to the subject's body are electrodes measuring heart rate and skin conductance. Fast brain waves (EEG) are characteristic of an emotional state. On the screen can be seen a computerized EEG reading that is being recorded from the subject.

The polygraph, above, shows a readout. The top tracing monitors heart rate. Below this is the trace from a timer which registers one-second intervals. The next two traces are measurements of skin conductance. The rapid fall which is shown on the readout accompanies an emotional response.

Cross-reference	pages
Cerebral cortex and thalamus	**80-81**
Consciousness	**106-107**
Hormones	**70-73**
Hypothalamus	**112-113**

The intellect

The human mind, the source of facts and fantasies, ideas and ideologies, faculties and feelings, has fascinated men throughout the ages. For centuries, the search for knowledge of the mind—and of its relationship to the body—was a philosophical, and even a spiritual, matter. Only recently, with the rise of psychiatry and psychology, has the quest become a scientific one as we seek to define and to quantify such abstract entities as intelligence and personality, emotions and creativity.

Indeed, great advances have been made in relating parts of the brain to particular functions of the mind. It is now known, for example, that the two halves of the brain—mirror images of each other—are involved in different kinds of activity. The left hemisphere is specialized mainly for inference, deduction and language, while the right brain furnishes our visio-spatial skills, our creativity and our appreciation of form and color.

There are still, however, many uncharted areas. Specific anatomical relationships between intellectual functions and clusters of brain cells have yet to be defined. Memory, self-awareness and thought processes, for example, are so complex that they are believed to require the involvement of the whole brain rather than one or two restricted regions.

That the study of man's mental processes remains so engrossing—and many of the answers so remote—is, in itself, an indication of the phenomenal powers of the cerebrum, the seat of human consciousness and of all learning, speech, thought and recall. This is the most challenging field of exploration of all.

The acquisition of symbolic language has been a major factor in man's development of rational thought. The extent of subtle phonetic combinations, as shown by the voiceprint, left, ensures a wide variety of words to communicate a vast range of ideas.

Frontal lobe and personality

The word "personality" comes from the Latin *personare*, to speak through. It derives from the practice, in classical times, of actors speaking through masks which they held to their faces to denote the character they were playing. But an adequate definition of personality, of the nature of man underlying the mask of behavior, has proved elusive.

It is difficult to say where, physiologically, personality originates. Since the frontal lobe is known to be especially significant in higher mental activity, it is believed that this is the segment of the brain which is mainly responsible for personality, for the combination of qualities which makes each one of us a unique being. The largest part of the cerebral cortex, and still the least understood, the frontal lobe is the great mass of tissue which extends from the region behind the forehead to the central sulcus, or groove.

At various times in medical history the frontal lobe has been considered the center of emotional control as well as the seat of intelligence, an assumption undoubtedly based on the fact that the most intelligent animals have the largest frontal lobes. This view foundered with the discovery that patients with frontal-lobe injury achieve normal scores on most standard intelligence tests.

Today all that is definitely known is that the frontal lobe is responsible for planning and judgment and for giving us the ability to form concepts, and to modify these concepts, by the use of information from other areas of the brain, including the memory stores.

The idea of a correlation between physical type and personality is an ancient one. It was Hippocrates who, in the fifth century B.C., first categorized individual temperament on the basis of four body substances, or "humors"—blood, phlegm and black and yellow bile. According to his theory, a preponderance of blood produced sanguine individuals; phlegmatic types were sluggish; those with an excess of yellow bile were choleric; and the type described as melancholic—from the Greek *melan coln*, black bile—tended toward depression. Hippocrates' descriptions still linger in the language of personality, and although inconsistent with present-day knowledge of physiology, his concept was the forerunner of modern theories correlating temperament with glandular activity.

It was not until the advent of scientific psychology and psychiatry at the end of the nineteenth century that intricate theories began to be formulated on clinical observation and research. There arose many rival personality theories—attempts to explain in scientific terms what human beings are really like and why they behave as they do.

In the early 1900s, Sigmund Freud, a Viennese neurologist and the father of psycho-

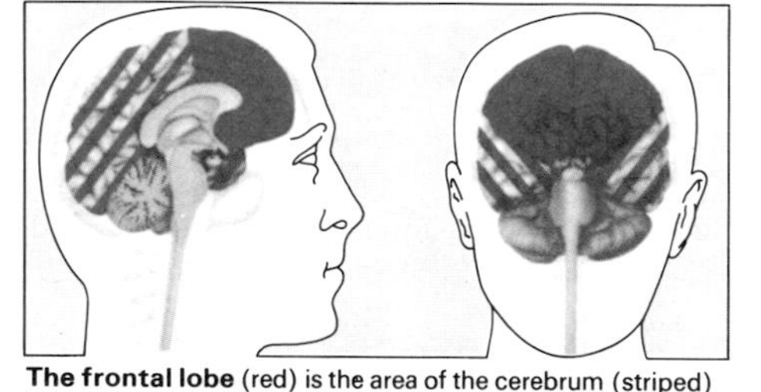

The frontal lobe (red) is the area of the cerebrum (striped) which has the greatest influence on personality development.

Frontal lobe damage can be detected by the formboard test, which is similar to a child's "letterbox" game. When this test is administered, the patients are blindfolded. Those with undamaged frontal lobes will correctly locate the shapes, top, those with damaged lobes will not.

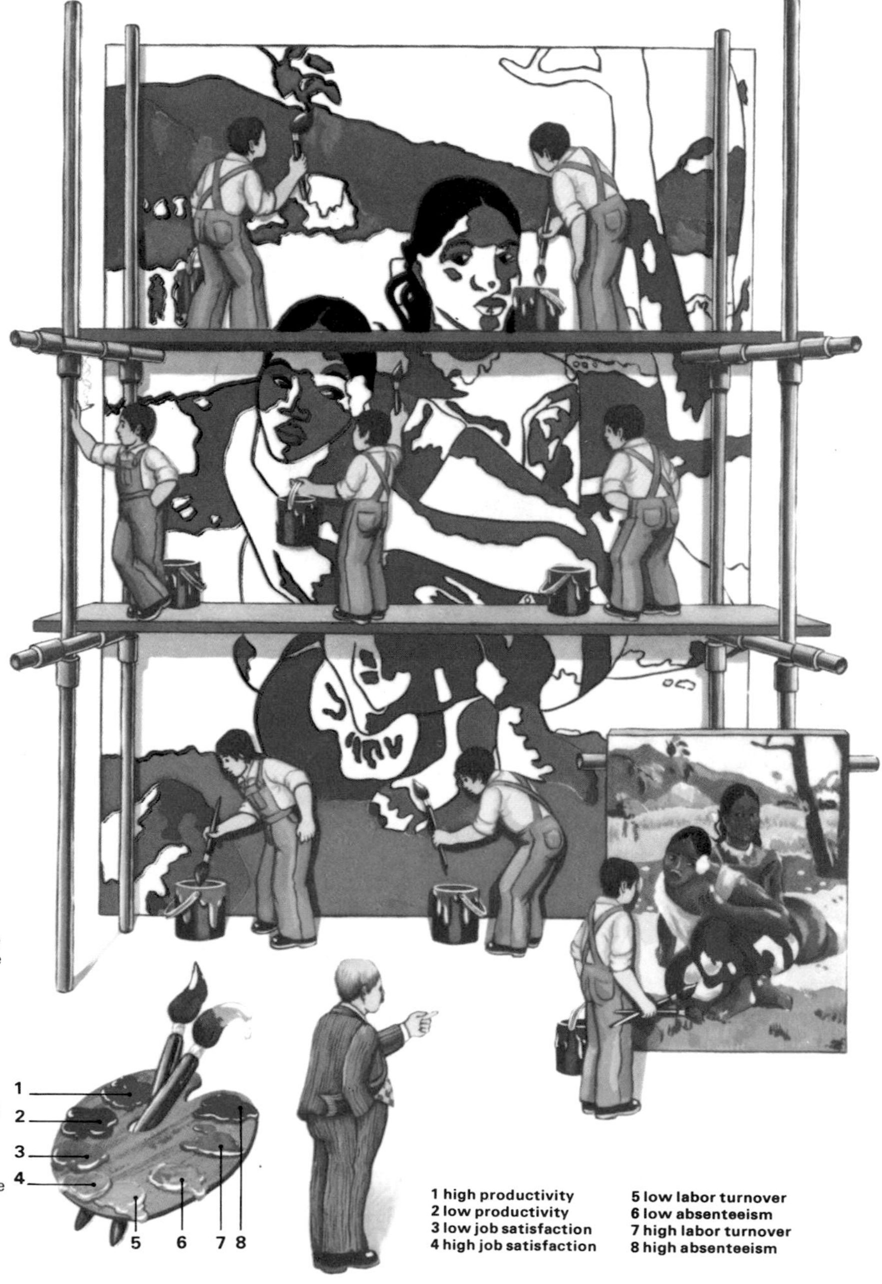

Authoritarian versus democratic leadership

A leader is usually described as "a person who has more influence over the behavior of a group than anyone else in the group." Leadership style is one way of expressing personality. Here are two contrasting leadership styles, authoritarian and democratic, organizing and directing the painting of a set. The authoritarian style, right, is rigid, with the uniformed workers dictated to by their director. The democratic style, far right, retains such elements of individualism as dress, and the workers communicate freely with their director. A leader is judged as successful or unsuccessful in terms of his effectiveness in job completion. This effectiveness is usually measured in terms of productivity, job satisfaction, personnel turnover and absenteeism, qualities indicated by the colors each group has filled in on the pictures. The workers under authoritarian direction have painted the picture in the colors that correspond, on the palette, to high productivity, low job satisfaction, high labor turnover and high absenteeism, while those under a democratic style of leadership have colored in those colors that represent low productivity, high job satisfaction, low labor turnover and low absenteeism.

analysis, developed a detailed theory of personality, based on fifty years' experience of treating patients. According to Freud, personality consists of three separate but interacting systems; the id, the unconscious part of the psyche, or mind, which comprises needs, drives and instinctual impulses; the ego, the executive of personality, which handles transactions with the outside world; and the superego, the moral aspect of personality, obtained from childhood conditioning by the parents, which maintains traditional values and ideals and is constantly in conflict with the id. Freud, who stressed the significance of the early years of childhood, was the first theorist to elaborate the developmental aspects of personality.

Carl Jung, a Swiss psychiatrist, as a young man worked with Freud and later formed his own school of psychoanalysis. According to Jung's complex and rather mystical theory of personality, there are two levels in the unconscious. The "personal unconscious" contains experiences the individual has had which have been repressed or forgotten. The "collective unconscious," common to everyone, contains behavior patterns and memories derived from man's ancestral past. Jung also grouped individuals under the headings of introversion and extroversion, a distinction later adopted and modified by other theorists.

In contrast, Alfred Adler, also an early associate of Freud, regarded consciousness as being the center of personality. He believed that human beings are aware of the reasons for their behavior, and that they are motivated primarily by social urges. Part of Adler's theory of personality, which distinguishes it from Freudian psychoanalysis, is his insistence on the uniqueness of the individual personality.

Another method of defining personality involves the pinpointing of traits, acquired through heredity, upbringing or environment, which characterize the individual. Although common traits may be shared by many people, there are endless combinations, yielding "trait profiles" which are almost as individual as sets of fingerprints.

Along with typologies, psychologists have been concerned with developing techniques for exploring the characteristic variables that are at the core of personality. The standard assessment device is the personality inventory, a test which requires subjects to respond to a series of statements or questions. One of the most widely used is the Minnesota Multiphasic Personality Inventory (MMPI), developed originally as a tool to assist diagnosis of psychiatric disorders.

Principal exponents of the quantitative and objective approach are the American psychologist R. B. Cattell and the British psychologist Hans Eysenck. Cattell has pioneered, among other tests, the 16PF, which uses sixteen different source traits to measure personality. Eysenck measures personality along two principal dimensions—introversion-extroversion and neuroticism-stability.

More impressionistic are projective tests, which probe for personality characteristics by confronting subjects with ambiguous or meaningless material for comment. The idea behind these tests is that the subject projects something of his own personality into his interpretation and response. The best-known projective test is the set of ten inkblots devised by the Swiss psychiatrist Hermann Rorschach. It consists of a series of ten carefully selected standard cards, where the image on each looks as if it had been made by spilling ink onto a piece of paper which was then folded, creating a symmetrical inkblot. There are no right or wrong answers to the Rorschach test. What the individual describes as seeing in each inkblot is the product of his imagination and draws on his capacity to fantasize. Psychologists believe that the answers are more revealing of personality than those given to direct questions asked in personality inventories.

Today there are so many different and often contradictory views of personality, each backed by a fund of evidence, that it is doubtful whether there will ever be one comprehensive theory to account for the complex nature of man. A simple definition of personality, on which everyone can agree, remains one of psychology's greatest challenges.

Cross-reference	pages
Cerebral cortex	**80-81**
Intelligence	**132-133**

Left hemisphere: the logical brain

Although they may look identical, the twin halves of the brain perform totally different functions. For many years there has been evidence of the specialization of the two brain halves, which came mainly from observing patients with brain lesions, brain damage caused by injury or disease. As long ago as 1874, the British neurologist John Hughlings Jackson described the left hemisphere as the seat of the "faculty of expression." He reported, too, that a patient with a tumor in the right side of the brain "did not know objects, persons and places."

Almost a century later, knowledge of hemisphere function advanced dramatically. Influenced by laboratory research on split-brain animals carried out in California, two American neurosurgeons pioneered a radical new treatment of epileptics who were disabled by frequent major seizures. They cut through the nervous pathway which links the two halves of the brain and thus unites the mind. Known as the corpus callosum, this link is a thick band of nerve fibers—about four inches long in humans—which penetrates deeply into the hemisphere in each side and transmits memories and learning from one hemisphere to the other. It was believed that cutting through the corpus callosum, the main line of communication between the two halves of the cerebral cortex, would reduce the number of cells involved in a seizure.

As a result of the operation the condition of the "split-brain" patients was improved and they showed little change in their day-to-day lives. But the subtle effects of surgery only became fully apparent when Roger Sperry, the psychologist who had done the original work on laboratory animals, followed through with a battery of evaluative tests. It was soon obvious that the operation had completely isolated the specialized functions of the two hemispheres. Here were people each of whom was in effect endowed with two minds—one verbal, analytical and dominant, the other artistic, but mute.

Simple verbal commands like "raise your hand" or "bend your knee" could only be carried out with the right side of the body, under the direction of the left hemisphere. Evidently the right brain could not comprehend these instructions. Similarly, if the patient had an object hidden from his view in his right hand, he could say what the object was. When he held it in his left hand, he could not say what it was, but he could subsequently distinguish it amid an assortment of items. It was clear that he could recognize the object

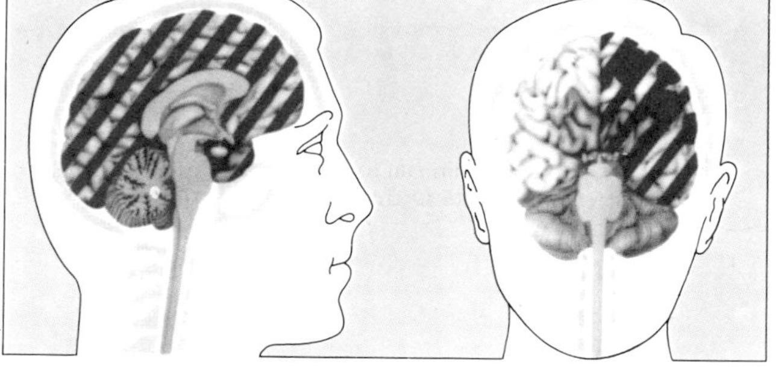

The left hemisphere of the human brain from three different views. Top left, the brain is seen in section from the side with the right hemisphere removed. Top right is a frontal elevation of the hemispheres. The solid red patches indicate some of the areas involved with the control and production of language. Language areas on the surface of the left hemisphere are shown right. Broca's area is involved in the motor control of speech, and Wernicke's area in more general language control. Naming and writing are controlled by other areas.

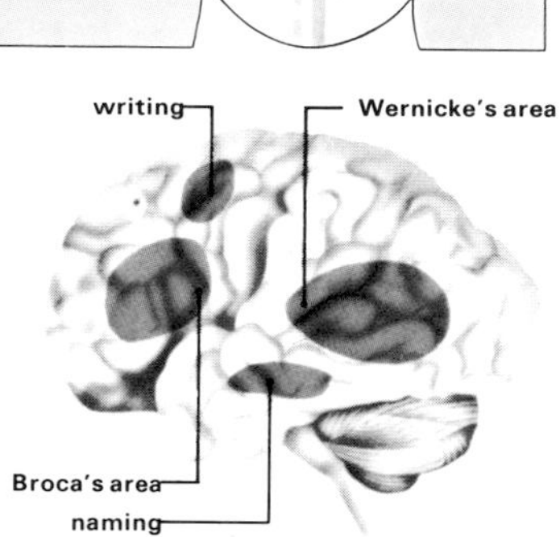

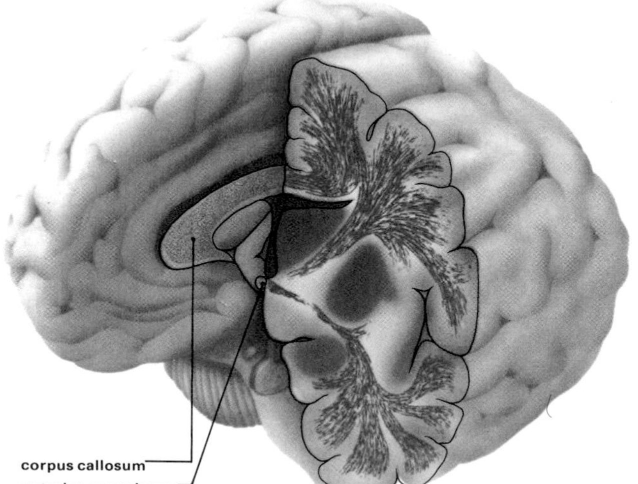

The corpus callosum, about four inches long, connects the two cerebral hemispheres. Thousands of nerve fibers disperse from this link into the white matter of the cerebrum. The anterior commissure, a smaller connection, lies in front of the corpus callosum. If the corpus callosum is cut, the two hemispheres become functionally independent.

The left hemisphere and language

The exact mechanism by which the mind assembles words into grammatical sentences is not completely understood. Psycholinguistic theory, one school of thought, believes that a number of ideas or principle mechanisms interact to produce a sentence. The transport system illustrated here incorporates psycholinguistic principles into its design, with the flow and direction of the passengers and trains arranged to appear analogous to the order of events in the production of a sentence from its basic units. The main terminus, in the foreground, is served by several smaller stations. Each subsidiary station and train represents one of the stages of psycholinguistic theory. The passengers continue to interchange until they arrive at the main terminus, where they form a regular sequence to exhibit the features of correctly constructed words. From the main platform they are transferred to the main-line train, which represents a sentence ready for dispatch as a comprehensible verbalization which also follows the rules of syntax and semantics.

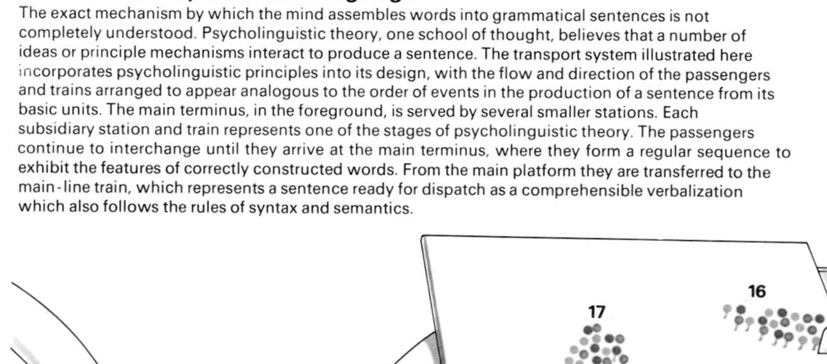

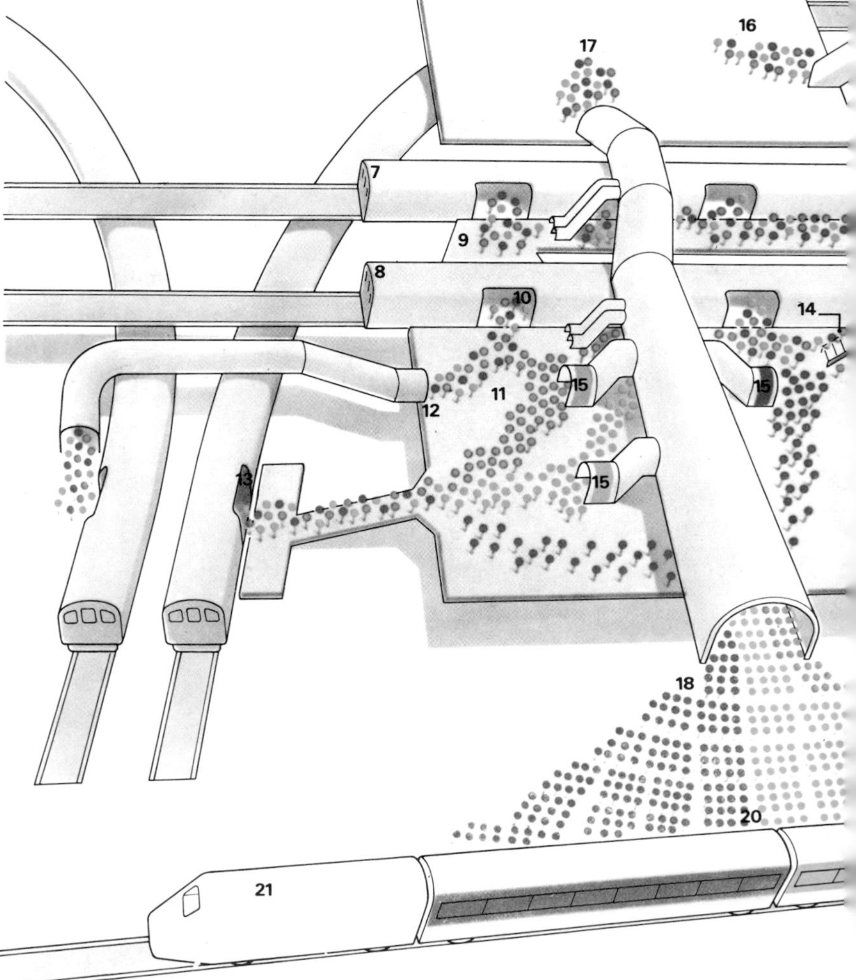

encountered by his left hand, but he was not able to talk about it.

Curiously, even when the corpus callosum is severed, the two halves of the brain continue to interact. One hemisphere will often "cross cue" the other. If, for example, the patient is confronted with a red or green light, only the right hemisphere, which is specialized in color discrimination, can interpret it. But although it can recognize the color of it, it lacks the verbal ability to name it. The left hemisphere can name the color, but is prevented from seeing it, so it guesses. When the right hemisphere hears its verbal twin making a wrong guess, it notes the error and at once causes the forehead to crease and the head to shake. The left brain, using its power of logic to act on the reprimand, immediately changes its guess.

Tests have shown that a patient who has only a functioning left hemisphere, acts more like a total person than the individual with only a functioning right hemisphere. He can speak, and can write with his right hand. The left hemisphere permits him to describe feelings and sensations, although he cannot draw shapes and is poor in music. However, he is unable to get around in the three-dimensional world in which he lives. Damage to the left hemisphere often interferes with language ability. Such patients speak with difficulty and in some cases they cannot speak at all.

In most adults the speech centers are limited to the left hemisphere. About 15 percent of left-handed people, however, and about 2 percent of right-handed people have speech on both sides. Some left-handed people, however, develop speech on the left; less than half have it on the right. Although the right side of the brain primarily controls the left side of the body and the left side of the brain largely controls the right side of the body, being left-handed usually indicates that the two halves of the brain have not become as fully specialized as they are in those people who are right-handed.

In young children each side of the brain probably possesses the potential for speech and language. In the early years brain damage on the left side often results in the right side developing a facility for language. Dominance for speech, and probably for other skills as well, is firmly established in one hemisphere by the age of ten and cannot be transferred to the other thereafter.

The left cerebral centers seem to be involved in the ability to recognize groups of letters as words and groups of words as sentences as well as in speech and spelling. Numeracy, mathematics and logic, and abilities necessary to process a mass of information into words, actions and thoughts are also controlled by the left cerebral hemisphere, which functions as the logical brain. Intelligence tests which investigate vocabulary, verbal comprehension, memory and mental arithmetic all evaluate left-hemisphere activity.

The difference in abilities between the two cerebral hemispheres seems to be unique to human beings. It has been argued that our brains have become specialized in this way because language and logic call for more-ordered thought processes than the brain-power needed for spatial orientation. Perhaps it is simply that the two halves of the brain are complementary, giving us alternative modes of thought which we can utilize in an increasingly complex world.

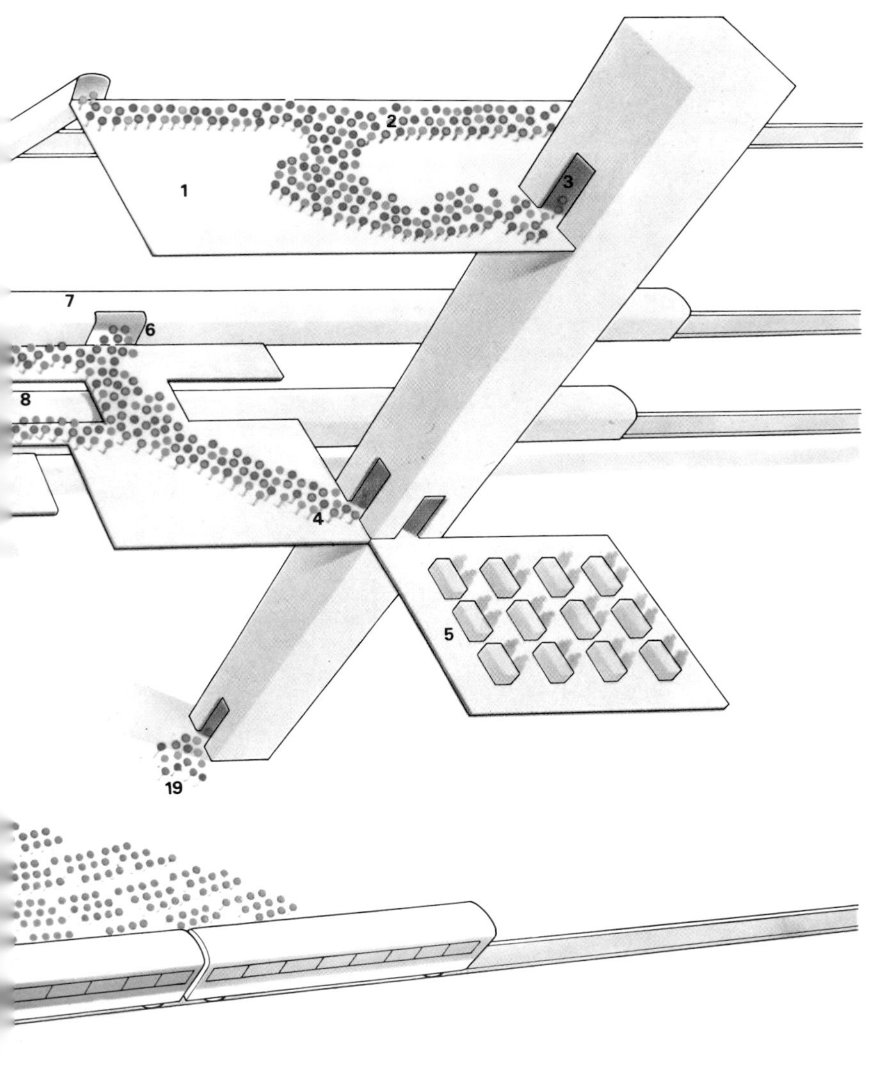

1. The assembly point of the lexicon store represents the mind's memory bank of words, or, since there is no evidence that words as they are spoken are stored in their entirety, features of words.

2. The assembly of words ready for the start of sentence formation.

3. The entry of words into the elevator leading to the deep structure which represents the part of the brain where the main rules of language are contained. These rules include both grammatical structure (syntax) and meaning (semantics).

4. Words enter the departure area for dispersal in the deep structure.

5. The overall control center symbolizes the brain's center for the organization of language, which is probably located in the left hemisphere.

6. Words enter the brain and proceed on their journey around the language center, where, by means of interchanges, they will be sorted according to the rules of syntax and semantics. This is perhaps the most important part of the process, for the syntax contains the correct grammatical rules for the construction of a sentence, while the semantics govern its meaning. The two are closely connected, for without syntax little meaning can be achieved.

7. The train bearing words which will be taken in the processing circuits of syntax and semantics.

8. The arrival of a train containing words that, during their journey, have acquired the attributes of meaning and the features of syntax.

9. An interchange point between trains of words which need to journey farther because more syntactical information is required.

10. Words that have completed their journey and now have a grammatical sense arrive in the transformational area.

11. In the transformational area, words with a grammatical constitution are reordered and combined for use in the sentence. They are now submitted to more rules, which govern such points as tense, mood and emphasis.

12. The arrival of words, such as negatives, with specific grammatical makeups at the transformational area.

13. More words—which could be in the passive mood—arrive at the transformational area.

14. The arrival at the transformational area of words that have escaped full syntactical and semantic processing. These could cause an ungrammatical sentence.

15. Words from the transformational area organize themselves to enter the surface structure.

16. The direct path of descent for words from the lexicon store allows words in common use direct access to the surface structure. These words already contain many syntactical features.

17. The entry of words in common use onto the surface structure, which contains the structural rules given to words governing presentation and pronunciation.

18. The emergence of words with correct grammatical makeup and pronunciation ready to take their place in a sentence.

19. Words that have fallen through the deep structure and which could emerge in the spoken sentence as totally garbled or ungrammatical language.

20. The organized features of a sentence lined up in the correct order to be used as comprehensible language.

21. A sentence ready for dispatch.

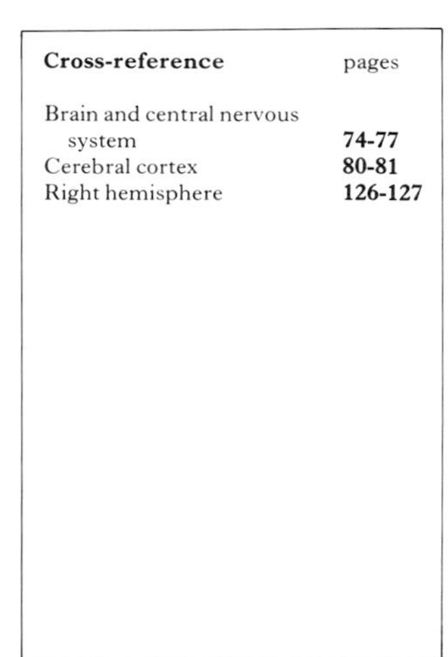

Cross-reference	pages
Brain and central nervous system	**74-77**
Cerebral cortex	**80-81**
Right hemisphere	**126-127**

Right hemisphere: the artistic brain

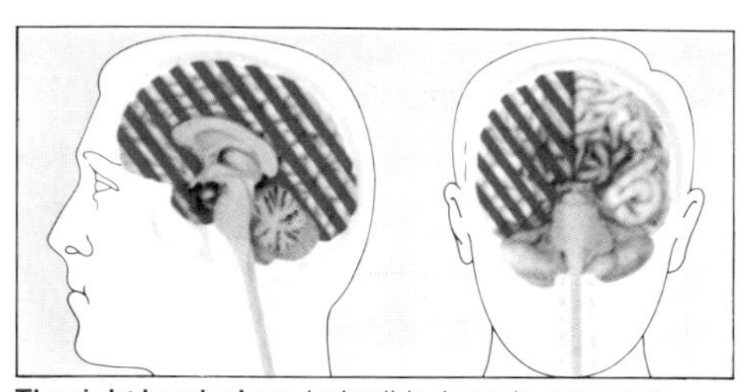

The right hemisphere (striped) is shown in cross section from the side with the left hemisphere removed and in a frontal view. This hemisphere has control over visio-spatial and musical abilities and abstract thought.

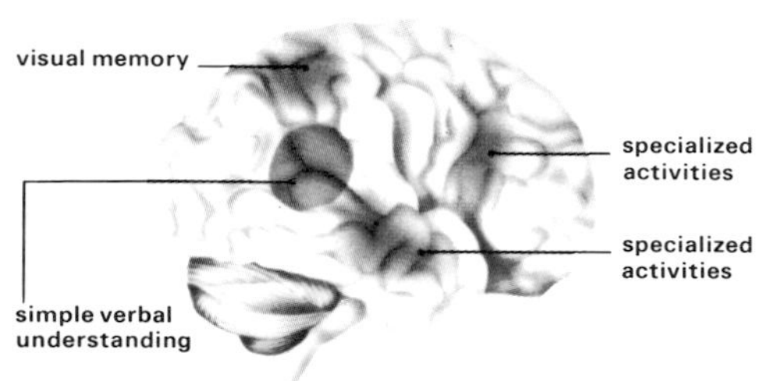

Visual memory has been successfully located on the lateral surface of the right hemisphere. Mapping produces areas for simple verbal understanding, present in the parietal lobe, while the specialized activities are located in areas similar to those on the left hemisphere.

In about 400 B.C. Hippocrates wrote of the possibility of the duality of the human brain, but it was not until 1684, when an Englishman, Sir Thomas Browne, suggested that the two halves of the brain might have a preference or "prevalency" for the behavior they control, that the significance of Hippocrates' observation became apparent. Browne's concept of the brain having two different functional halves was not, however, taken up and extended until 1874, when John Hughlings Jackson introduced the idea of the brain having a "leading" hemisphere.

Initially, the left hemisphere was regarded as the knowledgeable, vital side of the brain, eclipsing the right hemisphere in importance. Research since 1942 has, however, definitely shown that the right hemisphere is of equal importance and is principally responsible for visio-spatial ability, for artistic ability and for the understanding and appreciation of music.

A more complete understanding of the right hemisphere emerged as a result of the split-brain operation for epilepsy. This operation severs the bundle of nerve fibers, the corpus callosum, which joins, and serves as a link between, the two hemispheres. An exponent of the split-brain operation was the American Roger Sperry, and as a direct result of his investigations the true value of the right hemisphere has been revealed. Sperry's patients were operated on in an attempt to control severe epilepsy and in this the split-brain technique was found to be relatively successful. After the operation patients seemed to behave in a perfectly normal manner. Under closer examination, however, it was soon realized that something odd happened in their responses to psychological testing. It was clear that the two hemispheres of the brain were now functioning independently.

The right hemisphere was found to have as many specialized functions as the left hemisphere, and, perhaps more important, the method of information processing in each of the hemispheres was obviously different. The right hemisphere, or mind, does not use the conventional mechanisms for thought analysis as does the left hemisphere. Instead the nonverbal, visio-spatial right brain conceives situations, and strategies of thinking, in total. The left hemisphere processes information in logical, analytical stages, using its language power as the instigator and mediator, while the right hemisphere integrates several kinds of information quickly and then relays it as a complete whole. The method of processing used by the right hemisphere therefore lends itself to the special type of immediate answer necessary when dealing with visual cues and spatial orientation.

The right frontal lobe and the right temporal lobe appear to operate the nonverbal, specialized activities of the right hemisphere. This

The colorful right hemisphere

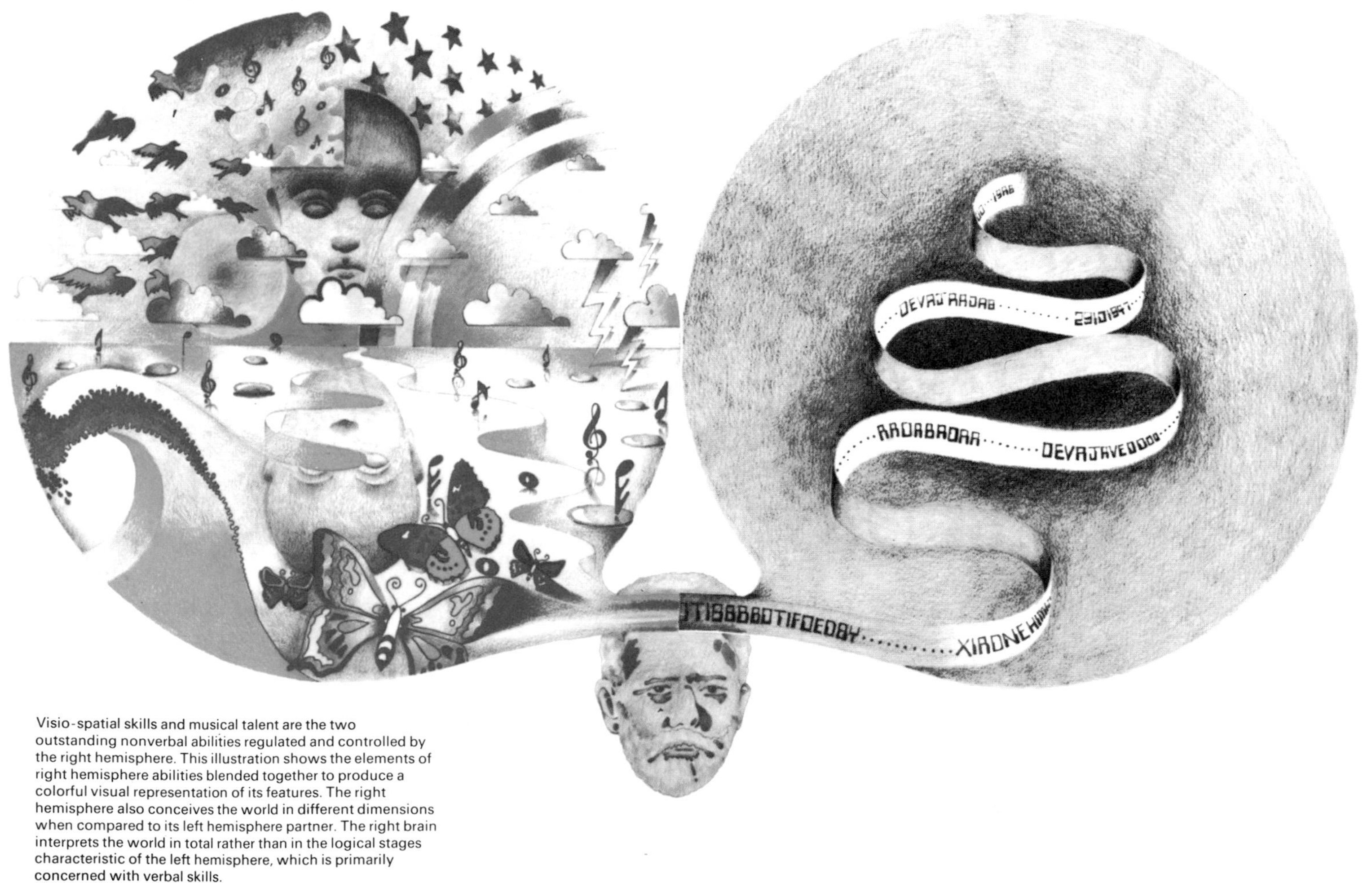

Visio-spatial skills and musical talent are the two outstanding nonverbal abilities regulated and controlled by the right hemisphere. This illustration shows the elements of right hemisphere abilities blended together to produce a colorful visual representation of its features. The right hemisphere also conceives the world in different dimensions when compared to its left hemisphere partner. The right brain interprets the world in total rather than in the logical stages characteristic of the left hemisphere, which is primarily concerned with verbal skills.

in many respects corresponds to the speech control functions of the left frontal and temporal lobes of the left hemisphere. The two other lobes of the right hemisphere, the parietal and occipital, appear to have less specific functions. However, as a result of studying split-brain patients, or patients with left hemisphere lesions, very simple verbal understanding has been found in the right parietal lobe, which has the capacity to comprehend a selection of simple nouns and verbs. In the same way, the left parietal lobe seems to possess limited spatial functions. So although the right hemisphere is quite distinctly specialized for nonverbal, visio-spatial abilities, there is accumulating evidence that the difference between left and right hemispheres is not one of simple differentiation.

The right hemisphere is, however, known to be the receiver and identifier of spatial orientation, responsible for the way in which we see the world in terms of color, shape and place. Using the skills of the right hemisphere we are able to find our way about. We know which street we are walking in simply by looking at the overall architecture—the shapes of the storefronts, of roofs and of doorways. When walking down the street, the face that stimulates recognition is also identified by the visual memory of the right hemisphere. The name that goes with the face is, of course, supplied by the left hemisphere.

Because of the evidence gathered from split-brain patients, it is now possible to test for evidence of right hemisphere specialization in normal people with more certainty of what to look for. One of the most obvious methods of investigation has been through the comparisons of electroencephalogram (EEG) readings. Experiments using the EEG to ascertain the properties and distinctive abilities of the right hemisphere have added weight to the argument of hemisphere specialization. By placing electrodes on the scalp it is possible to take a series of EEG readings for activities characteristic of each hemisphere.

When a person is writing a letter, for example, the left hemisphere should be suppressing the activity of the right. The EEG readings seem to confirm this assumption. From the left hemisphere the EEG had the characteristic fast waves of attention, or activity, while the EEG from the right hemisphere was slow, with high amplitude waves that occur in a relaxed state. The opposite effect was produced when a solely visio-spatial task was being done. This time the EEG from the right hemisphere was of the fast-wave variety and a slow wave appeared on the left hemisphere. Exactly the same EEG patterns are found even when volunteers are just thinking verbally or spatially.

Although this technique of investigating specific hemisphere activity is still very much in its infancy, these corresponding changes in EEG recordings would seem to indicate that the normal human brain does indeed have specialized thought processes.

More speculative research suggests that the preference a man shows for turning his head is related to his choice of work. If the head is habitually turned to the left, there is a tendency for nonverbal, right hemisphere activities including, for example, music or painting. If the head is habitually turned to the right, there seems to be a preference for activities which include language or logic.

Today anthropologists and psychologists believe that entire civilizations may have developed under the influence of the hemisphere which was dominant in the population. Western society is thought to be dominated by the left hemisphere and this is reflected in the design and values of that culture. Other cultures, however, have centered around a right-hemisphere-oriented mode of thinking. These societies are few, but one of the more extensively studied is that of the Trobriand islanders. Instead of considering information a step at a time, as is done in Western society, the Trobriand islanders process all of the available information at once. If a wind hits an islander's boat, for example, he does not see it as coming from a particular quarter, but describes it by the name of the part of the boat that it strikes.

Visual memory problems
Look at each of these photographs in turn and, as quickly as possible, identify what they are or what they are part of. The complete objects are shown on page 201. Correct identification draws solely on the right hemisphere's visual skills, with particular reference to its visual memory. If identification takes some time it is because the visual memory is searching for the correct solution to the puzzle, but has only been presented with a small part of the problem. Eventually, the right hemisphere's visual mechanism will place the object in the right category and identification will be possible.

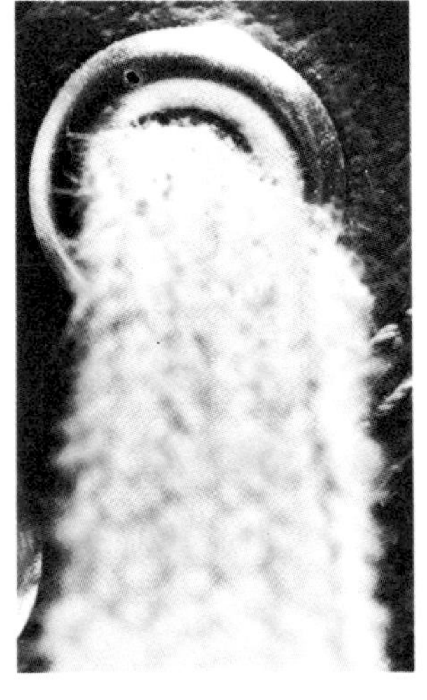

Visual memory is tested when identifying faces. Visual recognition is the property of the right hemisphere, while the addition of the name comes from the left hemisphere. Here are three internationally known faces of famous people past and present. The test is to find how quickly you can name and recognize them. The answers can be found at the back of the book on page 201.

Cross-reference	pages
Brain and central nervous system	74-77
Cerebral cortex	80-81
Frontal lobe	122-123
Left hemisphere	124-125

Learning: the acquisition of skills

With every passing moment of our waking lives we experience new things and store them in our memories for future use. As a result of past experiences, we respond in certain ways in the future. This is the process of learning. Reading a book, walking to work, taking a pill to cure a headache, simply wearing clothes—all are examples of learned responses, as are many other forms of behavior that we take for granted. Learning influences every aspect of our lives, our likes and dislikes, our opinions and beliefs, even the pattern of the society in which we live.

To understand what the learning process is, it is necessary to study behavior and to measure performance. Such factors as fatigue must be taken into account, as must the presence or absence of rewards for certain behavior.

At the beginning of the twentieth century, psychologists began to consider learning as a matter of associating a stimulus with a response, correct responses being "stamped in" by reward. A hungry cat, for example, in a cage that can be opened with a latch, will move about randomly, and will only by accident open the door and so reach the food that it can see. It will take several attempts before the cat's behavior changes. Then, instead of making random movements, the cat directs its attentions to the rewarding movement of pressing the latch. It is the gradualness with which the cat's behavior changes that suggests that learning consists of the stamping in of responses that bring a reward, and the stamping out of incorrect responses.

The Russian scientist Ivan Pavlov believed that all learning was built up from basic units of behavior called conditional reflexes. A reflex of this kind is set up when a neutral stimulus comes to replace one which has first produced a response. Thus if a dog hears a bell ring just before it receives a slight electric shock that makes it lift up one leg, it will, after the event has occurred several times, respond to the bell alone. This is a conditioned response.

Later theorists claimed that conditioned responses were only one special kind of learning, and that there is an important concept called "drive" at work. An animal deprived of food, for example, is in a state of hunger drive. Finding food and eating it is a form of reinforcement for behavior of a certain pattern; it meets and lessens the hunger drive.

If a rat is placed in a simple T-shaped maze, with a reward of food in only one arm of the T, it soon learns to run from the foot of the T to the appropriate food box, ignoring the empty arm. The sight of the right box becomes a reinforcement that strengthens the preceding habit—selecting the correct arm of the maze. When this behavior has been established, the rat can be said to have learned to run the maze.

When a rewarding pattern of behavior has been established, the rat will still continue to select that arm of the maze for a while, even if now there is no reward of food. If the food

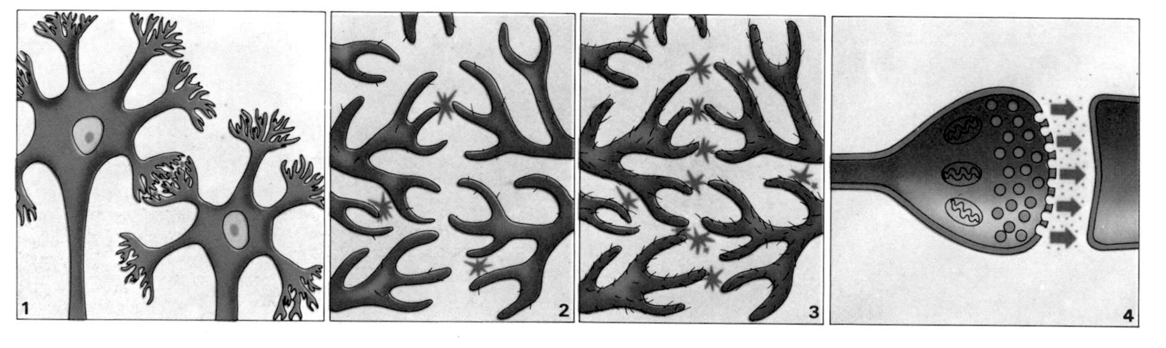

The anatomy of learning
When we learn something, new pathways are established in the brain's nerve network. Between the fine branches of adjacent nerve cells (**1**) are synapses, which transmit information. In an enlargement (**2**) there are at first few such junctions (red) between the dendritic spines, but during learning new spines grow, enabling new synapses to develop (**3**) so that there are now firmly established routes for the transfer of certain information. The brain's activity has altered subtly. (**4**) Arrows indicate the one-way flow of information at a synapse.

Classical conditioning

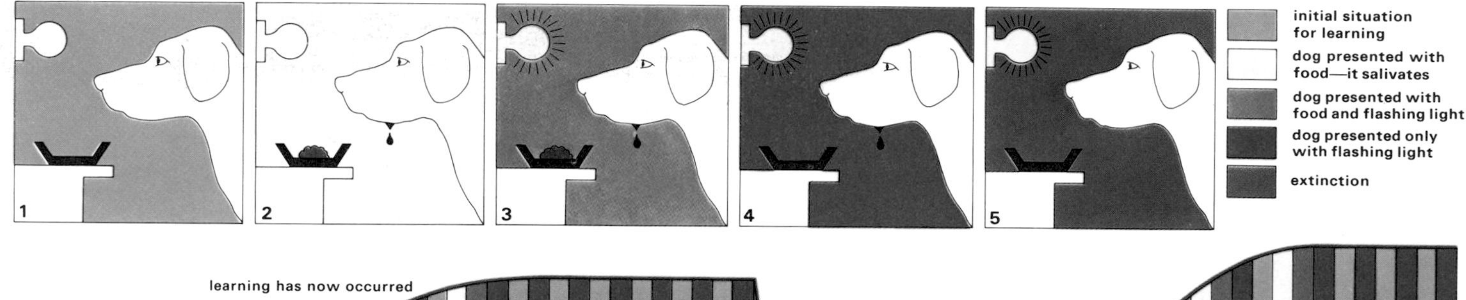

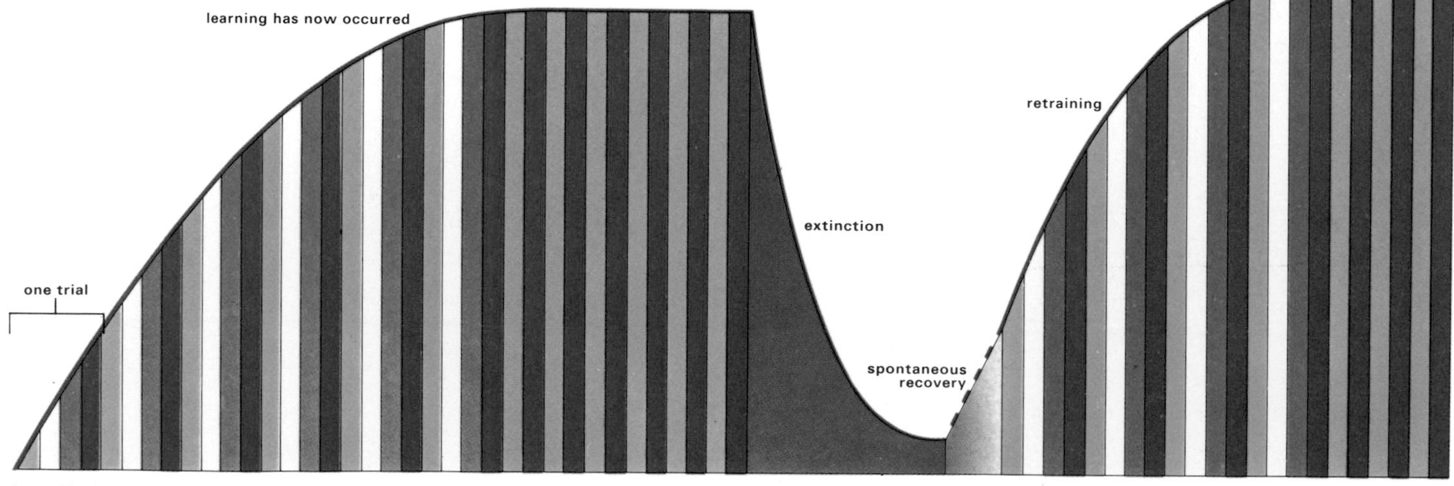

A graphical representation of the classical conditioning phenomenon illustrated by I. P. Pavlov's experiment with dogs. The learning principle involves making a dog salivate when a light flashes. For learning to occur, the dog is put through a number of trials. One trial consists of the dog starting in the training situation (**1**), then food is presented (**2**) and this becomes the unconditioned stimulus. The dog's reaction, salivation, is the unconditioned response. The food is then presented simultaneously with a flash of a light (**3**). The light is the conditioned stimulus. The food is withdrawn but the light is flashed on in the hope of eliciting the salivation response. This sequence is repeated several times until the dog salivates to the flash of the light. (**4**) When this happens salivation is now the conditioned response. The graph goes on to show the dramatic fall of extinction (**5**). This occurs when the dog is continually presented with the conditioned stimulus, even when he is no longer hungry and the conditioned response no longer occurs. Retraining occurs at a faster rate.

is continually missing, however, the rat will eventually "unlearn" the behavior. This is called extinction. An animal that is punished with a shock for making a certain response will temporarily stop making it—but will not necessarily learn another, substitute, response. This is not unlearning, but is suppression of a certain habit.

In human beings, psychiatrists may attempt to cure undesirable patterns of behavior by using the same principles of conditioning and counterconditioning. An alcoholic patient, for example, might be given a drug that induces nausea whenever alcohol is taken. Eventually, the patient will be conditioned to associate alcohol with nausea, and will develop an aversion to alcohol, therefore changing his behavior. For a person with a phobia, such as an unreasoning fear of spiders, a different technique might be used, encouraging him to come to terms with his fears gradually, and to lose the immediate feeling of panic, by demonstrating the essential harmlessness of a spider. Behavior therapy, however, must usually be used in combination with other forms of psychiatric treatment to be really successful. By itself it is simply an attack on the symptoms of a condition which may have very complex causes.

In 1930, the American psychologist B. F. Skinner described the form of learning called operant conditioning. This differs from the classical conditioning described by Pavlov in that it refers to modification of voluntary behavior and not to reflex activity. Using rats in boxes which were equipped with levers that could be pressed in certain circumstances to obtain food, Skinner showed how behavior could be shaped by punishments and rewards, and what the effects of reinforcement were. Usually the rat was required to press a bar when a light was shown, or when a buzzer sounded, to obtain a pellet of food. The reward would be immediate if the rat responded to the stimulus correctly, and because of this such operant responses were not easily extinguished. After learning to press the bar and obtain food, a rat might press the bar more than fifty times, even when food no longer appeared. This is operant conditioning.

Skinner tried to explain human behavior in the same terms—money, for example, being seen as a reinforcement for certain activities. But man's motivation to learn cannot be explained so simply. Our intelligence makes learning and behavior far more complex than a matter of more rewards and punishments. Learning for its own sake is found not only in man, but in other animals to which we attribute curiosity.

Our ability to learn far exceeds that of any other animal. By learning language we have taken one of the greatest steps forward in evolution, for through language not only can we learn from our own experiences but we can also pass on this learning to others.

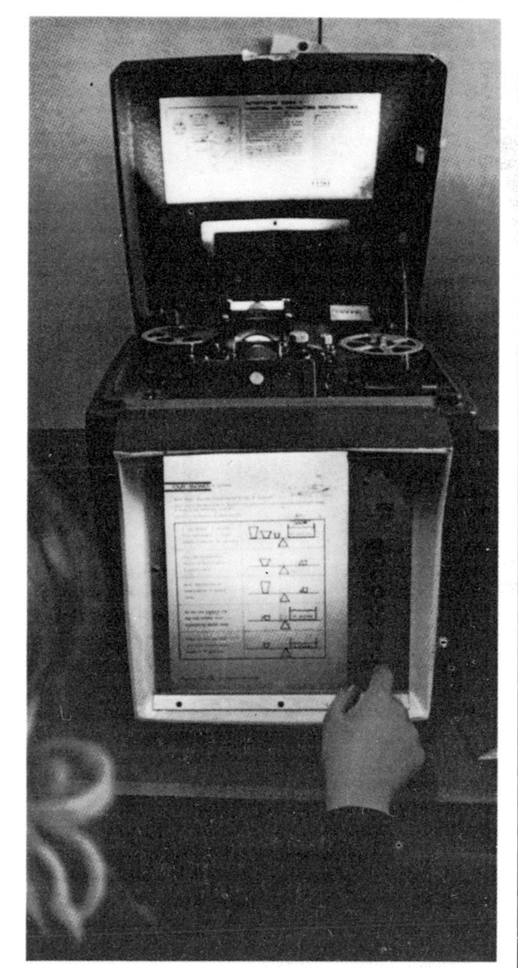

Teaching machines use theoretical principles of learning and apply them to real situations. In recent years they have been developed, under the guidance of applied and educational psychologists, for use in schools and other training establishments. The photograph shows a teaching machine being used, in a school, to teach a mathematical theory. The top of the machine is open, showing how the information is fed via a film onto the screen. The buttons allow the child to regulate the speed at which information is presented. Control over the learning situation is today believed to be the greatest advantage offered by teaching machines, as it permits the learner to progress at his or her own speed and to gain feedback to the answers in the shortest possible time. Teaching machines will probably become more widely used in the future.

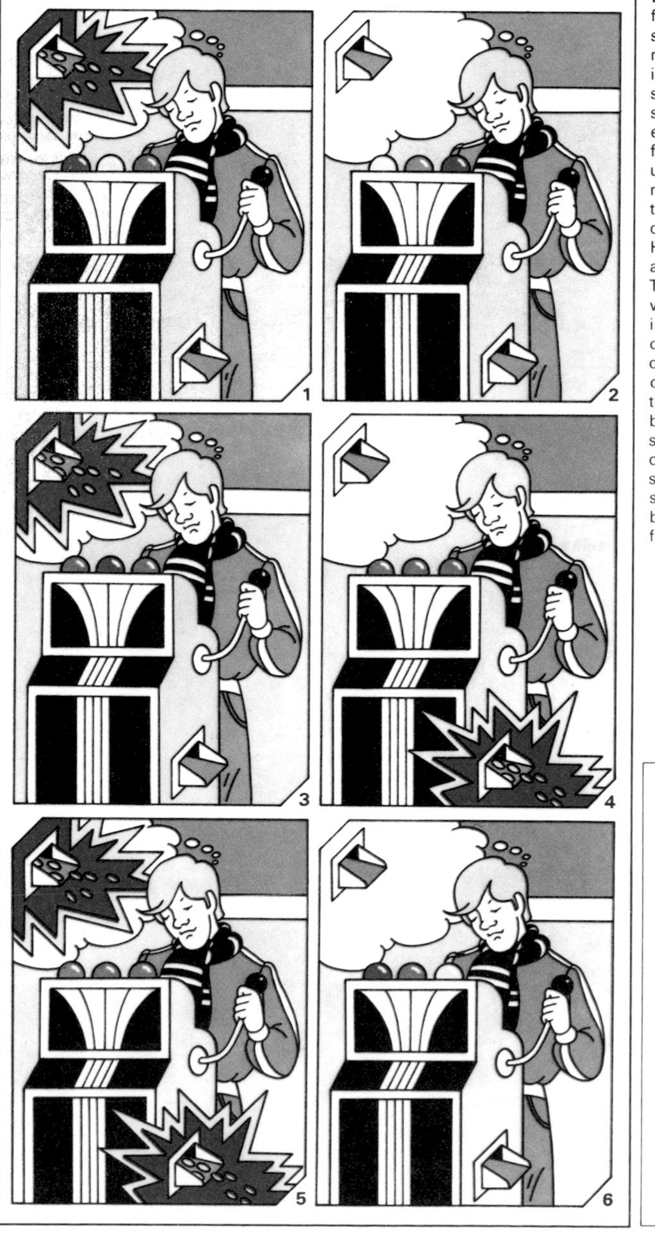

Partial reinforcement, or reward, for a behavioral action in a learning situation is one of the most effective methods of learning. The reinforcement is usually presented in a fixed schedule, or orderly way. Playing a slot machine is used here as a simple example of how the player himself forms a reinforcement schedule and uses it to keep gambling. To win he needs three of the same-colored lights to come up together; any other combination does not produce a win. His own schedule is set by his aspirations to win on every other play. The machine, however, is set to its own winning pattern and will pay out only intermittently. Out of six plays he wins only twice, and of these wins only one coincides with his schedule. The lack of correct forecasting does not deter the player, but serves to strengthen his behavior, for eventually he adapts his schedule to try to coincide with that of the machine's. Psychologists specializing in learning theory have shown that irregular reward for a behavior helps the learner to remember far better than continuous reward.

Cross-reference	pages
Brain and central nervous system	74-77
Hypothalamus and drives	112-115

Short-term and long-term memory

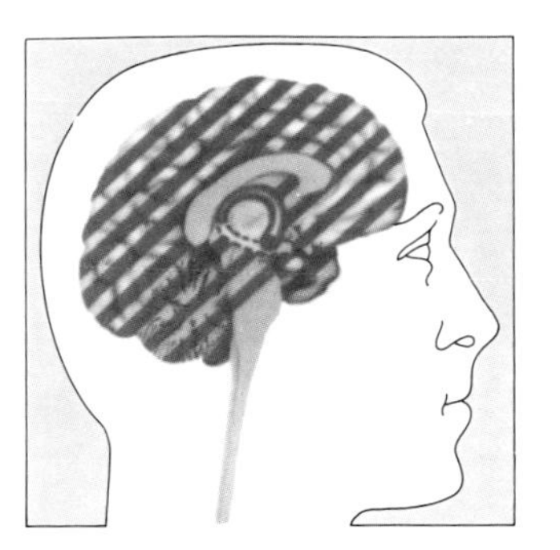

Memory involves the whole brain, but the indispensable regions include the hippocampi and the mammillary bodies.

A factory analogy to memory

This illustration fits together the major aspects of all the theories postulated to explain how man's memory may be organized. It makes an analogy with a vast factory storage system, showing the mechanisms by which memories may be selected, sorted and stored in the brain, and how they may be later retrieved when required for recall. There are two theoretical components of memory—primary memory and secondary memory. Primary memory has two stages, a precategorical stage and a short-term memory stage. Precategorical memory has no measurable capacity. Short-term memory has a very limited capacity, measured in terms of four to seven bits of information, and is useful only for immediate recall or rehearsal. From primary memory, information enters secondary memory, which is the storage system known as long-term memory. The characteristic feature of long-term memory is its seemingly limitless capacity, which makes it the permanent memory store. All the information being processed in the analogy is in a coded form.

We tend to talk of memory as though it were a distinct physical entity contained somewhere in our heads—a kind of scrapbook of mental pictures, perhaps, or a videotape to be played back at will on a slightly quirky recorder. Someone is likely to complain of a "bad memory" in much the same way as he might complain of a bad tooth or flat feet.

However, memory is nowhere near as tangible as a part of the body. It is better regarded not as a thing, but as a fundamental function built into the living brain. Brain tissue has the ability to change as messages pass within it. Its modification, whether for long or short periods, gives it the power of remembering, and gives us the ability to profit from past events. Without this ability we would be mindless, unable to learn, to read, to write, to talk or even to think. Lacking a memory we could comprehend nothing and communicate nothing.

Every fraction of every second, information is entering the brain and memory is at work, whether we are aware of it or not. The point of our attention, when reading a book, for example, is the meanings of the words on a page. But behind this the information that the print is black and the paper white is also being recorded, repeatedly, with each successive word we read. Otherwise, we would not notice if a specially important word were printed in red. Whether transiently or for long periods, the information from all our senses modifies the brain's activity in some way, so that events are "remembered" in its complex circuits.

Although we do not yet know the physical basis of memory, experiments indicate that there are quite distinct stages in the process of acquiring and storing information. During the phase of acquisition there is first an extremely brief stage which is called sensory storage, or sometimes immediate forgetting, and this is followed by short-term memory. Long-term memory, the storage of information for considerable periods, may or may not follow.

The stage of sensory storage lasts very briefly after an item of information has been received. During that time the sensory circuits involved remain activated, and a fleeting mental image is formed, only to die away. By far the greater part of the information stored in the image is simply forgotten. Anything that does survive for a particular reason passes to the stage of short-term memory.

A person might be asked, for example, to describe what he saw after being allowed to glimpse a photograph of twelve familiar objects for just a few hundredths of a second. It is likely that only four or so objects would be recalled, despite the fact that all twelve were perceived. The retention of the image in the sensory store is so transient that by the time four objects have been named the image has disappeared and, no longer being available for inspection, the others have been forgotten.

The very act of naming the four objects has, for a while, saved them from being forgotten. The information from the sensory circuits has been encoded into a verbal form, and reinforcement of the visual image is now given by hearing the names of the objects. They have entered short-term memory, and for a time can be recalled.

Short-term memory retains information and passes it into long-term memory only if the information is "rehearsed"—that is, thought about again and again by making a conscious effort. Otherwise the information disappears in probably less than half a minute. An example of rehearsal is repeating a telephone number to oneself after looking it up, in order to dial. Without rehearsal, a seven-digit number is forgotten in seconds. Given sufficient rehearsal, however, it may pass into long-term memory and be remembered for days or even years.

How information is retained in long-term memory is still a mystery. It may be through the occurrence of chemical changes in the neurons, or perhaps the establishment of new connections at their synapses, so that impulses travel more readily through certain circuits. It is not known. Nor is it even known whether one particular memory is stored in a specific part of the brain, or diffusely throughout it.

If chemical changes within the nerves are involved when information is stored, it could be that molecules of ribonucleic acid, RNA, take part. Of all the types of cell in the body, neurons are by far the biggest containers and producers of RNA. In experimental animals, when drugs that hinder the production of RNA are given, the process of learning is plainly disrupted. The passage of nerve impulses possibly influences RNA production during the acquisition of memory, and, therefore, the formation of new proteins in the nerve tissue. This would provide a link between the electrical activity of the brain and changes in its structure as memories were acquired.

Memory has a third component besides acquisition, or registration, and retention. This is recall. Again, it is not yet known how the process works, nor how, out of all the memories we store up in our lives, we can select certain ones at will quite readily, while others seem to remain temporarily inaccessible and yet others seem to disappear forever.

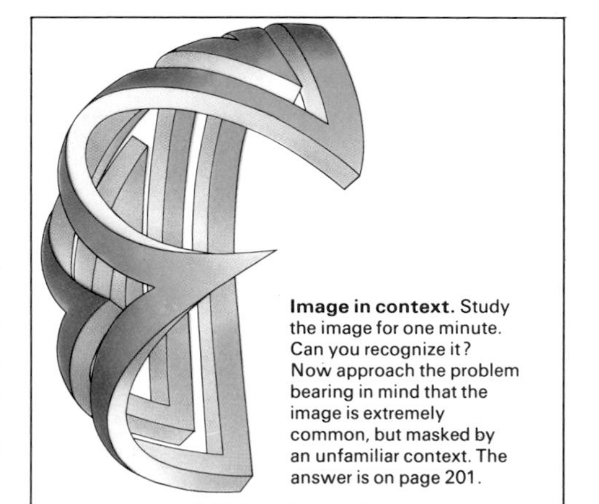

Image in context. Study the image for one minute. Can you recognize it? Now approach the problem bearing in mind that the image is extremely common, but masked by an unfamiliar context. The answer is on page 201.

primary memory—precategorical store
primary memory—short-term memory
secondary memory—long-term memory
input
retrieval
short-term recall center
decoding center

1. Information input from the sensory areas.

2. First stages of information coding.

3. Some information overflows off the main conveyor and is lost.

4. The passage of information from the precategorical store into the short-term memory.

5. Area of limited capacity characteristic of short-term memory.

6. Pathway taking information out of short-term memory and immediately transferring it to the recall center.

7. Access point for information from short-term memory to the decoding center for immediate recall or rehearsal.

8. Information that has either overflowed out of short-term memory, or has been ejected from short-term memory.

9. The transfer spiral for information leaving short-term memory and entering secondary memory for long-term storage.

10. The further coding area for information to be stored in long-term memory.

11. Information being selectively transferred to the memory banks.

12. The long-term memory banks containing the coded information.

13. Unpleasant memories that the mind wishes to eliminate are selectively disposed of down the chute out of the mind.

14. Top-priority bank containing highly specialized or important information.

15. New coded information being stored in a bank. All information enters the banks from the top. With time, information moves down the bank, being gradually distorted and suppressed by the weight of new information from above.

16. Information being taken out of storage in a coded form in the first stage of recall.

17. The retrieval mechanism collecting required information and transferring it up to the decoding center.

18. Access point for information to be conveyed from the long-term memory storage banks onto the retrieval mechanism.

19. The decoding center for all information required for complete recall.

20. The decoding and matching of recalled information in preparation for verbalization or action.

21. Information leaving the memory in decoded form.

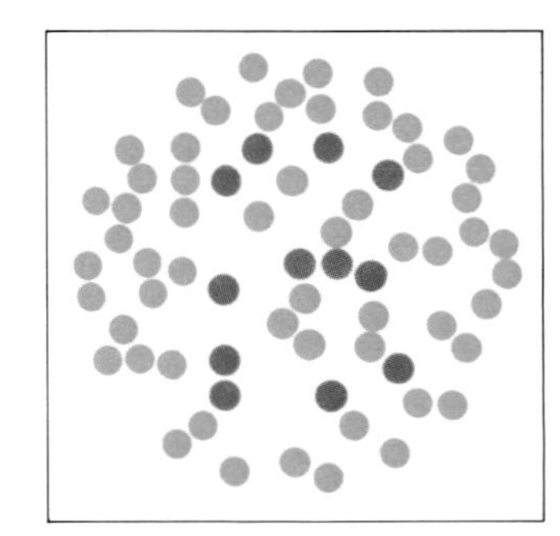

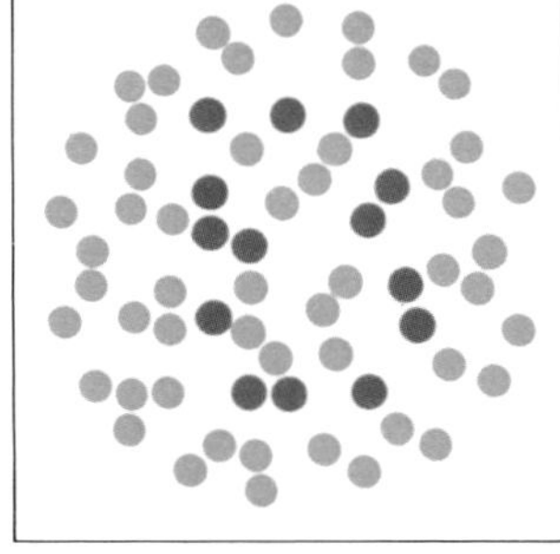

Eidetic imagery is experienced when an almost photographic image of a particular scene can be formed in the mind. A recalled eidetic image is a visual sensation and should be perfect. A very accurate description is not necessarily eidetic. Concentrate carefully on the first picture and try to remember it. Now try to superimpose this pattern on the second picture. If a sharp visual image then persists you have the ability to form eidetic images. The answer is on page 201.

Spatial memory is tested by this problem. Look at the eight shapes for one minute and try to commit them to memory. Now turn to page 201 and, without further reference, pick out the test images from the answer display. Correct recall of the shapes is not easy as the answer display includes similar and distracting shapes to try to eliminate or reduce any element of guesswork in choosing the correct answers. Most people's spatial memory is probably located in the right hemisphere.

Cross-reference	pages
Brain and central nervous system	**74-77**
Learning	**128-129**
Ribonucleic acid	**26-27**
Right hemisphere	**126-127**

Intelligence: mental skills

One of the most baffling abstracts for human beings to grasp is the size and scope of their own intelligence. For years, philosophers, psychologists and educators have grappled with the phenomenon of human intelligence, a resource so much greater than anything to be found in the animal world. Their concern has been to define human intelligence accurately and, in reaching such a definition, to find ways of measuring it and of comparing the amount present in different individuals.

Intelligence has, therefore, been described in many different ways. Perhaps the most relevant definition, even if it may seem somewhat facetious is "that which is measured by intelligence tests." In 1904, Charles Spearman, a British psychologist and statistician, wrote of intelligence as containing two factors. One he termed "general intelligence," which related to an individual's ability to solve such specific problems as jigsaw puzzles or mathematical equations. The other factor yielded the ability to solve general problems. His concept, however, was not acceptable to all psychologists, many of whom regarded intelligence as a collection of abilities all closely interwoven. The major reply to the two-factor theory came from an American, L. L. Thurstone, who produced a group factor theory, published in 1938, postulating seven primary mental abilities ranging from number and word fluency skills to reasoning and perceptual speed.

It is well recognized, however, that intelligence is a function of the entire brain and nervous system—right down to the receptors in the body which translate energy into a meaningful pattern. The brain as a whole seems to be implicated, rather than any specific part of it, although its two hemispheres are involved in different types of ability. Sometimes damage to one part of the brain can be offset by the use of other areas. Intelligence is not related to the brain's size.

It is thought that intelligence is 80 percent a product of inheritance and 20 percent of environment. Support for the prominence of the heredity factor comes from studies showing that, even if they are brought up separately, identical twins have almost the same intelligence quotient, or IQ. Birth order is a modifier of intelligence. Firstborn children tend to be more intelligent than their siblings, with a significant drop in IQ for fifth and subsequent children. But whatever the genetic endowment, a satisfactory environment is essential to the attainment of high intelligence levels. Psychological research has found that children reared in areas of cultural deprivation, offering perhaps only limited horizons, lack the stimuli necessary to develop their intelligence fully.

Intelligence tests arose from the need to assess children for education—specifically from a bid by the French authorities to screen out feebleminded children at an early age so that their special needs could be met. The pioneer was Alfred Binet, who, in 1905, devised a test involving measurable abilities that could be expected in children at each year. A child of six, for example, was given problems capable of solution by a normal six-year-old, and, depending on results, was assigned a mental age as distinct from his chronological age.

It was a German, Louis William Stern, who, in 1912, suggested that test scores should be expressed in the form of IQs—intelligence quotients—that would "show what fractional part of the intelligence normal to his age a feebleminded child obtains." In 1916, Lewis Terman, an American working at Stanford University, produced a revision of the Binet scale to provide, in addition, for the testing of children of normal and superior intelligence. The Stanford-Binet test is still in use for children up to the age of fifteen.

One of the most widely used tests for sixteen-year-olds and upward is the Wechsler Adult Intelligence Scale, introduced in 1955. The Wechsler test poses two groups of questions. One deals with language skills and the other with nonverbal or visio-spatial competence—the ability to cope with pattern and design. Ideally, intelligence tests should be free from cultural differences or constraints, especially as many nonintellectual abilities must affect human intelligence. Debates on racial differences in intelligence have had their roots in biases inherent in tests. To combat the culture-bias problem, attempts have been made to construct "culture-fair" tests. The underlying principle in the content of these tests is the avoidance of all reference to items that may be more familiar to one culture than another, so minimizing any advantage. Regardless, however, of the numerous approaches to finding a culture-fair solution, in the main their success is very limited.

Sex differences in intelligence are almost as difficult to isolate as racial differences. Controlled experiments in the 1970s, however, do show that some specific abilities favor one sex. Men generally tend to be superior in numerical reasoning, gross motor skills and mechanical aptitude, while women perform better on tests of detail, memory and verbal ability.

The scoring of intelligence tests has been standardized, with an average of 100 over the population as a whole. It has been found that 68 percent of people come within fifteen points on either side of the 100 IQ mark, and 95 percent fall between the range of 70 and 130. On the Wechsler scale, a score of more than 140 is rated as very superior; 120 to 139 is superior; 110 to 119, high average; 90 to 109, average; 80 to 89 low average; 70 to 79 borderline; and below 70 subnormal.

But to give a size and shape to the total resource—all that we bracket under the vague heading of intelligence—is a daunting task, and at best IQ scores can only be taken as a rough guide. True, it is relatively easy to pick out extremes—the severely retarded can be recognized and so can the intellectual giant—but in terms of his intellectual capacity, the average man is much more difficult to classify. In fact, many researchers have made the point that all that intelligence tests measure is people's ability to do intelligence tests.

Central to this whole issue is a tendency to consider intelligence in isolation from other closely related, if not integral, attributes. Creativity, for example, is a vital feature of intellect, and yet it is often treated by psychologists as separate from IQ, implying that IQ tests take no account of creativity. Particularly controversial was a suggestion, based on early studies into creativity and intelligence, that people with a high IQ could be divided into two categories, one creative and the other not.

In a class of their own are geniuses, most of whom share a close relationship with their parents, tend to develop early and to experience a very unusual childhood. Men like Francis Galton, an early pioneer in the measurement of intelligence, knew several languages and was reading Homer at the age of six. Mozart, who played the harpsichord at the age of three, was composing at four and touring at six.

While history has many examples of the way in which geniuses tend to change the whole direction of the subjects in which they excel, it is a fact that Olympian intelligence does not necessarily correlate well with success in life. It is difficult for psychologists to find with any certainty, for example, what proportion IQ may contribute to measures of academic successes, such as examinations, or of commercial or industrial triumph. Whatever its dimensions, therefore, intelligence is still only a part of what makes an individual an outstanding human being.

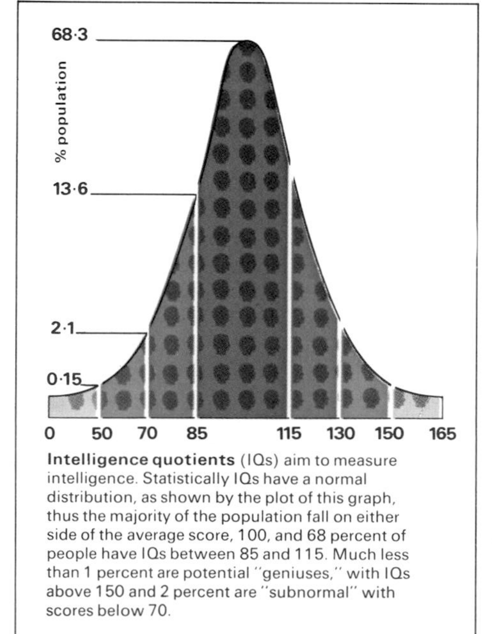

Intelligence quotients (IQs) aim to measure intelligence. Statistically IQs have a normal distribution, as shown by the plot of this graph, thus the majority of the population fall on either side of the average score, 100, and 68 percent of people have IQs between 85 and 115. Much less than 1 percent are potential "geniuses," with IQs above 150 and 2 percent are "subnormal" with scores below 70.

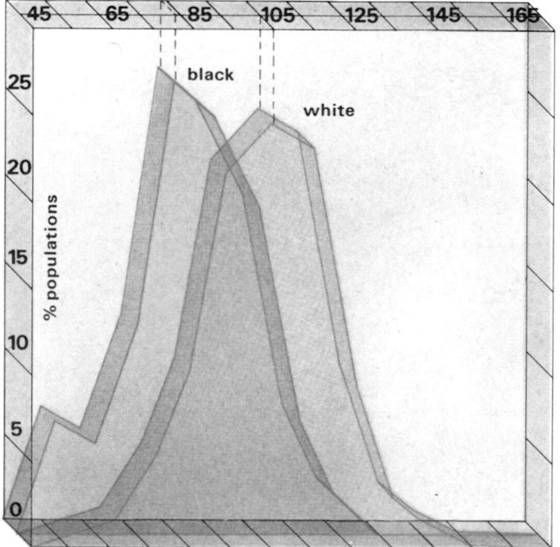

IQ tests tend to be dependent upon culture and are, therefore, biased and fallible, resulting in misleading comparisons of intelligence between ethnic groups. The two graphs above show the results obtained in Canada where, on the same tests, a white group scored consistently higher than a black group. Whether there is any difference between intelligence distribution in black and white populations will not be known until reliable culture-free tests are fully developed.

The measurement of intelligence

Intelligence quotients, or scores, are the products of numerous separate tests aimed at measuring the abilities, aptitudes and skills considered by psychologists to be the basis for overall intelligence. Intelligence tests are usually administered to individuals or groups in batteries, with each test measuring a specific aptitude. These batteries include tests of language skills, such as comprehension, vocabulary and verbal reasoning, as well as tests of nonverbal skills. The scores received from the tests are taken and treated by statistical analysis, which then results in the proper intelligence quotient. Examples of tests for such nonverbal skills as numerical, arithmetical and visio-spatial performance are given here.

Intelligence tests

Intelligence testing commonly involves the assessment of numerical and visio-spatial abilities. The first two tests evaluate numerical ability and the other two visio-spatial skills. (**1**) Find the concept which gives the missing number in the circle. (**2**) Use the first two completed examples to form a concept and then fill in the missing number at the top of the square. (**3**) Choose the odd shape. (**4**) Follow the sequence and then choose the corresponding shape from one of the four possibilities given alongside. The solutions are given on page 201.

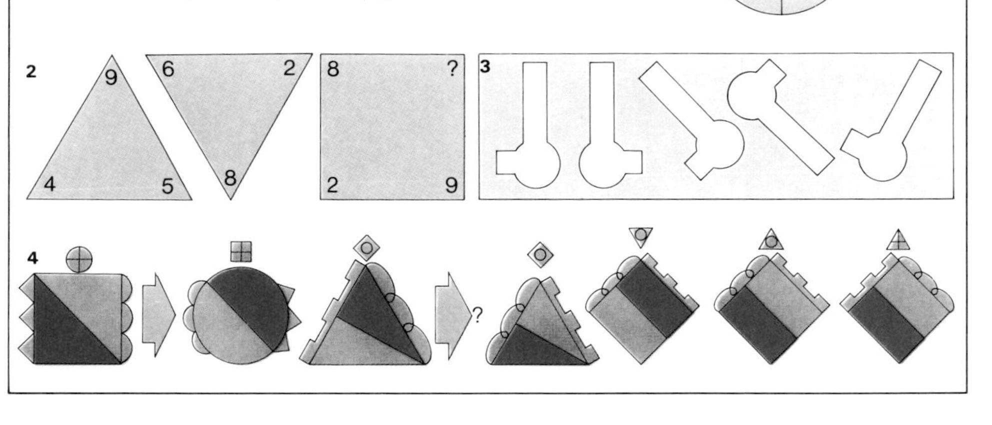

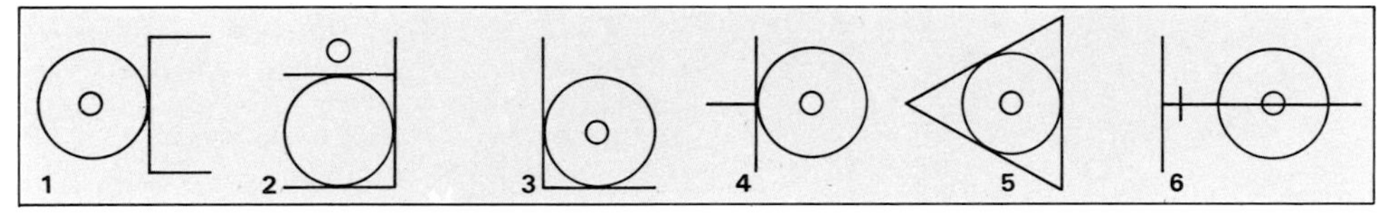

Items used in intelligence tests are designed to eliminate all possibility of there being more than one correct answer. Unfortunately, the creative abilities of the examinees often defeat the structure of intelligence test problems. The test above requires the odd shape to be chosen from the sequence. Which shape would you choose? The correct answer is 2. However, intelligent reasons can be found for leaving 2 in and excluding 5 or 6.

The two nonverbal performance tests above involve picture arrangement and block design. In (**1**) the bird must be correctly completed from the pieces presented. In the second test (**2**) the design must be correctly reproduced using color blocks. Similar tests to these are used in the Wechsler Adult Intelligence Scale.

A picture-completion test

At casual glance all the elements in this illustration probably seem correct. Study the scene in detail and discover how much of the illustration is incomplete or incorrect and what items should be added. (The answers are on page 201.) This type of picture-completion task is usually less complicated than it is presented here. When giving this type of test it is only the essential missing parts that are relevant, although astute subjects may find other items missing. The test normally has one item per test card and is presented to the subject by an examiner. There would be a number of test cards, each with one object or simple scene that needs completion. The instructions for the test are given verbally and no comment is made until all the cards have been presented. Picture-completion tests are one of a battery of tests from which a final IQ score is determined.

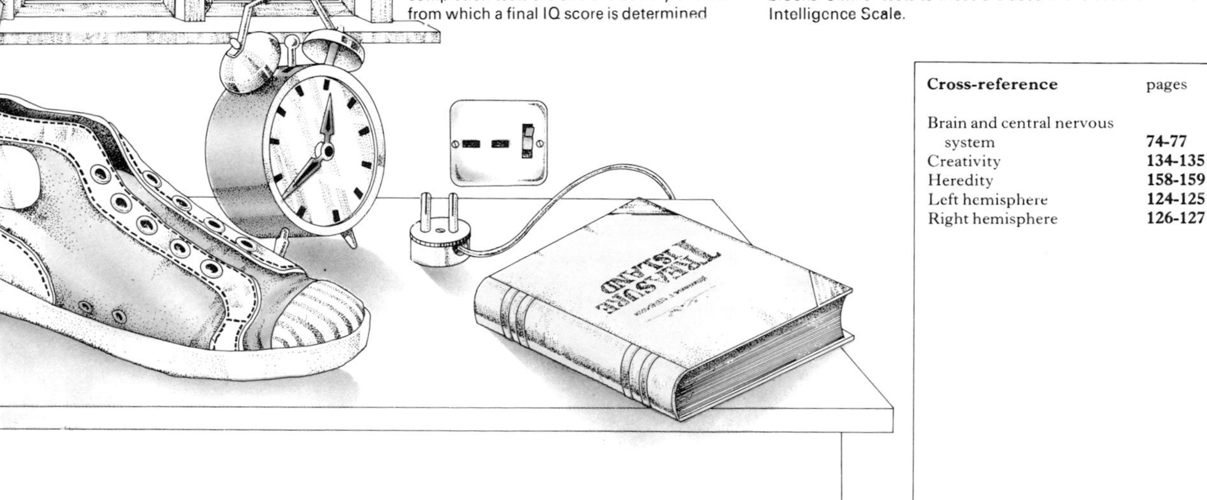

Cross-reference	pages
Brain and central nervous system	74-77
Creativity	134-135
Heredity	158-159
Left hemisphere	124-125
Right hemisphere	126-127

Creativity: novel thinking

Of all human abilities, the creative faculty has always been regarded as the most mysterious. In many cultures, creativity has been believed to emanate from a divine or at least an unconscious source, a power not to be summoned up at will or brought under control by ordinary conscious resources.

Descriptions of creative activity by those who have experienced it at firsthand do little to dispel this impression. Mozart, speaking of his musical ideas, said, "Whence and how they come, I know not, nor can I force them. . . . I really do not aim at any originality." Jules Poincaré described how an important mathematical insight came to him as he boarded a bus. "At the moment when I put my foot on the step the idea came to me, without anything in my former thoughts seeming to have paved the way for it."

Modern psychology is still far from explaining creativity in objective and logical terms, but in recent years some understanding has been gained of the personality types and the circumstances in which it is most likely to appear. There have been a number of studies in which creative individuals, selected on the evidence of their achievements and including architects, scientists and writers, have been compared with their less creative colleagues. The difference between the highly creative groups and the relatively noncreative groups did not lie in intelligence as it is measured by intelligence tests. The creative individual could, however, be sharply differentiated from the others on the basis of personality measures.

There are, of course, many exceptions, but on the whole it has been shown that highly creative people tend to be introverted, need long periods of solitude and seem to have little time for what they regard as the trivia of everyday life and social activity. They tend to be strongly intuitive and more interested in the abstract meaning of the outside world than in the way it appears to the senses. According to these studies, too, creative people are often poor at dealing with people and may avoid social occasions; there is often a tendency on their part to feel that most ordinary people are stupid, as well as a desire to dominate others rather than to relate to them on equal terms. Creative individuals seem, too, to be relatively free from conventional restraints and are not particularly concerned with what other people think of them. They have little regard for established tradition and authority in their fields, preferring to trust their own judgments. Creative men often score highly on a test of "femininity," indicating that they have more sensitivity, self-awareness and openness to emotion and intuition than does the average man in Western culture. An outstanding feature of the creative mind is the preference for complexity.

This creative personality can be subdivided, roughly, into two types—the artistic and the scientific. The basic characteristics remain the same in both, but, on the whole, the artist is more likely than the scientist to express his unconventionality in his life as well as in his work. The unconventional artist is commonplace, the unconventional scientist relatively rare. Musicians and creative scientists tend, if anything, to be more emotionally stable than the average individual, and when this is not so their instability emerges as anxiety, depression, social fears or excitability, rather than as full-blown neurosis. Among artists and writers, however, genius often appears to be akin to madness; serious neurosis, drug or alcohol addiction or insanity appear with abnormal frequency.

There is little relationship between creativity and IQ; it is quite possible to be highly creative but of average intelligence or to be of high intelligence but entirely unoriginal. It is difficult to trap creativity in a formal test situation and tests for creativity are by no means as reliable as are those for intelligence. Typical procedures involve seeing how many unusual responses the subject can give to such questions as "How many uses can you think of for a brick?" and "What consequences would result if all private cars were banned?" Such tests reveal the existence of two basic mental "styles," the convergent and the divergent. Convergent-thinking people tend to approach problems logically and seek conventional solutions. Divergent thinkers tend to use nonlogical or "lateral" thinking, seeking fresh and nonconformist approaches.

To a great extent, the modern Western educational school system favors the intelligent noncreative child—the converger—over the creative child. The creative child may not have a very likable personality. He may be shy, withdrawn, unwilling to take the teacher's word for anything, preferring to follow his own interests rather than those laid down in the curriculum. The converger, by definition, is likely to produce the kind of work the academic machine asks for, not doubting the textbooks.

This division between the creative diverger and the conventionally minded converger is not absolute, however. Convergers told to respond to tests as though they were divergers—to answer the questions as they imagine an unconventional artist might—can produce scores close to those of the "genuine" divergers. This would seem to indicate that while there may well be innate and unalterable individual differences in creativity, much conformist thinking is due not to an incapacity for original thought as much as to a fear of appearing odd or losing social approval, or to an unwillingness to trust to intuition instead of reason.

One important component of creativity is independence from the opinions of others. This may be why an above-average proportion of highly creative people are firstborn children, since this family position often results in an independent attitude. One test for social conformity involves asking the subject to say whether a line projected onto a screen is longer or shorter than a standard line projected earlier. The subject gives his judgment after a number of other people have given theirs. The others are in fact confederates of the tester and their judgments are wrong, sometimes absurdly so. Nevertheless, many people will echo the majority opinion although it means disbelieving the evidence of their own eyes. Those who stick to their own judgment regardless tend also to score highly

Creative thinking and reasoning

Creative thinking is a mental process that usually produces a novel solution to a problem. This illustration suggests the types of mental image that may be formed by two different thinking processes—conventional and creative. The initial problem is presented in the lower right-hand corner and consists of finding alternative ways of using three rectangles, each of which is made up from three small triangles. The vertical flow of images shows how conventional thinking might handle the problem. The three basic rectangular shapes are kept and are only arranged in different sequences, thus limiting the number of possible combinations and, therefore, the number of solutions. The lateral flow shows a succession of mental images that might be produced by using a creative thinking process. The first stage would be to reduce the rectangles to their basic units, small triangles. Then the shapes might be rearranged into three larger triangles. This results in a completely new solution. It is not known exactly how mental strategies differ in conventional thinking and creative thinking, but the quality of creativity can be measured by the end result.

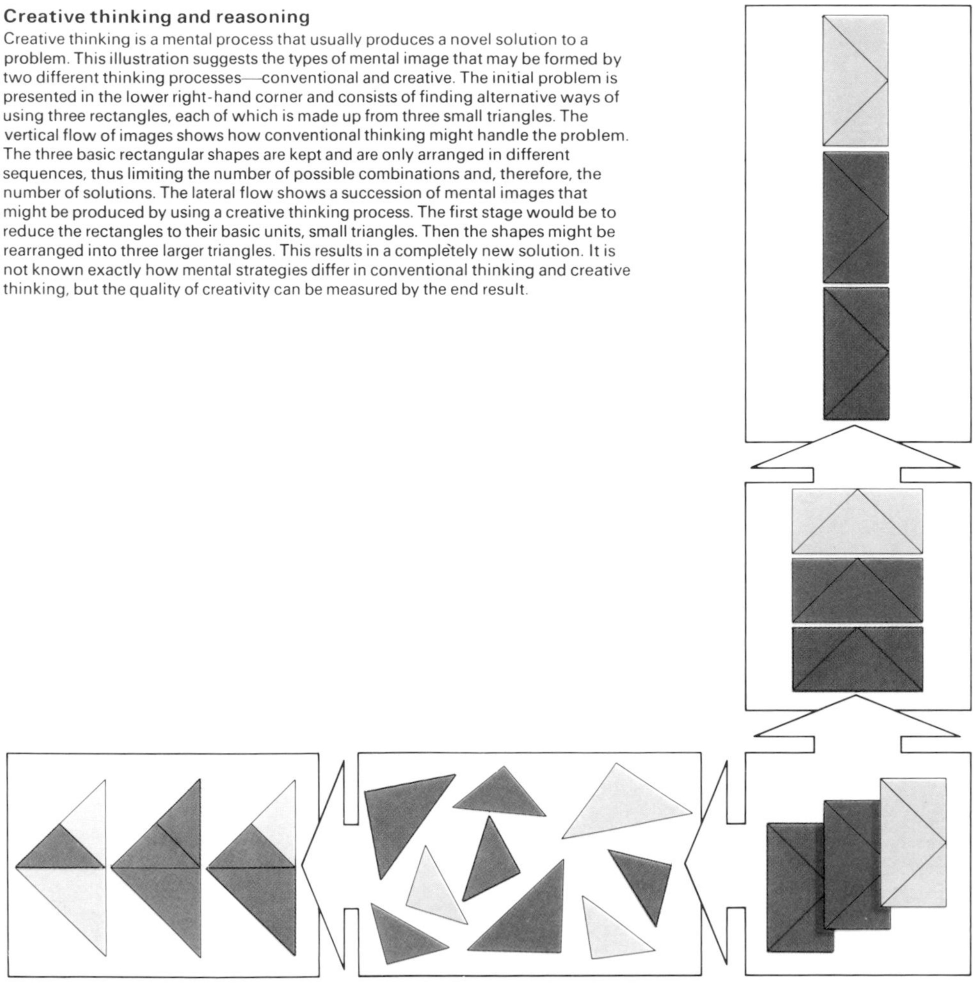

on tests of creativity. Such inhibitions can be overcome to some extent by modern brainstorming techniques, in which a group of people is brought together for the purpose of producing new ideas or problem solutions. Participants are encouraged to voice every passing notion, however wild, without self-censorship or criticism.

Other ways in which creativity can be increased have been suggested by studies of the states of mind in which creative people usually receive their insights. The pattern is almost invariable. The mind must be prepared by the collection of all relevant information. There are persistent attempts to come to terms with the problem in a logical fashion while being careful not to accept a conventional solution. But the answer itself, the creative idea, nearly always comes at a time when the individual is not concentrating on the problem, but is in a peaceful state of drowsiness, waking dream or reverie.

Creative insights, indeed, often seem to occur on train or bus journeys or in the bath, both situations which can induce, by their monotony, a trancelike condition. In these states of consciousness the barriers against the unconscious are lowered and fantasy and imagery are allowed free play. And it is from the unconscious, with its faculty for synthesizing across gaps which logic is insufficient to span, and its relative freedom from conformity and convention, that creativity ultimately springs.

Concept formation is one of the basic thinking strategies used when solving problems. Find the concept contained in this set of cards. The left-hand column of cards has the positive features of the concept, while the cards in the right-hand column do not. The problem is approached by testing the elements of the left-hand pictures and comparing them with the pictures on the right-hand side where the concept is not present. The elements should then fit into a general formula that can be identified as the concept. The answer to this problem is given on page 201.

Hidden in this scene are the outlines of seven triangles of varying sizes. See how quickly you can find them. (The answers are given on page 201.) This test is an adaptation of the hidden objects test developed and used by psychologists as one of the tests for creativity. The conventional test requires subjects to locate a regular geometric shape which has been concealed in a more complex geometric pattern. In this particular illustration, however, an everyday situation has been used as the background within which the triangular shapes have been hidden.

Logical thinking

True logic is not used by most people. Instead they deviate, and often the departure is informative and useful. This problem tests your power of logic. You are given the rule: "When traveling to Moscow I always go by train." Now look at the pictures on the cards. If each card has another picture on the reverse side which is consistent with the rule, then what is the minimum number of cards which need to be turned over to prove the rule? Now try the problem. It is only necessary to look on the reverse side of the picture of Moscow to ensure that there is a train, and to turn over the aeroplane in case Moscow is on the other side, which would violate the rule. The train can be traveling to any destination without disproving the rule so there is no need to turn it over. London is not mentioned and can also be ignored.

Can you solve this? Starting from any point you wish, all the dots must be joined together using four straight lines only and without taking your pen off the paper. Read no further until you have tried the problem for at least one minute. The dots are perceived as forming a square and therefore set up perceptual boundaries that are fixed in the mind as if they were drawn in position. To solve this problem, drop perceptual convention and look outside the boundaries for the solution. The answer is on page 201.

Cross-reference	pages
Frontal lobe and personality	**122-123**
Intelligence	**132-133**
Left hemisphere	**124-125**

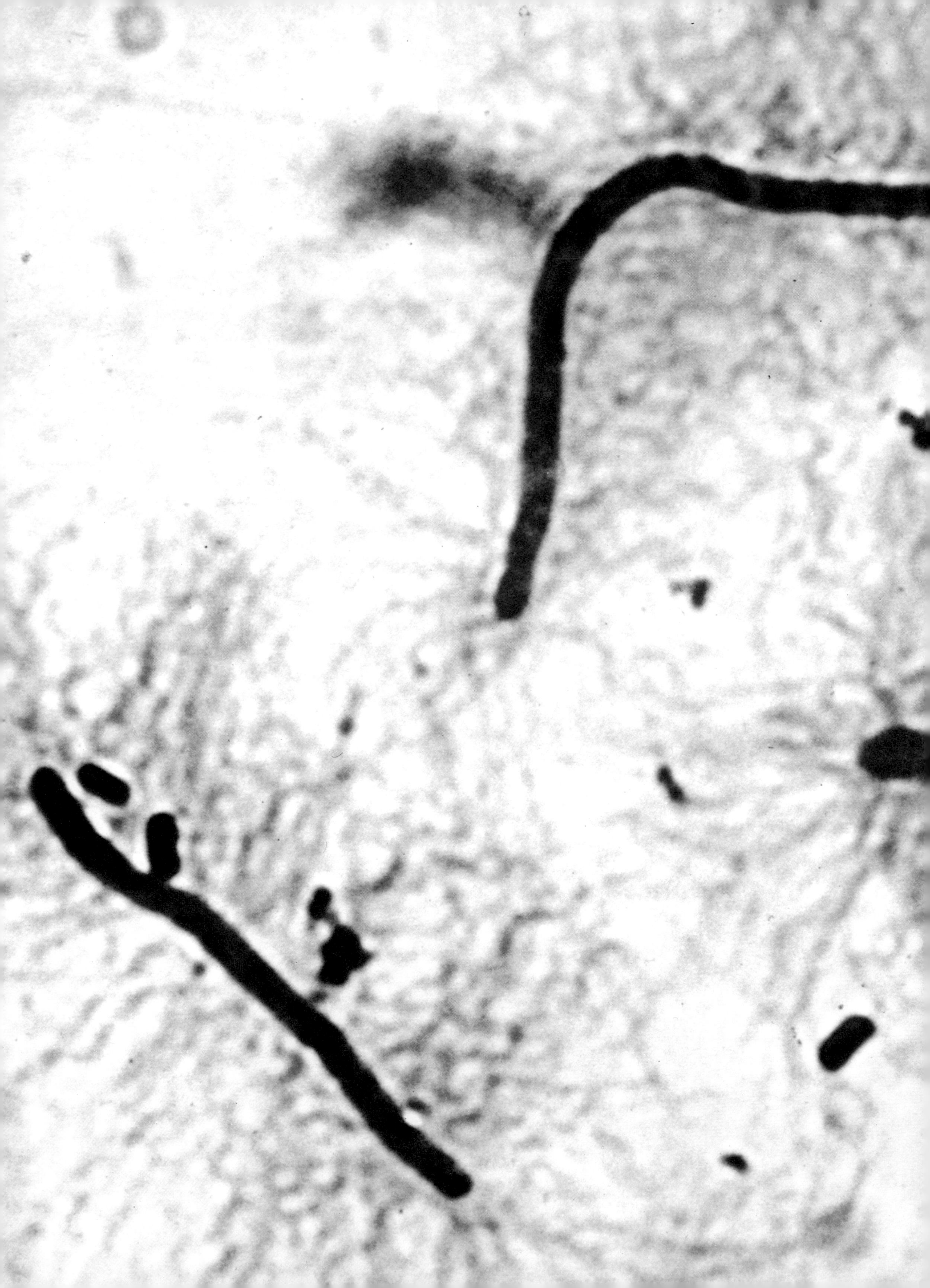

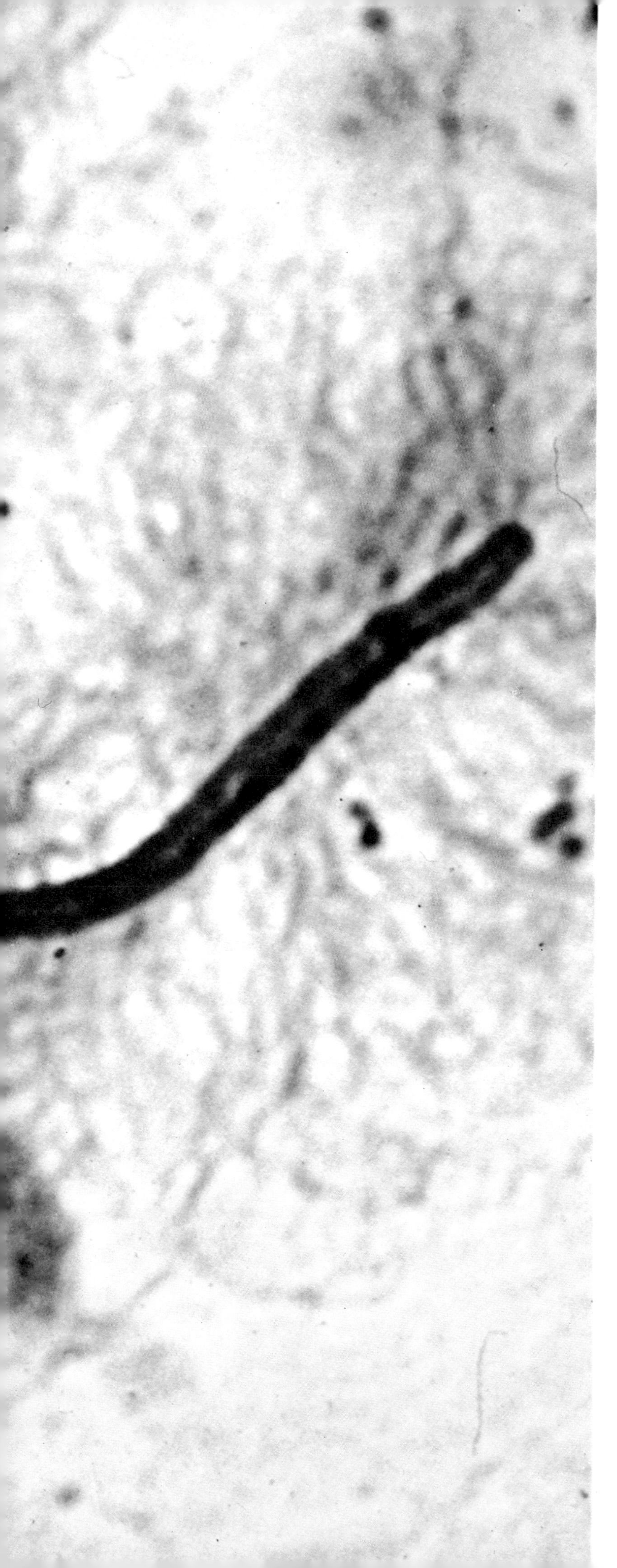

Defenses of the body

The human body is constantly faced with assault by disease-causing bacteria, viruses and parasites and with the threat of injury and damage to tissues. It is, however, remarkably resilient and has coped with its vulnerability by developing a range of integral defense and repair systems.

The first line of defense is provided by the body surfaces through which the hostile organisms may seek entry—the skin, and the alimentary and respiratory tracts—and here protective filter mechanisms and chemical action attempt to defy the invaders. If they do manage to penetrate the surface defenses and enter the bloodstream, they are attacked by specialized blood cells, which destroy and dispose of them. This process is backed up by a complex chemical system, the immune system, which deactivates the offending micro-organisms and neutralizes the poisons which they secrete.

The skin and other body organs, notably the liver, have the ability to repair damage that has occurred to them, but others, such as nerves, unfortunately lack this capacity and damage can have serious and permanent consequences. Most physical injuries result in blood loss, either internally or externally, but another effective defensive mechanism, blood clotting, keeps blood loss to a minimum. Understanding of all these natural defenses has led to advances in medicine which reinforce the body's own efforts.

The bacteria Bacillus proteus, *shown here ten thousand times life-size, infects the human urinary tract and is among the many microorganisms which attack our bodies.*

Barriers to infection

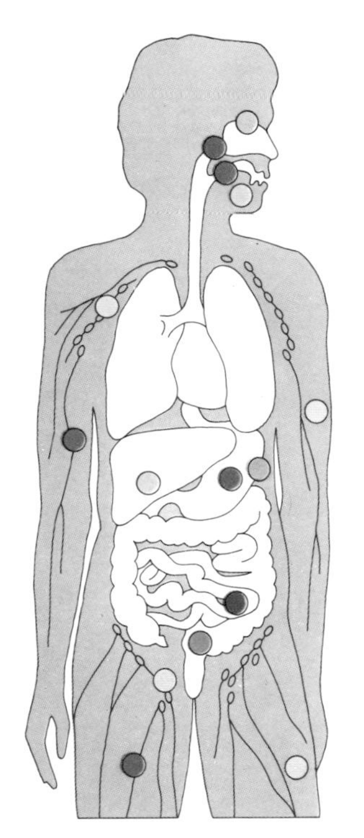

The defense systems of the body

Tonsils and adenoids are composed of lymphoid tissue. Located at the entrances to the throat, they act as barriers to bacteria and viruses.

Lachrymal glands, above the outer corner of each eye, secrete tears, which wash dust and dirt from the eyes. They contain chemicals that kill bacteria.

Salivary glands are found in the cheeks and under the tongue. The saliva they secrete contains substances which help to resist infection.

Lymph nodes are small glands which produce white blood cells. Some of these cells produce antibodies and others ingest bacteria.

The lymphatic system carries tissue fluid, or lymph, around the body. Lymph contains white cells and carries bacteria to the lymph nodes, where they are trapped and destroyed.

The liver produces some of the factors which make blood clot and so initiate wound repair.

The stomach produces hydrochloric acid, which sterilizes ingested food and kills bacteria.

The spleen, composed of lymphoid tissue, produces white blood cells and removes unwanted debris from the blood.

The small intestine is usually sterile, for bacteria have previously been killed by gastric acid and broken down by digestive enzymes.

The large intestine contains harmless bacteria which effectively exclude harmful bacteria from colonizing.

The skin is the body's main protection against disease. Invasion of organisms through the skin only occurs when it is damaged.

The three main routes of entry to the body for agents of disease are through wounds, by inhalation and by ingestion. The skin is the body's first line of defense and is remarkably impervious to harmful microorganisms and poisons, unless there is a break, such as a cut, in the epidermis. Such wounds are, however, quickly blocked by blood clots, which bar further entry of bacteria. The air we breathe is full of potentially harmful organisms, but the mucus and ciliated epithelium, lining the respiratory tract, are designed to trap and remove any invaders. The tonsils and adenoids form a protective ring of tissue around the entrance to the throat. Ingested organisms are usually destroyed by chemicals in the saliva or by the strong acid in the stomach. There is also a backup mechanism in the blood and in the lymphatic systems of white blood cells which destroys bacteria, fungi and viruses or neutralizes any toxins which are secreted by them.

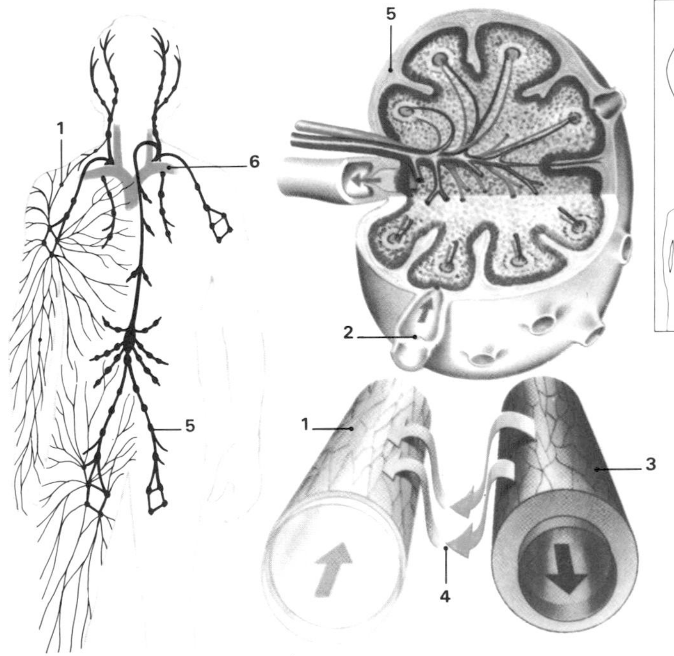

The lymphatic system, a network of vessels (**1**) which carries lymph from around the tissues to the blood, has valves (**2**) which prevent backflow. Lymph is formed when high arterial pressure forces fluid out of the capillaries (**3**) into the tissue spaces (**4**). From there it is taken up into the lymphatic vessels. The lymph carries harmful substances from the tissues to be removed by the lymph nodes (**5**) before the lymph is returned to the blood at the veins in the neck (**6**). In the lymph nodes are large numbers of white blood cells which produce antibodies and ingest any foreign material.

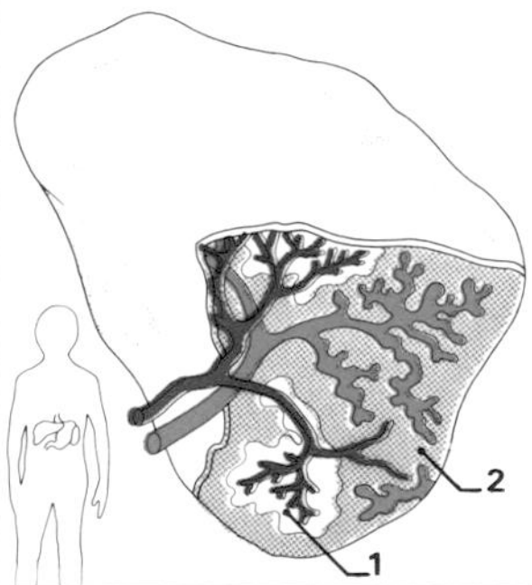

The spleen is a small organ located behind the stomach in the left, upper part of the abdomen. A mass of lymphoid tissue, it contains two distinct areas, the white pulp (**1**) and the red pulp (**2**). The white pulp is distributed, as tiny nodules of lymphatic tissue, among the red pulp, which fills all the remaining space. The white pulp is responsible for the production of lymphocytes. Some lymphocytes enter the blood, while others remain in the spleen, where they ingest and destroy foreign or damaged cells. The red pulp contains many blood channels, expanded capillary lakes, which act as filters when blood is forced through them, removing damaged or abnormal red cells and allowing normal ones to pass through. The lymphocytes then engulf and destroy these damaged red cells. The spleen also acts as a store for red blood cells.

The human body is a vulnerable organism living in an essentially hostile environment in which disease-causing bacteria, viruses, fungi and parasites abound. Without an efficient defense system we could only survive for a matter of hours before certain bacteria would begin to putrefy our body tissues. However, not all microorganisms are undesirable. Indeed many are permanent residents of our bodies. The bacteria which live on the skin, for example, are like a mercenary army assisting in the defense of the body. Their very density leaves no foothold for less desirable colonists and the compounds they produce are inhibitory to other microorganisms.

The body's primary defenses try to prevent possible infectious agents from penetrating into the body. If this is not successful then a secondary line of defense stops these organisms from entering the bloodstream, which would distribute them to every tissue. Those that do manage to enter the bloodstream can still be intercepted as the blood filters through the spleen, the liver and the bone marrow.

The skin is the first major barrier to assault by unwelcome invaders. The body's largest organ, it is tough, flexible, self-repairing, waterproof and almost all-enveloping except at the body's orifices, where it is modified to form mucous membranes. The mucus produced by the membranes is thick, slow moving and an effective trap for bacteria. This job of trapping or destroying invaders at the vulnerable body openings is a shared one. In the windpipe ciliated cells push mucus containing inhaled dust particles and bacteria toward the throat so that they can be coughed out or swallowed. In the eyes there are the washing properties of the secretions from the lachrymal glands. In the urinary tract there are the volume and the bactericidal action of urine. Hairs in the nose and ears and the cerumen in the ears have great trapping ability.

Despite these various mechanisms, however, many bacteria and viruses do penetrate into the body. Fortunately, there is another line of defense—white blood cells which can engulf and deactivate many foreign particles. These are circulating cells and together with other white cells, which are always resident in the lymph nodes, spleen, liver, bone marrow and lungs, comprise what is known as the body's reticuloendothelial system.

Lymph, the interstitial fluid which surrounds the body's cells, passes through the sinuses of the lymph nodes, the rounded structures scattered throughout the body. Lymph nodes contain large numbers of white cells which ingest microorganisms and dust particles to prevent them from traveling around the body. The lymphatic system thus provides a route by which lymph can pass into the lymph nodes, carrying with it any large particles such as bacteria, before it joins the bloodstream.

If the lymphatic system should fail to filter out all potentially harmful invaders before they enter the bloodstream, the body can fall back upon yet another organized line of defense within the spleen, liver and bone marrow. As blood circulates around the body it flows into the spleen. There whole cells pass out of the spleen's porous capillaries into the surrounding red pulp, which is densely packed with white cells, before passing into the venous sinuses. The white cells in the spleen specialize in removing unwanted debris.

The few bacteria that survive the hostile conditions of the alimentary tract, where there are stomach acid and digestive enzymes, are absorbed from the intestine along with digested food particles and pass into the portal vein, which supplies the liver. In the liver they are destroyed by reticulum, or Kupffer, cells, which are capable of ingesting invading bacteria. Similar cells in the bone marrow also remove and destroy harmful substances.

There is a mobile element to this versatile string of defenses—a variety of white cells called macrophages. In all the tissues of the body, for example, there are cells known as histiocytes which, when necessary, travel through the tissues and bloodstream to specific areas of inflammation, where they pick up and destroy organisms and foreign particles which are causing infection. Monocytes and lymphocytes, formed in the lymph nodes, and the granulocytes, formed in the red bone marrow, are also macrophages.

One way that bacteria succeed in causing infection is when tissue is damaged by chemicals or by physical force. When this happens the tissue cells release histamine, a substance which increases the permeability of blood vessels in the immediate area, causing blood plasma and blood cells to leak into the tissue. If bacteria seize this opportunity to invade the tissue they may multiply in great numbers and cause infection.

Within a few hours, however, circulating white blood cells are attracted to the site of the infection and begin to engulf the bacteria and any dead or damaged cells. Macrophages begin to arrive a few hours later as reinforcements to fight the invaders. All these defending cells ingest large numbers of bacteria and dead tissue cells and sometimes die themselves. After several days, when the battle is finished, and usually won by the body, a cavity is present within the tissue, filled with dead tissue cells, blood cells and macrophages—a collection commonly known as pus. This eventually empties and the tissue is repaired and heals over.

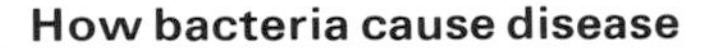

How bacteria cause disease

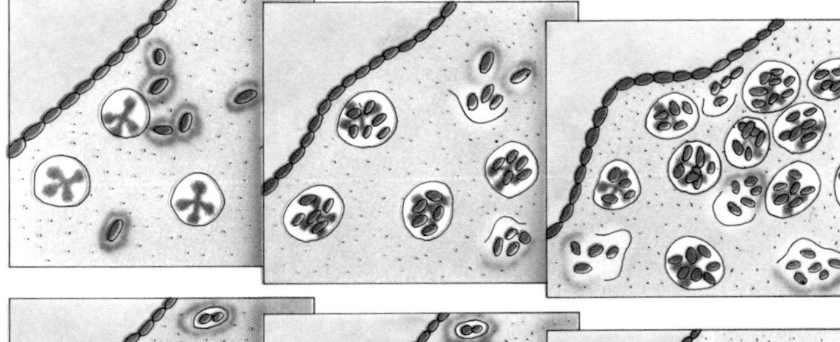

When a tissue becomes infected by bacteria, white blood cells migrate to the site of infection and begin to ingest the bacteria. The bacteria, however, may survive and multiply within the white cells, which then burst, releasing the bacteria into the tissues once again, causing severe inflammation. More white cells then enter the area to try to combat the infection.

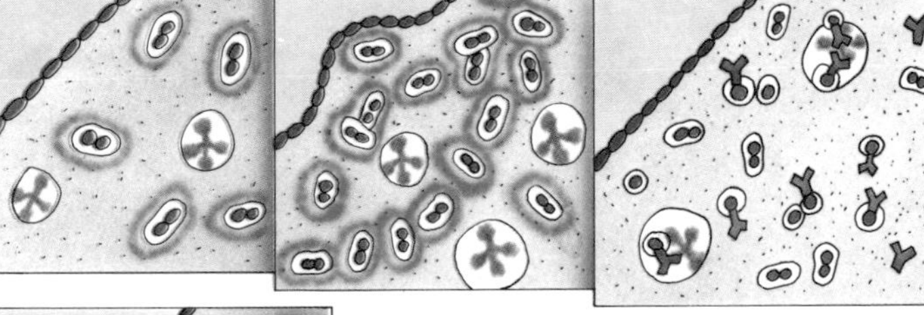

White blood cells are attracted to an infected site where bacteria are multiplying and causing inflammation. Some bacteria have thick capsules around them which prevents them from being engulfed; consequently, the bacteria increase. The body then manufactures antibodies which attach themselves to the bacteria, making them vulnerable to ingestion by the white cells.

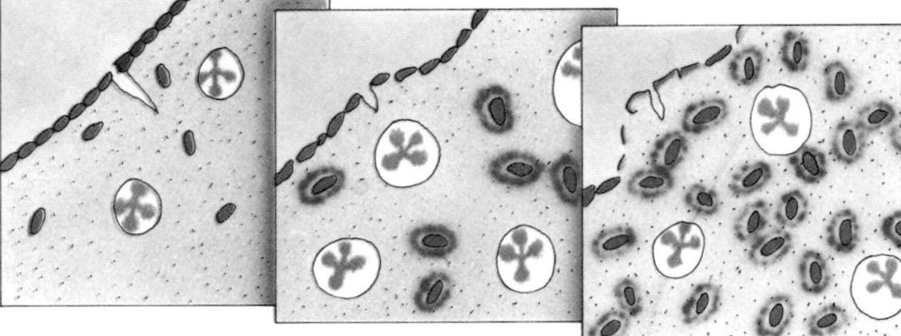

The infected site is supplied with white blood cells. The bacteria have gained entry by a small cut in the skin and are beginning to multiply and grow rapidly, secreting toxins which can act locally or can enter the general circulation to be widely dispersed, causing the death or malfunction of body cells. If the white cells fail to combat the bacteria in the initial stages, then tissue damage may occur.

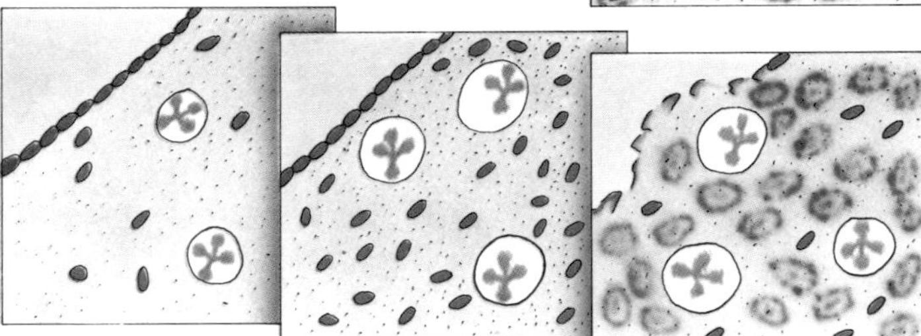

The white blood cells appear, ready to combat the infection. If they initially fail to destroy the bacteria, then the bacteria increase in number and finally begin to break up, releasing toxic chemicals, called endotoxins, which are contained within their walls. The released endotoxins may cause local tissue damage.

Bacteria usually gain entrance to the body either by inhalation, by ingestion or through a break in the skin. Once they have entered a tissue, the bacteria may produce a substance known as the spreading factor, which digests connective tissue barriers in the tissues, thus allowing further spread of the bacteria or their toxins. Failure to overcome these invasive agents can lead to severe tissue damage. Large numbers of bacteria may compete with the healthy cells for nutrients and oxygen. Toxins, on the other hand, disrupt normal cell metabolism by interfering with chemical reactions.

bacteria
encapsulated bacteria
disintegrating bacteria
bacteria-producing endotoxin
dividing bacteria
antibodies

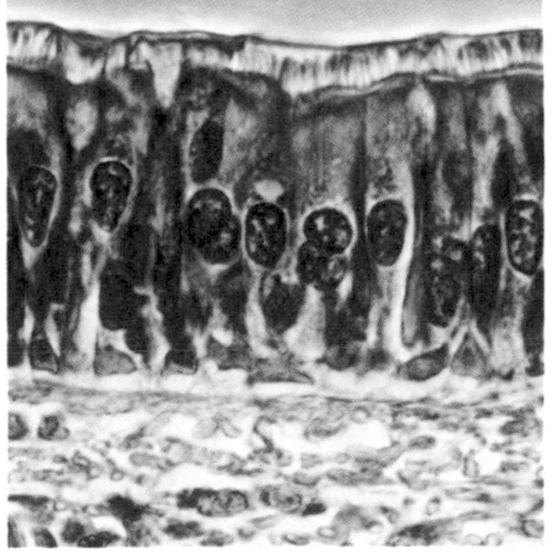

Ciliated cells line the passages of the respiratory tract. The cilia are mobile, hairlike structures. Inhaled dust and microorganisms are trapped in mucus, which is secreted by cells in the respiratory epithelium, and then is moved up toward the throat by the wavelike action of the cilia. Coughing dislodges the mucus from the throat, and it is swallowed, passing down the esophagus to be destroyed by the hydrochloric acid in the stomach.

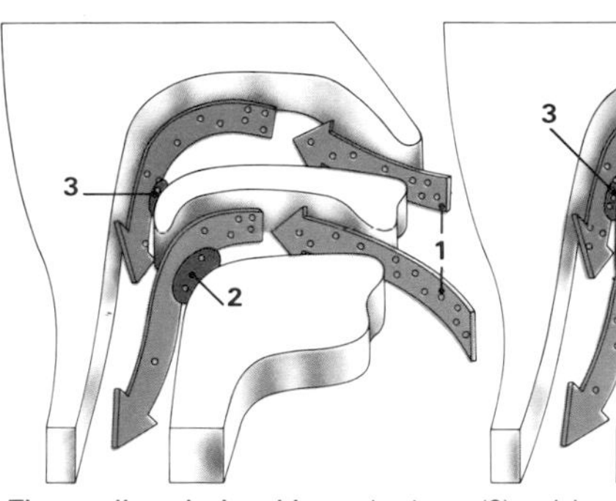

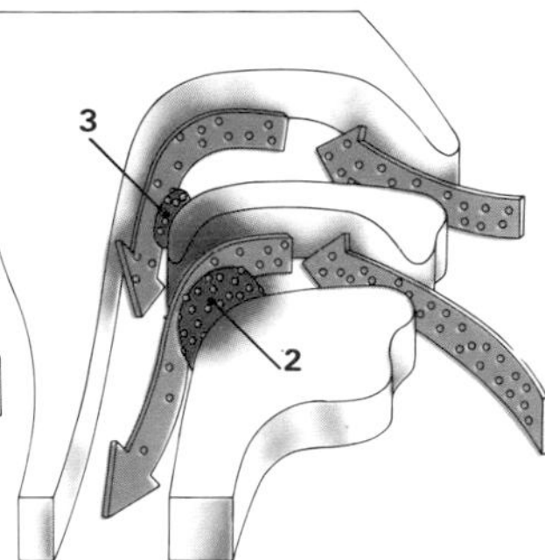

The tonsils and adenoids are glands composed of lymphoid tissue, which are found at the back of the mouth and the back of the throat. They form the first line of defense against airborne organisms. Bacteria and viruses in the air pass through the nose or the mouth (**1**). They are then carried to the tonsils (**2**) and the adenoids (**3**), where they multiply in number and cause inflammation. White blood cells are attracted to the area to combat the organisms, and cause the swelling and sore throat that accompany infection of these glands. Continued infection causes swelling of the lymph nodes in the neck.

Cross-reference	pages
Alimentary tract	**46-49**
Blood	**58-61**
Bone marrow	**30-31**
Circulation	**54-55**
Liver	**50-51**
Lungs	**64-65**
Skin	**28-29**

Immunity and repair

The development of immunity

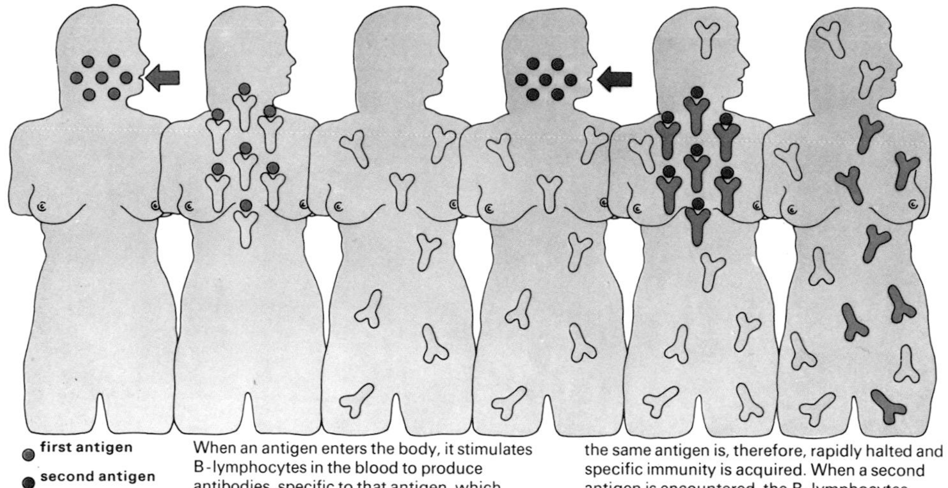

When an antigen enters the body, it stimulates B-lymphocytes in the blood to produce antibodies, specific to that antigen, which attack and destroy it. After the initial attack, the number of specific antibodies in the blood falls slowly, over several weeks, but the body "remembers" their structure, and can produce them at short notice. Subsequent invasion by the same antigen is, therefore, rapidly halted and specific immunity is acquired. When a second antigen is encountered, the B-lymphocytes must produce a new antibody, since the first antibody is specific only to the first antigen. The second antibody continues to circulate in the bloodstream and immunity to the second antigen is acquired.

The immune mechanism

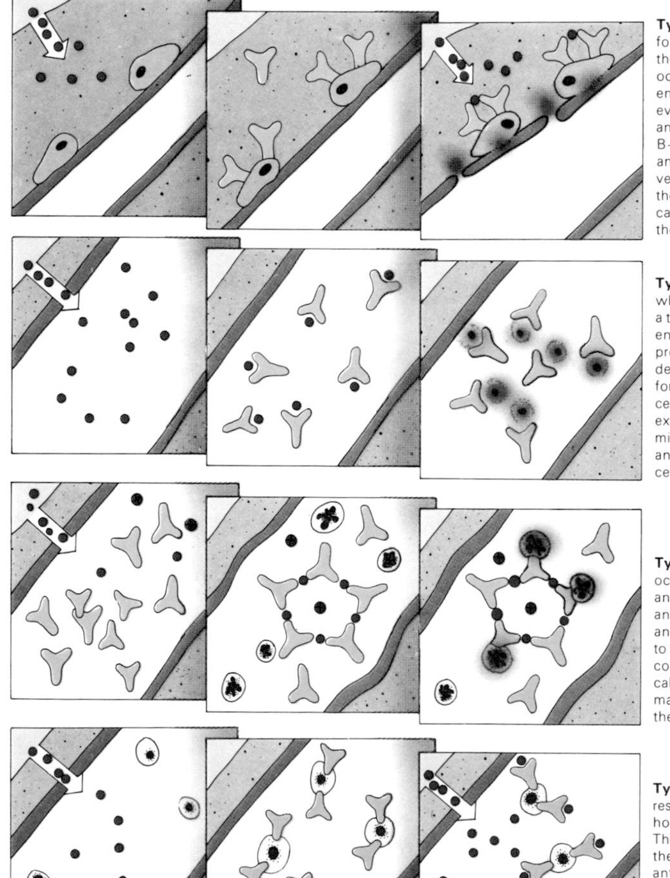

Type I reaction is an allergic response to foreign substances, usually proteins, entering the body. It is an immediate reaction which occurs within minutes or hours of the antigen entering the body. The diagram follows the events that occur in a type I response. The antigens enter the body and stimulate B-lymphocytes to produce antibodies. The antibodies then adhere to mast cells in the vessel wall. They neutralize the antigens and the mast cells release a chemical which causes, for example, the streaming eyes and the sneezing symptomatic of hay fever.

Type II reaction is initiated by antigens which are part of, or closely associated with, a tissue cell. The diagram shows antigens entering the bloodstream and invoking the production of antibodies. The antibodies destroy the antigens, but they may also cause, for example, a cross reaction with blood cells which can lead to cell damage. An example of this type of reaction is a mismatched blood transfusion, in which antibodies are formed against the donor red cells, which leads to their destruction.

Type III reaction is an immediate reaction occurring within a few hours of a small antigenic stimulation. The diagram shows antigens entering blood already filled with antibodies, formed during a previous exposure to these antigens. The antibodies form a complex with the antigen and a blood protein called complement. The complex so formed may damage tissue, such as the glomeruli of the kidneys, by blocking up the capillaries.

Type IV reaction is a delayed immune response which occurs more than twenty-four hours after the initial contact with the antigen. The antigens enter the bloodstream, where they stimulate T-lymphocytes to produce antibodies which remain attached to the cell wall. The antibodies then destroy the antigens. Once the T-lymphocytes have been sensitized by the antigen, they can produce antibodies and confer immunity. This is the basis of immunization against tuberculosis.

The immune mechanism is designed to protect the body against attack from invading microorganisms and foreign, potentially harmful molecules. There are four main types of immune mechanism. The nature of the invading antigen determines which type of mechanism is brought into action. Certain antigens promote an exaggerated response, called a hypersensitive reaction, or an allergy, which may be harmful to the body tissues.

The body's ingenious defense system of traps and defensive cells that ingest bacteria is backed up by a complex chemical system. There are constantly circulating in the blood chemicals which can react with some invading microorganisms and deactivate or destroy them by causing their cell walls to break down. The body's major chemical system, however, is the immune response. This consists of white blood cells, the lymphocytes, that are produced in the lymphoid tissue and are always in the blood. Lymphocytes are activated by, and will destroy, bacteria, viruses or their toxic products that are made up of proteins or complex sugars. Any substance or organism that induces an immune response is known as an antigen.

There are two different types of lymphocyte, T-lymphocytes and B-lymphocytes. They are identical in appearance, but function in different ways when exposed to an antigen. The T-lymphocytes are formed in the thymus gland during fetal and early childhood development, before the thymus shrinks to a fibrous remnant. Stem cells circulating in the blood are trapped in the thymus. There they rapidly divide to form lymphocytes, which pass into the bloodstream and circulate around the body. Eventually, the T-lymphocytes implant themselves in lymphoid tissue.

When a T-lymphocyte comes into contact with an antigen it divides rapidly and forms many daughter cells, all of which have identical properties—they are able to attach themselves to an invading microorganism and produce a specific substance, called an antibody. This antibody breaks down the external membrane of the microorganism, destroying it as the insides spill out. The T-lymphocytes can also release a substance which stimulates macrophages in the area to engulf the antigen.

B-lymphocytes do not pass through the thymus gland before they settle in the lymphoid tissue, but are believed to be formed in the gastrointestinal tract. First discovered in chickens, B-lymphocytes take their name from the bursa of Fabricius, a unique lymphoid tissue found only in birds.

Within the body's lymphoid tissues, the B-lymphocytes exist in isolated collections which continually divide and produce a stream of new lymphocytes that circulate throughout the body. The B-lymphocytes are the source of circulating antibodies—specifically structured proteins called immunoglobulins.

These antibodies can react in several ways when they come into contact with antigens. The antibody can attach itself to the antigen and neutralize it by covering on the membrane the chemical groups which have a specifically toxic effect. This combined particle is then ingested by macrophages. Antibodies are also capable of linking together many molecules of the antigens. This creates a mass which, again, macrophages are able to destroy.

Antibodies are capable of destroying bacteria, viruses and other microorganisms more directly. They attach themselves to the membranes of the microorganisms and make the membrane more permeable. A protein which constantly circulates in the bloodstream, known as complement, is then able to penetrate the cell membrane and cause it to rupture. Even if the antibody cannot penetrate the cell membrane, the presence of complement on its surface makes it more susceptible to being engulfed by the ubiquitous macrophages.

Certain antigens can kill the lymphocytes. This can have serious consequences, because each lymphocyte can recognize only one specific antigen. If all of the lymphocytes were eliminated then the body would no longer have the ability to recognize that particular antigen.

As antibodies are very specific in their actions, those formed in response to invasion by chicken pox virus, for example, are ineffective against the measles virus. In some cases, antibodies once formed remain circulating in the body for life. That is the reason that one attack by a particular virus can confer lifelong immunity to that infection.

The natural immunity conferred on the body by attack by an antigen can be supplemented, and even preempted, by vaccination. This usually involves introducing into the body small quantities of an antigen, but only enough to stimulate antibody production and not enough to produce disease. Another method is to use antigens that have been detoxified in some way but whose antigen structure remains unaltered and can still, therefore, trigger an immune response. Although purified antibodies can be injected directly into the bloodstream, the effects are transitory. Immunization techniques have, in recent years, been dramatically effective in controlling such epidemic killer diseases as smallpox, poliomyelitis and whooping cough.

One of the most remarkable aspects of the immune system is its ability to recognize body tissues and not attack them as if they were antigens. The mechanism of this immunological tolerance is not yet fully understood, but one theory is that during fetal and early life, before the immune system is fully mature and while the thymus is active, the body suppresses the production of lymphocytes which would attack vital tissues.

This is not the only way that the immune system can go wrong. Hypersensitivity reactions, often described as allergies, are caused when such normally harmless substances as dust, pollen or cat's fur trigger an immune response with resulting uncomfortable symptoms such as sneezing, itching and overproduction of mucus and tears.

Another aspect of the body's defensive system is that of repair to damaged tissues. Blood cells are replaced within hours, skin within days and bone within weeks. The healing processes in different tissues follow the same basic pattern. Blood clots and plugs an initial wound while macrophages remove debris from the site of injury. The local tissue immediately begins to divide and deposits the structural protein collagen into the spaces.

In bony tissue, osteogenic, bone-producing cells deposit bone on the inside and outside of the shaft until an excess of bone is present. The excess is then remodeled until the normal bone shape is restored.

Severe mutilation or burning of the skin may require a graft of healthy skin to rapidly repair the damage. In any form of tissue grafting, grafts from elsewhere in the body or grafts from an identical twin produce the best results, as these grafts are unlikely to trigger an immune response, unlike grafts from unrelated donors. The transplanting of corneas rarely promotes an immune response and subsequent rejection because they contain no blood vessels. With transplants of such organs as kidneys, in which there are many blood vessels, powerful drugs are required to try to suppress the immune system.

Repair of bone

A fracture of a bone is usually caused by direct violence or by a strong twisting strain. When a bone fractures the two fragments separate and in so doing tear the arteries in the Haversian systems that cross the fracture line. This damage results in the leakage of blood into the fracture, where it is trapped and soon clots. After a short time, the Haversian arteries go into spasm, causing the death of active bone cells not only at the fracture site, but also for some distance along the shaft. About two days after the break, the blood clot is invaded by capillaries and fibroblasts. The fibroblasts differentiate into bone-forming cells, or osteoblasts, and cells that form periosteal tissue on the outside of the bone. New tissue, called callus, surrounds the fracture and replaces the dead bone. The dead bone is absorbed and replaced by new bone, formed by the osteoblasts in the callus, which is remodeled by cells, called osteoclasts, to its original shape. This remodeling process is so effective that after a few months it is difficult to detect the fracture site.

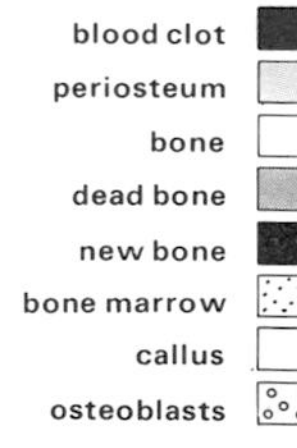

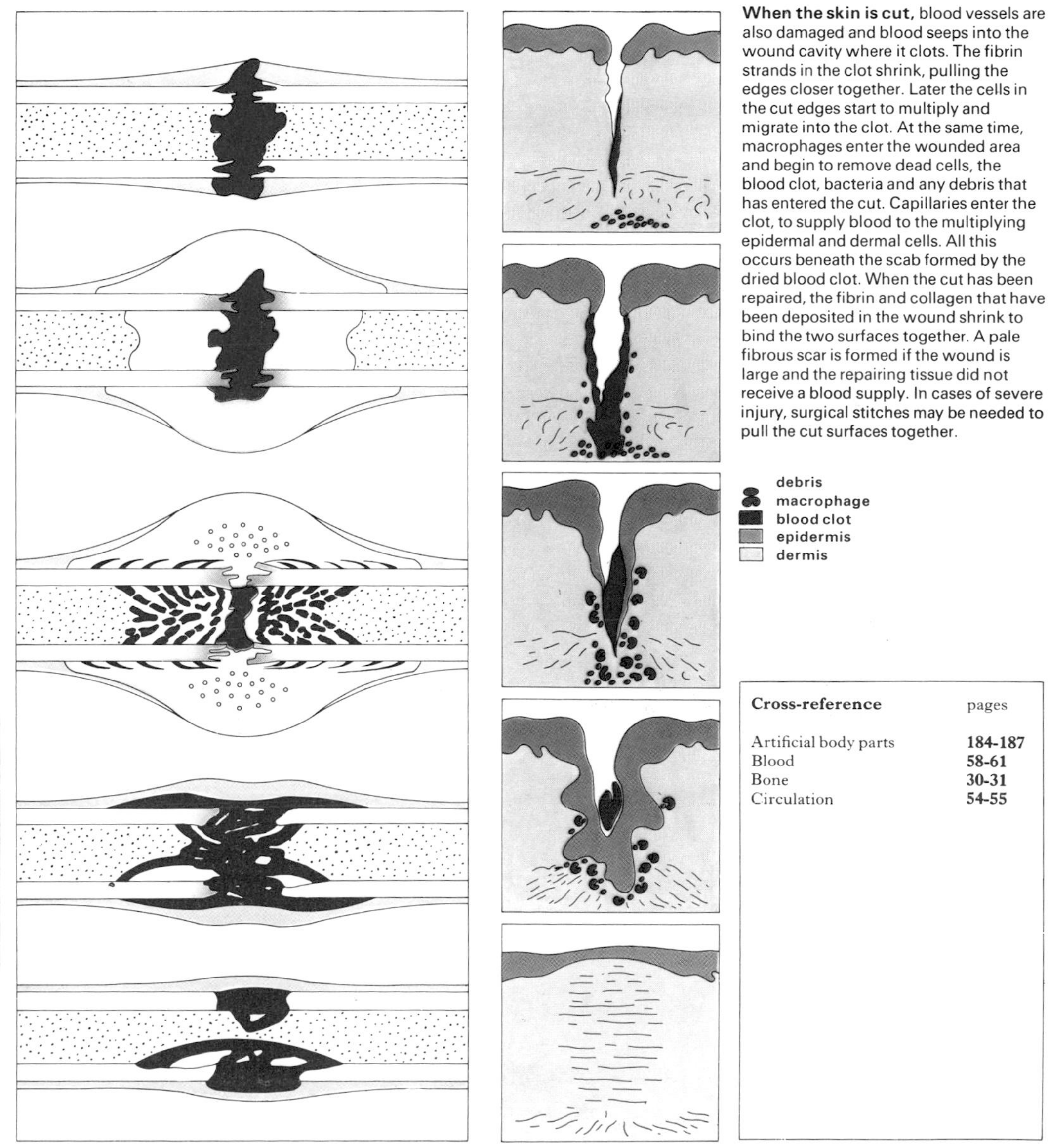

When the skin is cut, blood vessels are also damaged and blood seeps into the wound cavity where it clots. The fibrin strands in the clot shrink, pulling the edges closer together. Later the cells in the cut edges start to multiply and migrate into the clot. At the same time, macrophages enter the wounded area and begin to remove dead cells, the blood clot, bacteria and any debris that has entered the cut. Capillaries enter the clot, to supply blood to the multiplying epidermal and dermal cells. All this occurs beneath the scab formed by the dried blood clot. When the cut has been repaired, the fibrin and collagen that have been deposited in the wound shrink to bind the two surfaces together. A pale fibrous scar is formed if the wound is large and the repairing tissue did not receive a blood supply. In cases of severe injury, surgical stitches may be needed to pull the cut surfaces together.

Cross-reference	pages
Artificial body parts	184-187
Blood	58-61
Bone	30-31
Circulation	54-55

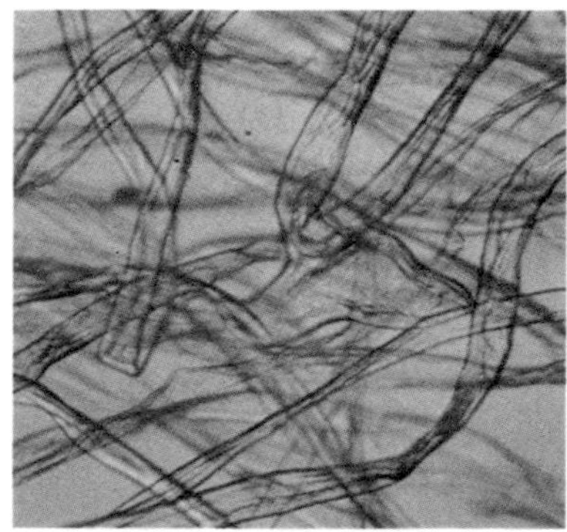

Strands of fibrin are deposited in a blood clot. They form a network binding the surfaces of a wound together and forming a matrix into which white blood cells can migrate during repair.

Reproduction

All life comes only from living things. This is one of the foundation stones of the life sciences. Because the individuals of a species grow old and die, or are killed by disease or accident, the very survival of this species depends upon the process of reproduction, or multiplication of the kind. It was once thought that simple creatures could be generated from dead matter. Worms were believed, for example, to arise from horsehairs lying in pools of water, and frogs were thought to be products of decaying meat. It was not until the seventeenth century that such spontaneous generation was disproved.

Simple creatures reproduce by simply splitting in two, the process of binary fission. This is asexual reproduction. In more complex creatures, sexual reproduction is the rule. This involves two individuals, each of which supplies a specialized reproductive cell, or gamete. In fusing together, the male and female gametes produce a fertilized single cell, the zygote, which is the start of a new individual. Sexual reproduction has great biological advantages, because it allows for the blending together of the inheritable characteristics of both parents. It also provides the possibility of the offspring being better adapted to survive than either parent.

In human beings, a male gamete, or spermatozoon, unites with a female gamete, or ovum, at the moment of conception, following the sexual act. From the zygote develops an embryo, from the embryo a fetus, and after some nine months a baby is born, helpless at first, but soon to develop all the physical and mental facilities of an independent, new member of the species. With the surge of hormonal activity at puberty that individual in turn becomes able to reproduce—and so the cycle of life continues.

To understand how human reproduction works, it is necessary first to understand a little of the structure of the reproductive organs, or genitalia, and how the male and female germ cells develop within the reproductive tracts. For a father, all responsibilities for the development of the unborn child are shed at the moment of conception. The mother, however, must protect and nourish the child within her through the long months of pregnancy. The changes in the bodies of both are remarkable and complex.

To understand how the new individual inherits characteristics from both parents, it is necessary also to understand a little of genetic mechanisms, and how traits can be mingled by the separation and recombination of genes.

Crystals of the sex hormone estrogen, left, photographed with polarized light. Estrogen is primarily involved in female fertility, but small amounts are also secreted by the male, for it is the balance between the male and female hormones which determines sexuality.

The male anatomy and physiology

Male reproductive organs

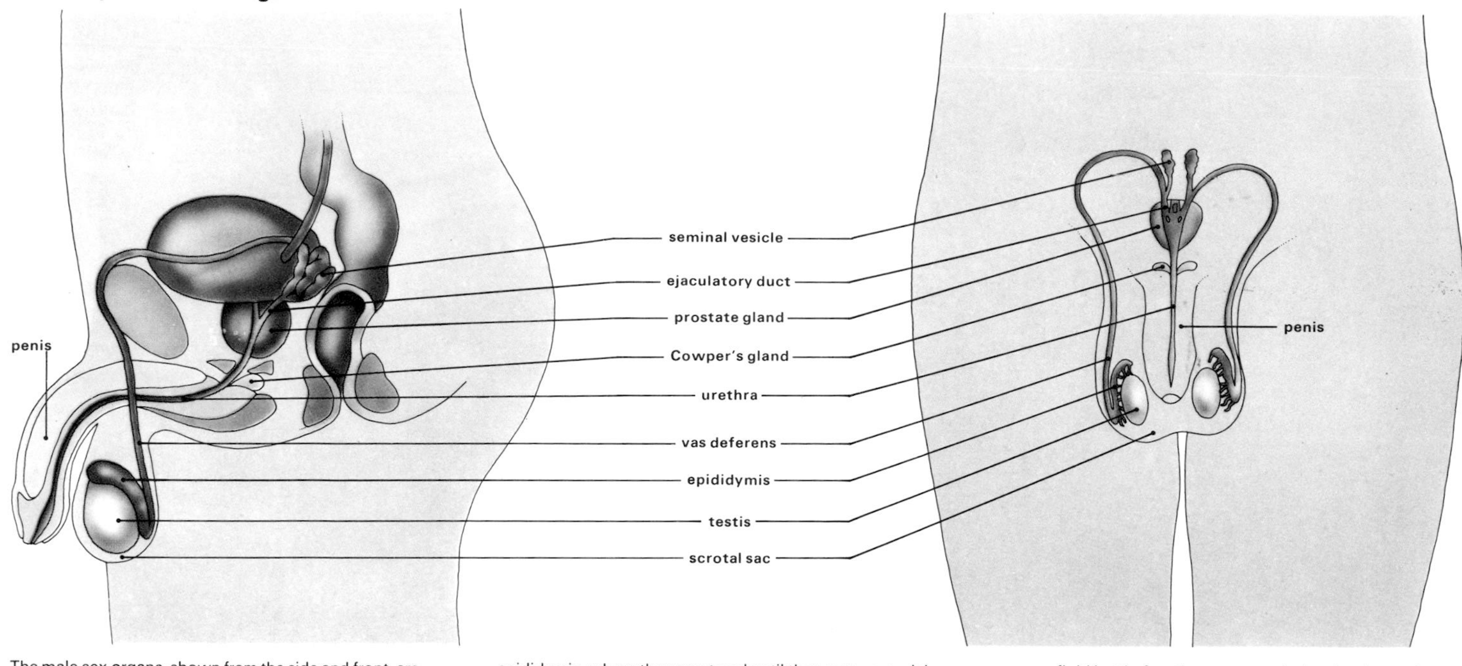

The male sex organs, shown from the side and front, are essentially two glands, or testes, producing hormones and sperm, and a system of tubes which enable the sperm to be transferred to the female. The testes are contained in the scrotum hanging outside the pelvis. From puberty, each testis continually manufactures sperm which are transported to the epididymis, where they are stored until they mature, and then to the vas deferens. Cells in the testes also secrete the hormone testosterone, which controls development of the male sex organs and the secondary sex characteristics. The rhythmic contractions force sperm into the urethra, which is also the route for excretion of urine. The paired seminal vesicles secrete a sugary fluid just before the sperm reach the ejaculatory duct. The prostate gland adds a milky, alkaline fluid, and a pair of Cowper's glands contribute mucus. The semen, which is ejaculated through the penis, is a mixture of sperm and these fluids. One ejaculate may contain as many as one hundred million sperm.

Unlike the female reproduction system, which is mainly concealed within the body, that of the male is for the most part external. The visible genital organs consist of the testes, or testicles, suspended in the baglike scrotum from the arch formed by the pubic bones, and in front of this the penis. After puberty the lower abdomen, from the navel to the region of the groins, and the scrotum are covered in pubic hair, which is usually slightly darker than head hair and is short, coarse and curly.

The penis has two roles to play. One is the routine conveyance of urine that has been stored in the bladder to the exterior—urination, or micturition. The other, more special role, is its function during sexual intercourse of depositing seminal fluid, or semen, in the vagina of the female genital tract—ejaculation. Semen contains the male sex cells, the spermatozoa, which are formed in the testes.

Linking the testes and the urethra, through which urine passes from the bladder to the tip of the penis, are ducts to convey the fluid, produced by the testes, containing the sperm cells. The ducts pass upward from the testes into the abdomen and join the urethra within the prostate gland, which encircles the exit from the bladder. The prostate gland is found only in the male, and is an "accessory gland" of reproduction. It makes its own contribution to the seminal fluid. From this point, semen takes the same pathway as urine to the tip of the penis. During sexual intercourse, muscles around the neck of the bladder contract, ensuring that no urine enters the urethra and that no seminal fluid enters the bladder.

In its normal limp, or flaccid, state the penis is not suitable for introduction into the vagina so that sexual intercourse can take place. During sexual arousal, however, it becomes erect and stiff and enlarges in size. This ability is due to the presence of three columns of tissue. At the top and sides are the paired corpora cavernosa, and underneath is the corpus spongiosum, which also forms the acorn-shaped glans at the end of the penis. These tissues are spongy and well supplied with blood. During sexual arousal, the return pathway of blood is restricted by muscular contractions at the base of the penis, and the columns, therefore, become engorged with blood and rigid.

In men who have not been circumcised, the sensitive glans is surrounded by a collarlike flap of tissue called the foreskin. During erection the enlargement of the glans normally causes this to retract, so that friction from contact with the vaginal lining can bring about orgasm and ejaculation.

The testes are two egg-shaped glands, each from one and a half to two inches long, which lie in the scrotum beneath the penis. They contain a large number of narrow seminiferous tubules, ducts in which the sperm are formed. It takes about forty-six days for a sperm to be produced and on average two hundred million sperm are made every twenty-four hours. The formation of sperm is a continuous process which is believed to go on, although with decreasing vigor, into old age and indeed sometimes right up to death. Sperm which are not ejaculated degenerate and are absorbed.

Between the tubules are the interstitial cells, which produce the male sex hormone testosterone. It is this hormone which, in balance with adrenal hormones, keeps the tubules functioning and which, from puberty, at the age of eleven or twelve, produces the male secondary sexual characteristics, including muscular development, growth of body hair and the breaking of the voice. The testes also produce some of the female hormone estrogen. The balance between estrogen and testosterone is essential to the development of the secondary sexual characteristics.

The ducts in each testis lead into the vasa efferentia, and so through the testicular wall to join the epididymis, a single coiled tube which runs down the side of the testis. This duct in turn opens into the thick-walled vas deferens. The vas deferens ultimately carries the sperm to the urethra.

During their passage through the epididymis, a journey taking about three weeks, the sperm undergo their final maturing stage. Exactly what happens in this final maturation process is not known, but sperm taken from the epididymis and used for insemination will rarely fertilize an egg.

If the vasa deferentia are cut, thus interrupting the flow of sperm, male sterility is the result, but the ability to have sexual intercourse is entirely unaffected. Since the mid-1960s this operation of vasectomy has become a popular method of birth control. Sperm continue to be produced in the testes of the vasectomized man, but they are reabsorbed by the duct system and never reach the penis.

The male accessory glands, the seminal vesicles, in which spermatozoa can be stored, the prostate and two sets of glands called Cowper's and Littre's glands empty into the upper part of the urethra and produce the bulk of the fluid of which semen is composed. Although they are the cells which carry the father's chromosomes to the ovum, the sperm make up only a very small part of the ejaculate.

The semen contains a complex mixture of chemical substances. It is rich in proteins and also contains fructose, a type of sugar rarely found elsewhere in the body, which, it is thought, the sperm use as a source of energy.

The scrotum is itself an organ of some

importance. As some male babies develop, one or both of the testes may not descend into the scrotum, as they normally do at about the time of birth, but remain in the body cavity. In such cases, only abnormal sperm, or none at all, are likely to be formed, because the normal body temperature is too high for the delicate process of sperm production. The temperature in the scrotum is one or two degrees lower than that in the abdomen, conditions better suited to sperm formation. An operation to bring undescended testes into the scrotum is often performed in such cases.

The production of sperm in the testis involves the spermatogonia, special cells in the lining of the seminiferous tubule. These cells can divide to produce more spermatogonia, or to produce sperm cells. Before sperm are produced, each spermatogonium divides to produce spermatocytes, which then divide twice more. These two divisions of one spermatocyte produce four sperm cells.

The first of these last two cell divisions is particularly important. Every nucleated cell in the body (except the sperm and the ova) has a double set of chromosomes, known as the diploid number. But a mature sperm and egg each have only one set. They are haploid. When they meet, the original diploid number is established again in the fertilized egg. The first division of the spermatocyte is, therefore, one of meiosis, which halves the chromosome number. All other cell divisions in the body, except egg formation, occur by mitosis, which involves no chromosome reduction.

The mature sperm, which is ejaculated, is about one five-hundredth of an inch long. It consists of a head, a midpiece and a tail. The sperm head, which is a flattened oval disk, contains the chromosomes which form half of the blueprint for the next generation. They are so tightly contracted, however, that they cannot be seen separately under the microscope. The front portion of the sperm head is covered by a tightly fitting acrosome cap, which detaches just before the sperm penetrates the egg. The midpiece of the sperm contains the mitochondria, originally present in the spermatogonium, which provide the energy necessary for its movement.

The tail of the sperm makes up the greater part of its length, and its characteristic wavelike side-to-side movements drive the sperm forward through its fluid medium. The sperm tail actually contains protein fibers which contract, first on one side and then on the other; the contractions result in the sperm's forward movement.

Not all sperm, however, have this normal healthy appearance. Even fertile men usually have some defective sperm in their ejaculate. Sometimes a sperm may have two heads or two tails. Occasionally, a group of them may clump, or aggregate, together. If a large number do so, infertility may result.

Men differ widely in the number of sperm they produce at each ejaculation. It may vary from as few as two million to as many as two hundred million. Most men produce about three milliliters of semen (a standard medicine spoon holds five milliliters) and there are about one hundred million sperm per milliliter. A concentration below sixty million per milliliter is the cutoff figure which may lead to relative infertility, but much depends on the proportion of the sperm which have a normal appearance, together with normal motility.

The site of sperm production

Each testis is an oval gland about two inches long, shown right in cross section, which produces male hormones and sperm, or spermatozoa. Within a tough outer capsule (**1**) the testis is divided into more than two hundred lobes, which contain four to six hundred minute seminiferous tubules (**2**) each about thirty inches long. Beginning at puberty, the cells lining the tubules produce about two hundred million sperm every twenty-four hours. Groups of cells between the tubules secrete male hormones called androgens. The tubules lead into a network of communicating tubes, known as the rete testis (**3**), which in turn branches into between twelve and twenty efferent ductules (**4**). These join a long coiled tube, the epididymis (**5**), where sperm mature and are stored for up to ten days on their way to the vas deferens (**6**), where they remain fertile for up to six weeks. Sperm are moved from the rete testis, through the efferent ductules and epididymis to the vas deferens on each side, by cilia and by the contractions of the smooth muscle which lines the tubules.

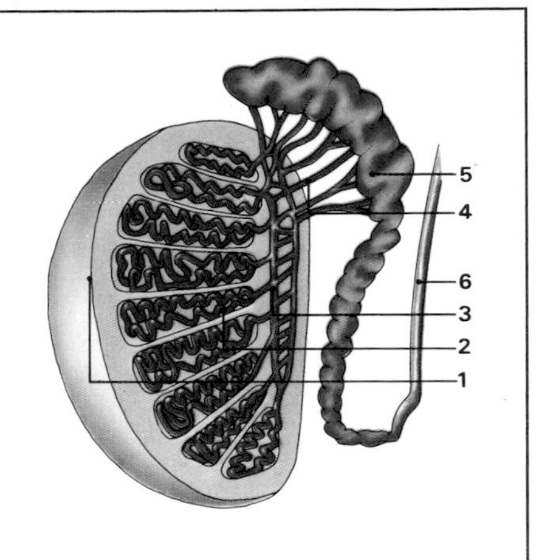

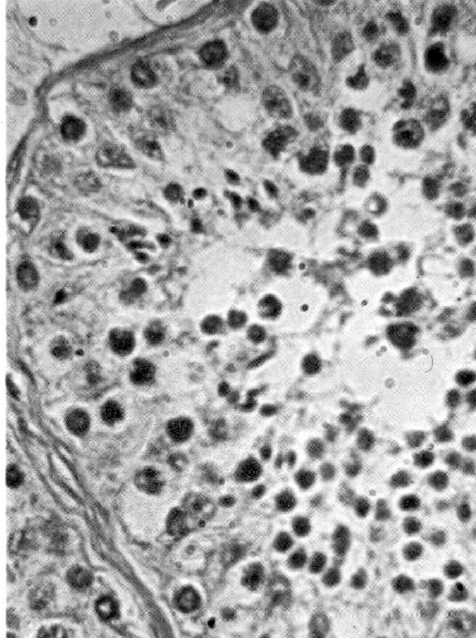

Seminiferous tubules, shown above, stained and in cross section, are the site of sperm production. The tubules are surrounded by a membrane, which provides an attachment for spermatogonia, and cells, which supply the developing sperm with nutrients. Over a period of about forty-six days, spermatogonia develop into spermatozoa. As they mature they move toward the center of the tubule.

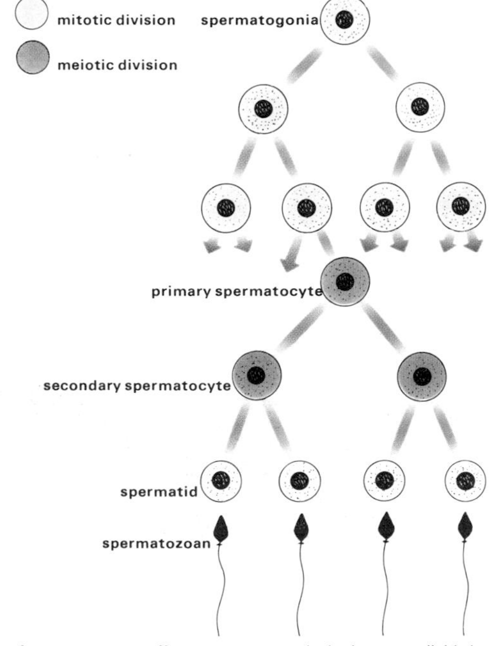

Immature sex cells, or spermatogonia, in the testes divide by mitosis and become primary spermatocytes. Each one then divides by meiosis, a cell division peculiar to the reproductive organs, to form two secondary spermatocytes, each containing half the full number of chromosomes. A second meiotic division splits each spermatocyte into two spermatids, which then mature into sperm.

1
2
3
5
4
6

Spermatozoa, above left, are magnified four hundred times. Above right, the head (**1**) has a large nucleus filled with genetic material (**2**). The neck (**3**) joins the body (**5**), which contains mitochondria (**4**) that provide energy for the tail's (**6**) motility.

Cross-reference	pages
Chromosomes	**26-27**
	158-159
	188-191
Coitus	**148-149**
Fertilization	**150-151**
Hormones	**70-73**
Secondary sexual characteristics	**170-171**
Sex drive	**114-115**

The female anatomy and physiology

Female reproductive organs

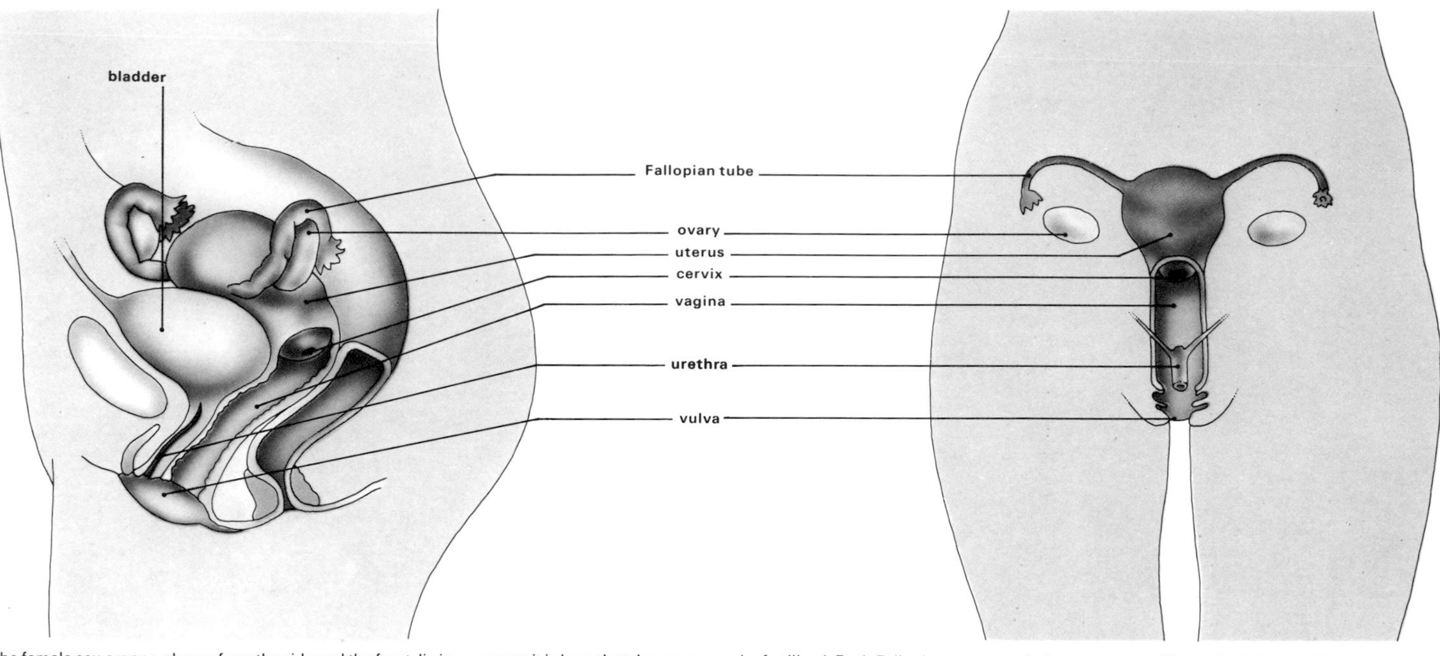

The female sex organs, shown from the side and the front, lie in the lower abdomen, where they are protected by the bony pelvis. At puberty, the two almond-shaped ovaries begin to produce the female hormones, estrogen and progesterone, and the sex cells, or ova. Each month, the ovary releases an ovum into the funnel-shaped opening of a Fallopian tube. If coitus occurs it is here that the ovum may be fertilized. Each Fallopian tube is approximately four inches long and is lined with ciliated cells. The two Fallopian tubes lead into the womb, or uterus, a three-inch-long pear-shaped organ lined with the endometrium, which is shed and built up again each month. It is in this endometrium that the fertilized egg implants and grows during pregnancy. The neck of the womb, or cervix, joins the uterus with the muscular vagina, which leads to the external genitalia, or vulva. The female urinary system, unlike that of the male, is separate from the reproductive system. The bladder empties into the urethra, which opens just in front of the vagina.

Whereas the male reproductive system is designed to produce male sex cells, spermatozoa, and deposit them in the female tract, the female reproductive system not only must produce egg cells, ova, and receive spermatozoa, but also must be adapted to nurture a growing embryo if fertilization should occur.

The outer region of the system is the vulva, consisting of the paired, liplike outer and inner labia, which between them conceal the opening of the vagina, the opening of the urethra, leading from the bladder, and the sexually sensitive clitoris. The clitoris is a small projection at the upper end of the vulva and, like the penis in the male, contains erectile tissue. Internally, the female reproductive system consists essentially of four parts—the ovaries, the Fallopian, or uterine, tubes, the uterus, or womb, and the vagina.

The ovaries are paired organs, about the size of walnuts, situated on either side of the uterus. One of the functions of the ovaries is to produce ova, or egg cells. When a baby girl is born, she has within her ovaries between forty thousand and three hundred thousand ova, of which she will shed a maximum of about five hundred. Ovum formation is, therefore, a ripening process rather than one of manufacture. On its release, an ovum moves through the Fallopian tube into the uterus. In the process it may meet sperm ascending from the vagina.

The Fallopian tubes, each about four inches long, connect the ovaries to the uterus. At their ovarian ends they have a number of fingerlike projections which extend over the surface of the ovary, apparently reaching out to the maturing follicle to ensure that its ovum will actually enter the tube and not be lost elsewhere in the abdomen. The tubes themselves carry the egg down toward the uterus. They are muscular organs, lined with many small folds and having cilia, waving hairlike projections, which help to carry the ovum onward.

Since the Fallopian tubes play such a vital role, obviously any interference with them will prevent the chance of conception. Today an increasing number of women are choosing to have their Fallopian tubes ligated, or tied, by a surgical operation. Such sterilization is a safe and effective means of controlling the size of the family.

The uterus is a muscular, hollow, pear-shaped organ a little smaller than a woman's fist. The ancient Greeks believed that the uterus had a life of its own and that it could wander freely in the abdomen, sometimes pressing on the diaphragm and producing breathlessness and anxiety—in short hysteria, a word derived from the Greek word for womb. The uterus is actually firmly anchored in the body by a number of ligaments. The organ itself consists of a thick muscular wall inside which is a thin lining, the endometrium. Muscular contractions of the wall, which increase during and just after coitus, help to move the sperm toward the ovum in the Fallopian tube. Muscular contractions are also responsible for forcing out the baby some ten lunar months later, at its birth.

The lower part of the uterus, the narrow cervical canal, opens directly into the region of the vagina where sperm are deposited during coitus. The cervical canal is usually full of a special mucus which, most of the time, prevents the movement of any small particles up into the uterus. At the time of ovulation, however, this mucus becomes thin and watery and will allow the passage of sperm.

The vagina is a channel about four or five inches long. At times other than during coitus it is "closed," its walls lying flat against each other. Because the uterus is usually tipped forward in the abdomen, the vagina and cervix lie at an angle to each other.

The functions of the various reproductive organs are closely coordinated. Each month from puberty, at about the age of eleven, to the menopause, at about fifty, a number of small swellings, or follicles, appear on the surface of the ovary. One of these continues to swell and eventually bursts, releasing the fully mature egg cell, or ovum, into the tube. As with sperm formation, this process of development involves the two divisions of meiosis, which reduce the number being restored at fertilization. Unlike sperm production, however, the first division produces one healthy pre-egg cell, or oocyte, together with a small disorganized envelope of chromosomes called the polar body. At the second division the polar body may divide into two. The oocyte itself divides, usually after the sperm has entered it, forming a third polar body. Thus one oocyte gives rise to only one mature egg; the rest of the chromosomes enter polar bodies, which eventually disintegrate.

But egg production is only one of the ovary's functions. Another is the production of the female sex hormones, estrogen and progesterone, which are produced cyclically. The pattern is caused by the stimulatory effect on the ovary of a further set of gonadotrophic hormones from the pituitary gland. During the first half of the menstrual cycle, as the ovum grows and develops, estrogen is the main hormone to be produced. Once the egg cell has been shed at ovulation, progesterone is also produced by the ovary. The two hormones together make the uterine lining develop in such a way that the fertilized egg can embed, or implant, itself—the first stage of pregnancy.

The production of ova ceases at the meno-

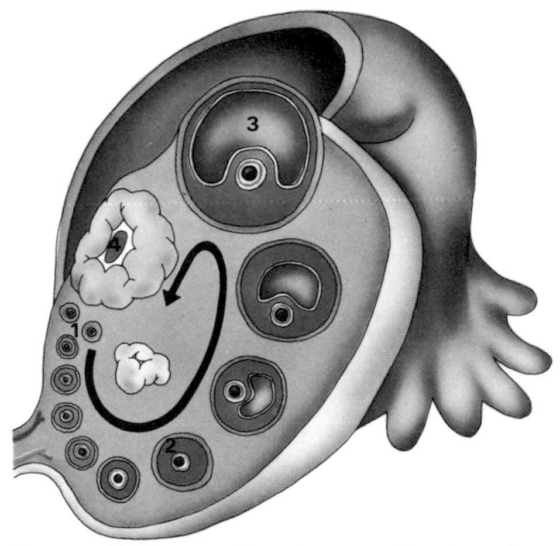

One ovum in an ovary, shown in cross section above, ripens each month. Around the ovum the saclike follicle (**1**), which absorbs fluid, grows and swells (**2**) for about fourteen days until it bursts onto the surface of the ovary and the mature egg, or ovum, is released (**3**). After ovulation, the empty follicle forms the hormone-producing corpus luteum (**4**), which helps to maintain pregnancy if fertilization occurs. If fertilization does not occur the corpus luteum simply withers away.

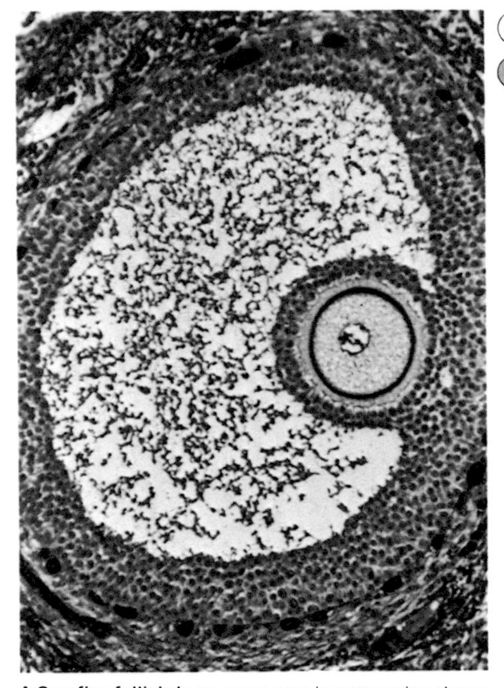

A Graafian follicle in an ovary, seen in cross section, shows the circular ovum surrounded by the pale zona pellucida. The ovum lies within a fluid-filled cavity and is attached to the inside of it by a ring of granulosa cells. The whole structure is surrounded by another layer of cells, which later secrete the female sex hormones.

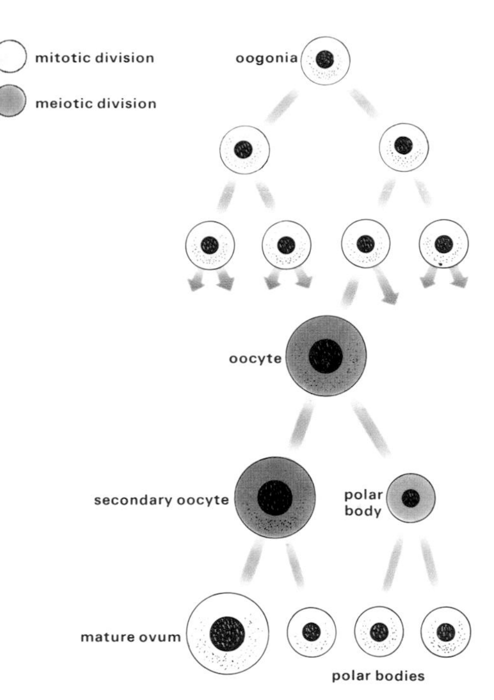

Immature sex cells, or oogonia, in the ovary divide by mitosis to produce oocytes. Before ovulation, one oocyte divides by meiosis to produce a secondary oocyte with half the full number of chromosomes, and a polar body. At fertilization, further divisions result in the production of the mature ovum and three superfluous polar bodies.

pause. Although the pituitary hormones are still produced in increasing quantities, to attempt to stimulate it, the ovary loses its ability to respond to them. The reason for this is still a mystery. The effect, however, is that estrogen levels fall in women after the menopause, and gonadotrophin levels rise. This leads to distressing symptoms like sweating and hot flashes. Estrogen tablets may be prescribed to restore hormonal balance.

By altering the cyclic production of hormones, ovulation can be prevented. Perhaps the most convenient of all contraceptive methods so far is the "pill," which contains combined synthetic hormones. Taken daily, the hormones override the ovary's natural rhythm and prevent an ovum from developing.

The hormonal cycle of the ovary has a profound effect on other parts of the body, particularly the endometrium. In most months no ovum is fertilized and the uterine lining has no role to play in implanting the early embryo. At the end of the month the ovary produces little of the hormone on which the endometrium depends for its integrity. It therefore breaks down and is shed from the uterus in the form of menstrual fluid. If, however, a fertilized ovum becomes embedded in the wall of the uterus, the situation is quite different. The young embryo produces a substance which stimulates the ovary to continue producing hormones, particularly progesterone. This in turn stimulates the endometrium and keeps it intact. Since it does not break down, no menstrual "period" ensues.

At the end of a normal cycle, however, when fertilization does not occur, the ovary stops providing both estrogen and progesterone. This change in hormonal balance, apart from the effect it has on the endometrium, is also believed to be responsible for premenstrual tension, which often involves changes in the woman's emotional makeup for several days preceding menstruation. Further evidence for this view is provided by the fact that premenstrual tension is often relieved if the cycle is overridden by the contraceptive pill.

The menstrual cycle

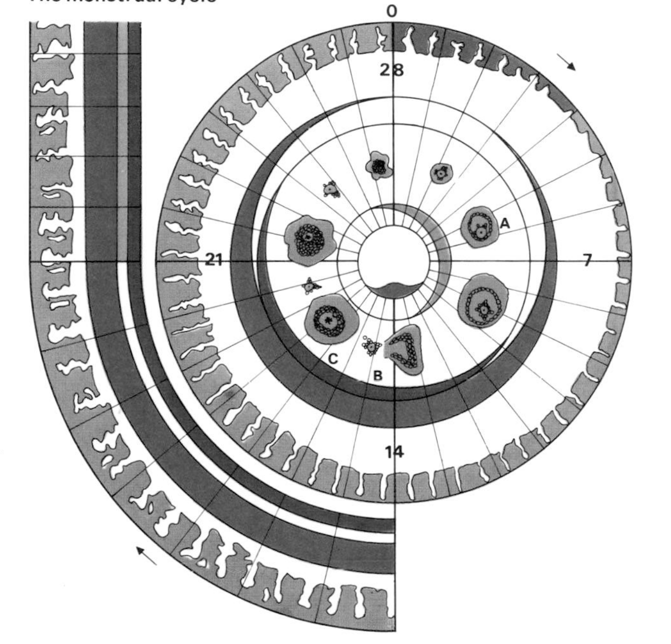

Controlled by hormones, the average menstrual cycle is twenty-eight days. From about day 5 pituitary follicle-stimulating hormone (FSH) promotes the growth of a follicle (**A**) in the ovary. Follicular cells produce estrogen, which builds up the endometrium, the uterine lining (light brown), in readiness for implantation of the fertilized egg. In midcycle, a surge of pituitary luteinizing hormone (LH) causes ovulation (**B**). The corpus luteum (**C**) secretes progesterone, which reinforces estrogen and builds up the endometrium to its maximum thickness. If the egg is not fertilized, the corpus luteum degenerates in the last few days of the cycle and the lining is shed during menstruation (light orange), which marks the start of a new cycle. If fertilization occurs, about day 14 the fertilized egg secretes human chorionic gonadotrophin (HCG). This maintains the corpus luteum and its output of progesterone and estrogen so that the endometrium is maintained during pregnancy and menstruation does not occur. The endometrium then nourishes the fertilized egg.

Cross-reference	pages
Coitus	148-149
Fertilization	150-151
Hormones	70-73
Menopause	178-179
Pregnancy	150-155
Puberty	170-171

Coitus and orgasm

Sexual intercourse, in all its detail, has never been a topic of polite conversation, and because it is such a personal and private affair, hardly suited to the overt scrutiny of inquisitive scientists, the exact nature of the physical changes that accompany this all-important process was shrouded in mystery until relatively recently. In the 1940s, original work by the American Alfred Kinsey paved the way for studies in the two subsequent decades by William Masters and Virginia Johnson, using techniques of direct observation, photography and laboratory measurements, in the United States. As a result, a great deal is now known about the physiological and anatomical changes that accompany sexual intercourse.

Sexual intercourse, or coitus, is the insertion of the penis into the vagina during sexual arousal, followed by movements of the bodies of both partners that bring about increasing sexual excitement until a climax, or orgasm, is reached. In a man, ejaculation of semen occurs at orgasm. In a woman, the changes are more subtle. Orgasm is an overwhelmingly intense feeling of pleasure, centered on the genital area, but from there pervading the entire body. During sexual intercourse there are increases in blood pressure, and in the rates of heartbeat and respiration as well as other physiological changes. With orgasm comes a relief of the tensions that have built up, and the body then reverts gradually to the normal state.

Precoital behavior that leads to sexual intercourse can take many forms, but usually begins with visual stimuli that trigger off the sex drive. Physical contact, the touching and petting of the erogenous zones, as well as verbal inducements, lead to an increase of excitement.

In men, the first sign of sexual arousal is erection of the penis. Within the length of the penis, and extending back into the pelvis beneath the arch of the pubic bones, are three columns of spongy, erectile tissue. On either side of the urethra are the corpora cavernosa, and beneath it the corpus spongiosum, which is continuous with the enlarged end of the penis, the glans. These tissues are well supplied with blood. During sexual arousal, which may be induced by a wide variety of stimuli, there is an increase in the amount of blood entering from the arterial system, but contraction of the muscles surrounding the base of the penis, the bulbocavernosus and ischiocavernosus muscles, restricts the return of blood in the veins. The tissues, therefore, inflate with blood under pressure, and the penis becomes longer, wider, stiff and erect.

Masters and Johnson found that the size of the erect penis is not related to the overall size or stature of the individual, nor is it closely related to the length of the flaccid penis. On average, penile length before erection is between one and five inches. At erection it increases to roughly six to seven and a half inches. Its length, however, bears no relation to the duration of intercourse nor to the amount of semen produced.

The first sign of sexual arousal in women, comparable to erection in man, is the secretion of a fluid which lubricates the walls of the vagina. This fluid originates along the whole length of the vaginal canal and not, as previously believed, from particular glands near its entrance. The walls of the vagina become congested with blood and redder in color. In response to initial sexual stimulation the breasts, the nipples and the clitoris increase in size. The clitoris, a small rod-shaped organ which lies close to the pubic arch in the midline, in a "hood" of tissue, is the female analogue of the penis. During the development of the embryo, tissues in the genital area, which contain a large number of nerve endings, develop either into a penis or a clitoris depending upon the sex.

Masters and Johnson describe this initial period of sexual stimulation as the excitement phase. It is followed after insertion of the penis into the vagina by the plateau phase, which precedes orgasm and which varies enormously in its duration, depending upon such factors as age, physical health and sexual experience. In men, the plateau phase is characterized by a further increase in the diameter of the penis, especially the nerve-containing tip, or glans. In women, a number of changes occur during this phase. The erect clitoris retreats somewhat into its sheath so that it is not usually in direct contact with the surface of the penis. The innermost two-thirds of the vagina expands both in length and in diameter, but the outermost one-third tightens somewhat around the penis. The uterus rises higher in the abdomen and the cervix begins to distend.

The plateau phase is followed by orgasm itself. In men, there are a series of violent contractions of the penile urethra throughout its length. The first few contractions occur at intervals of little more than one second, although with subsequent contractions the frequency decreases. Similar contractions also occur in the accessory glands, including the prostate gland, which produces the bulk of the seminal fluid. These coordinated contractions result in the ejection of semen into the upper vagina.

Orgasm in women has many similarities to that in men. The vagina, particularly in its more narrow lower third, also experiences powerful contractions lasting from two to four seconds, followed by shorter contractions about eight-tenths of a second apart. The uterus moves upward and it, too, experiences waves of contractions; during the intervening expansions it may temporarily become considerably larger than normal.

Following orgasm is the resolution phase, during which the genital organs return to their original prestimulatory condition. In men, this phase is twofold. There is an initial and rapid decrease in penile size, followed by a slower phase during which the penis returns to nor-

Sexual intercourse

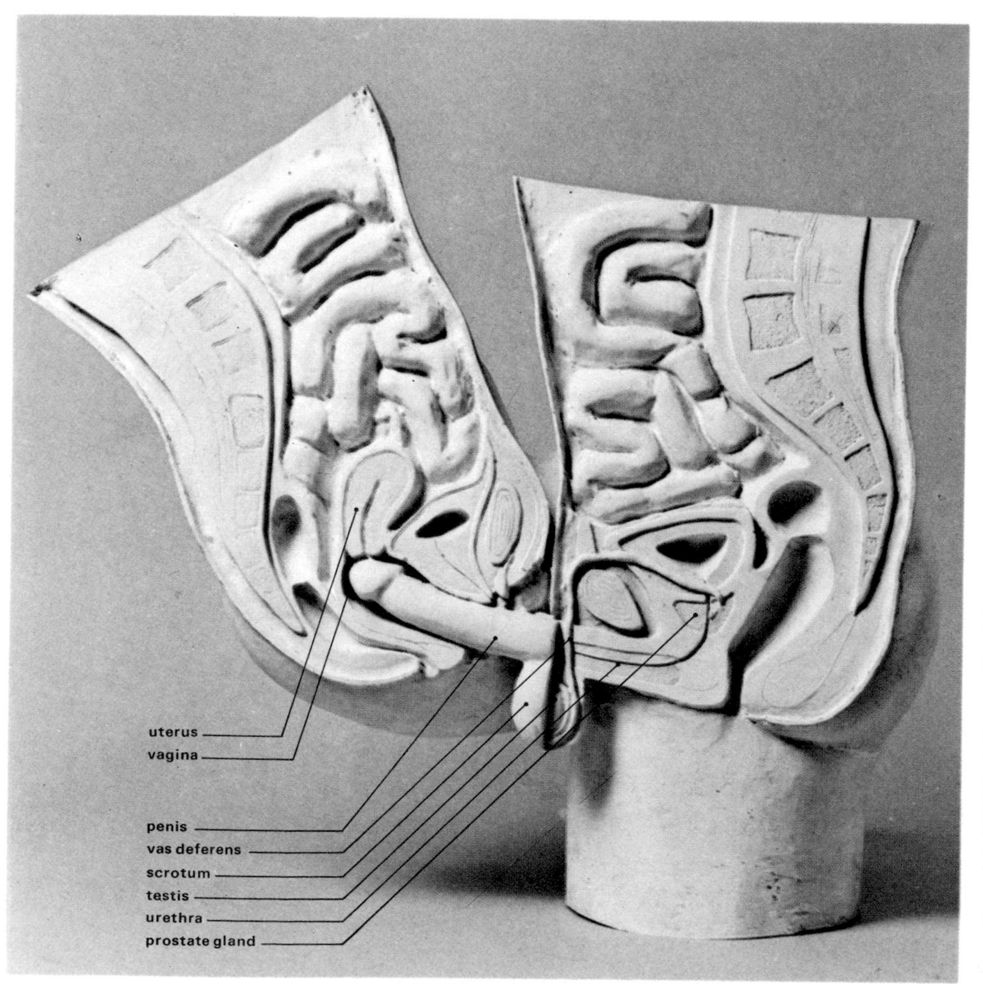

Coitus, or sexual intercourse, is achieved by the insertion of the engorged, erect penis into the lubricated and receptive vagina. This cutaway model shows how the vaginal muscles tighten around the penis. The movement of the penis within the vagina results in the male ejaculating and thereby depositing semen into the upper part of the vagina. The sperm then begin their upward journey through the cervix, uterus and into the Fallopian tubes.

mal. A similar two-phase return also occurs in the clitoris. Following orgasm, the vagina relaxes to regain its original size and the uterus descends back to its original position. The lower part of the cervical canal, however, remains open for as long as thirty minutes.

These are the resolution changes which occur if there is no further stimulation. But the woman may return only as far as the plateau phase if stimulation continues. From this level some women are capable of experiencing a second, third or even a multiple series of orgasms.

Indeed, there is no reason to assume that any woman is not innately capable of doing so. Whether she actually does achieve multiple orgasm depends on a large number of factors, including the duration and the intensity of the stimulus that she receives.

It is only during the resolution phase that the man's reactions differ significantly from those of his partner. For during this phase, for most men, there is a refractory period in which no amount of stimulation will induce a second erection. The duration of this phase varies with the individual and the occasion. He may not in fact revert to his normal unstimulated condition, but only to the excitement phase—that is, to a preparation for the next full erection and the next complete cycle of sexual response.

There is a wide variance in the frequency of sexual intercourse, depending on the needs and desires of the individual. According to Kinsey the average number of orgasms experienced during a week was three. However, in this study the extremes varied from one man, who had ejaculated only once in thirty years, to another who claimed to have experienced thirty orgasms a week over a period of thirty years.

Sexual capacity reaches a peak in the male during adolescence and then declines gradually into old age, while the sexual drive of the female increases in middle-age. Although sexual activity becomes less frequent as individuals become older, there is no physiological reason why it should not be continued into old age.

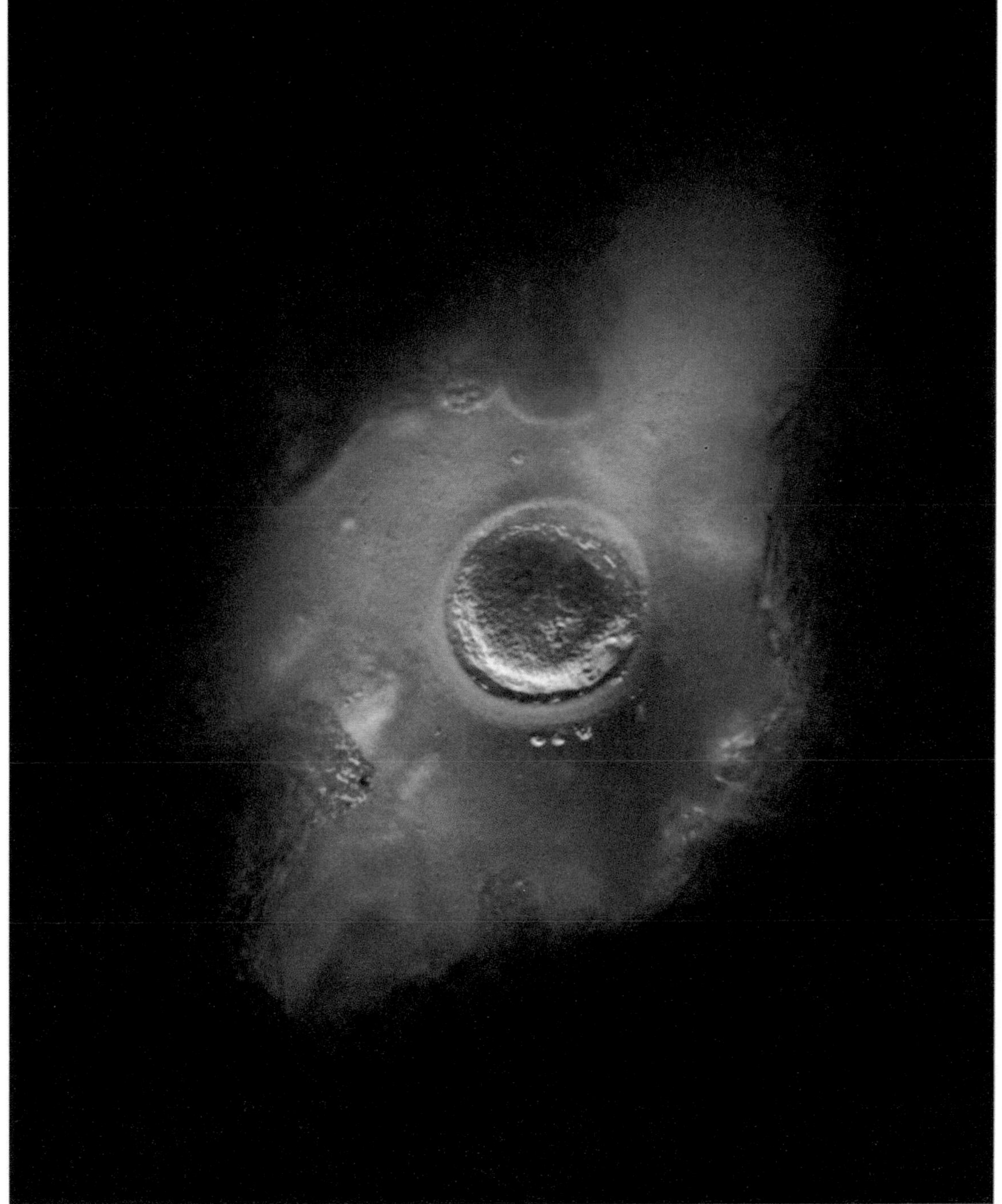

An ovum enters a Fallopian tube in the photograph, left. There it may fuse with a sperm. The few sperm which reach the ovum are only a small percentage of the total that were deposited in the vagina during sexual intercourse.

Cross-reference	pages
Female anatomy and physiology	**146-147**
Male anatomy and physiology	**144-145**
Sex drive	**114-115**

Fertilization: the beginning of life

Fertilization is perhaps the most important single event in the whole process of sexual reproduction. Through the fusion of genetic material from the father and the mother a new individual is formed who is genetically different from either of them. It is on this ability of the species to produce such genetically different individuals that the possibility of their evolution and, indeed, their ultimate survival must depend.

In human beings, fertilization occurs in the upper third of one of the Fallopian tubes. The ovum at ovulation is guided into the tube by the fingerlike projections in which the tube ends. The ovum moves slowly down the tube, probably propelled by contractions of the tubal muscles. Its journey before its fertilization is a comparatively short one.

Quite different, however, is the journey faced by the sperm. From the vagina it has to pass through the cervical canal, through the cavity into the uterus, through the narrow junction that joins the uterus to the Fallopian tube and, finally, through most of the tube's length until it reaches the ovum. This journey is thousands of times greater than the sperm's own length. Because sperm have lashing tails which drive them along, it was assumed by the microscopists of the seventeenth century, and by many people since, that sperm simply swim all the way to the egg. It is now clear, however, that for them to do so would take much longer than the twelve to twenty-four hours that they actually need to get there. During part of their journey they do, indeed, seem to swim; at other times they are passively carried along.

Their movement, for example, through the mucus which fills the cervix is probably due to their own efforts. Once within the uterus, however, they are carried along by waves of uterine contraction. In getting into the Fallopian tube their own movement is again important, but once inside the tube the regular beating of the small hairlike projections which line it possibly helps them along to the egg traveling toward them.

Since the journey is so long and so difficult it is perhaps not surprising that relatively few sperm reach their destination. At ejaculation, the sperm in the vagina can be numbered in hundreds of millions. Estimates of how many enter the uterus are difficult to make, but the figure is probably about one million. And only about one thousand may enter the Fallopian tubes. It seems that the male produces so many sperm because so few actually arrive at the ovum. The ultimate fate of the surplus spermatozoa in the female genital tract is still not known. Many simply leak out of the vagina, while others are removed by white blood cells in the uterus. Although this removal is rapid, some spermatozoa remain fertile and motile for twenty-four to forty-eight hours.

But the problem can be turned around, to ask why, since only one sperm is required for fertilization, so many rather than so few should get into the tube. One belief is that the cluster of sperm in the tube help the one that ultimately penetrates the ovum to do so. This view suggests that all the sperm around the ovum shed their acrosomal caps at the same time. The acrosome is believed to contain a substance which dissolves the outer membrane surrounding the ovum, thus making penetration by that one sperm much easier. After penetrating the ovum membranes and entering its protoplasm, the sperm loses its tail and midpiece. The sperm head begins to swell and, for the first time, its chromosomes become visible. In this form it is known as the male pronucleus. The female nucleus undergoes a similar increase in size, and chromosomes in this pronucleus also appear. Near the edge of the ovum protoplasm there may be seen a polar body, a clump of waste chromosomal material that results from the final division of the egg nucleus, and is extruded from the ovum.

The male and female pronuclei then fuse to restore the normal diploid number of forty-six chromosomes. Two of these chromosomes are sex chromosomes, and it is at this stage that the sex of the baby is determined. In the process of sperm production in the testes, two types of sperm are formed, one carrying a sex chromosome known as X, and the other carrying a Y chromosome. Ova, however, bear only sex chromosomes of the X type. If a sperm carrying an X chromosome fertilizes the ovum, the baby's chromosomes will be XX and it will be a girl. If, however, a Y-carrying sperm meets and fuses with the ovum, then the baby's

Sperm transport

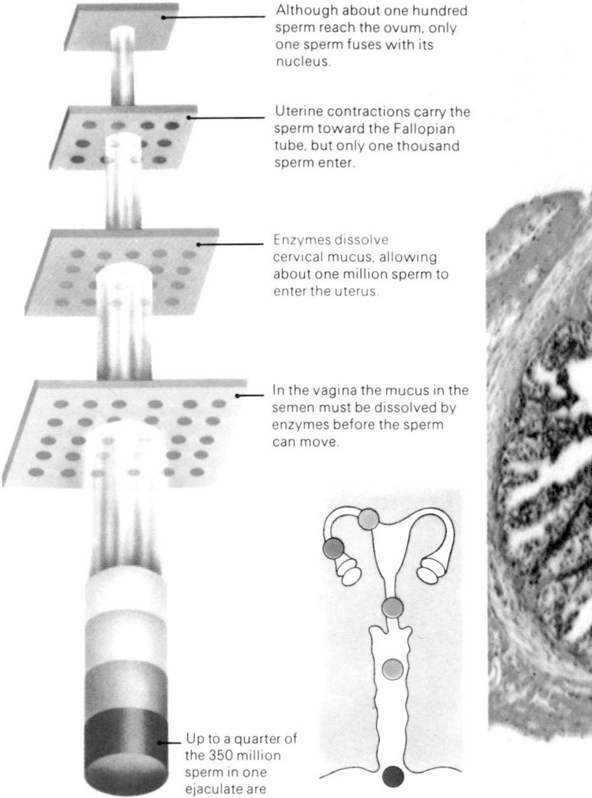

There are more than 350 million sperm in a single ejaculate, but only one sperm fertilizes an ovum. The initial reduction occurs at the cervix and about one million sperm enter the uterus. Of these only about one thousand enter the Fallopian tube and possibly only one hundred successfully reach the ovum in the tubal ampulla.

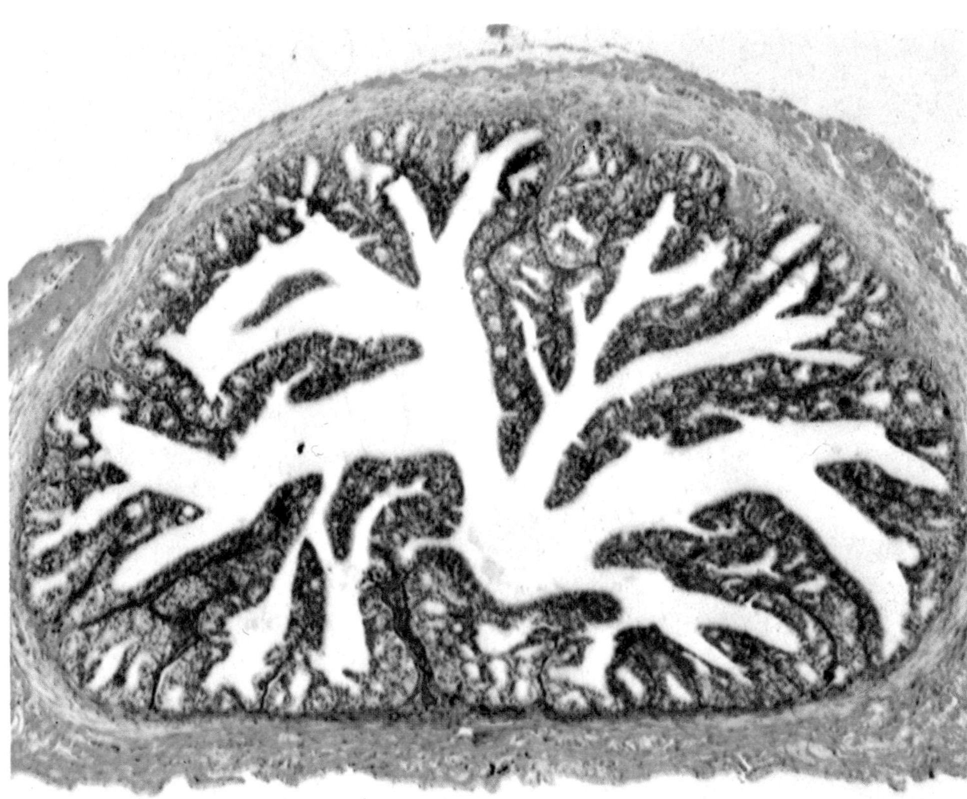

The Fallopian tube, seen here stained in cross section and thirty-five times life-size, has walls consisting of three layers—an external layer, a middle, muscular layer and a folded mucus-secreting layer, which extends into the center of the tube. This inner layer contains ciliated cells, which beat toward the uterus, carrying the fertilized ovum, or zygote, along in the fluid current that is set up. At the same time, the projections in the central canal of the tube slow the passage of the zygote, thus ensuring that several stages of division occur and it is sufficiently mature for implantation before it enters the uterus. The mucus secreted is important for the viability of the sperm, ovum and the developing zygote.

cells will carry the chromosome arrangement XY and the baby will be male.

Since this difference in sperm types was discovered, scientists have tried to exploit it by separating the sperm and using them for insemination. Many parents would like to be able to decide the sex of their children in planning a family. Because an X chromosome is some five times larger than a Y chromosome, it is possible to achieve some separation of the sperm by using a centrifuge.

Sometimes more than one sperm may penetrate the ovum, but usually only one is transformed into a pronucleus to fuse with the ovum chromosomes. If more than one did so, then the normal chromosome number would be exceeded by one-third and an embryo with such an abnormal chromosome number would die. Indeed, even penetration by more than one sperm is unusual. It seems that the first penetration activates a defense system in the egg which makes it repel any others.

Following fertilization the egg begins to divide into a number of cells. As it does so, it moves down the Fallopian tube into the uterus. The ovum enters the uterus about one week after fertilization, by which time it has developed into a ball of perhaps thirty-two to sixty-four cells. The ball fills with fluid and the cells separate and come to lie on its surface. It is at this, the blastocyst stage of development, that the young embryo first makes contact with the lining of the uterus—the first stage in the development of the placenta.

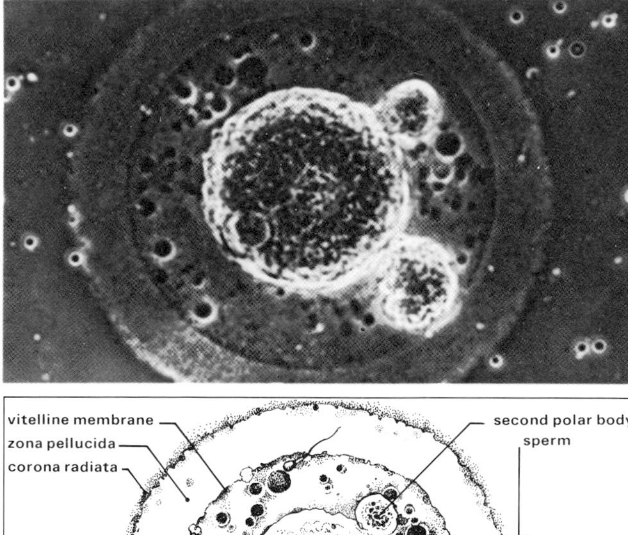

The immature egg is released into the Fallopian tube after undergoing the first meiotic division and formation of the first polar body. The egg consists of a nucleus and cytoplasm enclosed in the vitelline membrane and surrounded by the zona pellucida, which is itself covered by a layer of loose cells, the corona radiata. The sperm passes through all these layers to reach the nucleus of the egg, and on penetration of the egg it stimulates the second meiotic division of the ovum and the formation of a second polar body.

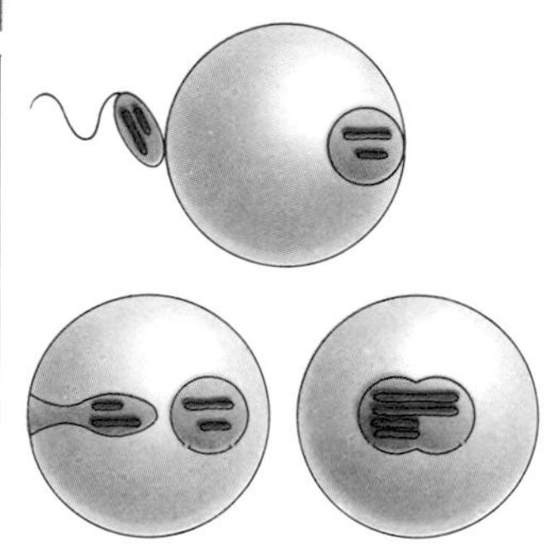

The nucleus of each sperm and ovum contains twenty-three chromosomes. The two nuclei fuse in the cytoplasm of the ovum.

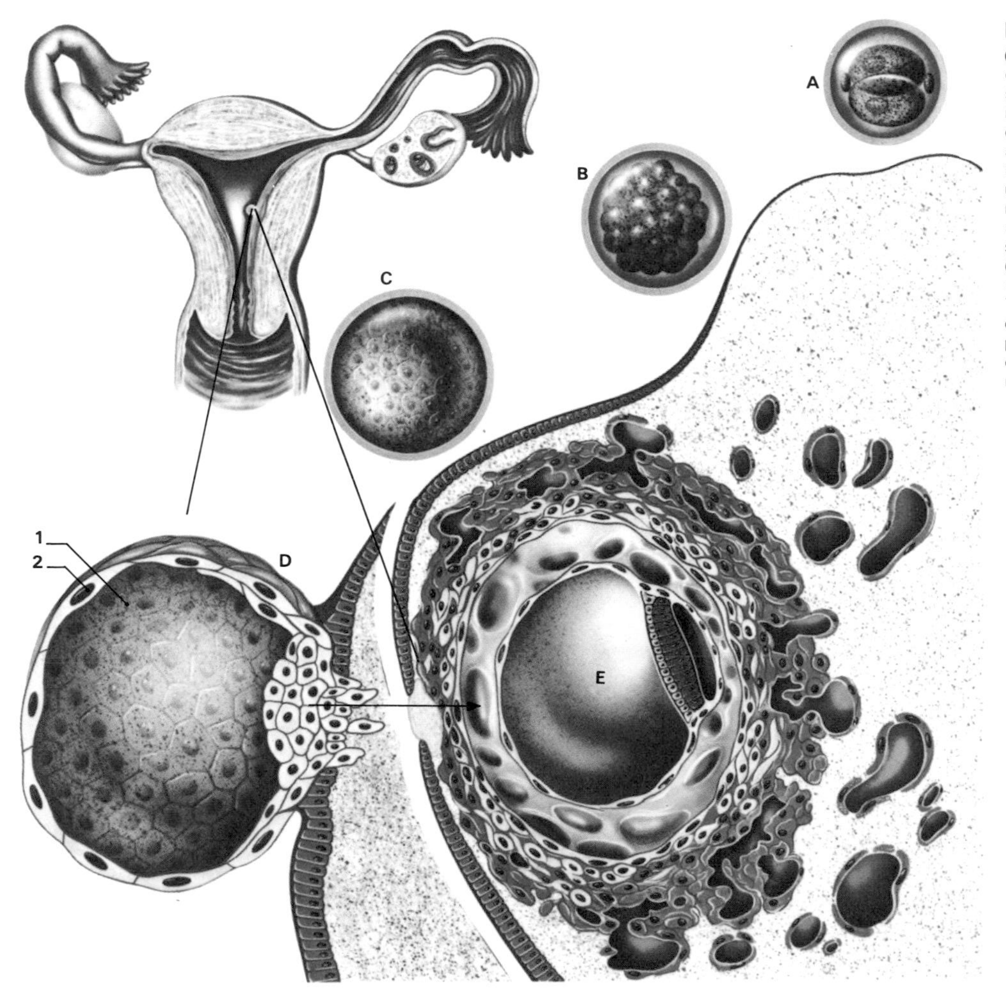

Implantation of the ovum

(**A**) Within a few hours of fusion of the sperm and the ovum cell division begins. The number of cells doubles with each successive division until a hollow ball of rapidly dividing cells is produced. (**B**) Throughout the process the ovum is moving down the Fallopian tube toward the uterus. Five to six days after the fusion, the ovum reaches the uterus and implants itself in the soft wall. (**C**) The hollow ball of cells now contains up to one hundred cells, arranged as an outer ring and an inner mass. The outer ring grows outward and penetrates the uterine tissue (**D**), while the inner mass divides into two separate layers, an inner endoderm (**1**) and an outer ectoderm (**2**). These two layers become separated from the outer ring and by nine days form two cavities within the hollow ball (**E**). The ectoderm later spreads around the periphery of the smaller cavity, forming a sac, the amnion, which will surround and protect the developing embryo. The endoderm similarly encloses the larger cavity to form a yolk sac, which will have a nutritive function.

Cross-reference	pages
Coitus	148-149
Female anatomy and physiology	146-147
Fertilization in the future	188-191
Heredity	158-159
Male anatomy and physiology	144-145
Placenta	152-157

Early pregnancy: one week to three months

From embryo to fetus

Twenty days after fertilization the embryo (**1**) is a three-layered disk of ectoderm, mesoderm and endoderm cells. Eight days later the yolk sac (**2**) has shrunk, the amnion (**3**) has enlarged and the embryo has recognizable features. (**A**) The embryo at three weeks, shown actual size in gray, has a primitive heart and rudimentary brain tissue. (**B**) One week later the heart is pumping blood around the body and into the placenta. (**C**) The embryo at five weeks has twenty-eight pairs of tissue blocks, which later form bones and muscles. (**D**) Limb buds and rudimentary eyes are visible and the tail is receding. One week later the hands have rudimentary fingers, the eyes have lenses and the liver and heart are visible. (**E**) At seven weeks the head grows rapidly, external ears are visible, the eyes are open and the tail has disappeared. (**F**) At eight weeks the embryo is now called the fetus. It has most of its tissues and organs and well-developed hands and feet.

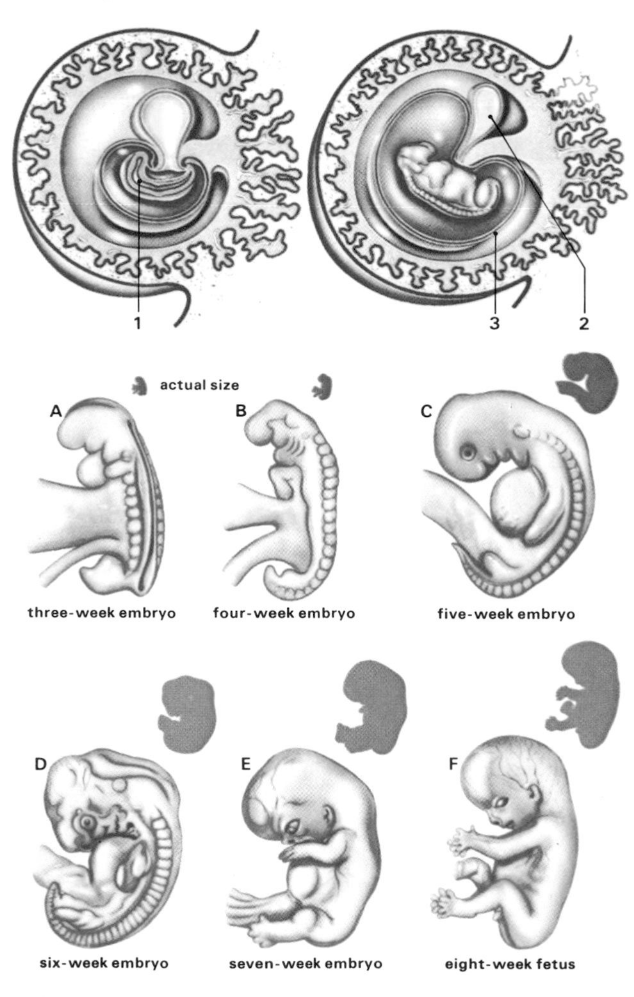

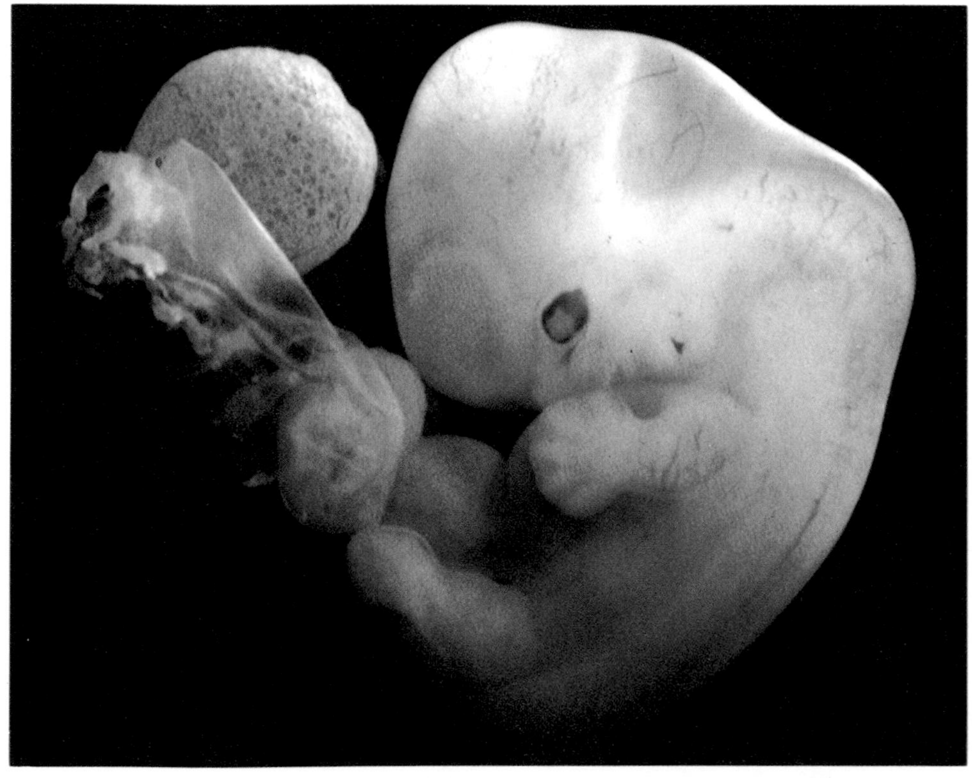

The six-week-old embryo is shown approximately sixteen times life-size. It has rudimentary limbs with the fingers more developed than the toes, well-formed, pigmented eyes and developing external ears which are clearly visible.

The human blastocyst, as it begins to implant in the lining of the uterus about seven days after fertilization, is made up of two distinct parts. Covering the whole surface is a layer of cells, known as the trophoblast, which never becomes incorporated into the embryo. Actively invasive cells, they erode their way into the tissue lining the uterus and in so doing anchor the blastocyst in place. By breaking down the maternal blood vessels, they also ensure that the embryo becomes bathed in maternal blood—its source of oxygen and dissolved foodstuffs.

Later, these trophoblastic cells develop into fingerlike projections, or villi, which contain their own blood vessels and are involved in continuous exchange of dissolved materials between the blood circulations of the mother and the fetus. So fine and delicate are these villi that at the time of birth, if placed end to end, it is estimated that they would stretch thirty miles.

The other function of the trophoblast in early pregnancy is the production of the hormone human chorionic gonadotrophin, or HCG. This hormone stimulates the ovaries to continue producing another hormone, progesterone. If the ovaries did not, as happens at the end of a menstrual cycle in which no fertilization has occurred, then the uterine lining, the endometrium, would break down as menstrual fluid. By producing its own special hormone, however, the trophoblast maintains the uterine lining and ensures that the developing embryo has a source both of anchorage and of nutrition. It is because of the early production of HCG that the first sign of pregnancy is a missed menstrual period, and nearly all pregnancy tests depend on the detection of HCG in the urine.

The embryo itself develops from a small disk of cells immediately below the trophoblast at the point of implantation. This disk is known as the inner cell mass. Almost as soon as implantation occurs, a fluid-filled cavity starts to appear between the inner cell mass and the trophoblast above it. This is the amniotic cavity, which will eventually surround the entire embryo until its birth, protecting it from physical shocks.

By sixteen days after fertilization, the inner cell mass can be seen as two layers. The upper layer forms the floor of the amniotic sac. The lower layer has cells which line the blastocyst cavity to form the yolk sac, a large fluid-filled space. At about sixteen days, the trophoblast linking the embryonic cells to the maternal tissue begins to elongate to produce the body stalk which will eventually develop into the umbilical cord.

In the third week of development an important rearrangement of cells takes place in the inner cell mass. The two-layered structure develops a third intermediate layer and in this three-layered sandwich the plan for all the tissues in the body of the fetus—and indeed the child and adult—is laid down. The uppermost layer is known as the ectoderm. From it develops the tissues of the nervous system, the skin, hair and nails and parts of such sense organs as the eyes. From the lowermost endodermal layer arise the tissues which line many of the body organs, for example, the alimentary canal and the lungs as well as such organs as the liver and the pancreas. From the mesoderm arise the blood system and the heart, the skeleton, the muscles and the

connective tissue—in fact the majority of the body's tissues.

In this third week very rapid differentiation of the rearranged cellular layers takes place. The heart begins to develop and the ectodermal layer folds itself in such a way that the neural tube is produced. This tube closes at both ends to develop at the front into brain tissue and behind into the spinal cord.

The mesodermal tissue becomes organized into blocks, or somites, which increase in number and give rise to regularly spaced blocks of muscle. Subsequent development overshadows the regularity of the somites, but in the embryo at least, the developing human body has a "segmental" pattern, as do many amphibians and reptiles.

By four weeks there are about twenty-five somites and the embryo, from its crown to its rump, measures about one-eighth of an inch—clearly visible to the naked eye. The first shallow pits appear, which are destined to become its ears. At the front end of the embryo, the bulge of the heart is clearly seen. Indeed, in proportion it is some ten times larger than the heart of the adult.

At five weeks the pigmented regions of the eyes develop, as does another pit, which will become the nose. At this stage, too, the first limb buds appear. The forelimbs are somewhat earlier in development than the leg buds. One week later the embryo is about five-eighths of an inch long with forty somites and it has begun to take on a more recognizably human appearance. The outside flaps of the ear, for example, can just be detected. The paddlelike hands are starting to have the outlines of the fingers and the rear limbs can just be distinguished as leg and foot.

This period, from the beginning of the fourth week until the end of the sixth week of pregnancy, is, therefore, a critical period for the development of the embryo. It is a period of organogenesis, during which the major organ systems of the body are being laid down. This is a period which is particularly sensitive to the malforming influence of drugs or natural infection.

By the eighth week of gestation the embryo still possesses forty somites, but its length has increased to about one inch. Yet this tiny creature is now quite definitely recognizable as a human being. It has lips, a nose, eyes and even eyelids. At this stage it is said to undergo the transition from embryo to fetus. Although in the latter six months of pregnancy a number of organ systems develop further, the greatest change in the fetus is one of sheer growth. During the third month, for example, the fetus doubles its length and its relative proportions become somewhat more like those of the infant at birth.

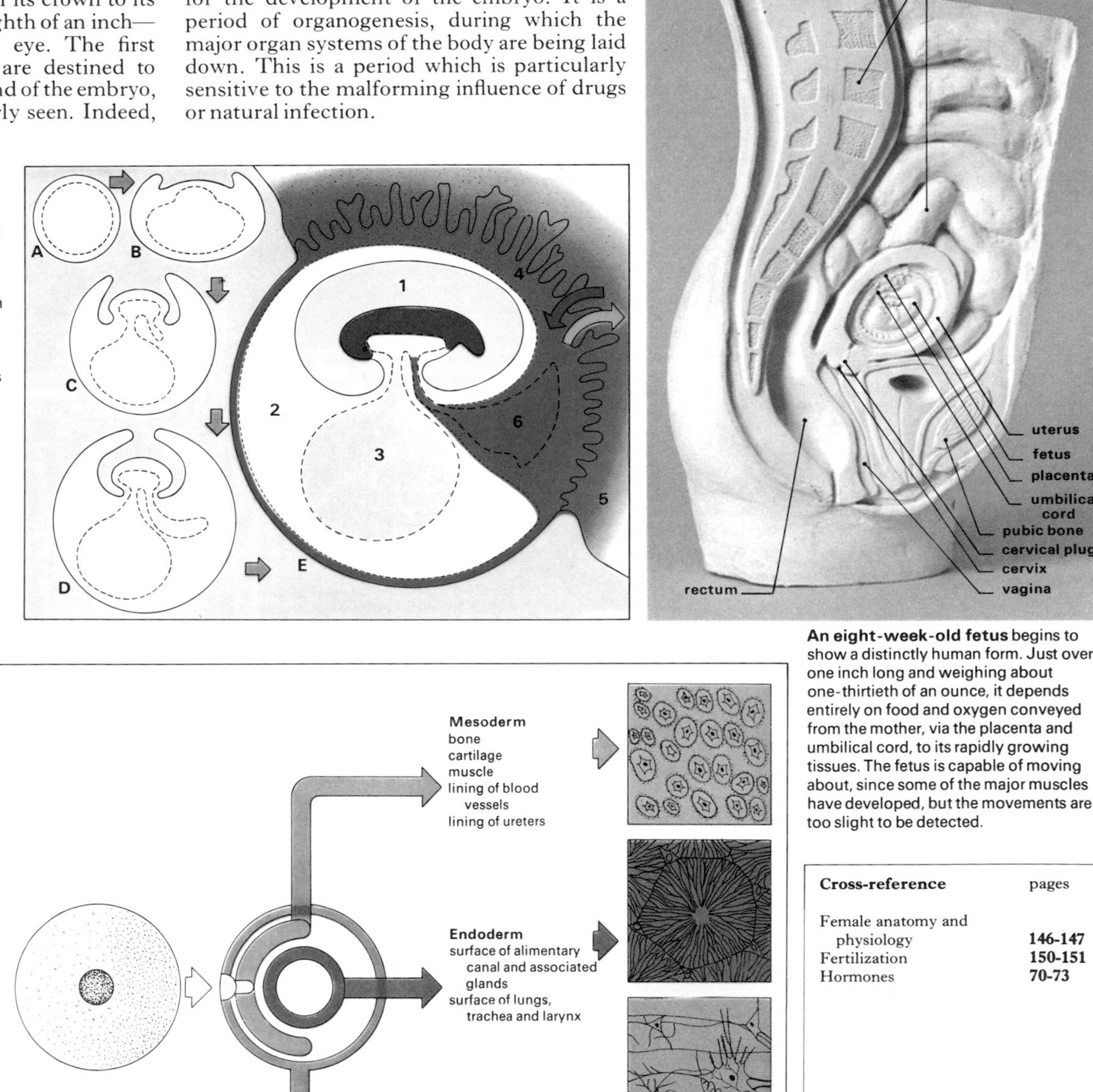

The growing embryo is kept alive by the placenta, an organ formed from embryonic tissue which connects it to the mother's uterus. Five stages (**A** to **E**) in the formation of the placenta are shown right. The embryo (**1**), protected by fluid in the amniotic sac (**2**), gains nourishment for the first few weeks from the yolk sac (**3**). Later the placenta, formed by projections of an embryonic membrane, the chorion (**4**), which penetrates the uterine wall (**5**), provides the embryo with food and oxygen. The blood vessels of the allantoic sac (**6**) later become part of the umbilical cord.

An eight-week-old fetus begins to show a distinctly human form. Just over one inch long and weighing about one-thirtieth of an ounce, it depends entirely on food and oxygen conveyed from the mother, via the placenta and umbilical cord, to its rapidly growing tissues. The fetus is capable of moving about, since some of the major muscles have developed, but the movements are too slight to be detected.

Three cell-producing layers

All the cells of the body are produced from the fertilized single-celled ovum. Although all the cells carry the same genetic information, different genes operate to produce each of the three cell types—ectoderm mesoderm and endoderm—of which the entire human body is composed. The embryo has an ectodermal, mesodermal and endodermal layer, and each layer produces particular kinds of tissue, examples of which are given in the diagram right.

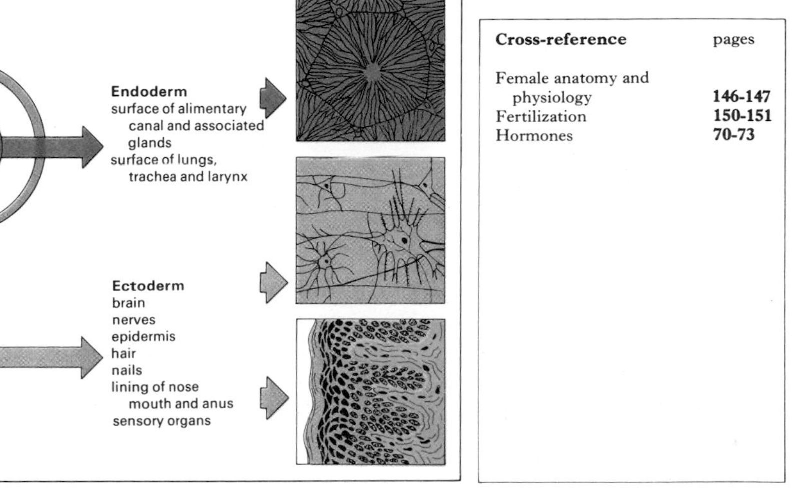

Cross-reference	pages
Female anatomy and physiology	146-147
Fertilization	150-151
Hormones	70-73

Late pregnancy: three to nine months

From embryo to fetus

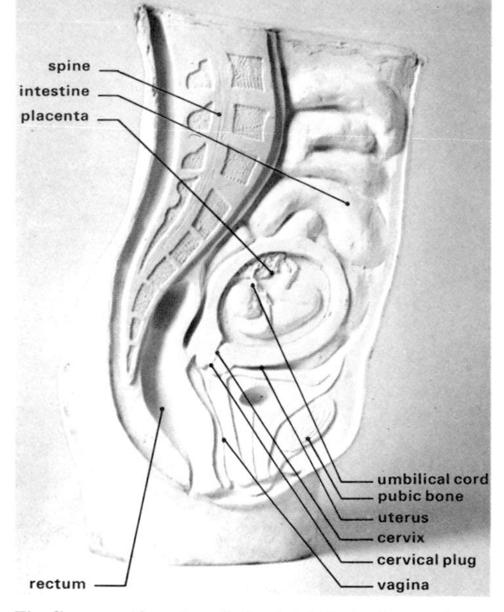

The five-month embryo is about eight inches long and weighs about half a pound. The lungs and digestive organs are not yet fully developed. The mother can now feel the baby as it turns and kicks.

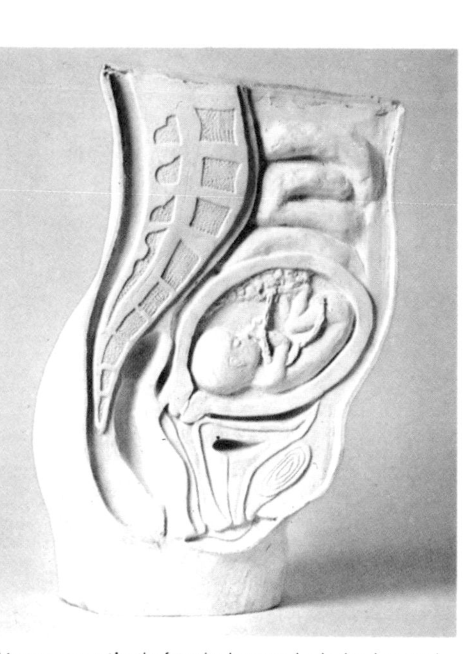

At seven months the fetus is about twelve inches long and weighs two to three pounds. Calcium and iron from the mother are being used in the final formation of the skeleton and the blood.

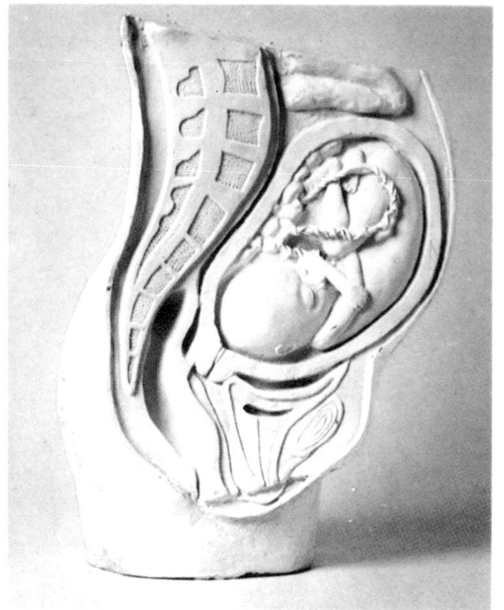

At nine months the fetus, now capable of an independent existence, moves into a vertical head-down position facing the mother's back. At birth the baby usually weighs between six and eight pounds.

By the end of the third month of pregnancy, although the fetus is only about three inches long, it is recognizably human. The systems of the body are already well developed, with many organs more or less complete. Nerves and muscles are working, and reflexes are becoming established. The heart pumps about fifty pints of blood through the circulatory system each day. Although the baby can move spontaneously, the mother is not yet aware of the movements.

During the remainder of pregnancy, the second and third trimesters, three related processes occur together. The fetus grows steadily in size and weight, and the internal organs become more elaborate. The placenta also increases in size, and, third, there are widespread changes in the mother's body as she adapts to the growing demands of the new individual within her.

In the fourth month of pregnancy, the fetus grows about five inches in length. At the same time, its weight increases from less than one ounce to more than a quarter of a pound. Suspended within the uterus in the watery amniotic fluid, and so in a state of near-weightlessness, the fetus can move freely, bending readily at the waist and hips, twisting and rolling and even performing complete somersaults.

During the third and fourth months, ossification centers, or sites of bone formation, which were just apparent at eight weeks, continue to develop in the cartilage of the face and the ribs, and appear for the first time in the vertebrae and in the fingers and toes. By fourteen weeks, the male fetus has a recognizable scrotum and the female has recognizable vaginal labia.

The scrotum and the labia, like the penis and clitoris, are equivalent, or homologous, structures. The genital organs of both sexes become differentiated from the same embryonic tissues under the influence of hormones. The development of the reproductive tract along male or female lines depends, of course, upon the incorporation of a Y chromosome from the sperm into the fertilized egg cell. But exactly how the Y chromosome exerts its effects is only now becoming elucidated.

Recent research indicates that the development of male characteristics in a growing embryo may be caused by the presence of just a single molecule on the surface of the fertilized egg cell. The molecule, produced by one gene on the Y chromosome, is thought to serve as a signal to the growing cells in the embryo to become male, so that the tissues of the reproductive tract turn into testes and not ovaries.

The embryonic testes secrete the hormone testosterone, and this is responsible for the growth of the male external genital organs in the embryo, and of course, much later, of the development of all the male secondary sexual characteristics at the time of puberty. The one X chromosome present in the male also plays a part, however. A gene on this chromosome provides the growing cells with the equipment to respond to the presence of testosterone.

In the fifth month of pregnancy, the fetus begins to make its presence more obvious. Fetal "quickening" movements may be felt as it moves and kicks within the amniotic sac. Early Christian theologians believed that these movements represented the true moment of conception—presumably the time at which the child's developing body received its soul. During the fifth month, and in subsequent months, the fetal heartbeat can often be heard with its rate of up to 140 beats per minute—twice the pulse rate of the normal adult.

At about twenty weeks, the mother's weight gain begins to become noticeable. It will probably have increased by eight to nine pounds, although the fetus itself weighs only ten ounces. The extra weight comes from a variety of sources. First, the amniotic fluid surrounding the fetus at twenty weeks will weigh about the same as the fetus itself. The placenta will weigh about half as much and the uterus twice as much. Altogether, however, the uterus and its contents only account for one-third of the mother's weight gain.

For many years it was believed, mistakenly, that the developing fetus acted in an entirely parasitic manner and removed foodstuffs from the mother's circulation, even to her own detriment. In fact, during pregnancy the mother tends to take in and retain throughout her body more extra nutrient than is necessary to support the added demands of the fetus. Much of this nutrient will subsequently enter the milk during lactation. Indeed, the breasts in midpregnancy weigh about seven ounces more than before conception. At the moment of birth they will be considerably heavier. In addition, the pregnant woman retains water and also lays down fat reserves. Her total weight gain at the time of birth will be about twenty-two pounds, of which the fetus and its associated structures account for less than half.

The placenta, later to be expelled from the womb as the afterbirth, is fully developed by midpregnancy. Thereafter it simply increases in size. The fully developed placenta is about the size and shape of a dinner plate, although the diameter may vary from eight inches to about fourteen inches. It is basically a membrane, but a very much folded one. Some estimates of the total surface area put it in the region of one thousand square feet. This is made possible by fold upon fold within the tissues; even the cell walls are thrown into microscopic folds of their own.

The placenta is attached to the wall of the uterus, firmly but not immovably. The attachments are bathed in the mother's blood, which is, however, kept separate from the baby's blood by cell membranes. The baby's side of the placenta is covered by the thicker fetal membranes. The outer membrane is part of the mother's body, while the inner membrane is part of the fetus. From the middle of the placenta emerges the umbilical cord, which

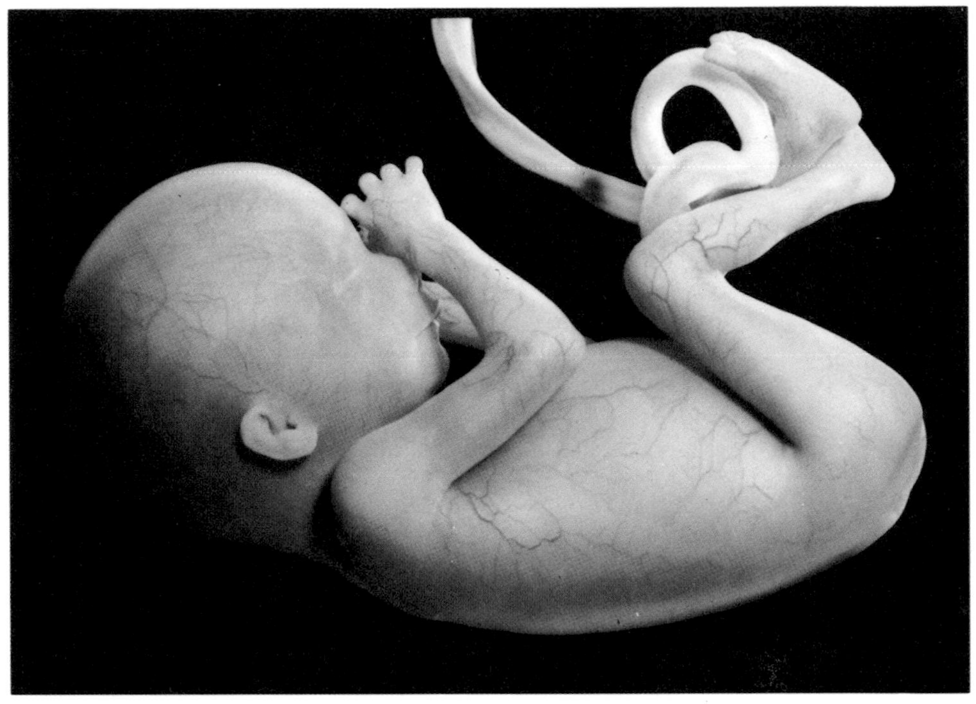

During the fourth month the baby grows so rapidly that it reaches a length of eight inches, half the height it will be at birth. The development of neck and back muscles is accompanied by the head being lifted off the chest, and the baby assumes a more upright posture. At this stage, blood vessels are clearly visible through the thin, loose skin.

is also covered by the fetal membranes. There is normally no mixing of the blood in the fetal and maternal bloodstreams, which are kept completely separate by the placenta unless there is an accidental hemorrhage that allows the blood to mingle.

The blood supply to the placenta comes from the umbilical arteries, running from the baby's heart along the umbilical cord. The placenta is divided into fifty or sixty subunits, called cotyledons, and each of these is supplied by a branch of one of the umbilical arteries. The branches of the umbilical arteries subdivide many times to form a dense network of capillaries. The blood is collected, when it has circulated through the capillaries, by branches of the umbilical vein, that also runs in the umbilical cord, and returns to the fetus.

On the mother's side of the placenta, spiral arteries from the wall of the uterus empty blood into a continuous space to give a large surface area for the exchange of dissolved gases and other materials.

Because the lungs of the fetus cannot function to exchange gases between the blood and the atmosphere, the placenta must act as a lung. Into it the baby's heart pumps blood to pick up oxygen and to eliminate carbon dioxide. The placenta must also act as a food supply for the fetus, providing protein in the form of dissolved amino acids, carbohydrates in the form of glucose, and other materials, including calcium, iron and other minerals.

The waste materials from the fetus flow the other way, and are transferred into the mother's circulation, to be excreted to the exterior with her own waste matter.

Although there are probably other reasons for the separation of the maternal and fetal bloodstreams, an obvious one is to prevent possible immunological reactions.

The fetus contains chromosomal material from its father. Thus it is immunologically different from its mother. In the same way that a skin graft would be rejected by the mother's body when her bloodstream came into contact with it, and antibodies were developed, so the fetus would be rejected as being "foreign." The fetus is in a sense a graft, but it is a foreign graft in the body that must not be rejected. Therefore the mother's blood circulation must be prevented from making direct contact with it. Occasionally, however, the placental tissue may be damaged so that a leakage occurs, and fetal blood cells get through. One of the most important consequences of this happens if the mother belongs to a different Rhesus blood group and develops antibodies to the baby's red blood cells. Although there may be no serious effects for that fetus, Rhesus incompatibility may require special precautions to be taken in ensuing pregnancies.

During the sixth month of pregnancy, the fetus begins to accumulate a layer of fat beneath the skin, and fine hair, called lanugo, covers the head and body, along with the white, greasy protective substance called the vernix. In this month the hands can be gripped tightly, and the eyes may even be opened and closed. At this stage, the baby is said to be "viable," which means that it stands a slight chance of survival should birth be precipitated.

During the last two months of pregnancy the fetus increases in weight by just under an ounce daily. The uterus expands accordingly and in the ninth month reaches up to the mother's breastbone. Movement inside the uterus is now much more restricted, and the baby settles into a curled position, usually head down, which is the most convenient position for the process of birth. If the baby settles bottom downward, this is called a breech presentation and is somewhat more awkward. If the baby lies transversely, every effort is made to turn it by pressing through the mother's abdominal wall; otherwise delivery may be extremely difficult.

Labor starts, on average, about two hundred and eighty days after the first day of the mother's last menstrual period. A day or so before labor begins, the uterus descends into the pelvic cavity, so that the baby's head, resting against the inner end of the cervical canal, is ready to be pushed through into the birth canal and so into the outside world.

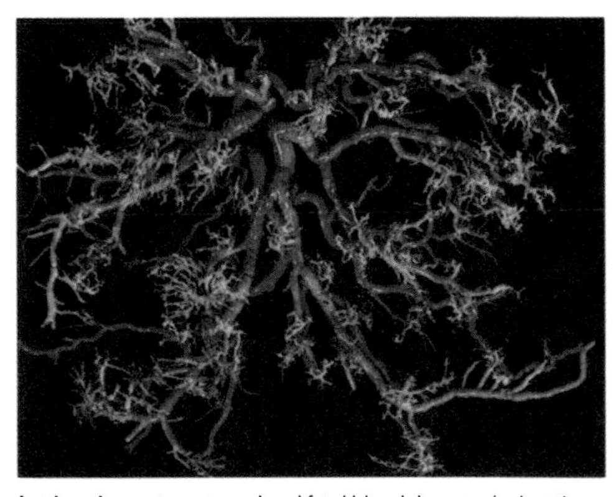

In the placenta maternal and fetal blood do not mix, but the blood vessels lie close together and materials pass from one to the other.

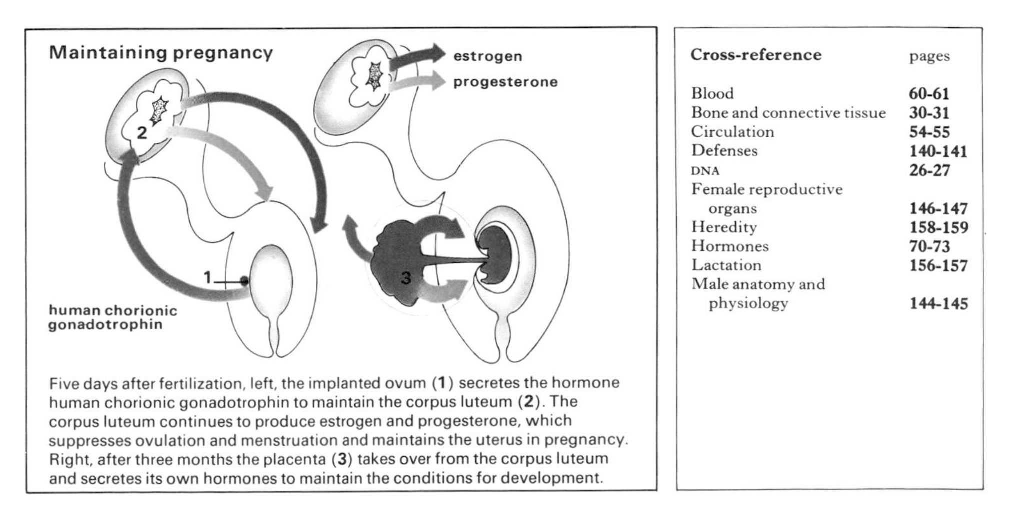

Five days after fertilization, left, the implanted ovum (**1**) secretes the hormone human chorionic gonadotrophin to maintain the corpus luteum (**2**). The corpus luteum continues to produce estrogen and progesterone, which suppresses ovulation and menstruation and maintains the uterus in pregnancy. Right, after three months the placenta (**3**) takes over from the corpus luteum and secretes its own hormones to maintain the conditions for development.

Cross-reference	pages
Blood	60-61
Bone and connective tissue	30-31
Circulation	54-55
Defenses	140-141
DNA	26-27
Female reproductive organs	146-147
Heredity	158-159
Hormones	70-73
Lactation	156-157
Male anatomy and physiology	144-145

The birth of the child

No one knows why (or more correctly how), after ten lunar months, or 280 days, in the uterus, the fetus, during the course of a few hours, is expelled from the uterus to take on its separate existence. But whatever the mechanism it works with very predictable frequency. Although pregnancies of 330 days or more have been reported, the likely date of a child's birth can be calculated with considerable accuracy.

Such consistent duration of pregnancy is, of course, in the interests of the fetus itself. Although babies delivered at about thirty weeks may survive under intensive care, their chances are not high without it. Conversely "postmature" babies, who remain within the uterus beyond their allotted time, may also have a higher mortality than normal. But such consistency does not, unfortunately, tell us much about the mechanism of birth. To the ancient Greeks, the problem was simple, they believed that the elastic-sided uterus simply expanded until it could take no more. At this point, like an elastic band, it snapped back again expelling the fetus in the process.

Today we are a little nearer a more scientific explanation, but many details still elude us. For example, although the uterus itself might set the general timing of birth, there is evidence that hormones from the fetal pituitary gland may interact with it and thus the fetus may set up its own "fine tuning" for the process.

Undoubtedly, however, the uterus plays a major part. Throughout pregnancy the uterine muscle undergoes small unnoticed contractions, but when birth is imminent their frequency and strength greatly increase. The question is what was holding them back earlier. The answer seems to be progesterone, which is produced in large quantities first by the ovaries and then by the placenta. Toward the end of pregnancy, progesterone levels in the blood begin to fall, and it is believed that this removal of a repressive influence may allow the uterus to begin its more extensive contractions.

But this is not the whole story. Coinciding with the fall in progesterone there appears to be an increase in estrogen, and there is certainly an increase in a hormone from the posterior part of the mother's pituitary gland. This hormone is known as oxytocin and, in the presence of estrogen, it stimulates uterine contractions so effectively that it may be used by a doctor to induce birth, or labor, if the uterus itself is slow to react at the end of pregnancy.

The process of labor is usually considered in three stages. In the first stage, uterine con-

Childbirth

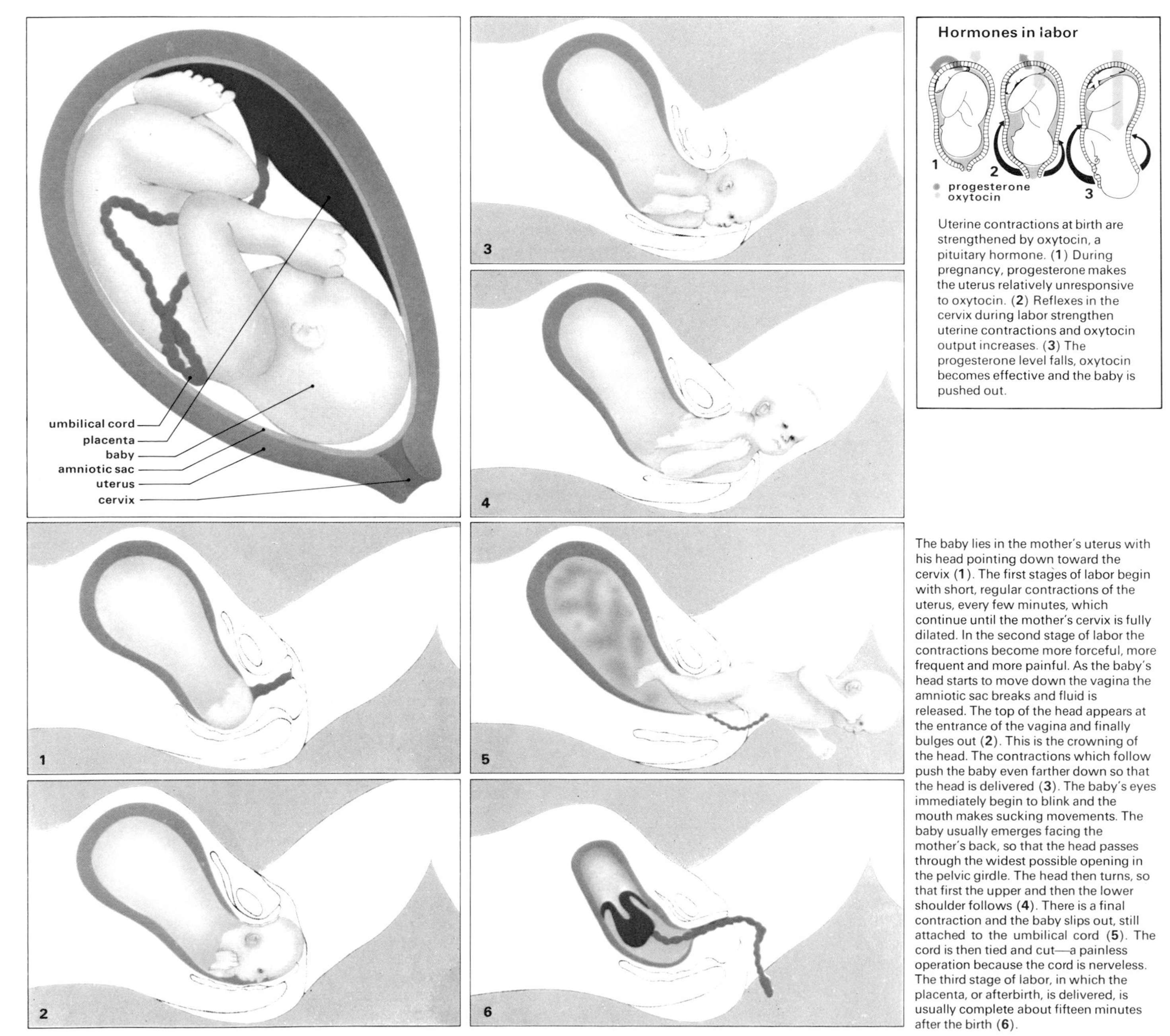

Uterine contractions at birth are strengthened by oxytocin, a pituitary hormone. (**1**) During pregnancy, progesterone makes the uterus relatively unresponsive to oxytocin. (**2**) Reflexes in the cervix during labor strengthen uterine contractions and oxytocin output increases. (**3**) The progesterone level falls, oxytocin becomes effective and the baby is pushed out.

The baby lies in the mother's uterus with his head pointing down toward the cervix (**1**). The first stages of labor begin with short, regular contractions of the uterus, every few minutes, which continue until the mother's cervix is fully dilated. In the second stage of labor the contractions become more forceful, more frequent and more painful. As the baby's head starts to move down the vagina the amniotic sac breaks and fluid is released. The top of the head appears at the entrance of the vagina and finally bulges out (**2**). This is the crowning of the head. The contractions which follow push the baby even farther down so that the head is delivered (**3**). The baby's eyes immediately begin to blink and the mouth makes sucking movements. The baby usually emerges facing the mother's back, so that the head passes through the widest possible opening in the pelvic girdle. The head then turns, so that first the upper and then the lower shoulder follows (**4**). There is a final contraction and the baby slips out, still attached to the umbilical cord (**5**). The cord is then tied and cut—a painless operation because the cord is nerveless. The third stage of labor, in which the placenta, or afterbirth, is delivered, is usually complete about fifteen minutes after the birth (**6**).

tractions begin to be noticeable at intervals of between twenty to thirty minutes. Progressively they get more frequent until they occur at two- to three-minute intervals and may last for a minute or more. During this time the fetal head is against the cervical entrance, and the cervical wall is becoming thinner as the cervical canal dilates to a diameter of about four inches. This first stage, which results in a cervix sufficiently dilated to allow the fetus to pass, and which usually involves rupture of the amniotic sac to release the amniotic fluid, may last for fourteen to sixteen hours in women having their first baby. In second and subsequent pregnancies the time is often shorter, perhaps eight to twelve hours.

In the second stage of labor, the continued uterine contractions expel the fetus, still connected to the umbilical cord, out through the cervix and the vagina. The duration of this second stage is also variable. In some women it may last only minutes; in others, several hours. In either event, it is likely to be the most painful phase of the birth process, although it is the most satisfying since it results in the appearance of a new human being.

The new baby is, however, a human being without its own separate existence until the cord which links it to the placenta is clamped and cut. During the third stage, which may take about twenty minutes, the placenta itself is delivered through the vagina. The separation of the placenta from the uterine wall is remarkable, considering the large quantities of blood which pass through it right up to delivery. Although a certain amount of bleeding accompanies its delivery, the uterus has a remarkable ability to "close down" its blood vessels and so prevent massive bleeding.

The newborn infant is capable of a number of reflex reactions, but because the cerebral cortex is not fully functional at birth, the range of the baby's reactions is extremely limited. One well-developed reflex is that of suckling. Indeed, swallowing movements of the mouth have been observed in the fetus even within the uterus. This instinctive reflex can be stimulated by any object placed near the infant's mouth, in particular, of course, the mother's nipple.

Psychiatrists have argued for years about the reason the human infant at birth should be so relatively inactive, compared to the young of other species that are capable of coordinated activities soon after they are born. While the answer is still not certain, it seems that this dependence of the human infant on others for his very survival over a period of years, provides him with an opportunity to undergo a long period of mental education.

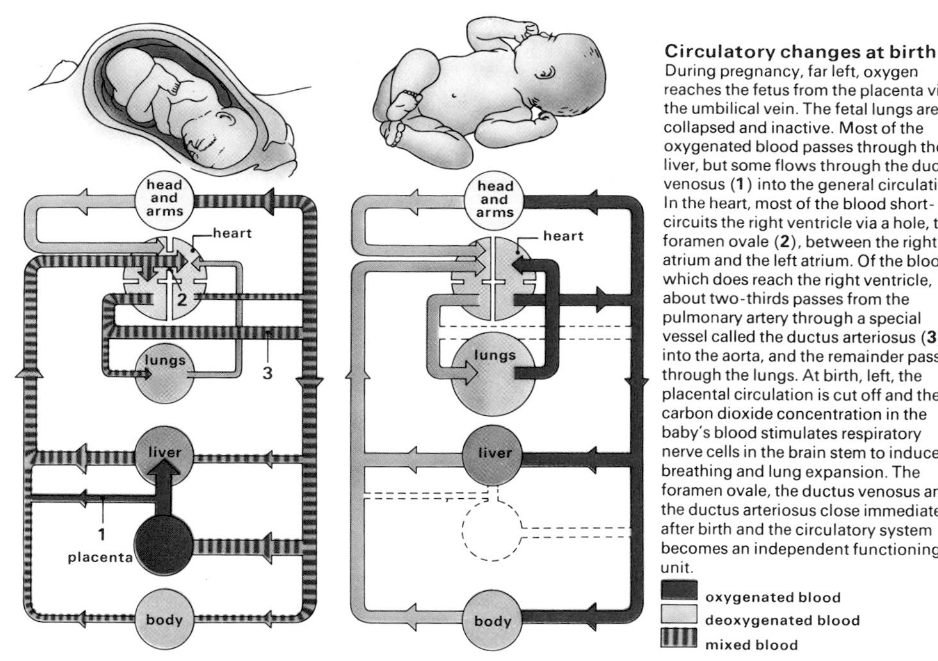

Circulatory changes at birth During pregnancy, far left, oxygen reaches the fetus from the placenta via the umbilical vein. The fetal lungs are collapsed and inactive. Most of the oxygenated blood passes through the liver, but some flows through the ductus venosus (**1**) into the general circulation. In the heart, most of the blood short-circuits the right ventricle via a hole, the foramen ovale (**2**), between the right atrium and the left atrium. Of the blood which does reach the right ventricle, about two-thirds passes from the pulmonary artery through a special vessel called the ductus arteriosus (**3**) into the aorta, and the remainder passes through the lungs. At birth, left, the placental circulation is cut off and the carbon dioxide concentration in the baby's blood stimulates respiratory nerve cells in the brain stem to induce breathing and lung expansion. The foramen ovale, the ductus venosus and the ductus arteriosus close immediately after birth and the circulatory system becomes an independent functioning unit.

oxygenated blood
deoxygenated blood
mixed blood

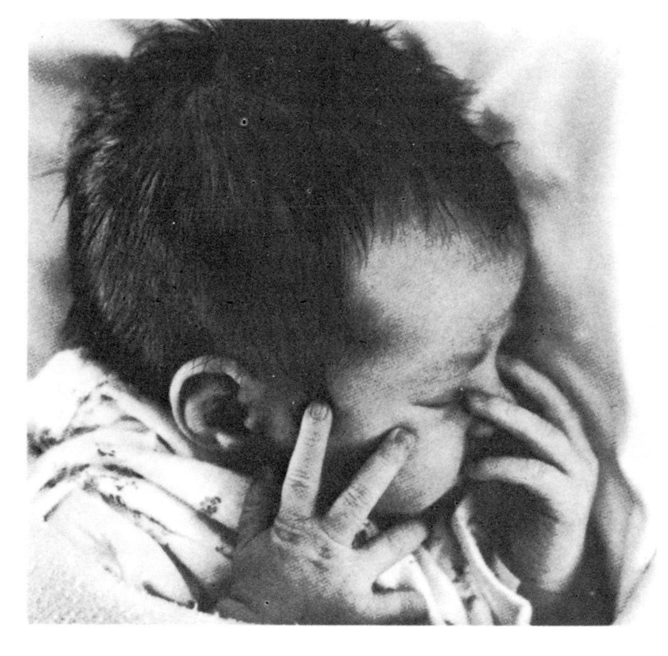

The newborn baby, cleaned and dried after leaving the comparative safety of the womb, spends most of his time sleeping. Awake he is capable only of reflex actions.

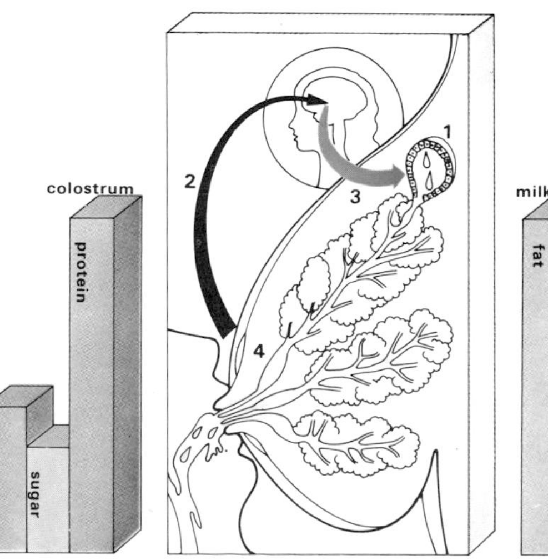

Milk let down is the term applied to the release of milk into the mother's breast ducts in response to the baby's suckling. At birth, milk is secreted into the breast alveoli (**1**). Suckling stimulates the passage of sensory impulses (**2**) to the hypothalamus, which orders the posterior pituitary gland to release the hormone oxytocin (**3**) into the bloodstream. Cells surrounding the outer walls of the alveoli are stimulated to contract, ejecting milk into the ducts. Thirty seconds after the baby begins to suck, milk flows into the reservoirs (**4**), or is let down. For a few days after birth the fluid produced is colostrum, which, although secreted in small quantities, is rich in protein and is well supplied with antibodies that protect the baby against infection. Milk begins to be secreted about three days after birth. The milk is almost twice as rich in fat and sugar as colostrum and is a complete food for the baby during the first months.

Cross-reference	pages
Circulation	54-55
Defenses	140-141
Female reproductive organs	146-147
Hormones	70-73
Neotony	16-17
Reflexes of the newborn	162-163

Heredity

Gamete cells are produced to transmit genetic information from parents to their progeny. The cells are formed by a special process of cell division, called meiosis, which results in the halving of the chromosome number in each parental gamete so that the full complement can be restored when sperm and ovum unite. First, the chromosomes appear in the nucleus (**1**) of the cell and then duplicate (**2**). Like chromosomes link together (**3**) and exchange different sections of their length before separating and lining up across the center (**4**). They are pulled to opposite poles (**5**) and eventually two cells are formed (**6**), each one containing one member of each pair of chromosomes from the original cell. Then a second division begins and, in the same way, the chromosomes arrange themselves across the center of the nucleus (**7**). This time one-half of each chromosome, a chromatid, moves to the poles of the nucleus (**8**). Within the four cells that are formed (**9**) each chromatid will eventually replicate to form a chromosome and each gamete cell will contain half the original chromosome number.

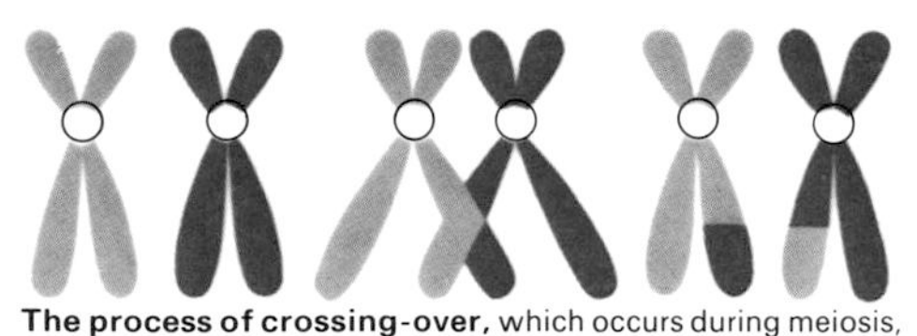

The process of crossing-over, which occurs during meiosis, is one of the most important genetic events, for it guarantees the individual characteristics of each one of us. When the two homologous chromosomes lie side by side, breaks occur in the chromosomes and corresponding lengths of the chromosome are interchanged. This crossing-over disrupts the original sequence of alleles and interchanges parts of both maternal and paternal chromosomes so that the possible number of gametes with different genetic information that any individual can produce is almost incalculable. This is why a child may inherit his father's nose and his mother's eyes.

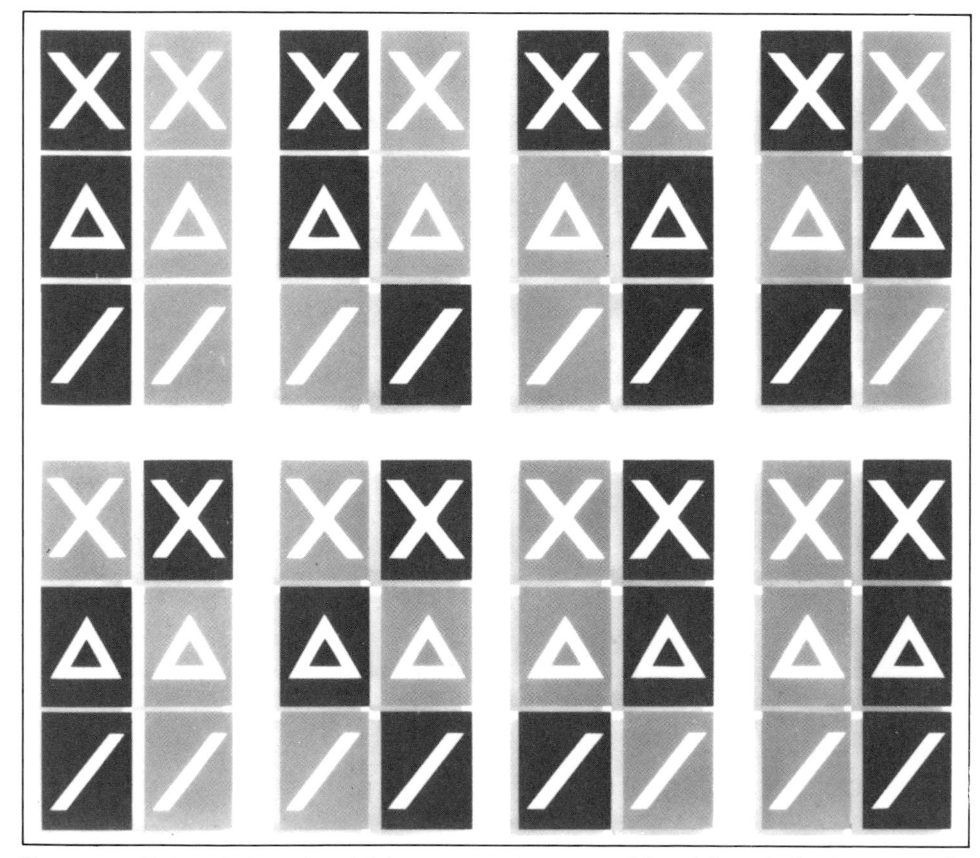

The process of independent assortment of chromosomes during the first division of meiosis ensures that each sperm and ovum contains slightly different information, and that all the progeny produced will have their own individual characteristics. In the diagram, dark brown cards represent maternal chromosomes and light brown cards paternal chromosomes. Cards with the same symbol represent the homologous chromosomes. Using only three sets of cards, one can see that it is possible to arrive at eight different combinations, or assortments, of chromosomes. Human beings have forty-six chromosomes arranged in twenty-three pairs within their cells. Independent assortment of these means that the possible number of sperm with slightly different genetic information exceeds eight million.

All of our inherited traits—the color of our eyes and hair, the size of our noses and teeth, even our basic temperaments—are passed in our chromosomes, which are built up of the units of heredity called genes. Even the most identical of identical twins show minute differences, although to a stranger they might appear to be perfect replicas of one another. Both will have the same color eyes and hair, the same shape of nose and ears and so on, and both will tend to behave in the same way. But their fingerprints will show differences, and on looking closely one will be able to spot other marks of individuality, perhaps moles, or a fractional difference in the spacing of the teeth. On the whole, however, identical twins seem to have inherited exactly the same characteristics from their parents.

Twins of this sort are so alike because they developed from the unusual division of a developing fertilized egg cell, or zygote, at an early stage. All the genetic material responsible for molding our characteristics is present in the zygote. Half has been given by the father, in the chromosomes contained in the sperm cell, and half by the mother in the ovum. The chromosomes are divided into genes.

When an ordinary tissue in the body grows, the cells within it divide into two by the process of mitosis. In this, each chromosome divides into two longitudinally, and half moves into each new cell. The normal number of chromosomes in a human cell is forty-six, and they are arranged in twenty-three pairs. But in the production of gametes—ovum and sperm—a special division occurs in the germ tissue that forms them. This is called meiosis, and results in each gamete having only half the full number of chromosomes, twenty-three. These twenty-three are the result of the separation of the original pairs of chromosomes, which are called homologous chromosomes. When an ovum is fertilized, the full number of chromosomes is restored.

Before meiosis occurs, the two members of each of the twenty-three pairs of chromosomes may come into contact with one another, and adhere at various points. At several different places along their length the chromosomes may then break, and sections are exchanged between the chromosomes. This process is called crossing-over, and allows for a certain amount of mixing up of genetic material before the gametes are formed. Each gamete now contains a random selection of the genes from the parent's forty-six chromosomes.

When fertilization occurs, the corresponding genes in the gametes come together, one from each parent. The pairs that are formed are called alleles. Many of the genes that are carried on the chromosomes are of course for very general characteristics—those common to all human beings, such as the possession of five fingers, two eyes and one nose. But there are probably as many as fifty thousand genes in each zygote, and some of these have the property of determining such characteristics as the color of the eyes, the blood group and the shape of the earlobe. In this case, the alleles may or may not consist of similar genes from each parent. When the two genes are identical, the new individual is said to be homozygous for that gene. When the alleles are dissimilar, the individual is heterozygous for that gene.

Whenever two dissimilar genes come together, there is one that will always assert its influence, and one that will "yield" to the

other's presence. The one that does show its presence is called the dominant gene, and the one that does not is said to be recessive. Examples of genes that are dominant are those that give a brown color to the eyes (B); blue eyes (b) is a recessive trait. Whenever an individual is homozygous for blue eyes—that is, has received genes for blue from both parents—then he will have blue eyes. In the same way, if he is homozygous for brown eyes, he will have brown eyes as a result. But if he inherits a gene for brown eyes from one parent and a gene for blue eyes from the other, the brown always wins, and the result will be brown eyes.

During meiosis, the alleles separate as the chromosomes separate and move to different gametes. So a person of the genotype Bb will produce equal numbers of gametes with genes B and b. The offspring of two such people result from a random combination of B and b gametes into the three genotypes BB, Bb and bb, in the ratio one : two : one. Half will, therefore, be of the same genotype, Bb, as the parents, and three-quarters will be of the same phenotype, BB or Bb. One-quarter will be bb, different from both parents.

It has been shown that the "elements" responsible for different traits are assorted and recombine independently of one another. This is true for genes that lie on different chromosomes, but, of course, is not true for genes that lie close together on the same chromosome, which can produce a linkage of certain characteristics, so that they normally occur together. However, the process of crossing-over can separate even these genes.

An interesting case of genetic linkage is found in the blood-clotting abnormality called hemophilia, where the gene responsible for the disorder lies on the X sex chromosome. It is a recessive gene, so that the trait is very rare in females, who have XX sex chromosomes. It can be passed on, however, to males, who have XY chromosomes and no dominant normal gene on the Y chromosome to annul the trait. Color blindness is also a sex-linked trait, and is ten times more common in males.

Most human traits are governed not by single genes but by the combined effects of several different genes, which form a polygenic system. The color of the eyes, for example, is not just a choice between blue and brown; there is a wide range of human eye colors. The genes in a polygenic system may be on the same or different chromosomes. Each contributes to a trait, but by itself has little effect.

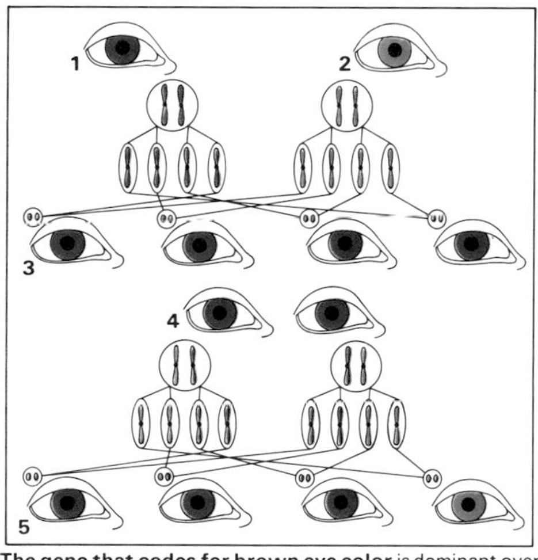

The gene that codes for brown eye color is dominant over that for blue. A man with two genes for brown eyes (**1**) and a woman with two genes for blue eyes (**2**) will produce all brown-eyed children (**3**), each of whom possesses one gene for brown eyes and one gene for blue eyes. If, however, two parents with brown eyes (**4**) each carry the recessive gene for blue eyes they will produce children in the ratio of one blue-eyed to three brown-eyed (**5**).

Albinism is caused by the absence of the pigment melanin, resulting in individuals with a light skin, white hair and pink or red eyes. The defect is carried by a single recessive gene, and one such gene must be inherited from each parent if the condition is to appear. In some cases, one or both parents are albino, but in other cases both parents may be normally pigmented, but still carry the recessive genes which results in an albino child.

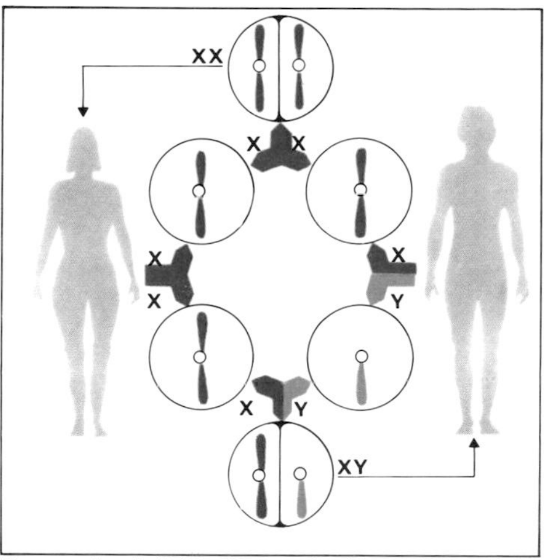

The sex of an individual is determined by a pair of chromosomes designated as XX in a female and XY in a male. Ova contain only X chromosomes, but sperm contain either an X or a Y in a one-to-one ratio. At fertilization, there is an equal chance of an XX or an XY pair being formed. Sex-linked characteristics, such as hemophilia, arise because there are regions of the X chromosome without corresponding regions on the smaller Y chromosome.

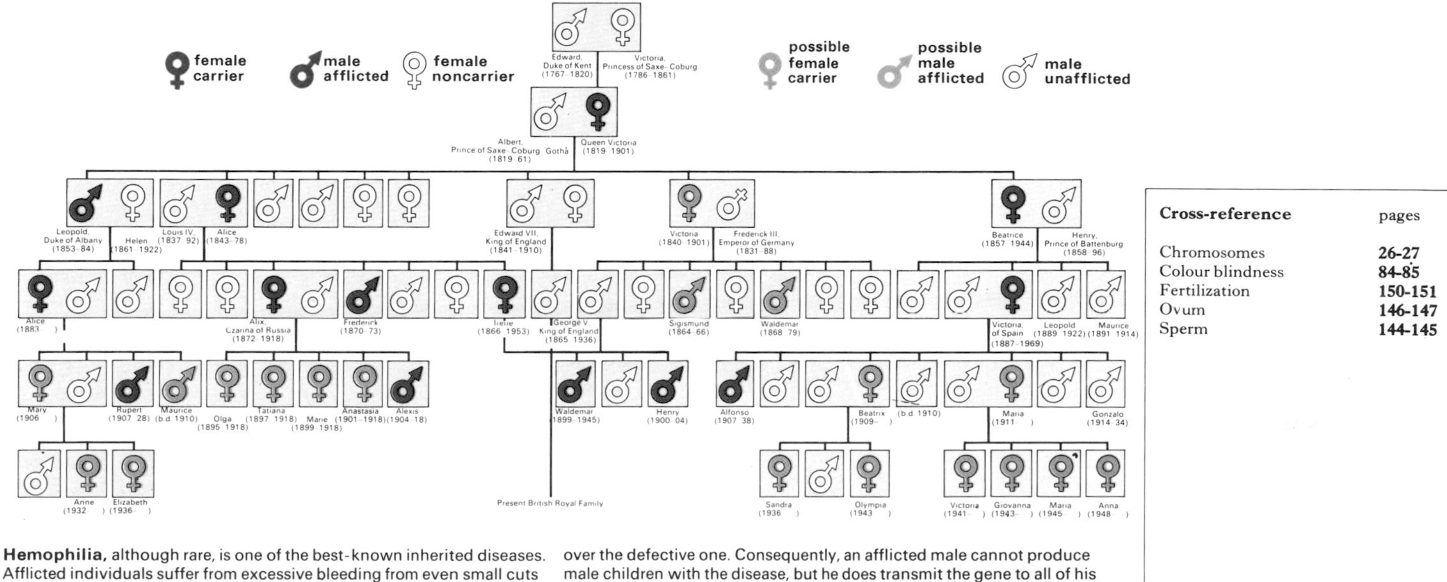

Hemophilia, although rare, is one of the best-known inherited diseases. Afflicted individuals suffer from excessive bleeding from even small cuts and bruises. This is caused by the lack of certain blood-clotting factors. The defective gene which fails to code for their synthesis is recessive and linked to the X chromosome, which means that only males (XY) can be afflicted. The Y chromosome has no corresponding gene to be dominant over the defective one. Consequently, an afflicted male cannot produce male children with the disease, but he does transmit the gene to all of his daughters, who will transmit the disease to half of their children, whether male or female. The best way to see how the gene for hemophilia is transmitted is to study a family tree. The illustration above shows the passage of this characteristic throughout the royal families of Europe.

Cross-reference	pages
Chromosomes	26-27
Colour blindness	84-85
Fertilization	150-151
Ovum	146-147
Sperm	144-145

Span of life

Shakespeare, speaking through the words of his character Jaques, sums up the life span of man as being seven ages—seven acts in a play in which we are all actors with an inexorable sequence of roles to play. The first role we play is that of the helpless babe in arms, the last that of the helpless senile. The wheel turns full circle, and we finish "sans teeth, sans eyes, sans taste, sans everything." As the plot of the play unfolds, between life and death we pass through childhood, adolescence, maturity, middle age and decline.

We can divide the span of life into as many ages, or sections, as we wish. We can use years, or decades or even milestones of physical development or physical and mental ability. But the exact pattern of any individual's role cannot be written until the play has actually been performed. Each stage in our lives is dependent on the previous one, on past experiences, memories, education, upbringing, diet, disease and other factors in our environment. Only the broad outline of events in the span of an "average" human being can be laid down.

About a quarter of our lives is spent in developing, in reaching physical and emotional maturity. Our bodies grow, and physical and mental skills are established. We become able to reproduce ourselves, and to take on the responsibilities that this ability confers upon us. And as soon as adulthood is reached, whether we like it or not, the process of physical decline starts, subtly, but immediately. The middle years, however, are accompanied by new satisfactions and new achievements that only accumulated experience and wisdom can bring. They include status in society, perhaps, and the pride of having done useful things, and of having brought into the world new individuals to perpetuate the cycle of life.

In the declining years, physical abilities inevitably diminish, but the intellect often stays bright. Grandparenthood brings its own flurry of pride and revived joy in existence. In old age, too, comes a merciful acceptance of the inevitability of death, and the disappearance of the fear of dying.

As medicine progresses, so the span of life is being extended. Disease is being conquered, and the quality of life in all its aspects is changing as our living standards improve. During the last century there have been changes in the "average" pattern of growth and aging; the ages of the onset of the menarche and the menopause have, for example, been changing. We can only guess what the future picture of a human being's life will be like, and can merely hope that we will avoid being overtaken by unforeseen dangers inherent in a society in which the rate of technological advancement is accelerating.

In the West, the expected human life span has increased from about fifty years at the turn of the century, when this family photograph was taken, to more than seventy years. Throughout the individual's life, the family plays an important although changing role.

The first year

The social attachment between mother and child is most important during the first year of life. The specific relationship begins to emerge at about three months, and is reinforced by feeding and cuddling, which bring mother and child into regular close contact. One of the major developmental aspects of this special bond is the ability of the baby to remember a familiar face. The attachment increases in intensity until the child is about one year old, when he becomes less specific in his response and begins to establish relationships with people outside his family.

For several weeks before birth, a child is in theory capable of a separate existence from his or her mother. At the moment of birth this theory is put to the test. Within minutes, the baby's lungs and circulatory system must adapt to the new, airy environment, and the other body systems must be brought into play to sustain life outside the womb.

But how really separate is this existence? A newborn human infant certainly has a life of his own, but he is absolutely helpless in his mother's arms. The only hope the infant has of survival lies in his mother's instincts and intelligence, and the responses she makes to his expressed needs. Crying may indicate discomfort or hunger, and both need attention. The baby's own reflexes of rooting—seeking for the nipple—and sucking, are of no avail if the mother does not proffer her breast.

For the infants of many animals, the first twenty-four hours are the most important, and instinctive behavior plays a large part in survival. Probably, because of the great capacity of the human brain to learn, and our fewer fixed patterns of instinctive behavior, this period may not be so important in man. What certainly is important, for satisfactory future psychological and physical development, however, is the bond that now must be created between mother and child.

Maternal instincts are recognizable throughout nature—a mother naturally tends, feeds and cares for her young. But in some animals, such as monkeys, this behavior may have to be learned. Young monkeys brought up deprived of a mother find it difficult to establish mature relationships with their fellows as they grow up. When they in turn have young they are unable to mother the infants successfully, and these young in turn are deprived, unsuccessful and so are themselves unable to mother.

There is now growing evidence that this may also be true of man. Mothers who find it difficult to care for their children have frequently themselves been deprived as infants and children. It would seem that in order for human children to be brought up successfully, their mothers must themselves have experienced during childhood the love, care and psychological well-being which is necessary for full emotional maturity.

There is no doubt that a close contact between infant and mother throughout the first year is essential for the child's normal development, but the culture into which the child is born determines the extent of this contact. Carrying a child around on the back, and breast-feeding, for example, are traditional in tribal societies, and result in constant contact. In an industrial society, the baby is often bottle-fed, left in a crib and frequently isolated from his mother for long periods.

Milk secretion by the mother begins immediately after birth. For the first few days only a yellowish fluid, colostrum, is produced by the breasts. Colostrum is of great importance to the infant because it contains maternal antibodies which help to protect him against infection for about three months, until his body can produce its own defense mechanisms, such as the antibody-forming thymus gland in the neck, which rapidly enlarges after birth. Other important systems within the infant's body rapidly mature after birth. Enzyme systems, for example, within the intestines and the liver are soon able to digest and utilize the mother's milk. The first year is

Reflexes of the newborn

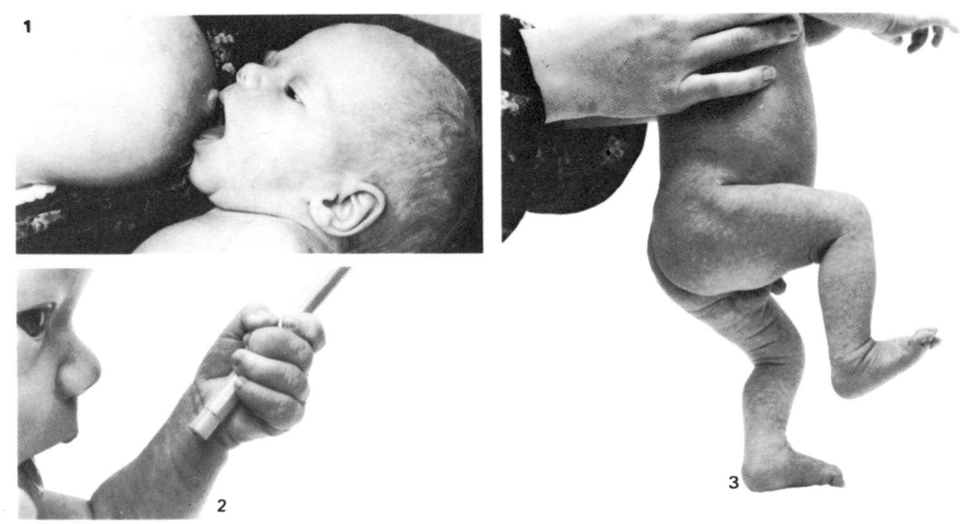

The newborn baby displays a variety of reflexes, some of which are necessary for survival and others which have evolutionary significance. Suckling is a combination of two reflexes, the rooting reflex (**1**), when the baby turns his head toward the nipple when it touches his cheek, and the sucking reflex, which initiates strong sucking when the nipple enters his mouth. With the grasp reflex (**2**), the baby will automatically grip any object that touches his palm. His grip is strong enough to support his own weight. The step reflex (**3**) causes the baby to lift one leg, or to step, when both his feet are put on a firm surface. He may also straighten his legs as if to stand. These reflexes disappear after a few weeks of life, when the nervous system matures and learned movements become possible.

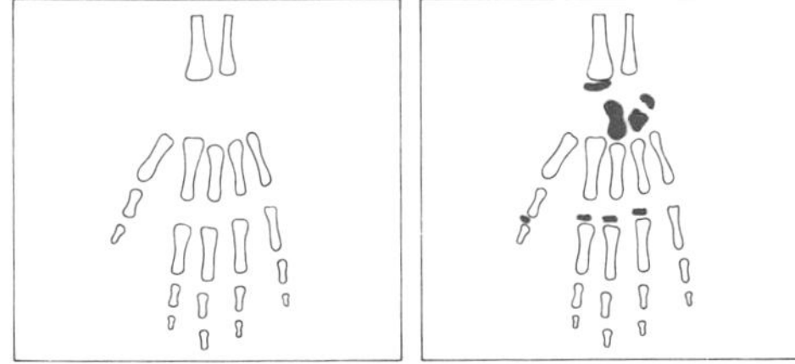

Most of the infant's skeleton is preformed in cartilage, which provides a model in which bone is laid down—a process called ossification. Some cartilage ossifies before birth (black) and some ossifies during the first year (red).

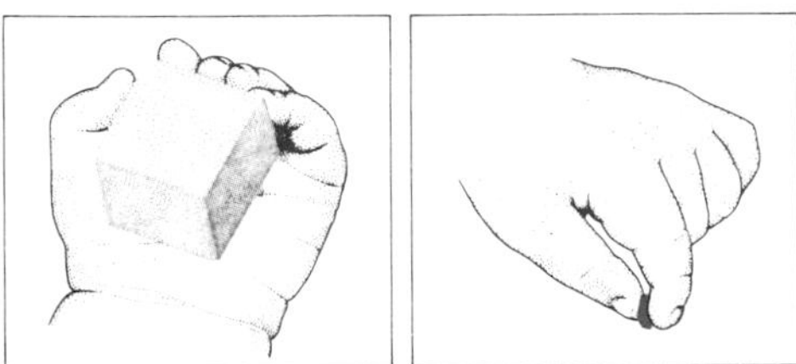

By six months the grasp reflex has disappeared and the baby can only hold large objects in his palm. At twelve months he is able to oppose his thumb and index finger. He can now pick up objects, both large and small, with delicate precision.

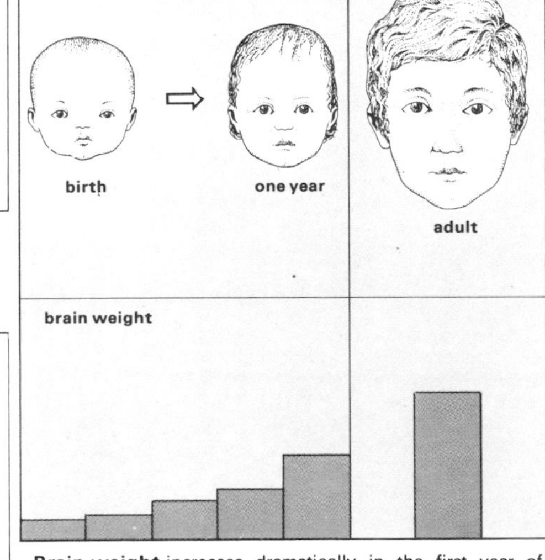

Brain weight increases dramatically in the first year of life. As the brain develops and the cerebral cortex increases in size, more areas of the child's body, as well as visual and auditory faculties, come under greater control. As the brain stem ceases to exert so much control, the reflex actions of the newborn disappear.

a time of rapid growth. The infant doubles his birth weight in six months and triples it within a year.

The newborn baby spends most of his time asleep. Up to the age of about four weeks the baby's main waking activity, when he is not feeding or otherwise occupied, seems to be long periods of unflinching staring at the objects around him. This is probably related to the slowly developing ability to focus the eyes and to control the numerous muscles which move them. The hands of the four-week-old infant are usually kept closed and are involved much less than the eyes in exploring the environment. By six weeks the baby can recognize his mother.

By sixteen weeks the baby begins to remain awake for much longer periods and displays more advanced behavior. His head is much more often brought to face the front, for example, and objects held in his hand are brought close to his eyes for careful scrutiny. He enjoys sitting up and looking around at what is going on. From now until the age of about twenty-eight weeks the most noticeable physical development is the acquisition of control over the muscles which support the head and move the arms.

By the age of twenty-eight weeks the baby begins to eat solids and may be able to sit upright without support for extended periods of time. He actively explores his surroundings with his hands. Full grasping, involving the use of the thumb, is developed, and objects can be transferred from one hand to the other. This is the age at which the first recognizable human speech sounds begin to be produced, by contrast to the gurgles and cooings of earlier weeks.

At about forty weeks, as a result of the rapid growth of his limbs and the musculature of his trunk, the baby is an adept crawler. The more distant parts of his body, such as the toes and the tip of the tongue, come under increasing conscious control, and the forty-week-old baby can pick up a crumb with some precision. He can respond socially by waving good-bye and showing various other responses to the people around him.

Recent studies have shown that in many ways the perceptual world of the infant is very similar to that of the adult, although it is much simpler. Experiments with young babies two to six weeks old show that they can track moving toys with their eyes and calculate, much as an adult would, the position of a toy when it emerges after being hidden by an object. They appear to have three-dimensional vision, and are able, even at this young age, to stretch out their hands in space to touch an object. However, if a young infant is shown a photograph, he is unable to judge depth because he has not yet learned the rules which govern perspective or shading and which to the adult eye indicate distance. This more subtle understanding must wait until the child is older and has learned by experience.

So the process of physical and psychological growth progresses. The cooing and smiling infant of three months, evoking in his mother the essential and necessary tender and protective feelings, changes slowly to the six-month-old who is more aware and becomes anxious when separated from his mother. By nine months to one year he is more unhappy when separated, and if this separation continues for any length of time he may become withdrawn. For such is the strength of the maternal bond that the infant is totally dependent on it.

Periods of perceptual development

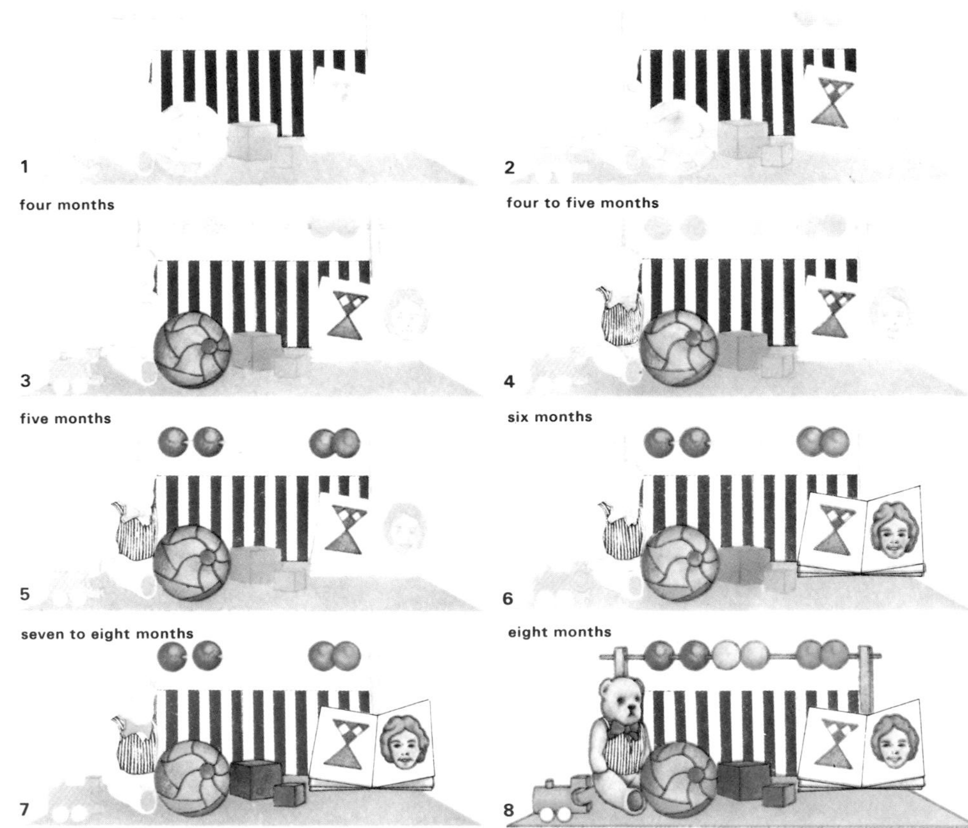

The baby's developing visual acuity allows him to discriminate between lines one-eighth of an inch apart (**1**). Later an interest in form develops and he prefers to look at complex geometric shapes (**2**). Gradually he prefers to stare at even more intricate patterns (**3**) and after another month can distinguish between lines one-sixty-fourth of an inch apart (**4**). Interest in color, primarily red and blue (**5**), occurs before an interest in faces (**6**). Awareness of size develops (**7**) in advance of shape and color discrimination (**8**). The child has now acquired most of the basic visual skills.

Locomotion and motor control

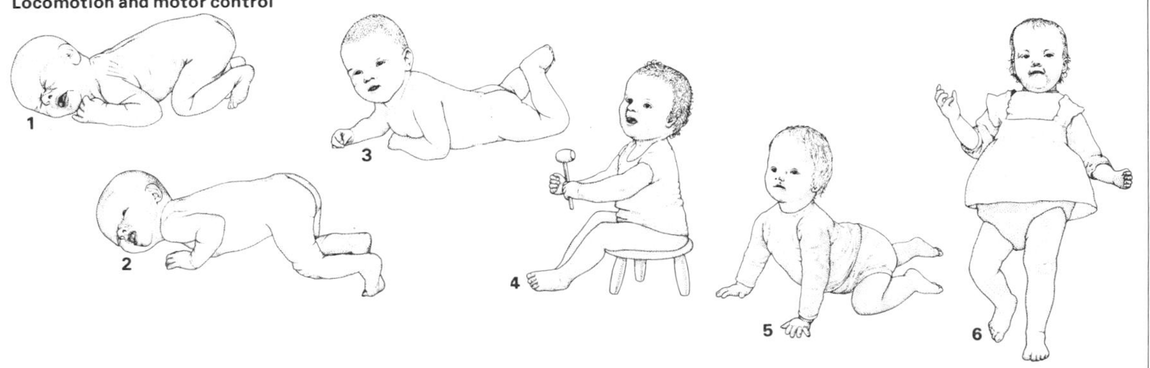

The newborn baby lies with his pelvis high and knees drawn up (**1**). By one month his knees are kept farther away from his body (**2**), and he can lift his chin and shoulders (**3**). Other areas of the body gradually become controlled so that by six months the baby can sit unaided (**4**) and by ten months he can usually crawl (**5**). At about one year the baby begins to take his first steps (**6**).

Cross-reference	pages
Birth	**156-157**
Brain and central nervous system	**74-77**
Lactation	**156-157**
Learning	**128-129**
Thymus and defense	**140-141**

One to three years

At thirteen months a child can climb stairs in an awkward fashion. By three years of age a child is able to walk up and down stairs alternating both feet. Although learning processes play a role in the improvement of such skills, the developing muscular and nervous control are also involved. Generally, by three years all trace of clumsy infant motor behavior has disappeared.

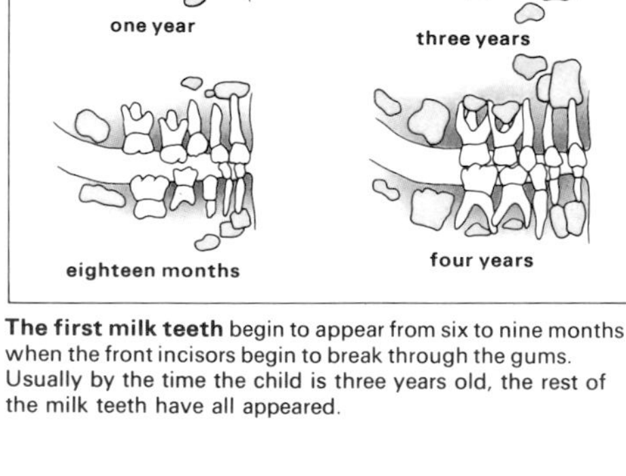

The first milk teeth begin to appear from six to nine months when the front incisors begin to break through the gums. Usually by the time the child is three years old, the rest of the milk teeth have all appeared.

Faced with a new social situation even the child of three, who no longer has the infant's fear of strangers, will become shy and display dependent behavior.

By the end of his first year the human child is sufficiently independent to begin to move about on his own and to start to explore his immediate world. The years from one to four are those in which both the inherited and the newly acquired facets of his character and personality begin to show. He is now becoming an individual in his own right, and establishing his own relationships both within and outside the family.

During this period the brain shows rapid growth and maturation. The most significant development is that of speech and language. It is language which sets man apart from the rest of the animal kingdom and which, in part, allows very rapid learning to take place as well as the full development of man's intellectual potential.

Play is of extreme importance during these years, for it is through play that the young child learns. By handling building blocks or colored solid shapes he learns about the properties of hard objects. Sand and water tell him about substances which flow. Dough and clay give him the feel for molding the world around him. It is through these primary and simple activities that he builds up his experience of what the world is like. The importance of play in these years cannot be overestimated. Several studies have shown that children who are not stimulated and who are denied the opportunity to play have difficulty in reaching their full intellectual potential. Equally important is the stimulation given by the parents. To succeed, children must be talked to, helped, kept interested and their experience widened.

The most significant hurdle for the growing child to clear is the loosening of the bond with his mother. By the age of one, the child becomes more aware of other members of the family. He begins to learn that sharing means giving up as well as gaining something. Children of this age frequently show open rivalry toward brothers and sisters. Although they may become openly aggressive, under the reassurance of the parents these feelings are normally brought under control. The young child also realizes that the parent to whom he is closest, usually the mother, is also shared with his father and he has to come to accept this, and to control his feelings of jealousy of the other parent, whose love and approval he also needs. These are the basic steps of psychological growth. The way the child copes with these will be reflected in the way he handles close relationships later in life.

The two-year-old, in becoming aware of other people within his family, starts to identify strongly with them. He imitates them and begins to incorporate their values and attitudes into his own developing personality. The values of his parents will form the basis of his future morality.

Two is the age of insistent demands, temper tantrums, jealousies and the ever-recurrent negativity. The two-year-old begins to feel himself a person in his own right, but his individuality is still fragile and easily swamped. Unless taken gently he finds it easier to say no than to go outside and explore. He is fearful in unfamiliar surroundings, particularly those outside the family.

By the age of three confidence is slowly building, and he begins to make friends of his own age. These friendships are still self-centered, but the child is gradually learning to

share and to cooperate with other children. This is the age of fears and nightmares. It is also the age when the child begins to realize that she or he is a girl or a boy, and objects if mistaken for the wrong sex.

This is an age, too, when overt sexual behavior may be observed. Until recently, Western societies have been disapproving of childish sexuality, but in more traditional cultures this behavior has always been accepted. The Western attitude stems from a mistaken belief that childish and adult sexuality are similar. The growing child, however, has no clear understanding of adult sexuality, or of its meaning, and although he may manipulate his genitals in sexual play because he finds it pleasurable, it has no direct sexual meaning. He becomes interested not only in his own genitals but also in those of the opposite sex, which are dissimilar and therefore arouse his curiosity and prompt him to ask questions.

This is the period, too, of rapid physical development. By the age of about fifteen months the average child can stand up independently and walk alone, although he may be a bit wobbly. He can put a pellet into a bottle and place one wooden block on top of another. He can scribble, communicate in gestures and use a few baby-talk phrases.

By eighteen months the child has probably reached a height of about thirty inches, weighs from twenty to twenty-eight pounds and may have as many as a dozen teeth. He is now a real toddler and can usually manage to maneuver himself into a child's chair. When asked, he can point out pictures of objects with which he is familiar and with luck and some effort he can build a tower of three blocks. He is less single-minded than the one-year-old, who tends to ignore everything but the one thing he is looking at. The eighteen-month-old seems to be interested in whole collections of objects. He may have a vocabulary of up to twelve words and shows full comprehension of a great many more.

The two-year-old shows evidence of continued rapid growth. The average child is now thirty-two to thirty-five inches tall, weighs twenty-three to thirty pounds, has about sixteen teeth and can chew efficiently. Although he still needs assistance on stairs, the two-year-old can get around most of the time on his own and is even beginning to run. He can kick a ball, he likes to chatter to himself and he can turn the pages of a book. As far as block-building is concerned, he can erect towers six blocks high. He can thread beads on a string and fit shapes into corresponding holes and he may have a vocabulary of three hundred words.

It is at about this time, early in the child's second year, that his nervous system matures sufficiently to allow him to become toilet-trained. This usually occurs at first during the day and later during the night, quite naturally, without pressure and with minimal training.

Between the ages of three and four the child ceases to be a baby. He can walk and run almost as well as an adult can, and is coming more and more to resemble physically a grown person in terms of body proportions and psychological responses. Mastery of language develops quickly and between the ages of three and four years many children may begin to read their first words. They also begin to draw simple pictures and to play games which frequently involve some element of play-acting or imitation of grown-ups.

Socialization begins at an early age when the child is forced to behave in a manner that is approved of by parents and society as a whole. Although she is capable of eating with a spoon, a child of two or three prefers to eat with her hands. With encouragement from her parents, however, she will learn to use the spoon because she seeks their approval.

The three-year-old has little difficulty in fitting the correct pieces of wood into the correct slots of the box. This shows developing hand–eye coordination.

A drawing of a man done by a two-year-old, left, and a three-year-old, right, shows the great development in motor control and cognitive ability.

Cross-reference pages

Brain and central nervous system **74-77**
Left hemisphere **124-125**

Three to seven years

A child of about four years old is able to build only a rough tower of six or nine blocks. The seven year old, whose sensorimotor and perceptual skills have developed, will select the specific shape and size of blocks and use them to build stable and more regular complex structures.

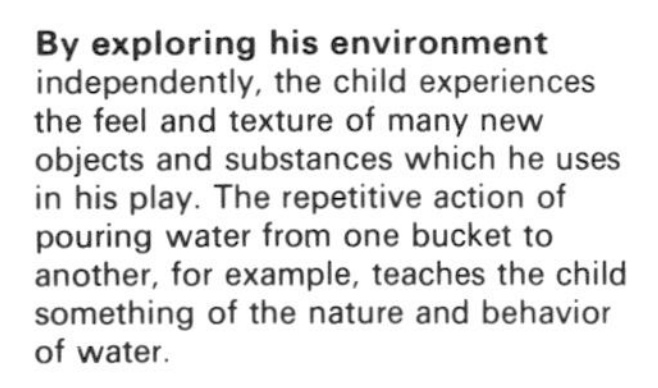

By exploring his environment independently, the child experiences the feel and texture of many new objects and substances which he uses in his play. The repetitive action of pouring water from one bucket to another, for example, teaches the child something of the nature and behavior of water.

Physical growth and increased motor ability give young children the opportunity to discover the limits and potentials of their physical skills. They prefer play involving other children and such games as follow-the-leader. Noncompetitive games involving such physical skills as climbing allow children to assess their own ability against that of others and also provide an outlet into a new social world.

During the period from four to seven years the major changes that occur in a growing child are psychological and social rather than physical. The pattern of growth continues—the brain grows more slowly and the limbs and body develop quickly. At the age of five the girl's ovaries have reached their adult size, but do not function until puberty when they are stimulated to action by hormones from the pituitary gland. Facts like this strongly support the idea that as a result of man's evolution the onset of physiological maturity has been retarded until a lengthy period of psychological and social development can be completed.

Between the ages of four and five the child makes a notable advance in independence. He becomes less clinging and positively enjoys opportunities to make wider social contacts with other children and adults. By the age of five the motor skills of walking, running and jumping have been completely mastered and the extent of control of body equilibrium is displayed by the five-year-old's ability to balance on one foot. He can, too, brush his teeth and comb his hair after a fashion. His perceptual and manipulative abilities allow him to draw a creditable picture of a man which, although still highly schematic, shows a great advance over the artistic efforts of the four-year-old. In play the five-year-old is more consistent and is more likely to finish any task which he has started.

The five-year-old can remember a long sentence and repeat the plot of a simple story. The 1,500 words of the four-year-old's vocabulary have become 2,200, on average, at five years old, and all the basic elements of grammar have been mastered.

From five to seven there is much less change in the child than can be observed, for example, between three and five. Now the pattern of growth is one of continuation of the trends established by the age of five. During these two years weight rises on average from about thirty-six to fifty pounds and height increases to an average of between three feet six inches and four feet two inches. Ossification of the bones becomes much more complete by the age of seven and replacement of the milk teeth can begin any time after five and a half and six, typically beginning with the eruption of the lower incisors.

Intellectually, this is a time of rapid development during which the child's knowledge and grasp of reality advances rapidly. This is particularly marked by his ability to understand abstractions and the beginning of his mastery of the fundamentals of reading, writing and arithmetic. This is also the period when the child will tax his parents and teachers with persistent questions.

According to Jean Piaget, a child psychologist, this time is characterized by a shift from direct perception of the world to one of representation. At the age of five, for example, the child will frequently confuse "b" and "d," "p" and "q." By the age of seven such confusion is less likely, for the child is beginning to understand fully such objects as being representations or symbols, in this case symbols of speech sounds, which are themselves organized into words which represent things. Obviously, this greater ability to approach things from the representational point of view goes along with the child's development of linguistic ability. The capacity to perceive things—words, signs, ideas, gestures—as

representations, which the child really acquires at this stage, is one of the first steps in the development of rational conceptual thought.

One of Piaget's best-known clinical experiments illustrates the acquisition of mental representations of a series of actions by a seven-year-old compared with that of a five-year-old. This involves the ability of the child to understand that the volume of a liquid or solid substance remains constant although the shape may change—an operation known as conservation.

If a typical five-year-old is shown two tumblers filled to the same level with water he will agree that they both contain the same amount. However, when the water from one glass is poured into a taller, thinner vessel, so that the water level is higher than in the tumbler, the child will then maintain that the tall container has more water in it, "because it is higher." A seven-year-old will insist that they must contain the same amount, "because you can pour the water back into the glass and it will be the same height." This shows that the older child is aware of the reverse operation, which will restore the original situation. This reversibility of operation is, according to Piaget, a most important developmental stage.

Psychologically this is a period when children begin to identify strongly with their sex and their various roles as adults in society. From time to time, the child will, for a short period, imitate the behavior of the opposite sex, but he or she will always come back again to identify strongly with his or her own sex, behaving very much as does the parent of the same sex within the family.

This is a time of friendships, of learning to leave the family, of going out to school and into the homes of peers and of seeing how other people live. Within the playground many groups emerge and dissolve. There is not one hierarchy, but many, each child learning how to play a different role and each child learning how to abide by the rules of the game. Most games appear to be as much self-competitive as they are peer-competitive. At this age children prefer games that restart automatically and in which everybody is given a new and equal chance.

While most children derive their pleasure more from competing than from winning and prefer games that do not directly compare abilities, some are fiercely competitive. But it is often adults who impose the idea of winning, and for this reason games organized by parents often fail to please.

Learning in the playground is as important to children as the formal learning provided in the classroom. It is through games that they learn the importance of social relationships, when to give and when to take, and they begin to understand the expectations of their peers.

Imitation of parental roles begins to develop during this period and boys will engage in activities which they associate with their fathers. Girls may find themselves imitating their mothers. Parents often reinforce such imitative behavior because they find it flattering to their own sense of identity. However, the changing attitudes toward the roles of both sexes allow for increasing freedom in the activities the children may select.

Growing children experience the delight of controlling body actions as they engage in boisterous play which involves running, jumping or bouncing. Girls enjoy physical games as much as boys, but also enjoy verbal-social interactions as much as physical ones.

a four-year-old's drawing

a seven-year-old's drawing

The drawing of a man more clearly resembles a human form as the child becomes older. The elements and details increase and more colors are used.

Cross-reference	pages
Bone	**30-33**
Hormones	**70-73**
Ovaries	**146-147**

Seven to eleven years

In the period which extends from the age of seven up to puberty, there are a number of profound changes in the lives of children in all societies. In the West, it is about this time that formal schooling begins and there is emphasis on scholastic achievement. Now the child's social horizons widen greatly and he begins to interact with many people outside the family, who come to exert almost as much influence on him as the more restricted sphere of family life did before.

In many nonindustrial societies, children now achieve a great measure of independence from adults, and become, in many respects, more or less wholly able to look after themselves. Many anthropologists have reported the way in which children of this age frequently take on some of the responsibilities of adults, and appear to have no difficulty in doing so. In many such societies, children are allowed much more freedom in later childhood than they will be permitted in puberty, when various taboos and restrictions will be brought to bear on them because of their approaching sexual maturity.

Emotionally and psychologically, this period is usually very much less eventful than the preceding one. Sigmund Freud called it the latency period, and modern research regarding the role of hormonal factors in suppressing sexual maturity at this age tends to bear out this theory. It is a period during which no great development takes place in the personality, and the identity established as a result of earlier experiences is confirmed.

The child of nine or ten has a well-developed conscience, or moral sense, as a result of adopting and identifying with the values of his parents. But he is also sufficiently independent to be able to practice subterfuge and to make a good show of lying.

In social development this period is one in which the trends established earlier become stronger. The peer group becomes, if anything, more important, more clearly demarcated on sexual lines. In nonindustrial societies the demarcation is usually even more evident, if only because by the age of seven or eight the little girl helps to look after other children in the often large family, and is actually doing some of the mother's work. The little boy begins to help the men in their particular activities and is learning to use bows and arrows or other items of masculine equipment.

At this stage the child acquires the ability to extend his intellectual horizons through the use of logic. The units of thought that he employs include images, symbols, concepts and rules. While the one-year-old, for example, perceives a circle simply as an image, to the four-year-old it may represent the letter O or zero, both of which are symbols. The eleven-year-old, however, may think of it as an image of the world and its contents. This is a concept and represents the method, which the child has grasped, of piecing together a collection of experiences that share one or more common characteristics.

At the age of seven auditory acuity is as good as it will ever become, although discrimination of pitch continues to improve for three or four years. At this time both girls and boys seem to show little interest in harmony, and melody, as a rule, is not appreciated as much as rhythm. In Western societies, boys tend to favor the march, or double time, while girls are more advanced and can quickly take to different rhythms, and by the time they are eight or nine begin to appreciate pop music.

Independence of parents and older brothers and sisters, who until now have provided security, is a major development of this period. The process of establishing independence begins early in life, and is greatly reinforced when the child leaves home each day to go to school. The nine- or ten-year-old is usually sufficiently independent to prefer to spend his free time with friends, involved in activities which do not require adult supervision.

Animals and pets provide an outlet for affection. The tolerance and care required to keep them is useful in teaching the child how to apply this thoughtfulness in relationships with other people.

Boys of this age on average are slightly stronger than girls and this often shows itself in their games. Although in finer finger work girls are superior to boys, interestingly enough, boys tend to draw better than girls up to about puberty.

The mental imagery of young children is very concrete. While adults tend to think in terms of scenes or words, children think in terms of actual sensations. Between seven and eight there is a changeover in children's imagery from the use of sensations and situations to a pattern more closely resembling the adult's, although this is still very limited. They tend to think either visually or auditorily since their thought content is mainly obtained from television, comic books and school.

The games that children of this age play provide an intriguing study in themselves, reflecting as they do the lore and tradition of their culture. From generation to generation, games such as leapfrog and tag have gone unaltered, and children have played with tops, hoops and kites throughout history and all over the world. In playground rhymes and singing games, important events of the past are often commemorated, although the significance of the words and jingles has long since been forgotten. Younger children learn from their immediate elders imitatively and without question, and so help to hand down games and rigmaroles from the remote past. Children must surely be the most traditional beings in the world.

By the age of ten many children are developing a first awareness of growing up. Girls may relinquish dolls that have been their constant companions and playthings since early childhood, and boys forget for long periods about cowboy outfits.

In the tenth and eleventh years girls particularly show an increase in maturity of attitude, and secondary sexual characteristics may appear as a prelude to the burst of puberty. The nipples may become more prominent, and the face and hips become slightly rounded as the typical female distribution of subcutaneous fat begins to become apparent.

In both sexes there may be an increase in height of between one and two inches during one year, the height for boys ranging from about four feet to over four feet ten inches, and for girls slightly, but not much, less. The mean weight for boys is about sixty-five pounds, with girls lagging a pound or so behind.

The ten-year-old is developing confidence and ease in the environments of both home and school, and usually gets on well with other people—except perhaps younger brothers or sisters, with whom there may be displays of impatience. Stresses and different kinds of problems at school or in the home may cause temporary disturbance, but overall the picture is one of poise, consolidation of what has gone before and preparedness for the next stage of growing up—adolescence.

Group games in later childhood fulfill a variety of functions. The rules of the games become less flexible and more important and these rules also begin to govern social relationships within the group. While older children generally adhere to the rules, younger ones will tend to drift away if they do not like them.

Parental control exerts a strong influence on the social and moral character of children. The child depends on his parents to fulfill such material needs as clothing and food as well as such psychological needs as affection and attention. Lack of parental affection and attention may lead to antisocial behavior.

Formal schooling leads to the child increasing her knowledge of the world around her. At the same time it establishes the foundation for the child's developing cognitive ability and powers of reasoning.

The drawings done by an eight-year-old, left, and by a nine-year-old, right, show the continuing development of perception and of motor skills.

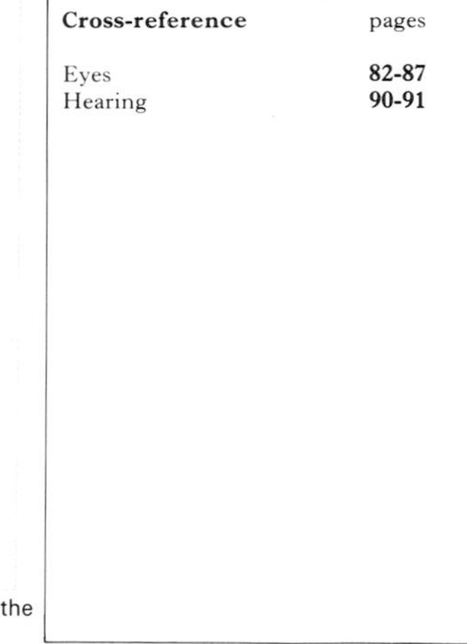

Cross-reference	pages
Eyes	82-87
Hearing	90-91

Puberty: physical changes

Puberty marks the beginning of the transition from child to adult. It is the time when a sense of identity begins to develop, and the adolescent is often torn between the childhood world of dependence and submission to authority and the need to rebel against adults in order to establish independence. Much of the emotional turmoil of puberty, as well as the physical development, is the result of changes in the hormonal balance.

Girls and boys at puberty

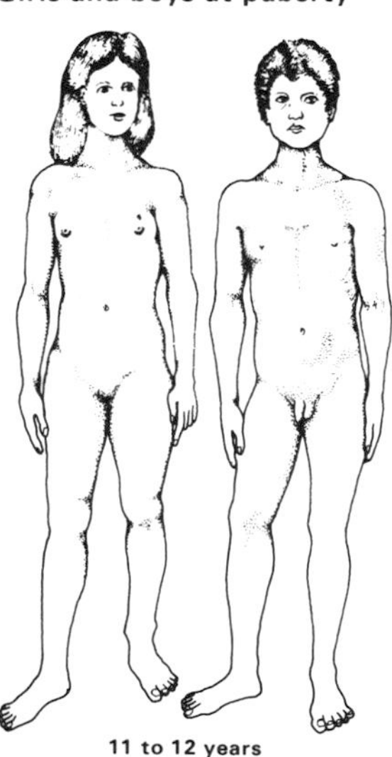

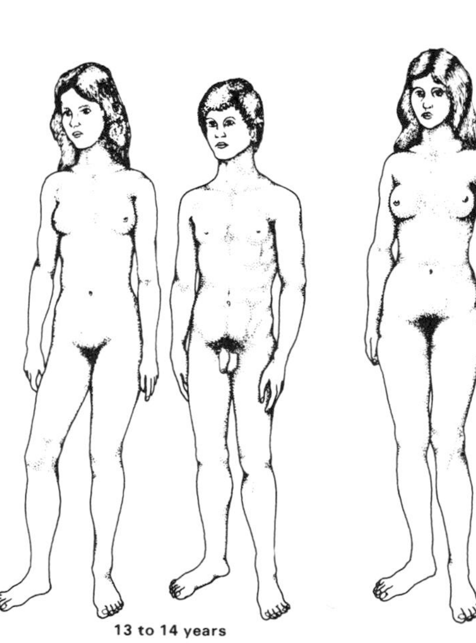

Puberty occurs about two years earlier in girls than in boys. Girls between the ages of eleven and fourteen tend to be taller, heavier and more mature than boys. At the ages of eleven to twelve secondary sexual characteristics start to develop in both sexes. From thirteen to fourteen boys grow rapidly, their voices break and pubic hair appears. Most girls begin to menstruate at or before this time, and their growth slows down.

The terms puberty and adolescence are often quite unnecessarily confused and misused, although they have reasonably precise meanings. Puberty, strictly, is the attainment of that degree of sexual maturity which makes reproduction possible; it refers to a stage in the body's anatomical and physiological development. More loosely, it is used of the whole period during which there are changes in the body leading to sexual maturity. Adolescence, on the other hand, is a much broader term. It refers not just to physical and sexual changes but to the whole range of bodily and emotional and intellectual changes that accompany the transition from childhood to adulthood.

In both sexes the physiological changes leading to puberty occur over a period of at least two years. In Western industrialized societies, noticeable changes usually begin in girls sometime between the ages of ten and sixteen, about two years earlier than in boys. There is a general spurt in growth as a result of suddenly increased activity in the anterior pituitary gland, which secretes growth hormone. At the same time there is increased secretion of hormones by the adrenals, the thyroid and the ovaries or testes. Under the influence of these hormones, sexual development accelerates.

In girls, the clearest indication of imminent reproductive ability is of course the onset of menstruation. The time at which a girl has her first menstrual period is known as the menarche, marking the beginning of her reproductive life, the other end of which will be marked by the cessation of menstruation at the menopause.

Menstruation is controlled by a roughly twenty-eight-day cycle of changes in the secretion of the hormones estrogen and progesterone by the ovaries, under the influence of the pituitary gland. In the first months of the menarche the loss of blood may be variable, and the ovaries may not in fact be releasing fertile ova.

Puberty in girls is also marked by the development of secondary sexual characteristics. There is an increase in subcutaneous fat, which is laid down in the characteristic female pattern—in the pads over the hips, thighs and buttocks and upper arms, and beneath the nipples of the breasts. At the same time, the glandular tissues of the breasts and the erectile tissue of the nipples begin to enlarge. Other developments are the appearance of armpit and pubic hair and the enlargement of the sex organs, the production of lubricating secretions by the vaginal glands and changes in the skin as the result of increased activity of the sebaceous and sweat glands.

In boys, there is a similar growth spurt heralding puberty. The increased activity of the pituitary leads to stimulation of the testes and secretion of increased amounts of the male hormone testosterone. There is marked enlargement of the penis and scrotal sac under the influence of testosterone and they darken in color.

At the same time circulating testosterone leads to the acquisition of the male secondary sexual characteristics. The shoulders broaden and the arms and legs become particularly well-muscled; there is hair growth on the face, pubis, arms, legs, armpits and on the chest, and, possibly, on the back as well. There is also growth and development of the larynx which results in deepening or breaking of the

voice. There is in the male no dramatic milestone in puberty such as the menarche in the female. The testes begin to produce spermatozoa, and there may be nocturnal emissions of semen. The first ejaculation is frequently experienced as a result of masturbation. Masturbation is a natural consequence of increasing sexuality in adolescents—both boys and girls.

The emergent sexuality of pubescent girls leads to their awareness of developing emotions which have consciously sexual overtones. The strength of these emotions both excite and disturb them. They become self-conscious, bewildered by their own feelings and by the intensity of the moods these feelings produce. Their labile emotions can lead to strong, passionate sexual crushes or to intense dislikes. Not yet of the adult world, but no longer children, they band together with their peers. In groups, they can get carried away, as at pop concerts, on waves of emotion, when the sheer intensity of their feelings may lead to hysterical reactions.

Girls of this age are characteristically passionate and loyal to their friends, yet they are frequently rebellious and hostile to their parents and to parental values. This period, when the intensity of the sexual feelings is frequently so strong as to seem unmanageable, is the time of puppy love, of admiration from a distance, of idols. The idol, however, must be unattainable and much time is spent alone and in fantasy. Overwhelmed by their feelings, but unwilling to admit their sexual nature, pubescent girls may often displace their feelings onto another interest, horses perhaps, or even religion.

In the case of girls, development of the breasts frequently carries with it a renewal of the long-standing identification with the mother. Now that the girl is developing the characteristic sexual features of the mother, she feels even more justified in identifying herself with her mother. Puberty also usually produces in girls an intense and sometimes obsessional preoccupation with appearance, which heralds the beginning of the typically feminine desire to be attractive to, and noticed and appreciated by, men.

In boys, such overt but disguised reactions to the onset of sexual maturity are not so common although they are not unknown. One of the reasons for this is that masturbation, the primary sexual practice of puberty and early adolescence, provides for the boy a more direct and less aim-inhibited outlet for his newfound sexual desires.

Nevertheless, the marked ambivalence toward the parents, which is typical of later adolescence, begins to appear in boys, perhaps even more than in girls, and this is frequently a period during which the boy or girl, stirred up by and unable to cope with their turbulent feelings, is moody or secretive, rebellious or depressed.

In non-Western societies, and even in the West only a century ago, puberty came much later. In nineteenth-century England, for example, the average age of the menarche was about eighteen. Today it is about eleven. Earlier menarche (and correspondingly later menopause) is known to be directly linked with the standard of living, but the exact causal mechanism is unclear. Puberty now seems to be a longer and more problematic process and only a prelude to adolescence.

Hormones at puberty
A boy (**1**) reaches puberty when the hypothalamus directs the pituitary gland to increase the secretion of the trophic hormones—luteinizing hormone, follicle-stimulating hormone and adrenocorticotrophic hormones. These act on the testes to start the production of sperm and on the adrenal cortex and testes to produce male hormones, or androgens. Testosterone, the most important androgen, promotes and controls the development of the sex organs and such secondary sexual characteristics as pubic and facial hair. In a girl (**2**), the trophic hormones stimulate the ovaries and adrenal glands to secrete androgens that control the shape of the female body. The ovarian hormones, progesterone and estrogen, initiate menstruation.

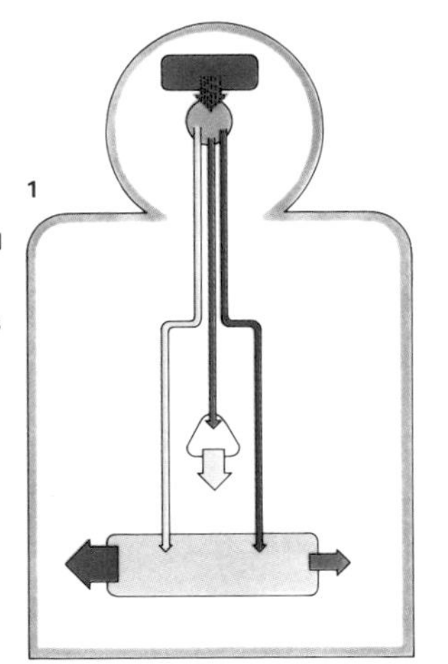

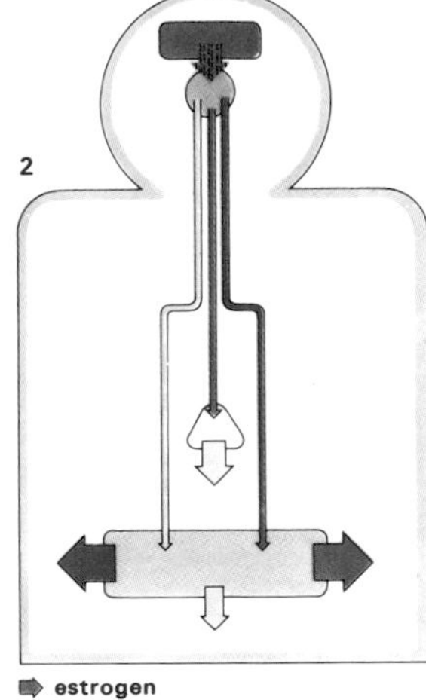

adrenocorticotrophic hormone
hypothalamic stimulus
hypothalamus
pituitary
adrenals

ovaries and testes
luteinizing hormone
follicle-stimulating hormone
ovarian androgens

estrogen
progesterone
testosterone
adrenal androgens

The physical changes of puberty coincide with a widening and deepening of intellectual capacity. The individual has now accumulated enough experience and has the mental flexibility to be able to reason in well-thought-out steps from which his or her own conclusions can be drawn. Each adolescent, however, advances intellectually at a different rate. As a result, material that challenges one student may bore another.

Puberty in primitive societies is clearly defined, particularly when it involves initiation ceremonies. In Sierra Leone, West Africa, for example, thirteen-year-olds are taken away from their village and for one month are instructed in customs, hygiene and self-discipline. After this they are considered adult.

Cross-reference	pages
Adolescence	**172-173**
Female anatomy and physiology	**146-147**
Hormones	**70-73**
Larynx	**62-63**
Male anatomy and physiology	**144-145**
Menopause	**178-179**
Skin and hair	**28-29**

Adolescence: emotional changes

Adolescence is the period during which the young person, no longer a child and already more or less physiologically mature, must gradually loosen the ties which have bound him to the home and to his parents. His social, intellectual and sexual interests expand and broaden. He must now begin to establish himself as an independent human being.

In primitive societies, adolescence seems to be of much shorter duration and a less troublesome experience than it is in the industrialized countries of the world. But what it lacks in length it makes up for in drama. In primitive societies, adolescence for both boys and girls, but especially for boys, is the occasion for a dramatic, sudden and often harrowing experience of transition to adult life. This comes about in the form of the initiation ceremony. Such ceremonies vary, but most of them contain many of the same ingredients.

First, the adolescent boy is ceremonially taken away from his mother and female relatives, often with the pretense of a fight or, at least, with a great show of sorrow on the women's part, and led away to an enclosure reserved exclusively for men. This symbolizes the breaking of the bond which until now has held the boy close to his mother and which was the original foundation of his social and emotional life. In some societies he may also transfer from the women's to the men's communal sleeping house.

Then the boy undergoes some ritual of initiation which is usually frightening, sometimes dangerous, often painful, and which, in many parts of the world, culminates in circumcision. The psychological effect of this initiation is to make the young man feel that he has undergone a profound and far-reaching inner transformation which has made him ready to assume the rights and obligations of adult life. Finally, there is often a period of sequestration during which the newly initiated boy must avoid all women and await the completion of the ritual transition to adulthood, an event which is signaled among the Australian Aborigines, for example, by the healing of the circumcision wound.

In technologically advanced cultures, such initiations do not occur, but adolescence is no easier because puberty comes earlier and, due to the demands of higher education for example, or military service, the final assumption of mature status comes later. The transition from childhood to adulthood, which is usually of very short duration in traditional societies, can last for anything up to ten years.

In the West, the boy who reaches puberty at eleven or twelve may not finish his formal education until he is twenty-one or twenty-two and may still not be in a position to marry and establish his own home for several years. Consequently, adolescence is a long and difficult experience for the young and looms large in the life of society as a whole. It is necessary for industrial societies to adjust somehow to having within them a large number of people who cannot be considered fully adult, but who are, equally, no longer children. The individual adolescent, too, finds the long and unsatisfactory period of transition taxing and alienating, since society gives him little help in establishing his new social identity as an adult. He is left to try to find this identity himself. As a result, the adolescent is likely to find this transitional stage far more difficult to cope with than does his equivalent in a primitive society.

The problem is equally difficult for girls. While in most nonindustrial societies a girl undergoes the menarche at about the age of sixteen, becomes initiated at once, or soon afterward, and is married and raising children soon after that, in an industrialized community

Adolescents have left childhood, but have not yet been fully admitted to the adult world. They need to belong to a group of people which they can understand, to which they can relate and which can provide the status denied them by adult society. The nature and the social structure of the group changes from the preadolescent one-sex gang, to a mixed clique which centers on verbal interaction and group activity and, finally, to a group which consists of couples with specific emotional attachments.

a girl is expected to wait for up to ten years before she can marry and have children. Today, too, in the West girls are able to compete on an equal footing with boys and have as many options open to them. As a result, the question of personal, social and sexual identity is frequently left a long time without a solution for girls in the same way as it is for boys. It is, therefore, not entirely surprising that in industrialized societies adolescents frequently resort to extremes of behavior or embrace revolutionary ideologies as a means of resolving these profound psychological conflicts. This immersion in a different way of life allows the adolescent to free himself from parental regulation, but at the same time imposes upon him new controls mediated by the group with which he is identifying.

For the adolescent in any society, sex is one of the greatest problems confronting him. In most primitive societies, sexual intercourse is practiced by young people as soon as they are physiologically capable of it. Even in those societies in which a girl is expected to be a virgin at marriage, the short time that typically elapses between puberty and marriage makes the intervening period of abstinence not too difficult to bear. In industrial societies, however, patterns of sexual activity among adolescents vary from place to place and between social groups.

A study made of sexual attitudes and behavior of adolescent girls in England, for example, showed that upper middle-class girls were the most sexually experienced social group and regarded the years of education and working, which typically precede marriage in the mid-twenties, as an opportunity for gaining wide sexual experience. Working-class girls, on the other hand, were less sexually experienced, but tended to embark on sexual intercourse earlier and to marry younger. Middle middle-class and lower middle-class girls showed the strongest preference for virginity at marriage and the greatest inhibition about premarital sex. Nevertheless, the percentage who married without previous sexual experience was small.

But no matter how adolescence is organized in a society, the patterns of transition and readjustment to adulthood remain largely the same. Only the time span and the means adopted for this process vary.

A wish to do well in sporting activities is often an expression of the adolescent's need for approval by his peers and by adults. Boys will, for example, train and compete to gain a coveted place in a national youth soccer team. Here the youth teams of West Germany and England are playing in a large public stadium.

Adolescent girls of the Berber tribe in Marrakesh, Morocco, are expected to undertake responsibilities associated with adulthood. Unlike adolescents in many other societies, who are free from any responsibility toward their brothers and sisters, Berber girls take complete charge of younger children.

The intense emotions of adolescence are often released as an obsessive or hysterical hero worship, particularly of such culture heros as sportsmen and pop stars. Screaming at pop concerts, for example, is one way the adolescent can release pent-up emotion.

Cross-reference	pages
Hormones	70-73
Puberty	170-171

The young adult

As the individual emerges from adolescence, his physical condition is at its peak. Muscles are well developed for strength, speed and agility and reaction times are fast. For these reasons people up to the age of twenty-six excel in many competitive sports, such as football, while other sports, such as golf or shooting, which rely less on strength and more on concentration, can be played successfully well into middle age. But even though there is an external appearance of physical perfection in the young adult, the aging process has begun, although it is imperceptible.

Loss of hearing, for example, begins in adolescence, height begins to decrease at the age of twenty-five, muscle strength begins to diminish at thirty and although body proportions remain the same, the percentage of fatty tissue increases. However, the condition of the body rarely concerns the young adult and social, emotional and psychological development provides the focus for his attention.

Although the young adult is faced with basically the same pressures as he was as an adolescent, for the first time in his life he finds himself totally independent, both economically and socially. The loss of dependence leads to an increase in insecurity. The generalized feelings of adolescence, such as searching for identity and new values, give way to a more specific definition of desirable goals within society.

In most societies, adulthood is represented by marriage and by the individual's establishment of a new home and a new family group. In many primitive societies the transition to adulthood, which is signaled by initiation, involves the newly initiated individual being indoctrinated in the values of his society and sometimes in its secret mythology. Even in industrial societies the roles that young adults assume, both within marriage and within the community, depend largely on the way in which training has molded their abilities and their emotions.

During the early years of adulthood the individual's personality and special skills are developed, and his or her work potentials and satisfaction begin to be realized. Training for work may take many forms, such as an apprenticeship in a craft, a vocational training for a career or it may involve academic achievement. But whatever its form the result should ideally involve some personal assessment of the individual's capacities, limitations and an evaluation of future goals.

The place which an adult takes in the community depends on his occupation and on his social life. Both are closely linked and both involve membership of groups. At work; group activity involves cooperation and commitment to a particular task, and may require obeying rules which the young person finds distasteful. He can approach the problem in different ways, either by coming to accept the rules, by changing his occupation or by joining an organized group within the group, such as a trade union, to help to bring about the alterations he desires.

Unlike industrialized societies, in traditional and more primitive societies the early adult life of the individual is more like that of his

Being in love involves sexual attraction, the sharing of private feelings, dreams and ideas, giving support and encouragement and experiencing a togetherness which helps to cushion the couple from the harsh realities of the world. During courtship, the young adult couple test each other, discovering whether they agree in their general outlook on life, whether their values are the same and whether their relationship has a strong enough basis for marriage and, eventually, for parenthood.

parents, and so acceptance of the adult role is easier. In many primitive societies, the transition to adult life is gentle. It is common, for example, for marriage to be regarded as fully consummated only after the birth of the first child. Prior to this, separation or divorce are relatively easy. In an industrial society there is frequently a period of waiting in early adulthood before the individual is able to marry. Although this sometimes happens in less advanced societies, too, it is usually the result of polygamy or because the older, richer and better-established men tend to attract many of the eligible girls.

In industrial societies, however, the period of waiting before marriage is usually the result of the need for individuals to achieve some degree of economic independence before they shoulder the financial burdens of married life. And even after marriage both husband and wife often continue to work for some time before raising a family. This is not only for economic reasons but also because the young wife may want to continue working.

In industrial societies, biological and social considerations are not in harmony, for, physiologically, the best years for child-bearing—the early twenties—are the years when young couples in the West frequently postpone having a family. This is, of course, a phenomenon which could only come about as the result of the widespread use of readily available contraceptives, which typifies such industrial societies.

In more traditional societies, however, this cannot happen because wives do not have separate occupations, apart from that of wife and mother. As a result of the extended family structure, too, which is typical of such societies, the young wife is supported and assisted in her role by her relatives and those of her husband. Many of the problems which are resolved by nursery schools in the West do not exist in traditional societies because the young wife with young children finds herself neither alone nor in an unfamiliar situation. Instead, she is supported by those around her, and is doing that for which she has been prepared by example and by helping out in her own home since early childhood.

As for the man in any society, much has to be achieved in the years between seventeen and thirty. In the West, this is the period during which the individual must further his career or occupation, and in which progress can be rapid if he is successful. The stresses and strains brought to bear on young men in such a situation are often considerable. The individual finds himself simultaneously subjected to demands from his home and his work. At home he must satisfy the emotional needs of his wife, and must carry the responsibility for the material well-being of his family. At work he must do all he can to achieve success, especially if he is in a middle-class profession which demands much of the young but tends to postpone rewards until later. In traditional societies, the young man is still subjected to comparable if somewhat different stresses. No matter what his culture, these years are crucial for laying the foundation for the rest of a man's life.

About one in twelve young people, in technologically advanced societies, go on to higher education before embarking on a career and marriage.

Today young women all over the industrialized world have the option of going to work and postponing marriage.

Young adults may experience a period of turmoil as they define their positions in society. They may reject many of the conventions of society as a protest against becoming immersed in the "establishment."

In most societies, getting married, setting up a home and raising a family marks the transition from adolescence to full adulthood and the young man and woman are then considered to be mature and responsible individuals. With the birth of the first baby the young couple becomes a family. In the West, both parents usually share in the upbringing of children.

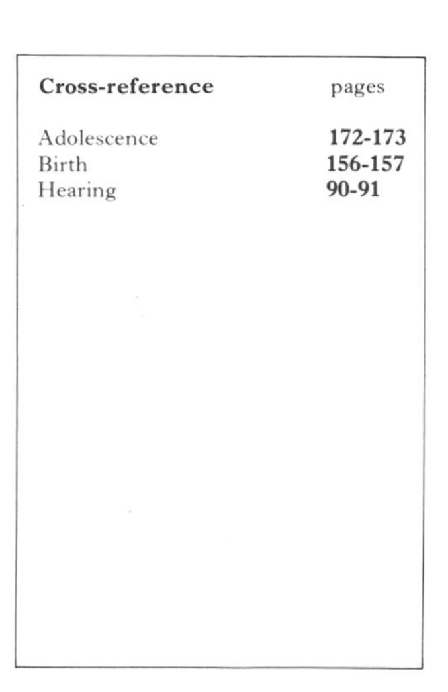

Cross-reference	pages
Adolescence	172-173
Birth	156-157
Hearing	90-91

Early middle years

In almost all societies, the years from thirty to forty-five are among an individual's best, for this is the period during which there is usually a consolidation of a person's public and occupational roles, as well as his family life. In these years an individual usually reaches his peak of physical attractiveness, physical agility, speed and strength. Although the process of aging is going on, for most people its effects are of little consequence. During these years, too, sexual and intellectual development will often reach their highest level of expression.

In an industrialized society, it is the period during which a man has achieved some kind of status at work and is probably beginning to enjoy the financial and social rewards which that brings. In less industrialized societies, it is the period during which a man joins the group of those who have passed through the stages of initiation, marriage and raising a family and have fully established themselves in the community.

During these years, women in advanced societies watch their children grow up and, hopefully, their material and social status is confirmed or enhanced. For women in traditional societies it is the period when they are raising and schooling their young and, therefore, making what is their greatest contribution to social life.

This is a highly productive phase of an individual's life. It is often the period of greatest output, in terms of physical work, emotional commitment and creativity. According to the classic American study of H. C. Lehman, the years from thirty to thirty-five frequently show a peak in intellectual, scientific and artistic achievement. He found, in a study of the life histories of chemists, that it was between the ages of thirty and thirty-four that they tended to make their greatest contributions to their field before falling off thereafter at a steady pace. A study of classic books for children written by women also showed a peak of achievement at this age, although this may be related to the pattern of childbearing among the sample of women authors used. Thomas Edison patented more inventions when he was about thirty-five than at any other period of his life.

Although such studies as Lehman's are open to all kinds of methodological and statistical criticisms, they do appear to confirm the general principle that these years are often the most productive in an individual's life. Even the peak year of athletic achievement, something which many might expect to come in the preceding period, according to Lehman is the thirty-first year.

Nevertheless, these are the years when, physiologically, decline begins. The weight of the brain and the number of nerve cells in it begin to decline during a person's twenties. Conduction velocity along nerve fibers declines steadily from the age of twenty. So do the basal metabolic rate, the strength of the grip, the cardiac output, liver weight and, in men, frequency of sexual intercourse. All these factors, however, operate only very slowly, and their cumulative effect, although considerable by the time a person is sixty or seventy, does not count for a great deal in early middle age. Many of them are, in any case, misleading.

Individuals of forty-five are not, for example, at a great intellectual disadvantage as compared with those of twenty-five. On the contrary, so great is the contribution of experience and memory to human intellectual skills that at forty-five a person's intellect can be expected to be as good, and often better, than it was at twenty-five.

Again, although the male frequency of sexual intercourse declines steadily after twenty, it is not true to conclude that men of thirty-five are sexually in decline. Often, a man in his thirties can sustain sexual intercourse for much longer than a man in his early twenties, and this is often the important factor for the achievement of full sexual satisfaction for both partners.

It is the latter part of this period of life that usually coincides with the growing up of the last children in the family and their departure from the household. The sudden sense of loss, coupled with awareness that her menopause is impending, sometimes makes a married woman depressed and anxious. She may feel that the best years of her life are now over. The future seems as empty as the house. This is particularly true of women who married early, without having sampled independence in young adulthood.

This problem was more likely to arise in years past when, except in the lower social classes, married women were not expected to go out to work. Today many women in industrialized societies find a new interest to alleviate the potential boredom. Some return to a previous occupation; others get training for a new occupation; and some devote themselves to a neglected hobby or go back to college.

The departure of the children may leave a couple alone together for the first time in a great many years, and this privacy can prove a source of relief and pleasure for both partners. It does not, however, always work this way for the wife. Her husband is quite likely to have established interests outside marriage, even if only his career, in which the wife can play little part. Far from promoting a "second honeymoon" situation, the absence of the children can in fact precipitate the wife into greater loneliness, unless the husband realizes the help that he can give. Without mutual understanding this is a time when marital problems may become acute. Now that there are no children to consider, separation or divorce all too often result.

Nevertheless, for the majority of men and women of all societies, the growing up of their children and the greater material comfort and security which early middle age brings make it a welcome and satisfying time of life.

Physical endurance and success

Edmund Hilary was thirty-four and Tenzing Norgay was thirty-nine when they conquered Everest. All in their late thirties, the crew of Apollo 10 traveled over 24,000 miles per hour. The peak physical condition, mature judgment and stable personality found in men in their late thirties is necessary to achieve such success.

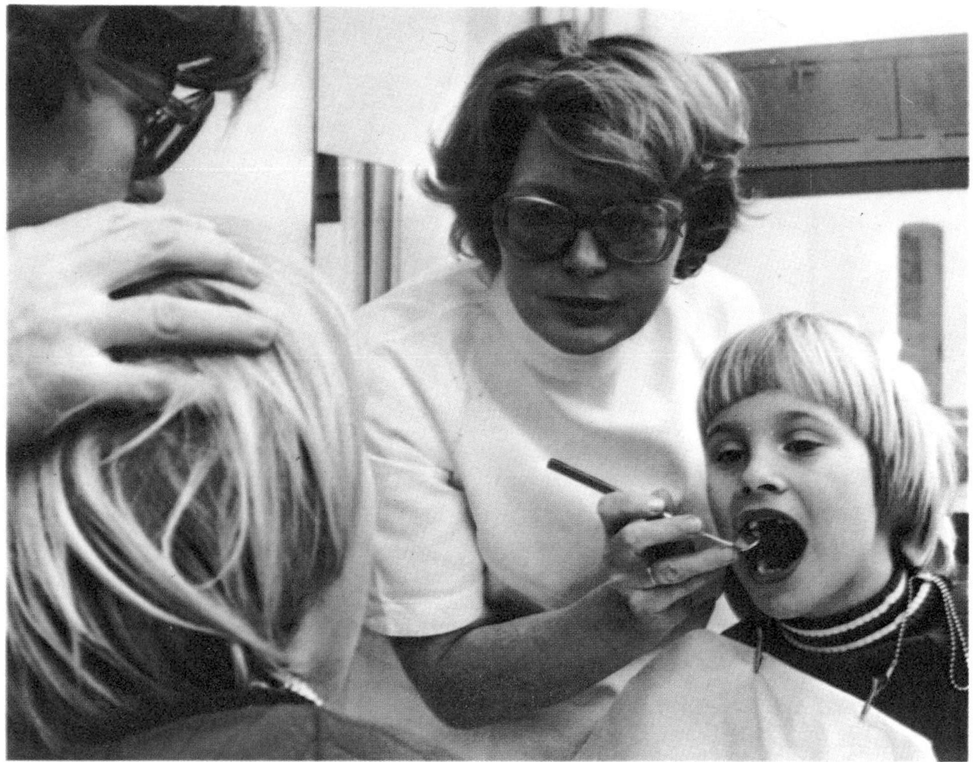

The father and son relationship often becomes stronger and closer as both grow older. The adolescent boy and middle-aged parent share common interests and spend more time together as the father teaches his son to become an independent and responsible adult. Through his son, the father often develops new interests and new viewpoints.

In their middle years, after their children have grown up, many women, like the dental hygienist, left, go back to work. In addition to their skills, they now have the assets of stability, self-confidence and experience.

Young grandparents, after they have reared their own children and had the satisfaction of seeing them develop and mature into responsible adults, get great emotional and psychological satisfaction from their grandchildren.

Cross-reference	pages
Brain and central nervous system	74-77
Intelligence	132-133
Sexual intercourse	148-149

Later middle years

In industrial societies people now live longer and have fewer children. Consequently, the postparental period makes up one-quarter to one-half of a person's married life. The transition to the childless home is an important one. For women, it is usually the time when they are relatively free from household responsibilities and are able to continue pursuing their own interests. Generally, too, they feel more satisfied with their marriage and the renewed companionship in the relationship. Old age is still sufficiently distant for life to be uncomplicated by the effects and most couples enjoy this new period of freedom.

Physiologically, aging, between the ages of forty-five and sixty, is such a slow process that it often seems to pass almost unnoticed to the people concerned, as well as to those around them. This is because we are all accustomed to seeing the effects of aging slowly appearing in later middle age. It is also because persons passing through this phase tend to make continual small adjustments in their behavior, in their style of life and in their attitudes toward themselves. Consequently, in many cases the slow but relentless effects of aging are mitigated. Sudden aging or abrupt changes in health during these years is usually the result of serious illness, bereavement or some other major psychological or physical shock.

Between the ages of forty-five and sixty, however, ill health does become more common, and the death rate for these years doubles, compared with that of the preceding period from thirty to forty-five. The principal causes of death are cancer, heart disease, circulatory disorders, strokes and diseases of the respiratory tract.

For women, the medical marker of these years is the menopause. During the menopause, which typically occurs between the ages of forty-five and fifty-five, ovulation, menstruation and reproductive capacity cease. This is the result of a change in hormonal secretion, principally the reduction of the

The later middle years are often a time of enormous creative energy and a great sense of purpose. Experience and the accumulation of knowledge and wisdom allow people of this age to successfully compete with those who are younger. These well-known people illustrate the success that can be achieved later in life.

The prima ballerina Dame Margot Fonteyn continued dancing into her forties and fifties. She is seen here dancing the principal role of the Cinderella ballet at Covent Garden, London. At the age of forty-six she formed her famous partnership with the young Russian dancer Rudolf Nureyev.

Frank Sinatra, who began his career when he was nineteen, and achieved fame not only as a singer but as an actor and film producer as well, made a successful comeback when he was in his fifties.

Helena Rubinstein reached the peak of her career in her later middle years as head of a world-famous perfume and cosmetics company.

When he was fifty-one, Richard Adams wrote his first book, *Watership Down,* and became a bestselling, widely acclaimed author.

Arnold Palmer, who became a professional golfer when he was twenty-five years old, was still successfully competing as one of the world's top golfers at the age of forty-six.

body's production of the sex hormones estrogen and progesterone. This terminates menstruation, and can cause numerous irritating physical symptoms, including hot flashes, excessive fatigue, emotional irritability, headaches and pelvic pain. The administration of synthetic sex steroids can alleviate these symptoms, which, although varied in length, are in any case transitory.

As important as the physical symptoms of the menopause, however, is the effect of the menopause on the psyche and on sexuality. Two quite opposite responses are common. For some women the menopause signals the effective end of their sex lives, because they regard themselves as being physically unfit for further sexual activity. For most women, however, the menopause produces an increase in sexual drive and pleasure which is in large part the result of change in hormonal balance. In many cases it is also due to the removal of fears of pregnancy which may have reduced a woman's sexual responsiveness during her fertile years.

The reason that some women lose their interest in sex at the menopause may be traced to long-term sexual inhibitions and maladjustments. Some authorities, although this is not generally accepted, have argued that all the symptoms of the menopause are basically neurotic, and proceed from a psychological rather than a physical cause.

For both men and women one major factor in maintaining both sexual appetite and performance into late middle age, and even into old age, is frequency of sexual activity. The decline of sexual responsiveness in some males after forty-five or fifty is related only in part to a slow decrease in sex-hormone secretion. Frequently, preoccupation with work, general physical unfitness and excessive indulgence in food and alcohol are far more to blame.

The late fifties are, however, often a notably rewarding and comfortable period of people's lives, both in industrial societies and elsewhere, and fifty-three to fifty-eight are the peak years as far as earnings are concerned. These are the years in which men have achieved much of that which they are going to achieve and, as many studies show, are getting the reward for it.

Studies of mental abilities show a slight decline in some areas throughout this period. Vocabulary and the knowledge an individual possesses, however, seem to remain constant or even to increase. More basic components of intellectual performance at tasks typically presented in nonverbal intelligence tests, such as forming cubes into patterns, seem in some cases to decline. When considering the effect of aging on intellectual capacities, however, it is necessary to distinguish between two kinds of intelligence and to separate unspecialized, innate, general cognitive ability from specialized, acquired cognitive ability, which is based on knowledge or experience. Considered in this way it seems that it is specialized intellectual ability which remains constant with age and it is only the unspecialized, general cognitive ability which declines.

If mental ability is broken down into its constituent parts, effects of aging can be seen in the way in which performance changes or declines. For persons over fifty, for example, tested on the Thurstone Primary Mental Ability Scale, the largest age-correlated decline was found in what is called "logical reasoning," and the next largest in "spatial reasoning." Other elements, such as verbal meaning, or comprehension, and word fluency, or articulateness, declined less rapidly. This means that people well into middle age are likely to display improved thinking, comprehension and informational skills.

But whatever such laboratory tests may show, it is, nevertheless, likely that the intellectual and psychological capabilities of individuals in later middle age are strongly affected by the use to which those abilities are put, and the amount of practice an individual gets in using them. Certainly, too much should not be made of findings about the psychological and mental abilities of aging people, which are too restricted in their scope to convey much information about the overall picture. The overall picture is, indeed, often far more positive. The older person displays all the advantages of long experience of life, wisdom rather than cleverness, and a wealth of accumulated knowledge which more than compensates for whatever slight loss there is due to neurological and physiological changes.

Many people reach a peak in their earning capacity in the later middle years and have an increasing amount of comfort and leisure time in which to pursue recreational pastimes and hobbies.

The later middle years are a time when the children have grown up and the married couple can savor their newfound freedom. They now have the time to enjoy life together and to plan activities which include both of them. As a result they often reestablish a close relationship and become more dependent upon one another than at any other time in their married life.

Cross-reference	pages
Heart	56-57
Hormones	70-73
Intelligence	132-133
Menstruation	146-147
Sex drive	114-115

Old age

The physiological changes which occur in old age are largely an accelerated continuation of those of the preceding years. The senses, particularly sight and hearing, all decline in acuity. Motor activity becomes impaired so that walking becomes slow. Handwriting becomes spidery and shaky because the harmony between hand, eye and brain has lost some of its precision. Also seen in old age is a marked impairment of memory, which paradoxically manifests itself as an inability to recall recent rather than long-term memories. Loss of weight and a reduction in height, partly caused by the adoption of a less erect posture, are also characteristic physical symptoms of aging. So, too, are a lowered resistance to disease and a reduced efficiency of the circulatory and respiratory systems.

Aging is believed by some researchers to have genetic causes. This theory is supported by the fact that longevity and susceptibility to certain diseases show distinct family patterning. It is also probable that the span of life is related to developmental speed, which would explain why man, who has a prolonged infancy, lives longer than the anthropoid apes, for example, which reach adulthood much more quickly.

More light may be shed on the mystery of aging by the study of cell biochemistry. There is some evidence that in the process of aging, errors gradually become more common in the transcription of genetic information into the proteins which are manufactured in the cell, and that this is one of the major sources of the progressive rundown in the body's renewal program, which is what aging essentially is. There is further strong evidence that failure of the immune system is also involved in aging. In time, the body's antibodies may become altered and unable to distinguish between cells belonging to its own organs and those from outside.

According to an American study of aging and social achievement, power and leadership in industry reaches a maximum for those in the age group of sixty-five to sixty-nine. This is in accord with the situation in non-Western societies. In more traditional cultures, most individuals reach the peak of their power and influence in old age. Usually, large, extended family structures are typical of such societies, and within such a family the old have great moral and political authority.

Indeed, in all societies it is the rule that where traditional or ritual knowledge is concerned it is always the older members of the community who are the authorities. And since traditional knowledge is the only kind of knowledge which many societies have, they are frequently ruled by a committee of elders, recruitment to which is principally on the basis of seniority. It is not, therefore, surprising to find the old accorded a respect and deference by the young which may seem exaggerated to Western eyes.

In the West, and more and more in all industrialized societies, the enviable position of prestigious authority which the old occupy in more traditional societies has been seriously undermined. The principal reason for this is that in such modern societies traditional

The achievements of these well-known people who, like many old people, successfully continued working when they were well beyond the accepted retirement age, serve to illustrate that old age is not a time when people physically and mentally deteriorate and can no longer make a worthwhile contribution to society.

Pablo Picasso, the Spanish artist, continued to explore new art forms and ideas when he was in his eighties. Between the ages of eighty-five and ninety he produced more than four hundred drawings and engravings.

At eighty-seven the English actress Dame Edith Evans appeared in a starring role on the London stage.

At sixty-nine Sir Francis Chichester made the remarkable four-thousand-mile journey from Portuguese Guinea, Africa, to Nicaragua, Central America, sailing his ketch *Gipsy Moth V*, singlehanded.

Dame Agatha Christie, the well-known English writer and playwright, wrote her seventy-seventh crime story at the age of eighty-five. Born in 1890, she spent much of her life writing her involved mystery stories with their famous detective characters. With her amazing capacity to produce new mystery plots, as well as poetry, plays and novels, she is one of the world's most widely read authors.

Although George Bernard Shaw, the famous British playwright, produced most of his plays in the early twentieth century, he was still writing with the same lucidity, humor, morality and satirical wit on his ninetieth birthday in 1945.

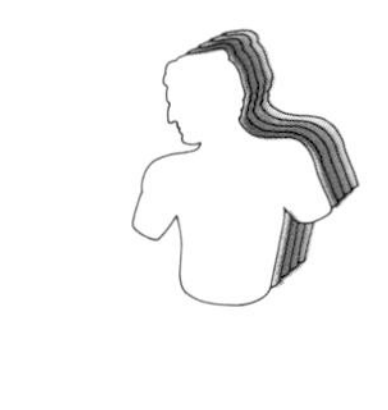

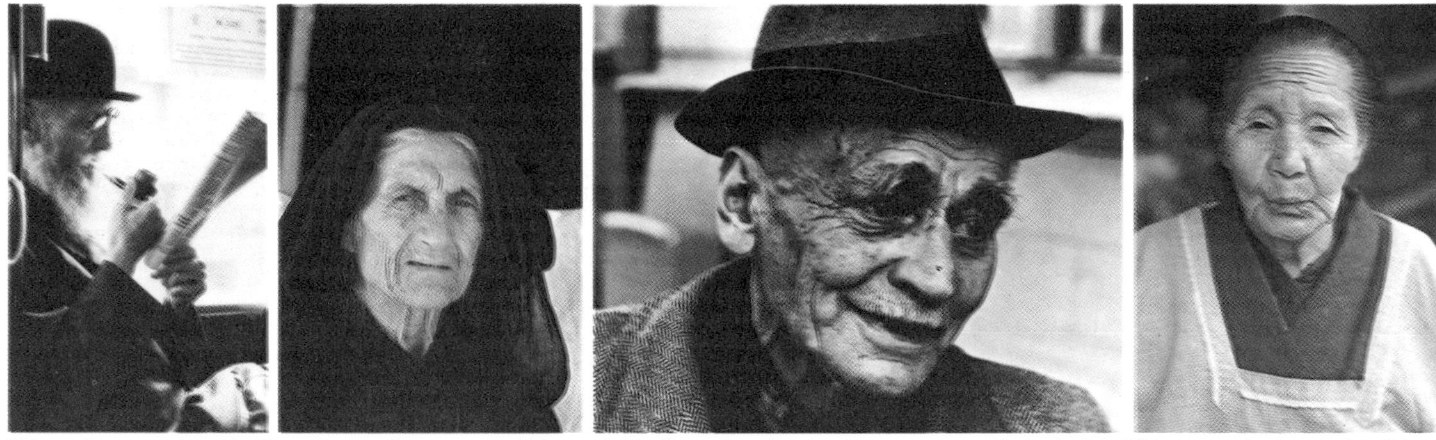

Old people are regarded with different degrees of respect and esteem in different societies. The venerability of old people in many traditional societies, as well as in societies in which extended families are the rule, is based on a respect for the knowledge they have accumulated during their lifetime. In Western industrial societies, however, where the old, despite their abilities, are permitted to contribute little to the economy, they are often considered unproductive and a burden to society.

knowledge and the wisdom of age are not of great significance in economic and scientific matters. These are societies where "new" is often equated with "better"—a proposition exactly contrary to traditional wisdom, but held in industrial societies because of the rapid progress of technology, which tends to make traditional knowledge seem obsolete.

A direct effect of this, and one also related to the work ethic of the West, is the phenomenon of retirement. In traditional societies an individual does not "retire." He or she may do a little less work in the fields because of advancing age or may not participate so widely in social events for the same reason. But the individual does not retire in the sense of giving up his occupational role in society, because he has no such clearly distinct role. Furthermore, because the aging person in the more traditional society is living in the same house as relatives, or at least close to them, he or she does not experience any abrupt change in social status comparable to that which afflicts the retired person in industrial societies.

Studies of retirement in the West show that the worst period of despondency and discomfort about retirement tends to occur just before it happens. Once the transition has been managed, it often turns out that retirement is not so bad as the person may have feared, and, such studies suggest, if he was looking forward to it he will find it even easier to adjust to than he may have imagined.

Another of the major psychological difficulties to which the older person is particularly prone in any society is that associated with the death of his friends and near relatives, and with the periods of bereavement and mourning which result. Again, traditional societies seem to cope so much better with this problem by having elaborate rituals through which the associated emotions can be expressed.

Psychological studies of mourning show that a major part of the process consists of turning aggressive and resentful feelings, which are unconsciously directed against the person who has died, back against the subject's own self. This accounts for the depression of the mourner and the exaggerated insistence on the excellence of the person who has died. This is a defense against the reproachful feelings that the mourner might have felt for the person who has died, but which he must now deny for fear of feeling guilt and implication in the death.

But the greatest and most intractable problem of old age is death itself and the response which the old person manages to make to it. Perhaps the best attitude to death—the one event which is a certainty of life—is reflected in the remark of Leonardo da Vinci, who said that death should come to a man after a full life as sleep comes after a hard day's work.

The face of old age bears its characteristic lines and wrinkles as the skin loses some of its elasticity, and the body stoops as the muscles and bones weaken, but the person inside the body is as young as ever.

Cross-reference	pages
Cell functioning	**24-25**
Circulation	**54-55**
Hearing	**90-91**
Heredity	**158-159**
Memory	**130-131**
Sight	**82-85**

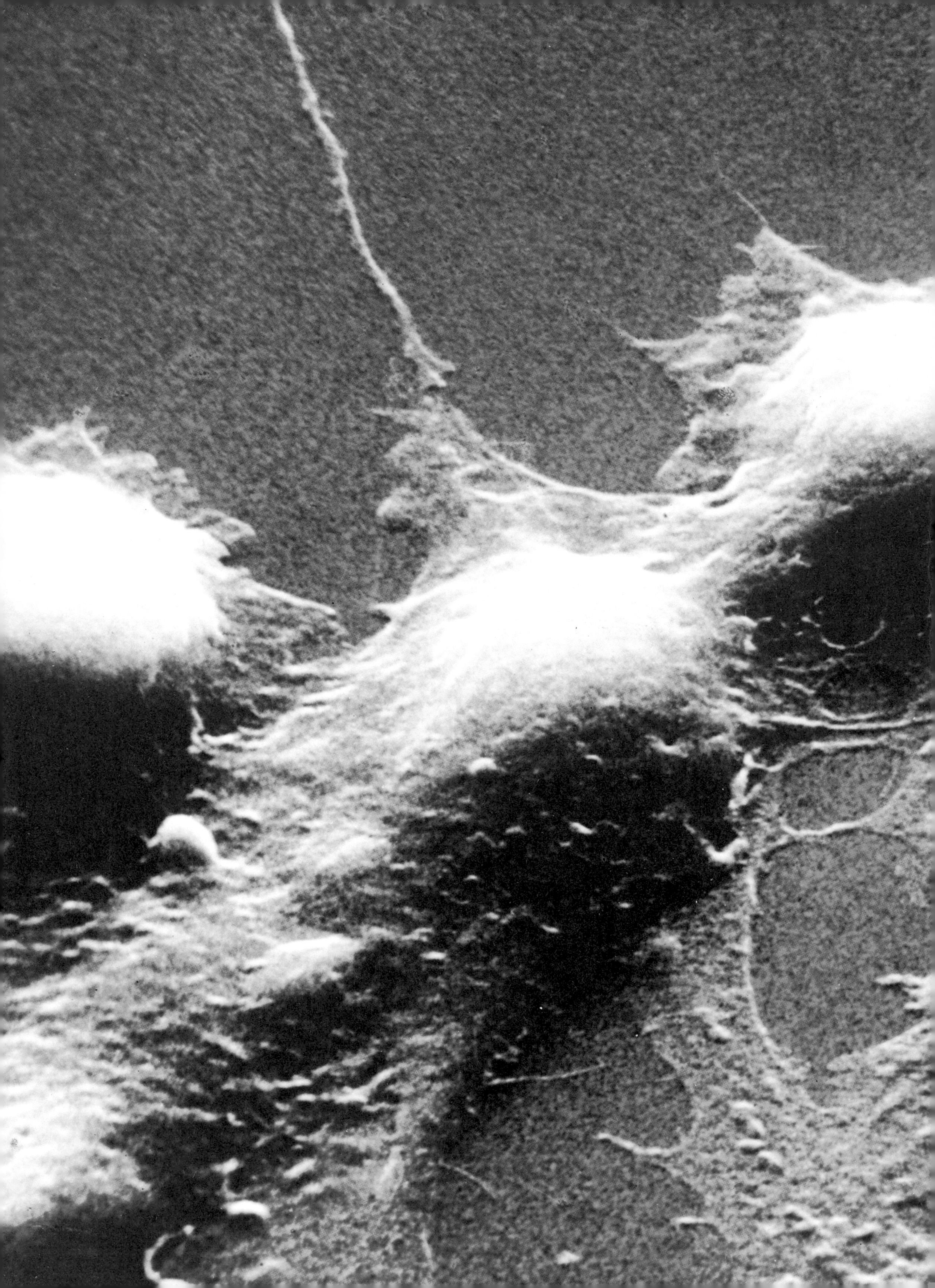

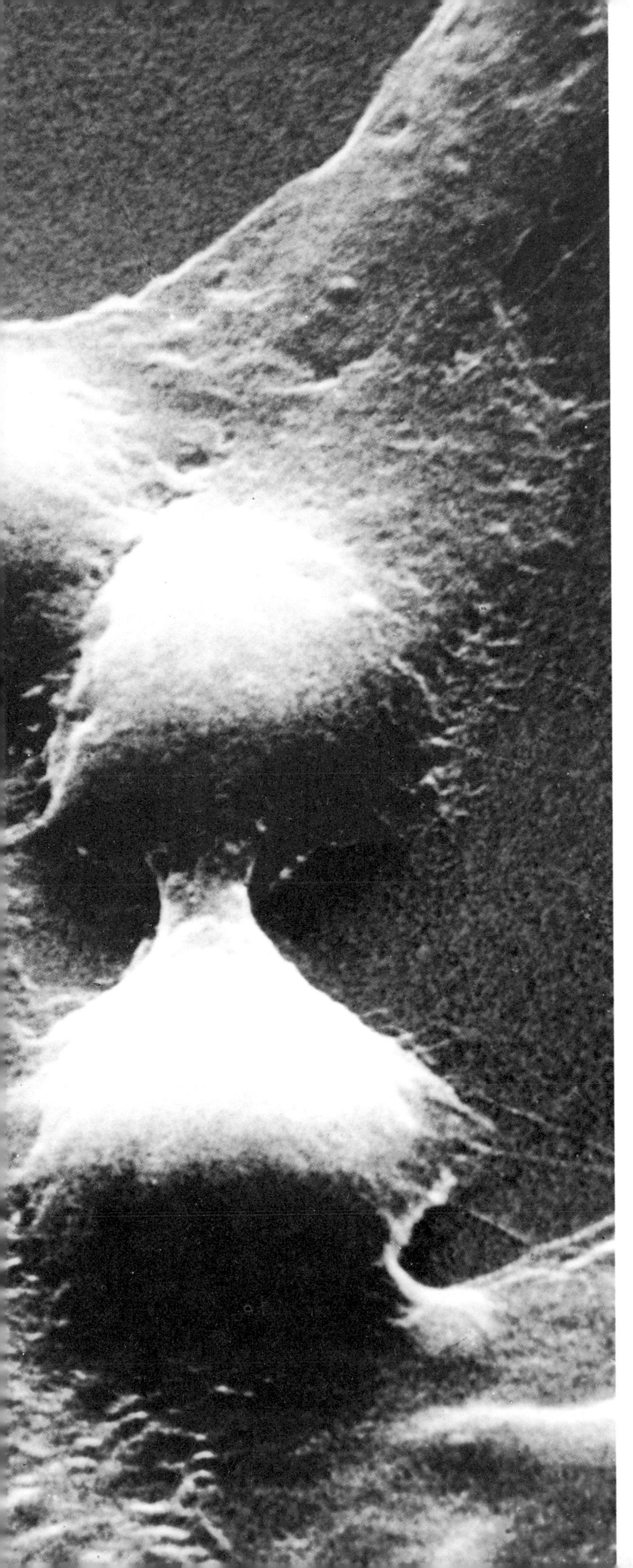

The future

Foretelling the future has always been a hazardous occupation. One can readily sympathize with the gods of ancient Greece, who, when consulted through oracles, would protect their divine status by making predictions that were always totally right because they were also totally ambiguous. And if even gods found it difficult to foresee the outcome of a simple battle or a love affair, how much harder it is for mortals to see the future of the whole human race. Speculation, however, is a fascinating pastime.

We regard ourselves as being at the top of the evolutionary tree. We dominate, manipulate and all too often despoil the world in which we live. Every branch of science sees a "breakthrough" in one direction or another as the weeks go by. A new insecticide is invented, crop production is increased and then we wonder why some species of birds have started to disappear. A new lifesaving drug appears, and a new contraceptive—and we are faced with the problems, in a few years, of an aging population. We are learning how to undertake genetic surgery—and are suddenly confronted with the uncomfortable prospect of producing man-animal hybrids, and clones of identical human beings.

It would seem that in many respects the grip that we have on our destinies is looser than we would like to think. It can only repay us if we occasionally take a critical look at man's achievements so far, and try to predict the future dangers along the pathways we are treading. We should remember, too, that what was yesterday's science fiction has uncannily often become today's matter of fact.

Man's future ultimately depends on the characteristics and properties of the cells of his own body. Seen left, magnified more than five thousand times, is the division of one cell.

Spare-part man

With our ever-widening knowledge of the intricate anatomical structure and the physical and chemical workings of the body, we have become accustomed to thinking of ourselves as exceptionally complex pieces of machinery. We are quite able to think of the food that we eat as being fuel to keep the machinery going, and to regard many of the processes of aging in the body as being the "wearing out" of its components. Indeed, to a truly remarkable extent, our bodies maintain themselves without outside help, for we have our own built-in repair and servicing systems.

However, everything cannot be left to nature. Through accident or disease, bodily components fail or are irreparably damaged. As good as our defenses may be, they are sometimes inadequate to cope with the onslaughts of bacteria and viruses. Over a period of years tuberculosis can destroy a lung. In a matter of days, poliomyelitis can devastate the nervous system leaving limbs useless. In a few hours, poisons can put the kidneys out of action forever. And remarkable as the body's powers of regeneration may be, they cannot supply a new hand to replace one lost after a second's contact with a power saw.

Modern medicine is gaining tremendous powers as new ways are being discovered, not only of postponing death, but also of fighting disability. If a component of the machine fails, then it is often possible to make an attempt to replace it. In some cases whole organs can be transplanted, or substitute parts, or prostheses, can be fashioned. With the present armory of drugs, surgical instruments and skills, and the aid of special machines, such as the "kidney machine," precious human life can be preserved and precious human abilities can be restored in ways that seemed impossible at the beginning of the twentieth century.

Human "spare parts" are of course nothing new. Probably the first prostheses were artificial legs. There is a representation of a person with an artificial leg on a Greek vase that is dated from the second century B.C. Quite clearly shown is a prosthesis of the type that today we would call a pylon prosthesis, from the right amputated knee joint to the ground, with a supporting flange extending up the inside of the thigh, and a strap reaching over the opposite shoulder. The problems of stability in an artificial leg of this kind are the same today.

In the sixteenth century, the French surgeon Ambroise Paré was able to design, and watchmakers and armorers were able to produce, quite serviceable artificial hands of metal. The steel fingers could even be made to open and close, with joints manipulated by buttons, springs and ratchets. Today artificial arms are made of metal and plastic, and they are activated by electricity and miniature motors. The electrical activity in the muscles of the amputation stump can be employed to direct the movement of the parts in some cases. Special plug-in extensions can be exchanged, as necessary, for a variety of functions—forks, screwdrivers, scrub brushes. Although it is possible to design quite natural-looking hands with a certain amount of function, it is generally better to sacrifice appearance for practicality.

False teeth and false eyes are not new ideas either. For teeth, many substances have been employed in the past, including bone, ivory, metal and even wood. For eyes, glass has naturally been the first choice. With these, as with all other prostheses, the main advances in the last few years have come about with the development of new materials, new plastics and new metal alloys.

The main problem with artificial body parts arises when the prostheses have to be inserted within the body's tissues. The body tends to react automatically to foreign material thrust within it, and to reject it. The rejection is very strong when the material contains foreign protein, and the body mobilizes all of its defenses; this is the reason that successful transplantation of body parts is so difficult. But even substances that might be thought of as being harmless to the tissues turn out in many cases to cause severe local reactions and irritation.

At the beginning of the twentieth century, stainless steel was widely employed when new parts were needed within the body. Plates and struts and wires of the metal can be used to join difficult fractures, and shaped pieces are used to patch up holes in the skull. But even stainless steel, like gold and silver and ivory, can cause irritation, and today special alloys, usually containing cobalt and titanium, are used because they are so chemically inert.

There are also suitable high-density polyethylene plastics that give rise to few problems. Probably the greatest use of such materials today is in the construction of whole joints to

Spare parts for man

Future man will benefit from increased surgical and technological skills. In many cases it will be possible to replace malfunctioning or damaged parts of the body with either newly designed artificial equipment or organic transplants. Shown here are some techniques that are already being used or might be possible in the future.

Restoring sight to the blind may be possible by implanting manufactured retinas which absorb light energy and relay the impulses to the brain. Mechanisms that produce images of objects as a series of dots are now being investigated.

The tympanic membrane and ossicles are being replaced with artificial components. Other methods of restoring hearing include amplifying sound waves that then vibrate facial bones and transmit these vibrations to the auditory nerves.

Food can be condensed to the minimum volume possible and still contain all the essential nutrients and vitamins necessary to keep our bodies healthy. Water intake cannot be reduced very much as it is necessary for many metabolic functions.

Surviving on a diet of pills and capsules could be very dull, but it is possible to provide taste and smell substitutes to stimulate both taste buds and olfactory organs.

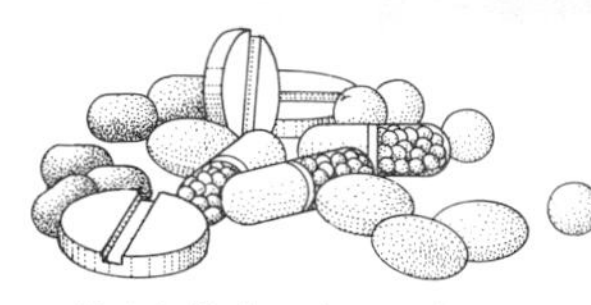

Metabolic function may be modified in the future by using capsules and pills which release a necessary metabolite, such as a hormone or coenzyme, over a set length of time or under particular conditions. The lack of feedback control is, however, a major disadvantage.

replace those damaged by arthritis. It is possible, for example, to construct hinge joints to replace elbows and knees, and metal ball-and-socket joints to replace those at the hip and shoulder. In the case of the hinge joints there is no need to imitate too closely the complexities of the real thing, because adequate movements can be attained more simply. In the case of the hip and shoulder, the construction could not be much simpler anyway.

The main difficulty with artificial joints is getting them to join firmly to the other bones, which may already be weak. Special cements to bind plastic to bone are being developed that will permit the living bone to grow into the new material and add strength.

Far in the future, scientists may discover how one cell taken from a particular tissue, such as the liver, can be cultured and grown outside the body until a new organ is produced. This organ could then be implanted into the original donor with little danger of rejection from the immune system.

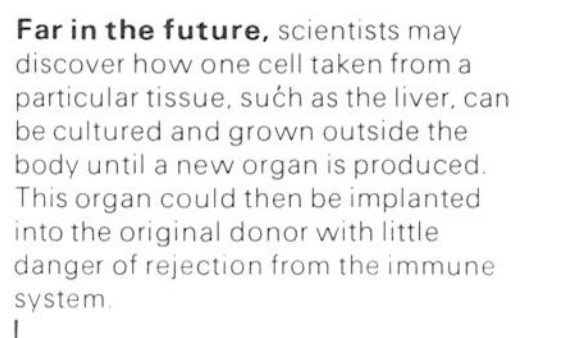

Heart pacemakers are used to mediate a regular heartbeat for damaged hearts. In future, it might be possible to attach auxiliary pumps to the aorta to aid failing hearts.

Damaged alveoli of lungs might possibly be replaced by new materials which would allow gaseous exchange.

Exercise while asleep could save the time and effort involved in keeping our muscles working. As our lives become more sedentary, it might be possible to have special attachments fixed to our limbs that will flex the muscles while we sleep.

Cross-reference	pages
Defenses	**138-141**
Diet	**52-53**
Joints	**34-35**

Maintenance and repair

To replace damaged blood vessels, knitted or woven tubes of synthetic fibers are used. If necessary, a great part of even the aorta can be replaced. Sections of the femoral artery, behind the thigh, or the popliteal artery, lower down, are sometimes replaced when there is danger of gangrene developing in a limb as a result of arteriosclerosis.

Replacement of damaged valves within the heart with valves of human design is prolonging the lives of many thousands of people. The most common replacement is a device of metal, like a cage with three bars, inside which a plastic or silicone rubber ball moves with the blood flow. This is used as a substitute for the three-cusped valve at the base of the aorta.

When the heart is not functioning properly because of the condition called heart block, in which normal electrical impulses are not reaching the heart muscle, an electronic pacemaker, with wires inserted into the muscle, can take over. As time goes on, such pacemakers are gradually being reduced in size. The main part, containing the batteries, which need occasional renewal, is implanted in the abdomen for fairly easy access.

The day when a completely artificial heart can be implanted may not be too far off; experiments with various models in animals have shown great promise. There is also the possibility of using, someday in the future, atomic power to run such devices.

The artificial kidney has certainly not reached the implant stage yet. This is one field in which miniaturization is desperately needed. At the moment the patient's blood has to be taken from an artery and passed through a piece of equipment, about the size of a large television set, for filtration of impuri-

Bones and joints made of inert plastic and metal alloys are now used to replace bones which are shattered or diseased. Substitute bones will only act as support and will not have a marrow which produces blood cells.

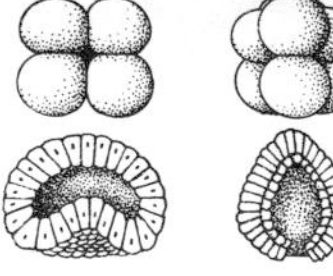

A sperm from the male and an ovum from the female could possibly be fused and the growth of the embryo continued outside the womb until maturity. In this way women would be able to produce children without the need to carry them for nine months.

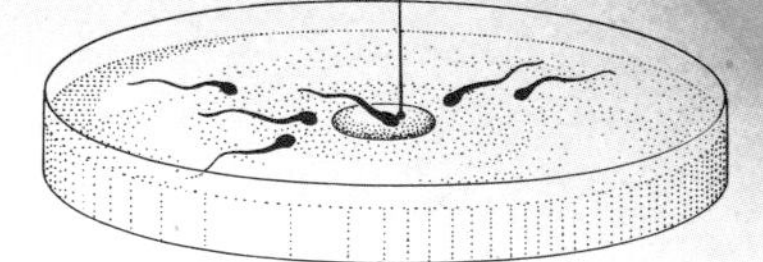

Fertilization and reimplantation are already being undertaken, and in the future, when the ability to separate X and Y sperm becomes more accurate, it could be possible to select the sex of the unborn child.

ties, before it is returned into a vein. It is fortunate that treatment with such a dialysis machine can be carried out at intervals, and does not have to be undertaken continuously. It is not too farfetched to hope that one day, however, an artificial kidney may be designed that can be carried about within the body.

Experiments with equipment to take over the functions of the liver are still at a very early stage. The problem here is that there are so many functions, of such a wide range, that have to be reproduced. It is rather like trying to fit into a small handbag all of the equipment of a laboratory staffed by a thousand or so busy scientists each doing something different, and fitting in the scientists as well. Nevertheless, no such problem should be dismissed as being completely insoluble. There may be very useful compromises.

In cases in which designing and putting into use working artificial body parts is particularly difficult, the other avenue open is to use transplants of living tissues or whole organs. Although it would seem that the use of spare parts that have been especially designed to perform functions efficiently by evolution over tens of millions of years must be better than using cumbersome and inefficient prostheses, transplantation does raise its own problems. The main problem, from the technical point of view, is that of rejection, the stumbling block that leads to the failure of so many transplant operations. Another major problem is the ethical question. If one remembers that even today there are people who object to such commonplace techniques as blood transfusion, then the very much greater problems of transplanting a whole heart without offending somebody's sensibilities are clear. There is also the delicate problem of deciding the moment of death of the donor when it comes to such vital organs as the heart or the liver.

There are good prospects of conquering the problem of tissue rejection with the use of immunosuppressive drugs and antilymphocyte serum. Kidney transplants are now relatively common, as the result of the development of these substances. But kidneys are simpler than hearts inasmuch as we each have two kidneys, and are able to survive on one alone. A kidney donor does not, therefore, have to be dead for the operation to be performed.

Advances have been made, too, in the typing of tissues, in much the same way that blood can be divided into compatible and incompatible groups. With this knowledge, the problem of tissue rejection can be reduced at the outset, by not attempting transplantation when the incompatibility is great between a donor's and a recipient's tissues.

The keeping alive of organs taken from dead persons, in organ "banks," will greatly improve the outlook for live transplantation, and we can perhaps look forward to the days when even such organs as the liver, lungs and heart, as well as minor organs and parts, can be kept in cold storage for future use.

Today kidneys are the most successful transplants, but there are no reasons why eventually other organs should not be transplanted with as much success. Hearts, with modern surgical techniques, are not really very much more difficult to transplant, and there may be advantages in transplanting hearts combined with lungs, although the thought seems more daunting. As transplants, lungs alone have not yet proved very successful. The liver poses a formidable problem, too, although attempts have been made. The way ahead is probably through the use of animal livers, kept alive within the animal, into which patients can be "plugged" in much the same way as a patient with kidney damage can be plugged into a dialysis machine. So far no proper liver transplant has achieved any more than temporary success.

If there is one transplant that really is impossible, despite the imaginings of science-fiction writers, it is that of a brain. For one thing, there could be no donor, dead or alive. The brain of a dead person is by definition dead beyond reversal, and there are no living donors on the market. Technically, the feat is impossible because of the task that would be involved of joining together the ten million nerves that run from the brain to the spinal cord, and the many millions more that link the brain to the sense organs of the head.

The use of transplants and man-made substitute organs can certainly prolong life usefully, but, perhaps mercifully, there are no prospects of achieving everlasting life by the continuous replacement of each body part.

Miniature metabolic machines for implanting into the body will possibly be produced as our technological skill increases.

The loss of the use of a limb, due to the damage of the nerve supply to a muscle, may be eliminated in the distant future by linking up sophisticated electronic circuits to the undamaged nerves of the area. These would be able to receive and transmit the nerve impulses from the brain and restore normal functioning to the limb.

Metabolic machines, designed to exercise minute-by-minute control over the availability of essential metabolites, have been linked to the venous system.

Cross-reference	pages
Heart	56-57
Kidneys	66-67
Liver	50-51
Tissue rejection	140-141

Continuing evolution

For many thousands of years, the human body and the human mind have been evolving in partnership. We have been changing both in physical appearance and ability, and in mental prowess, adapting to our environment as we find it, and learning also to bring about changes in that environment to suit ourselves. But having come thus far, what can we foretell for ourselves in the future?

It would seem unreasonable to believe that evolution of either body or mind should stop suddenly in our own particular era. Certainly, our growing control of the environment means that the pressures on us as living creatures to evolve physically in especially suitable ways are less. The "survival of the fittest" concept takes on a different significance when year by year we discover new means of ensuring that everybody has a reasonable chance of adequate "fitness."

But as each new breakthrough is made in science and in medicine, influencing through technology our standards of living and our human and social values, we are simply introducing different kinds of pressure, not removing them. We will certainly continue to evolve. The problem is that we now have a hand in shaping this evolution, and we must be sure that we are achieving the best results.

The science-fiction writer is all too often prone to assume that the physical characteristics which the human race has developed over perhaps the last two million years, and which distinguish humankind from the other primates, will be progressively exaggerated as time goes by. Because man's brain has steadily enlarged, it must therefore continue to enlarge. Because we have lost most of the hairiness of the apes, so in the future our bodies will be completely hairless.

Conceding that technology could possibly influence evolution, on the other hand, such a writer might also foresee that because we delegate transport to trains and cars, and walk far less than our even quite recent ancestors, our legs will wither beneath us. And because we now use machinery to undertake what would otherwise be manual tasks, our hands will dwindle into things suitable only for pressing buttons.

Such extrapolation leads to a bizarre future man, huge brained, totally bald and with a matchstick body; a nightmare being with vast mental powers, but little physical ability. It is extremely unlikely, however, that such is the fate of man in the centuries to come. With more thought, it would be realized that a huge brain needs an appropriately proportioned circulatory system and respiratory system to keep it going, and a supply of nutrients from an adequate digestive system. A brain to be powerful needs a healthy body.

Again, our limbs are not merely appendages, but organs that by their movement aid the circulation of the blood and also provide a means of getting rid of surplus energy, burning up materials that might otherwise lead to obesity and the attendant risks of arterial and heart disease and high blood pressure.

Trying to set aside insufficiently informed, if imaginative, guesswork, we should examine how much, so far, our activities have managed to upset the natural processes of evolution that shaped our present human characteristics. From there we could begin to speculate on the directions and pathways open to us for control of our own physical destinies.

When, in the middle of the nineteenth century, Charles Darwin first explained convincingly the mechanism by which animal populations diverge, develop and advance—the process of natural selection—he knew nothing of the units of inheritance called genes. It was not until the end of the century,

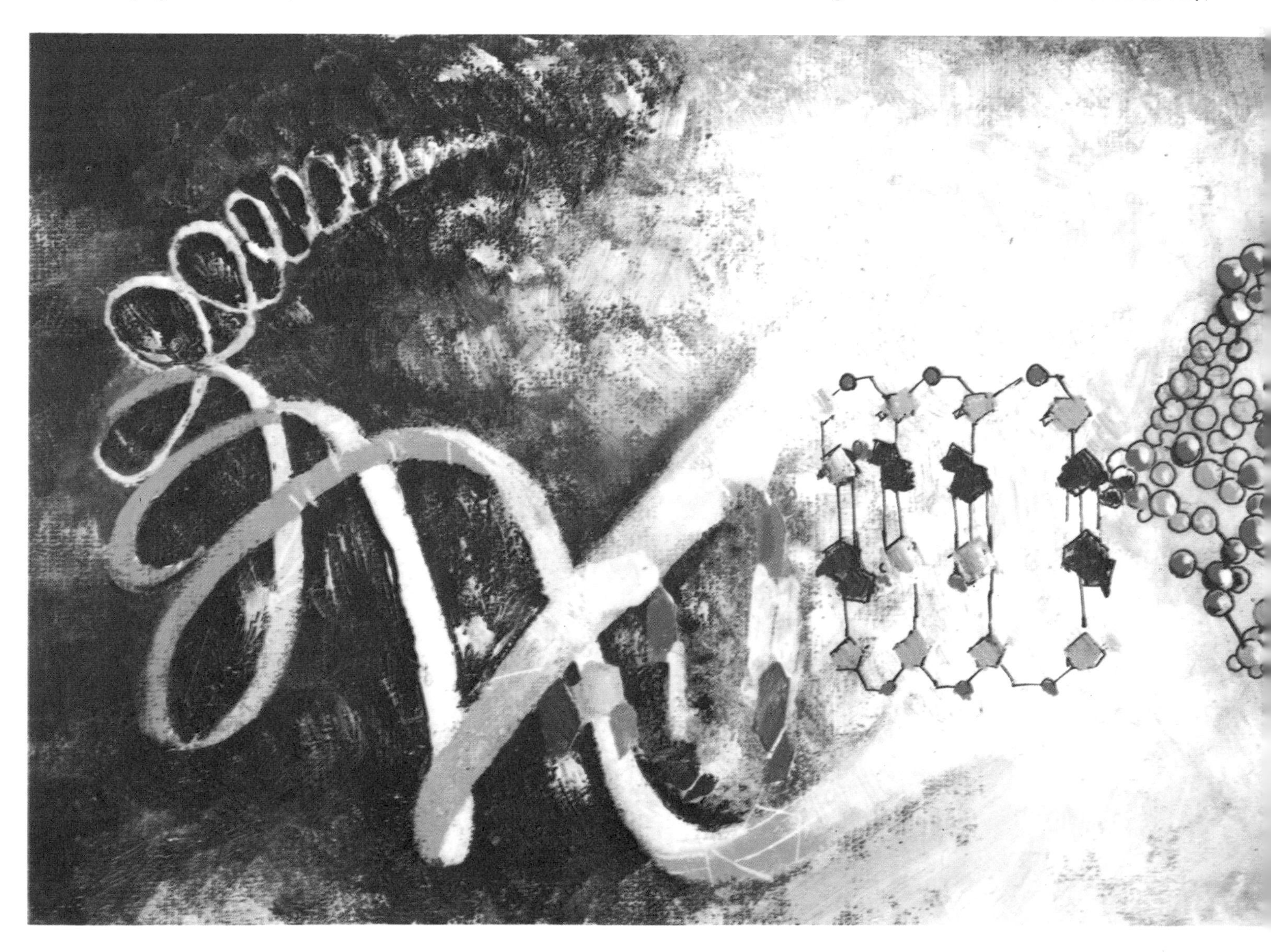

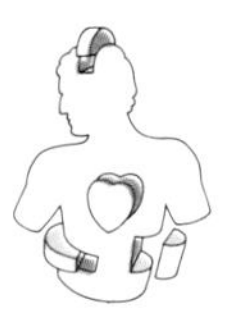

when the pioneer work of Gregor Mendel was brought to light again and properly examined, that the fundamentals of genetics became even partially understood.

Had Darwin been in possession of Mendel's experimental results, it is probable that his famous work *On the Origin of Species* would have been twice as thick a volume, and would have appeared twice as convincing to his more-than-skeptical readership. Had he known about genes he would have seen that here was a ready-made basis for natural selection—the elimination of less useful genes from a population, allowing more beneficial genes to survive. Natural selection could be seen as a gradual change in the overall gene content of the population.

The human population contains a mixture of both good and bad genes, giving rise to a mixture of people that are less well adapted for survival than normal and people who are well adapted, and perhaps distinctly advantageously adapted. Because of our medical and social advances, the pressures that would normally lead to the elimination of inferior individuals have been annulled.

The elimination of whole slices of humanity does still, of course, occur. Famines and floods and droughts kill vast numbers of people. But this is a random affair, and nothing to do with genetic makeup or survival of the fittest. The total pool of genes available to the human race is not significantly changed in this way. Again, the artificial pressures of civilization are certainly leading to the disappearance of certain vulnerable races, such as the Australian Aborigines, the Indians of South America and the Bushmen of South Africa, but it is not evolution at work, despite the fact that their particular genes are being lost from the total human pool. It seems extremely likely that racial differences will gradually disappear in the future, as rarer races are eliminated, and those left will mix and mate with one another.

Certain new characteristics are inevitably going to be added to the total human genetic pool because of mutations, spontaneous alterations in genes from a variety of different chemical and physical causes. But this in no way implies that we shall develop further along the lines that we have followed so far, with exaggeration of our characteristics to the point of caricature.

The pursuit of eugenics, or breeding people for certain characteristics that seem advantageous, raises very serious philosophical and moral problems. Who is to say which characteristics are desirable and which undesirable? It is rather arrogant even to state that exceptionally high intelligence makes a person any "better" than his normally intelligent fellows, and it is much like stating that people with blue eyes and blond hair are superior to all others.

Nevertheless, most people agree that there are certain genes that the human race could well do without, and that if there are ways of eliminating harmful genes, such as those responsible for incapacitating metabolic disorders in their recipients, then they should be closely studied. Advancing medical knowledge is making it possible to keep alive people who would, without modern therapeutic regimes, have gone to a very early grave. But such treatments as are available to combat the symptoms of, for example, cystic fibrosis do not tackle the basic cause. A child born with cystic fibrosis would, a short time ago, have lived for a few years at the most. Now, with extremely expensive treatment, he can be helped to survive past adolescence when he will be capable of reproducing—and so can live to pass on the genetic mistake to his offspring.

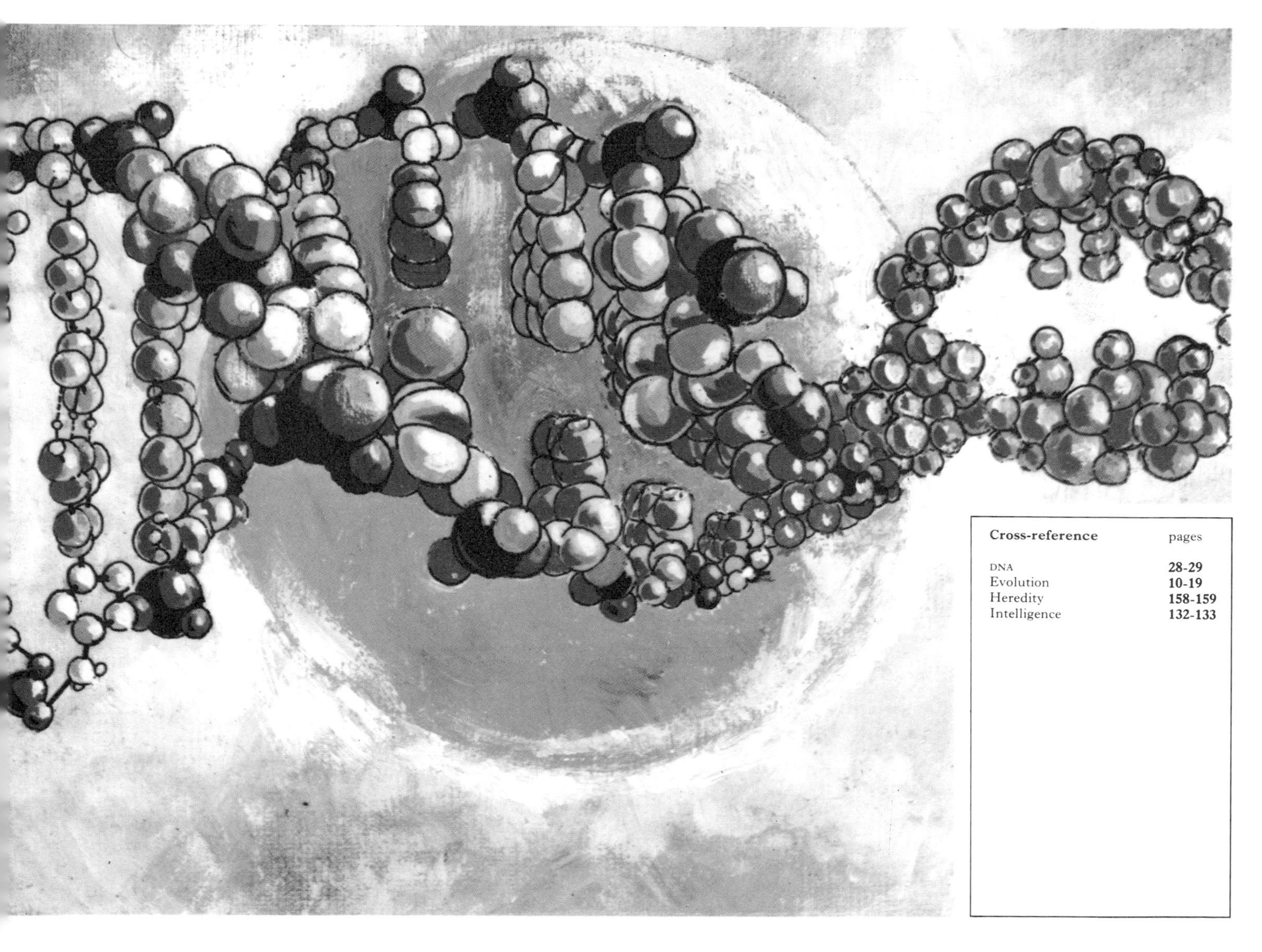

Cross-reference pages

DNA	**28-29**
Evolution	**10-19**
Heredity	**158-159**
Intelligence	**132-133**

Genetic engineering

Until we reach that stage where we can take ourselves to genetic "engineers" and have them put right errors in genes that we have inherited, we can only assess the actual disability of genetic defects, and calculate the risks of them being passed on in given circumstances. A genetic "counselor" may be able to save much heartbreak in parents by advising them of the risks of passing on bad genes to their children. This is, of course, a form of eugenics, but it is "negative" eugenics, in that the aim is to breed out faults rather than breed in doubtfully advantageous qualities. Again there are moral questions likely to arise; should parents who refuse to be advised, or are incapable of understanding what it is all about, be forcibly restrained from bearing children—perhaps sterilized? And should the abortion of high-risk fetuses be undertaken if there is just a chance that the child might be normal?

The number of defects that are brought about by dominant genes, and that will almost certainly be passed on, is far outnumbered by defects due to recessive genes, which will normally only produce an effect when both parents are affected or are carriers. In the case of both parents being carriers, the likelihood of bearing an obviously affected child is only one in four, so that even if every obviously affected child were to be eliminated, or sterilized, then only a fraction of the problem has been solved. The contribution toward getting rid of the unwanted gene from the population as a whole has been negligible.

In the case of the rarer dominant genetic disorders, the risk of passing on the defect is quite plain, and at first it might seem that some progress might be made if all people with such disorders were to be sterilized. In fact this is not so. It is the dominant genes for disorder that are mainly the lethal ones—that is they tend to cause death in the bearer probably at or soon after birth. How then could such a disorder be passed on to a future generation anyway? The answer is that these disorders are continually arising by mutations in generation after generation. There is little that can be done to actively prevent mutations from occurring.

Another unfortunate facet of dominant genetic defects is that sometimes the defects that are not immediately or quickly lethal may not manifest themselves until middle age—when they are likely to have already been passed on to the parent's offspring quite unknowingly. Trying to eliminate unwanted genes, therefore, seems a thankless task. It has been calculated that each person probably carries genes for some half a dozen different disorders of a major or minor degree, anyway. This means that to make a thorough job of eliminating bad genes we would have to destroy or sterilize the whole human race.

The possibility remains that in the future scientists will be able to undertake some genetic engineering. Certain, and some people would say certainly risky, steps in the direction of manipulating chromosomes and the genes they bear has already been made. In simple creatures it has been possible to alter the structure of chromosomes by ingenious methods, and so to alter the qualities of the new cells resulting from cell division.

This has not been done by unraveling the DNA threads of which the chromosomes are made, and snipping out lengths and adding others with ultramicroscopic surgical instruments—such procedures seem way beyond man's present capabilities—but by employing living surgical tools, bacteria and virus particles, which have the power to transfer genes from one cell to another. In simple animal, plant and bacterial cells it has proved quite

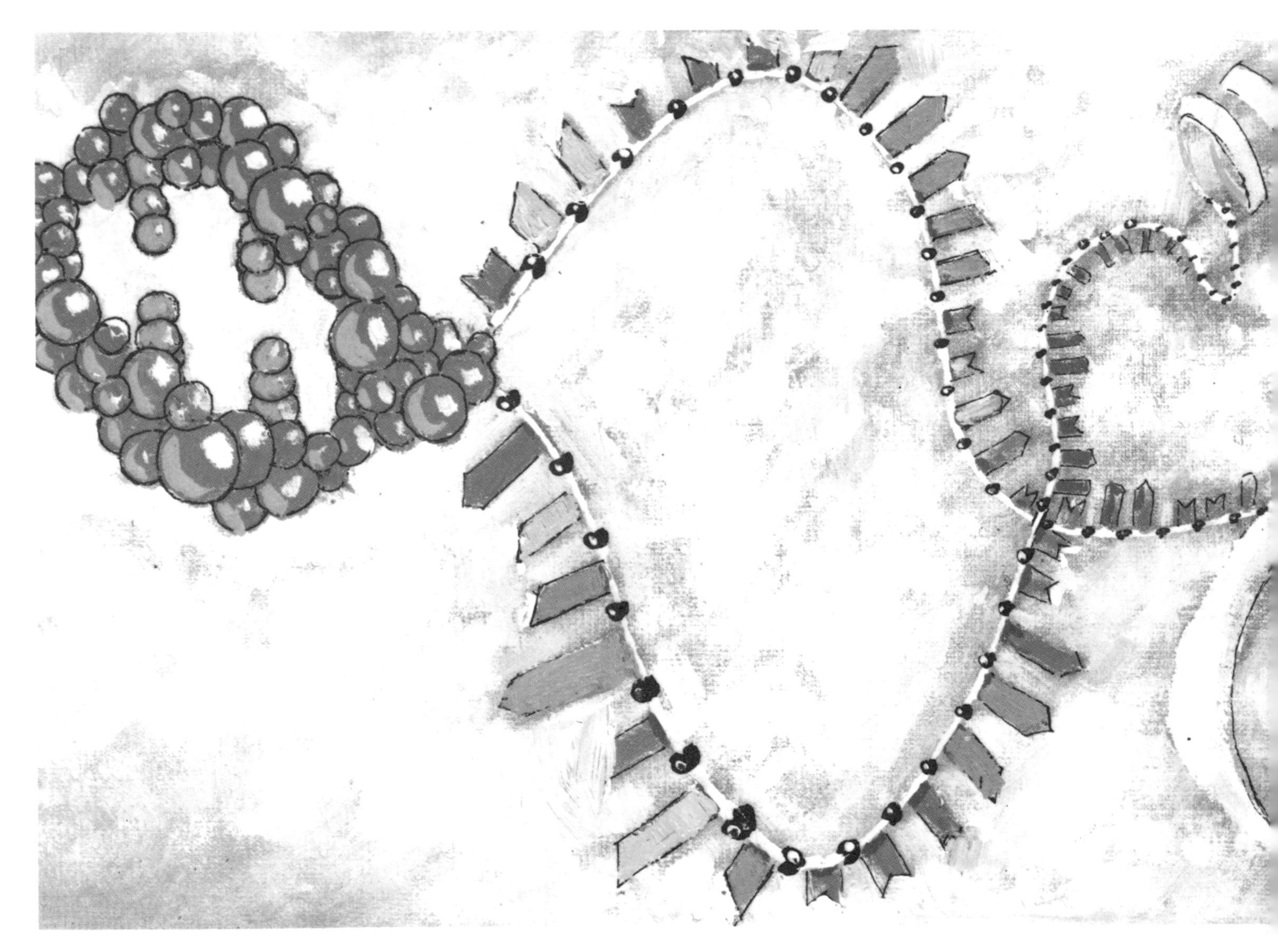

possible to alter, delete and replace whole lengths of DNA by the use of selected viruses deliberately inoculated into the cytoplasm and nucleus.

Because genes are responsible for the production of enzymes, this means that one day it may be possible to transfer whole genes into cells where they are deficient for some reason, and so perhaps cure diseases due to genetic mistakes.

The first steps in human genetic surgery of this kind will clearly be with the body cells, so that any alteration will not be apparent in every cell of the body. Only when the stage is reached of being able to manipulate ova and sperm cells prior to conception, or operate on a newly formed zygote, and tinker with the genetic mechanism, can we look forward to altering the basic anatomy and physiology of a whole human being.

Those days may not be all that far in the future. We are already able to store sperm for use in artificial insemination, and experiments with animals have shown that it is possible to separate X-carrying and Y-carrying spermatozoa and use them to produce either male or female offspring as we choose. This is certainly one form of controlling the chromosomal makeup of creatures yet to be born; even if we are dealing with just one chromosome, it is a very important chromosome, determining sex. From being able to dictate the sex of offspring to being able to alter other chromosomes cannot be a very large step.

Another point to be seriously considered, and one which has raised great controversy, is the possibility of cloning. Cloning is the production of new individuals that have an identical genetic makeup to one of the parents. By taking the nucleus of a mature cell, which has the full number of chromosomes, it is possible to form a zygote that will develop into a replica of the individual from whom the replacement nucleus was taken. So a mother could produce a child that would be her own identical, but clearly younger, twin, or the identical twin of the "father," or another "mother," depending upon where the nucleus for the zygote was taken from.

Using such a technique, it would be possible to raise whole families of carbon-copy people in unlimited numbers—a thousand Einsteins or Picassos. Clones of Olympic hurdlers or rugged soldiers or brilliant scientists could in theory be produced as though from a factory production line.

Cloning, of course, is no more than what a gardener does when he propagates a plant with particularly desirable characteristics by grafting a bud from it into the tissues of another plant. However, the thought of propagating human beings in such an impersonal manner must make most people feel slightly uncomfortable.

If we admit that in the future our bodies and brains are likely to remain more or less as they are, as regards evolutionary change, what then are the prospects for the mind? The idea that because we use our brains a lot we will develop bigger brains is regarded as unbiological, but our thinking has developed over the ages in a remarkable way. It is not a matter of passing on from generation to generation the accumulated wisdom of the past in the form of subtly coded genetic material, but a matter of improving the communication of our learning as time goes by.

Certainly as time goes by there will be successive breakthroughs in every field of knowledge. It is our duty to make sure that this knowledge is widely imparted, and wisely used, rather than hoping that nature will eventually turn us all into superbeings living in dream Utopias.

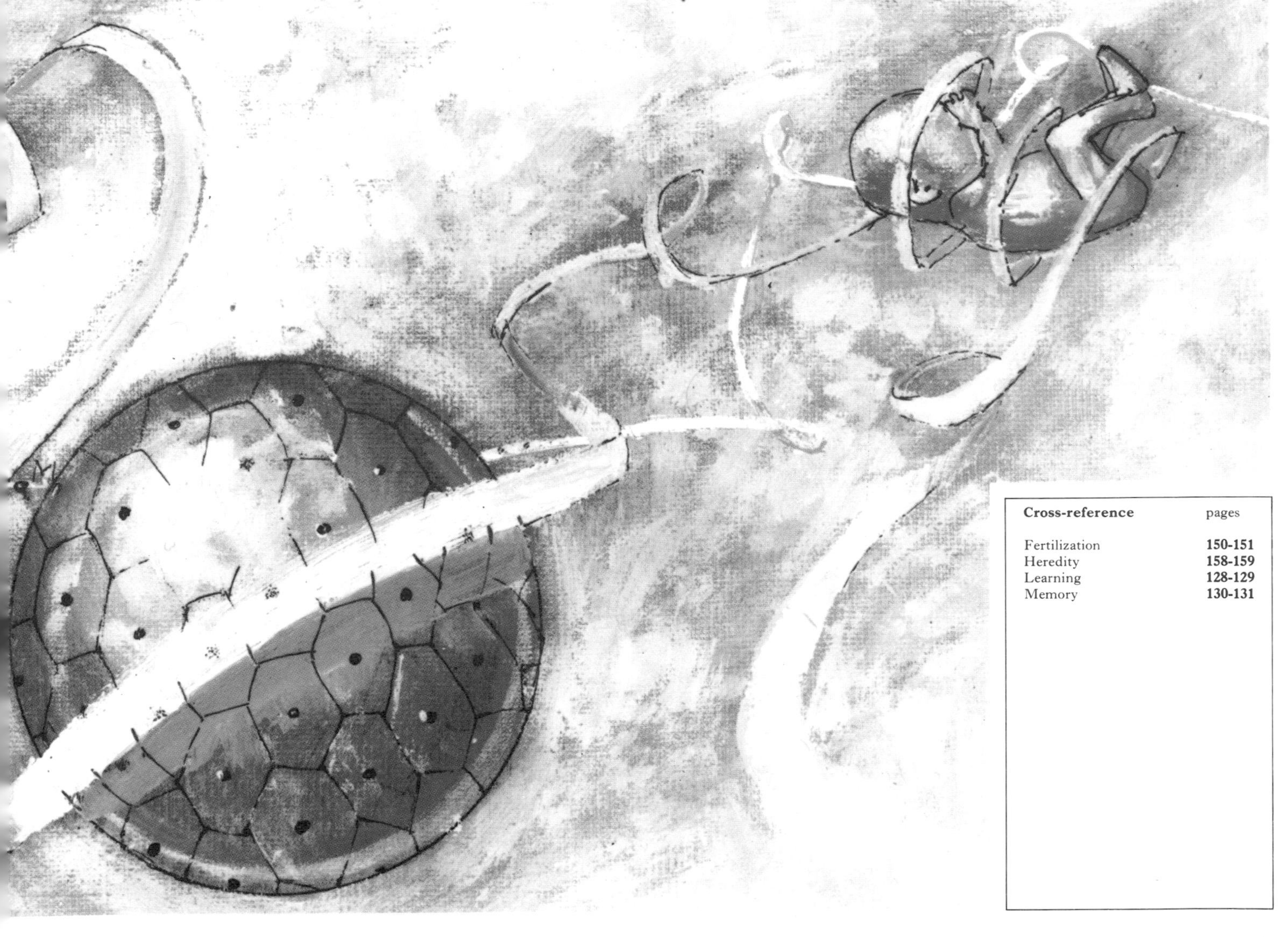

Cross-reference	pages
Fertilization	**150-151**
Heredity	**158-159**
Learning	**128-129**
Memory	**130-131**

Glossary

Abdomen. The name given to the lower part of the trunk between the diaphragm and the pelvic girdle. Some of the important organs found in the abdomen are the kidneys, stomach, intestines, liver and pancreas.

Absorption. The process of taking in materials from the environment. Movement of the products of digestion from the alimentary tract into the bloodstream occurs by absorption across the intestinal wall.

Accommodation. In vision, adjustment in shape of the lens of the eye in order to focus images of objects at different distances on the retina at the back of the eye.

Acetylcholine. A chemical released at the end of certain nerve fibers to set off impulses in nearby cells or to start the contraction of muscle fibers.

Acoustic. An adjective used of any structure or process associated with the sense of hearing. The ears, which are sensitive to sound waves, are the body's acoustic organs.

Actin. A protein molecule found in muscle tissue which, on interaction with the protein myosin, produces contraction of the muscle fibers.

Adaptation. In visual perception, the alteration of the sensory receptor cells in response to an increase or a decrease in light so that a clear image is formed.

Adenoids. Lymphoid tissue at the back of the nose which plays a defensive role against inhaled bacteria.

Adenosine triphosphate (ATP). A chemical that supplies energy for the reactions which occur within cells. Energy is released when one or two of the three phosphate components are removed, to leave adenosine diphosphate (ADP) or adenosine monophosphate (AMP). ATP is the major storage molecule of energy for all the body's activities.

Adipose tissue. A type of connective tissue with many fat-containing cells. It forms a thick layer beneath the skin which acts as a fuel store and helps to insulate the body against cold.

Adolescence. The period of life, between puberty and adulthood, when secondary sexual characteristics develop.

Adrenal glands. A pair of small, hormone-secreting organs, also called the suprarenal glands, which lie over the upper pole of each kidney.

Adrenaline. A hormone, also called epinephrine, secreted mainly by the medulla of the adrenal gland to prepare the body for physical action. This it does by increasing the pulse rate, diverting blood from the intestines and skin to the muscles and by converting the glycogen stored in the liver to glucose and the triglyceride stored in adipose tissue to fatty acids, both easily used sources of energy. Adrenaline (and a related compound called noradrenaline) is also released from certain nerve endings in the sympathetic nervous system.

Adrenocorticotrophic hormone (ACTH). A hormone, secreted by the pituitary gland, that stimulates the cortex of the adrenal gland to produce other hormones, called corticosteroids, which help to control the body's metabolism.

Agglutination. The clumping together of suspended particles in a fluid. For example, the red cells of a blood donor will be agglutinated by the serum of a recipient if the blood groups are not compatible.

Aldosterone. A hormone secreted by the cortex of the adrenal gland. It regulates the excretion of salts by the kidneys and, therefore, the overall balance of salts in the body.

Alimentary canal. The digestive tract, which includes the mouth, the pharynx, the esophagus, the stomach and the small and large intestines.

Alkaloid. The name given to a group of related chemical compounds that are commonly found in plants and may have medicinal or poisonous effects on the body. Common alkaloids include morphine and codeine from the opium poppy, nicotine from tobacco, quinine from the bark of the cinchona tree and strychnine, derived from the seeds of an East Indian tree.

Allergy. An excessive reaction of the body's immune system against such normally harmless substances as pollen, animal fur or dust, which cause such unpleasant symptoms as inflammation of the skin, eyes and nose, wheezy breathing and intestinal upsets.

Alpha wave. One type of brain wave, or variation in electrical activity of the brain, which occurs characteristically when the eyes are closed and a person is awake but relaxed.

Alveolus. A general term for a small cavity, or socket, like a socket of a tooth in the jawbone and a small air sac in the lungs.

Amino acids. These are the basic building blocks of proteins. Human proteins are made up of twenty-two different amino acids. Eight or nine, the "essential" amino acids, cannot be produced by the body itself, but must be provided in the proteins of the diet.

Amnion. One of the membranes surrounding the fetus before birth, enclosing about two pints of the watery fluid called amniotic fluid.

Anabolism. The building up in the body of complex chemical substances from simpler ones, for example the formation of proteins from amino acids. The opposite of catabolism.

Androgen. A general term for any chemical that acts as a male sex hormone.

Antibiotic. A general name for substances, produced by fungi or other microorganisms, that have the ability to destroy or halt the growth of disease-causing organisms. Penicillin and streptomycin, for example, are extracted from fungi. Many antibiotics can now be produced synthetically.

Antibodies. Proteins formed by certain body cells in response to the intrusion of foreign materials, or antigens, that enter the body. Antibodies play a major role in the body's defenses against infection by attacking and destroying bacteria.

Antidiuretic hormone (ADH). A hormone, produced by the posterior pituitary gland, which acts on the kidneys to reduce the volume of water lost in urine.

Antigen. A general term for any foreign material that, entering the body, provokes the production of antibodies. Bacteria, viruses and all allergy-inducing substances, such as pollen dust, are classed as antigens.

Antiseptic. Any chemical used to destroy harmful, infective microorganisms while not damaging the body's tissues.

Antitoxin. A type of antibody produced by the body to neutralize the harmful effects of bacterial poisons, or toxins.

Anus. The last inch or so of the alimentary canal, which opens to the outside of the body.

Anxiety. An emotional state that gives rise to feelings of apprehension. Anxiety may occur in response to imagined or real dangers.

Aorta. The largest artery in the body, measuring a little more than one inch in diameter. It leads directly from the left ventricle of the heart and branches to form the rest of the arterial system.

Appendix. A tubular cul-de-sac, usually about three inches long, but showing considerable individual variation, that projects from the cecum, or first part of the large intestine. Although the appendix is now usually regarded as a redundant organ, it may possibly have a defensive role.

Appetite. The desire to eat stimulated by the sight, smell or thought of food. Appetite is very much linked with hunger, which is a series of physical responses to a fall in blood sugar levels.

Areola. The dark, pigmented area around the nipple. It becomes darker in women who have borne children.

Areolar tissue. A form of loose connective tissue that binds organs and tissues together to keep them in place in the body. It also acts as packing material in and between organs.

Arterial system. The portion of the circulatory system through which blood passes away from the heart to be distributed to all parts of the body.

Artery. A large vessel, or tube, through which blood is conveyed away from the heart to various parts of the body. Its smaller branches are called arterioles.

Association area. An area of the cortex of the brain which is not primarily motor or sensory. The association areas probably combine information relayed from different sensory organs, such as the eyes, ears and nose, to give an integrated awareness of what is happening in the outside world.

Astigmatism. A defect of vision caused by the curvature of the transparent cornea being different in different planes, resulting in the distortion of the images of objects. It can usually be remedied by the use of corrective lenses.

Atlas. The first vertebra, or bony unit, of the spinal column, on which the occipital bone of the skull rests at the atlanto-occipital joint.

Atrium. One of the two smaller chambers of the heart, often incorrectly called auricles. The left atrium receives oxygenated blood from the lungs and the right atrium receives deoxygenated blood from the rest of the body.

Auto-immunity. A malfunction of the immune system. Normally only foreign proteins entering the body are attacked, but in some individuals the immune system turns against certain of the body's own proteins, causing damage to tissues and organs.

Autonomic nervous system. The "self-governing" part of the nervous system that controls such bodily activities as digestion and blood pressure, which are beyond conscious control. Its sympathetic and parasympathetic divisions work in balanced opposition to regulate all involuntary actions.

Axis. The second vertebra in the neck, below the atlas.

Axon. The portion of each nerve cell, or neuron, which conducts nerve impulses away from the cell body. At the end of the axon impulses are chemically transmitted to neighboring nerve cells, or effector organs such as glands or muscles.

Bacteria. Microscopic organisms that have the ability to live in a wide range of environments, including the human body, and which reproduce quickly. Although some bacteria are invasive and disease-causing, or pathogenic, the majority are harmless and many are beneficial. Bacteria can be broadly classified by their shapes when examined under the microscope. Thus bacilli are rod shaped, cocci are round, vibrios are comma shaped and spirilla are corkscrew shaped. Streptococci grow in chains, and staphylococci form groups like miniature bunches of grapes.

Barbiturates. A group of chemically related drugs that act as depressants on the central nervous system and are therefore widely employed as anesthetics, sedatives and hypnotics (sleeping pills).

Basal ganglia. A group of nerve cells in the thalamus, deep within the forebrain, which act as relay stations for nerve impulses involved in the control of posture and muscular movements.

Basal metabolism. The minimal rate at which the body consumes energy, when at rest, for only essential processes and tissue maintenance.

Bile. The thick, greenish-brown fluid secreted by the liver into the biliary tract, where it may be stored in the enlarged side branch called the gall bladder.

Bilirubin. The major pigment of bile, derived from the breakdown of hemoglobin, the red pigment in red blood cells.

Bipolar cell. A nerve cell with extensions in two opposite directions. Bipolar cells in the retina of the eye, for example, transmit messages arriving from the light receptor cells through one extension, into their cell bodies, and out through another to the optic nerves.

Bladder. A distensible muscular bag in which fluid may be stored temporarily. The urinary bladder, in the pelvis, receives urine from the kidneys via the ureters, and stores it until it may be discharged conveniently through the urethra to the exterior. The gall bladder stores bile secreted by the liver until it is needed in the digestive tract.

Blastula, blastocyst. A very early stage in the development of an embryo, consisting of a fluid-filled hollow cluster of cells.

Blind spot. The small area at the back of the retina where nerve fibers meet and leave the eye to become the optic nerve. The blind spot lacks rods and cones and so is insensitive to light.

Blood. The red fluid contained within the vessels of the circulatory system, consisting of a straw-colored liquid, plasma, rich in proteins, in which are suspended red blood cells, or erythrocytes, white cells, or leucocytes, and particles called platelets.

Blood coagulation. The chemical change in soluble blood proteins and platelets that produces a meshwork of insoluble fibrin threads, which then traps blood cells to form a clot. This local transformation of blood into a semisolid mass protects against loss of blood after injury, and is the first stage in wound healing.

Blood groups. Different types of blood, decided by the presence or absence of certain proteins on the surfaces of the red blood cells and in the plasma. Because the mixing of noncompatible blood groups causes the red cells to clump together, or agglutinate, the donor and recipient in a transfusion must have blood groups that match. The two most familiar and important blood group divisions are the ABO and Rhesus systems.

Blood pressure. The pressure of the blood in the body's major arteries, normally measured in millimeters of mercury by means of an instrument called a sphygmomanometer. Two measurements are taken: one of the highest pressure reached when the ventricles of the heart contract (systolic pressure), and one of the lowest pressure between contractions (diastolic pressure). The results are recorded as a ratio of systolic to diastolic pressures. For a young adult, resting measurements in the region of 120/80 are regarded as normal, but there is considerable variation between individuals, and a gradual rise with increasing age.

Bone. The substance that forms the framework, or skeleton, of the body. Most bone starts as cartilage, and is a matrix of connective tissue that is made rigid by the laying down of mineral salts, such as calcium phosphate, and yet retains resilience because of the presence of protein fibers. In the long bones, such as the femur, a load-bearing, hard cylinder, the cortex, surrounds a spongier interior medulla, which may contain marrow. The surface of the bone is covered by a thin, tough membrane, the periosteum.

Bowman's capsule. One of the million or so microscopic cup-shaped structures in each kidney. Each capsule has at its center a tuft of capillaries, or glomerulus. It collects the watery solution filtered under pressure from the blood, and passes it to the tubules where urine is formed.

Brain. The largest and main organ of the nervous system, surrounded and protected by the skull. It regulates all the body's responses, from simple movements to emotional behavior and intelligence.

Brain stem. The central "core" of the brain, linking the top of the spinal cord with the hindbrain and the midbrain.

Brain waves. The continually fluctuating electrical activity of the brain as recorded by an instrument called an electroencephalograph through electrodes in contact with the scalp. There are variations in the patterns and frequency of the waves traced on a moving tape (the electroencephalogram or EEG), corresponding to different states of sleep and wakefulness.

Broca's area. Part of the frontal lobes of the forebrain concerned with the muscle movements involved in speech.

Bronchus. The tube through which air passes into and out of each lung. The windpipe, or trachea, divides into two main bronchi just above the heart. Inside the lungs each bronchus divides and subdivides into smaller bronchioles, and these lead eventually into the minute air sacs called alveoli, where gaseous exchange takes place during respiration.

Bundle of His. The bundle of specialized muscle fibers which conducts electrical impulses from the atria to the ventricles of the heart.

C

Calcitonin. A hormone produced by the thyroid gland, and involved, with other factors, in the control of the level of calcium in the blood.

Calcium. A chemical element vital to the formation and growth of teeth and bones, and the proper functioning of nerve and muscle fibers and many cell enzymes. The main sources of calcium in the diet are milk, cheese and bread.

Calorie. A unit of energy defined as the amount of heat required to raise one gram of water through a temperature of one degree centigrade. The calorific value of food is the number of calories it would yield if completely utilized. As a calorie itself is so small, food values are usually expressed as Kilocalories, or Calories with a capital C. One Calorie is a thousand calories. The average adult requires about 2,000 or 2,500 Calories a day.

Cancer. A malignant growth caused by the disorderly and uncontrolled multiplication of cells in a tissue.

Capillaries. The smallest blood vessels in the body, penetrating and permeating all the body's tissues. They join the smallest branches of the arteries to those of the veins, and carry nutrients and oxygen to the cells, and waste products away from them.

Carbohydrate. One of the basic kinds of food material, containing only carbon, hydrogen and oxygen. The smallest carbohydrate units are simple sugars, or monosaccharides, which may link together to form the large molecules, or polysaccharides, found in starches. The body utilizes glucose, or "blood sugar," as a major source of energy during metabolism. Quantities of glucose in excess of the body's immediate requirements may be stored in the liver and muscles when linked together to form "animal starch," or glycogen.

Carbon dioxide. One of the waste products of metabolism, released by the body's cells during the chemical breakdown of food materials to provide energy. Carried in the bloodstream from respiring tissues to the lungs, it is released as a gas in expired air.

Carcinogen. Any chemical substance known to cause cancer.

Cardiac muscle. The special type of muscle found exclusively in the heart. Cardiac muscle has a built-in rhythm of contraction and can function without stimulation from the nervous system, although normally controlled by it. The regular, powerful and coordinated contractions of cardiac muscle are necessary to maintain the circulation of blood.

Carotid arteries. The two arteries running from the aorta upward on either side of the windpipe, or trachea, which through their internal and external branches supply blood to most of the head, neck and brain.

Cartilage. A tough, glistening form of dense connective tissue found throughout the body having many different roles. It covers the ends of bones, cushioning them and allowing easy movement at joints; disks of cartilage lie between the vertebrae of the spine and bands of cartilage link the ribs and sternum. C-shaped pieces of cartilage keep the trachea open and cartilaginous struts give shape to the nose and ears.

Catabolism. The chemical breakdown into simple substances of more complex materials. Thus the splitting of proteins into amino acids is a catabolic process; the building up of proteins from amino acids is the opposite—an example of anabolism.

Cecum, caecum. The saclike first portion of the large intestine, into which the ileum of the small intestine empties, and which leads to the ascending colon. The appendix forms a blind side branch from its lower part.

Cell. The basic structural and functional unit of the body and of all living things. Every tissue and organ is composed of aggregates of cells, each of which is to some extent a self-contained unit, but relies upon others for its continued existence. Most cells have a jellylike cytoplasm enclosed by a cell membrane and itself surrounding a nucleus that orders nearly all of its activities.

Cellulose. A complex carbohydrate that forms the basic structural material of all plants. Unlike starches, cellulose cannot be broken down by human enzymes, and so passes through the digestive system almost unchanged, as the "roughage" of the diet.

Central nervous system (CNS). The brain and spinal cord, which communicate with and regulate directly or indirectly nearly all body activities by means of messages passed through the branches of the peripheral nervous system.

Cerebellum. One of the major divisions of the brain, consisting of an oval structure with two lobes, situated beneath the larger, overhanging occipital lobes of the cerebrum, and attached to the brain stem. The cerebellum is responsible for integrating nerve impulses from the cerebrum to produce balanced and coordinated muscular movements.

Cerebrospinal fluid (CSF). A watery liquid that circulates both within the linked inner cavities of the brain (the ventricles) and the central canal of the spinal cord, and also outside the brain and spinal cord. It has a mainly shock-absorbing function, although it does contain a little nutrient glucose.

Cervix. The Latin for "neck," applied both to the upper region of the spinal column, as in the "cervical" vertebrae, and to the narrow muscular portion of the uterus protruding into the vagina.

Cholesterol. A chemical necessary for the formation of bile salts and a number of hormones.

Cholinesterase. An enzyme found throughout the body that breaks down acetylcholine, the chemical transmitter released at many nerve endings by nerve impulses. The rapidity of its action ensures that one nerve impulse cannot have too prolonged an effect.

Chorion. The outer of the two membranes that surround a developing fetus.

Choroid. The layer of the eyeball between the tough, white sclera outside and the light-sensitive retina within. It contains a network of blood vessels to supply the retina with oxygen and nutrients.

Chromosomes. Tiny strands in the nucleus of every cell which carry genetic information in the form of units of DNA called genes. In human beings there are twenty-three pairs, of which one pair are called sex chromosomes and the remainder autosomes. Each reproductive cell contains twenty-three single chromosomes, twenty-three pairs being reestablished when an ovum is fertilized by a sperm.

Chyme. The partly digested and semi-liquid food which passes from the stomach into the small intestine.

Cilia. Any small, hairlike structures on the outer surface of cells. Those lining the respiratory tract have a rhythmic beating action, and sweep mucus and dust particles out of the air passages. The eyelashes are also termed cilia.

Ciliary muscle. The muscle surrounding the lens of the eye, to which it is attached by suspensory ligaments. Its fibers contract and relax to adjust the focal length of the lens.

Circadian rhythm. A recurring twenty-four-hour cycle of activity found in many of the body's processes, meaning literally a rhythm of "about a day."

Clitoris. The small protrusion of erectile tissue at the front of the female genitalia, sensitive to sexual stimulation and analogous to the penis in the male.

Coccyx. The lower end of the spinal column, consisting of a triangular bone made up of four small vertebrae fused together. A vestigial structure, the coccyx represents in humans the remains of an ancestral tail.

Cochlea. The coiled tube of the inner ear, shaped like a snail shell, which contains receptor cells for the sense of hearing.

Coitus. Sexual intercourse.

Collagen. A tough protein that occurs in fibers as a major component of connective tissue. About one-third of the body's protein content is collagen; tendons are almost pure collagen.

Colon. The main portion of the large intestine, linking the cecum and the rectum and divided into ascending, transverse and descending thirds.

Color blindness. An inability to distinguish colors, more correctly termed color-defective vision. The most common defect is an inability to distinguish between red and green. Men are affected about ten times more frequently than women.

Colostrum. The yellowish fluid secreted by the mother's breasts for the first few days after childbirth. Colostrum is rich in proteins, antibodies and minerals, thus providing essential nutrients for the baby. The breasts later secrete milk instead of colostrum.

Conception. The moment of fertilization, when a (male) sperm unites with a (female) ovum to form a single cell, or zygote, which will develop into an embryo.

Cone. One of the light-sensitive receptor cells in the retina of the eye responsible for distinguishing colors. Cones are not stimulated by light of low intensity, and in conditions of poor illumination only the other light/dark receptors, called rods, are used for vision.

Conjunctiva. A delicate, transparent membrane covering the front of the eye and lining the inside of the eyelids. Continually moistened with the secretions of the lacrimal gland, its smooth surface allows easy movement of the eyelids.

Connective tissue. The material which in various forms protects, separates and supports the organs of the body. Bone, cartilage, ligaments and tendons, adipose tissue and loose, fibrous areolar tissue are all connective tissue.

Consciousness. The state of being awake and aware of oneself and of the environment.

Contraception. The use of one or more of the several methods available for ensuring that coitus does not lead to pregnancy.

Convolutions. The irregular folds on the surface of the brain which serve to increase the area of the cerebral cortex within the skull.

Cornea. The bulging transparent front part of the eyeball, continuous at its edges with the opaque sclera, or "white," of the eye. Within it may be seen the colored iris with its central aperture, the pupil.

Coronary arteries. The arteries which, branching directly from the aorta, spread over the heart to supply the cardiac muscle with blood.

Corpus callosum. The thick bundle of nerve fibers which joins together the two cerebral hemispheres of the forebrain, allowing interchange of information between one side and the other.

Corpuscle. Any small, usually rounded, structure. The term is most often used to refer to blood cells, but the minute tactile corpuscles in the skin are touch receptors.

Corpus luteum. The "yellow body," a mass of cells within the surface of an ovary, formed at the site of a released mature ovum after ovulation. The corpus luteum secretes progesterone, one of the hormones which prepares the uterus for pregnancy, but degenerates and becomes smaller and paler unless the ovum is fertilized.

Cortex. Any outer covering or layer, such as that of the cerebrum (the cerebral cortex or "gray matter"), or of the kidney (renal cortex).

Corticosteroids. Hormones secreted by the cortex of the adrenal glands, and classified according to their main actions. The gluco-corticoids, including hydrocortisone (cortisol) and cortisone, influence glucose, protein and fat metabolism; mineralo-corticoids, the main one of which is aldosterone, regulate the salt and water balance of the body.

Corticotrophin. A hormone secreted by the pituitary gland that stimulates the secretion of hormones by the adrenal cortex. It is also known as adrenocorticotrophin, or ACTH.

Cowper's gland. A pair of pea-sized glands opening into the urethra of the male. Their secretion makes a small contribution to seminal fluid.

Cranial nerves. The twelve pairs of nerves emerging directly from the brain, with both sensory and motor functions.

Cranium. The portion of the skull surrounding and protecting the brain.

Cytoplasm. The jellylike substance of a cell. The cytoplasm is enclosed by the cell membrane and contains all the intracellular units, such as the mitochondria and ribosomes.

D

Dark adaptation. The eye's adjustment to poor illumination. Dilatation of the pupil allows more light to enter the eye, and chemical changes in the rods and cones of the retina render them (particularly the rods) more sensitive to light of low intensity. Increasing adaptation may take place over several hours.

Dead space. The volume of air in the passages between nose and lungs that takes no part in respiration and has to be moved in and out with each breath before gaseous exchange can occur.

Decibel. A measure of the intensity of sound.

Decidua. The lining of the uterus; it gradually increases in thickness during each menstrual cycle and is shed monthly (menstruation), unless pregnancy occurs.

Dendrites. The many short fibers radiating from the body of a nerve cell through which impulses enter from other nerve cells.

Dentin, dentine. The dense white material in teeth, beneath the visible thin layer of enamel, and surrounding the soft central pulp.

Dermis. The inner layer of the skin, lying beneath the epidermis. It is also known as the corium.

Dialysis. A form of filtration in which a fluid comes into contact with a membrane that will allow only molecules below a certain size to pass through. Thus large protein molecules may be retained, while small salt molecules will diffuse through. In an artificial kidney, dialysis is used to rid a patient's blood of waste materials, but to retain plasma proteins and blood cells.

Diaphragm. The sheet of muscle between the thorax, or chest cavity, and the abdomen. The diaphragm stretches up in a dome shape; it is attached to the lumbar vertebrae behind, and the lower end of the breastbone and the lower ribs in front. Contracting the diaphragm enlarges the chest cavity and draws air into the lungs.

Diastole. The resting period between heart beats, when the cardiac muscle relaxes and blood flows into the ventricles.

Diencephalon. Part of the brain, sometimes also known as the interbrain, which includes the thalamus and hypothalamus.

Diet. The food and drink taken into the body to provide the energy and the raw materials for growth and maintenance.

Diffusion. The process in which molecules dispersed in a fluid move from areas of high concentration to those of lower concentration.

Digestion. The chemical processes brought about by the enzymes secreted in the stomach and intestine which break down foodstuffs from large molecules into smaller molecules. These small units can be absorbed through the intestinal wall, and transported in the blood to the liver and other organs for utilization.

Dilatation. The process of widening or expanding as, for example, when the pupil of the eye opens up in dim light. The term is sometimes shortened to dilation.

Disk, disc. A shock-absorbing and slightly flexible pad of cartilage between two vertebrae of the spine. The optic disk is the region of the back of the retina insensitive to light—the "blind spot."

Diuretic. The name given to any substance which increases the output of urine.

Dorsal. A term that describes anything related to the back, or on the back surface of an organ.

Dreams. Sequences of images passing through the mind while asleep. Dreaming appears to be concentrated into twenty-minute periods, during which the brain becomes more active and "rapid eye movement" (REM) occurs, separated by an hour or more of dreamless sleep.

Drive. A term used by psychologists to describe the motive force, often unconscious, prompting an animal to behave in a certain manner. The hunger drive, for example, prompts animals to seek and eat food.

Ductus arteriosus. A blood vessel that in the fetus allows blood to bypass the lungs, which have not yet a respiratory function. After birth the ductus constricts, and blood is directed through the pulmonary circulation, for oxygenation by the now air-filled lungs.

Duodenum. The first, C-shaped, part of the small intestine between the stomach and jejunum, or, as the Latin name implies, the width of twelve fingers. Into the duodenum, through a single opening, are secreted the digestive juices of the pancreas and bile from the liver.

Dura mater. The outermost of the three protective membranes, or meninges, which enclose the brain and spinal cord.

Ear. The complex structure, set mostly within the bone of the skull, which acts as a sensory organ, not only for sound but also for posture, balance and movement.

Ectoderm. The outer layer of cells in the growing fertilized ovum. From it develops not only the skin of the embryo but, by a process of infolding, the tissues of the nervous system.

Eidetic image. An intense memory image, of an object or scene once perceived, so vivid as to be almost photographic in nature, and "seen" in the mind as though still before the eyes.

Ejaculation. The forceful expulsion of seminal fluid through the penis, occurring during orgasm in the male.

Elastic tissue. Connective tissue containing a high proportion of fibers of the protein elastin, which recoils after distortion.

Electrocardiogram (EKG) (ECG). A recording made with an instrument called an electrocardiograph, of the sequence of electrical changes in the heart muscle during each heartbeat. Such recordings may be useful in the diagnosis of heart disorders.

Electroencephalogram (EEG). A recording of the continuous changes in the electrical activity of the different parts of the brain, made with an instrument called an electroencephalograph, employing electrodes attached to the scalp.

Embryo. An unborn animal in the early stages of development. In human beings an embryo has become recognizably human, although less than two inches long, by the end of the eighth week of pregnancy; after this time it is known as a fetus.

Enamel. The thin, hard layer covering and protecting the crown of a tooth. Enamel is composed mainly of calcium phosphate and fluoride, crystalline in structure and similar to, but much denser than, bone. It is the hardest substance in the body.

Endocrine gland. A gland that produces hormones and secretes them directly into the bloodstream, as opposed to an exocrine gland, which has a duct to carry secretions to where they are needed.

Endoderm. The innermost of the three layers of cells in a developing embryo. From it the tissues of the digestive and respiratory systems develop.

Endolymph. One of the fluids which fill the membranous canals of the sense organs in the inner ear.

Endometrium. The tissue lining the uterus, within the myometrium, or muscle layers.

Enzyme. A substance which speeds up chemical reactions in the body. Each enzyme is a protein; each functions in only one specific reaction, and may require the presence of additional substances called coenzymes before it will work. Many thousands of different enzymes, inside and outside cells, are needed in the interlinked processes that go to make up the body's metabolism.

Epidermis. The outer layer of the skin.

Epididymis. The system of coiled tubules associated with each testis, in which spermatozoa are stored and mature before being released during ejaculation.

Epiglottis. A flap of cartilage at the base of the tongue which, during swallowing, prevents food from accidentally entering the trachea instead of the esophagus.

Epiphysis. The end of a growing bone, separated from the shaft, or diaphysis, by a disk of cartilage. At different stages of development, epiphyses in different parts of the body unite with the diaphyses as the cartilage ossifies, or hardens, into bone.

Epithelium. A covering tissue, such as skin or mucous membrane, forming surfaces inside and outside the body.

Erection. The stiffening of the penis and clitoris when filled with blood during sexual excitement.

Erogenous zones. Those parts of the body, including the genitalia, lips and breasts, which are sensitive to sexual stimulation.

Erythrocyte. One of the disk-shaped red blood cells responsible for transport of carbon dioxide and oxygen around the body. Erythrocytes have no nuclei; their color is due to the oxygen-carrying pigment hemoglobin.

Esophagus, oesophagus. The muscular tube, about ten inches long, that carries food from the back of the throat to the stomach, the esophagus runs between the windpipe and backbone and then through the diaphragm to the stomach. Muscular contractions push food and drink along the esophagus.

Estrogens, oestrogens. One group of the female sex hormones which control ovulation, the release of an ovum by an ovary, and influence the growth and development of both the internal reproductive organs and the secondary sexual characteristics which appear at puberty. The most powerful of the naturally occurring estrogens, estradiol, is secreted by the cells in the ovaries. The corpus luteum, which forms from a Graafian follicle after ovulation, produces estrogens, as do the adrenal glands and the placenta, the organ that connects a growing fetus with the wall of the uterus.

Eustachian tube. A fine tube connecting the middle ear with the back of the throat, with the function of equalizing the air pressure on either side of the eardrum.

Excretion. The elimination of waste products of metabolism from the body. Examples are the exhalation of water vapor and carbon dioxide from the lungs and the formation of urine, containing urea, water and salts, by the kidneys.

Exocrine gland. Any gland, such as a salivary or sweat gland, which secretes its product into a duct leading to a site of action.

Eye. The organ of sight. Each eyeball is nearly spherical, less than one inch (2 cm) in diameter.

F

Fallopian tubes. The paired tubes extending from the uterus toward each ovary. Ova released from the ovaries enter the branched, open endings of the tubes, and are impelled by ciliary action into the uterus. Fertilization of an ovum by a sperm swimming in the opposite direction in one tube, following coitus, leads to conception.

Fascia. Any thin sheet of fibrous connective tissue, such as that covering the thigh muscles, the fascia lata.

Fat. One of the three main types of food in the human diet, the others being protein and carbohydrate. Fat provides a richer source of energy than carbohydrates, and is also an essential structural component of cell membranes. Fat is stored in the body in adipose tissue.

Fauces. The upper part of the throat, from the root of the tongue to the opening of the esophagus. Lymphoid tissue in this area, the tonsils and adenoids, provides a line of defense against infection.

Feces, faeces. The semisolid contents of the large intestine, consisting mainly of indigestible cellulose from the diet, mucus and dead bacteria (about one-tenth by weight). The color depends upon bile pigments. The expulsion of feces through the anus is termed defecation.

Femur. The thighbone, the longest bone in the body, articulating with the pelvis at the hip joint and the tibia at the knee joint.

Fertilization. The union of a male gamete, or sperm, with a female gamete, an egg cell or ovum, to produce a single cell, a zygote, from which an embryo will develop. Fertilization normally occurs in one Fallopian tube, after which the zygote passes into the uterus to become implanted in its wall.

Fetus, foetus. An unborn child, after the eighth week of pregnancy, before which it is known as an embryo.

Fibrin. An insoluble blood protein formed from a precursor called fibrinogen during the clotting of blood. Threads of fibrin, forming a mesh across a wound, trap blood cells to form the basis of a clot.

Fibroblasts. Cells concerned with making the fibrous elements in connective tissue, especially that found in healing wounds.

Fibula. The slender outer bone, sometimes called the splint bone, of the lower leg. Its lower end forms part of the ankle joint, but its upper end, bound to the tibia, does not form part of the knee joint.

Fissure. The general term for a cleft, or opening, separating parts of the body.

Follicle. A small cavity or sac. Hair follicles in the skin each produce a hair shaft. In the ovary, Graafian, or ovarian, follicles produce the female gametes, ova.

Follicle-stimulating hormone (FSH). A hormone, secreted by the pituitary gland, which causes an ovarian follicle to ripen and release a mature ovum.

Fontanelle. One of the gaps between the growing bones in the skull of a newborn baby. The fontanelles gradually narrow during development, and are normally closed by about eighteen months.

Forebrain. The part of the brain at its upper and foremost end, above the midbrain. It consists of the two cerebral hemispheres, the thalamus and hypothalamus and part of the pituitary gland, and is concerned with perception, emotions and drives.

Foreskin. The short, retractable sleeve of skin surrounding the glans of the penis in uncircumcised males. Also known as the prepuce.

Fovea. The shallow, depressed area at the back of the retina, containing only cone cells, which corresponds to the center of the visual field, and is responsible for the sharpest vision.

Frontal lobes. The foremost lobes of each cerebral hemisphere. The posterior regions of the lobes control muscular activity. The anterior, or prefrontal, areas are involved in learning, behavior and personality.

G

Gall bladder. The small, pear-shaped sac beneath the liver in which bile produced by the liver is stored and concentrated, until poured into the duodenum to aid the digestion of fatty foods.

Gamma globulins. Kinds of protein found in the blood plasma, including the antibodies responsible for defending the body against infection. They are produced by the lymphocytic system.

Ganglion. A cluster of nerve cell bodies forming a swelling in the course of a nerve.

Gastric. The adjective for anything relating to the structure or function of the stomach.

Gene. One of the basic units of heredity, responsible for the development of a specific trait, or characteristic, in an organism. Genes are short sequences of different units which are linked together to form the long helical molecules of DNA in the chromosomes of every cell nucleus. Each gene is responsible for directing the production of a specific kind of protein by a cell, and so contributes to the overall pattern of growth in the organism.

Genital. The adjective for anything to do with reproduction. The genital tract is thus another name for the reproductive system. The genitals, or genitalia, are the external reproductive organs.

Genotype. An individual's genetic makeup either as a whole or referring to a particular gene, or unit, of inheritance.

Gestation. The development of a new individual between the times of conception and birth. The gestation period in human beings is about thirty-eight weeks.

Gland. An organ that produces and releases a secretion. Exocrine glands such as sweat glands release their products through a duct. Endocrine, or hormone-secreting, glands release material directly into the bloodstream.

Globulins. One of the kinds of protein in the blood, the others being albumins. Antibodies, vital in the body's fight against infection, are globulins of the type called gamma globulins.

Glomerulus. A knot of tiny capillaries enclosed in each cup-shaped Bowman's capsule in the kidneys. As blood passes through the glomerulus at high pressure, waste materials and water filter out into the tubule and pass on from there to form urine.

Glottis. The entrance to the windpipe, or trachea, from the pharynx, the cavity behind the mouth.

Glucagon. A hormone secreted by the pancreas that increases the level of sugar in the blood.

Glucocorticoids. A group of hormones secreted by the cortex of the adrenal glands that are involved mainly in growth and in handling of food materials. In higher concentrations, the glucocorticoids call into action the repair mechanisms and anti-inflammatory responses of the body and help to make available adequate supplies of glucose to provide energy.

Glucose. A simple sugar and a major source of energy for cells. It enters the bloodstream directly from the small intestine or can be formed from other carbohydrates in the liver.

Gluteus maximus. The largest of the buttock muscles and the most powerful skeletal muscle in the body.

Glycerol. One of the constituent molecules of fat.

Glycogen. The form in which carbohydrate is stored in the body, being a large-molecule "animal starch," built up from units of glucose.

Gonad. One of the primary sex organs — the testis in the male, the ovary in the female.

Gonadotrophins. A group of hormones, produced by the anterior, or frontal, lobe of the pituitary gland, which act on the ovaries or testes and influence their growth and development. Gonadotrophins are also produced by the placenta during pregnancy.

Graafian follicle. Another name for an ovarian follicle, in which egg cells, or ova, develop in the ovary, to be released at ovulation.

Granulocytes. One of the types of white blood cells, or leucocytes, so called because of their granular appearance when stained and examined microscopically.

Gray matter. Nerve tissue, gray in color, composed mainly of cell bodies, in contrast to white matter, which is composed mainly of nerve fibers with the insulating fatty material myelin. Gray matter is found in the outer layer, or cortex, of the cerebrum, and in the inner layers of the spinal cord.

Growth hormone (GH). The hormone secreted by the anterior, or frontal, lobe of the pituitary gland which promotes body growth.

H

Hallucination. An apparent sensory perception that has no basis in external reality.

Hand. Eight bones called carpals form the wrist and base of the hand and five metacarpals support the phalanges, the three bones of each finger and the two bones of the thumb. Muscles in the forearm control the fingers through tendons running in front of and behind the wrist. Other small muscles within the hand help to bring about precise movements of the fingers.

Heart. The muscular organ, which is a double pump for forcing blood through the circulatory system, lying between the lungs slightly to the left of the center of the chest.

Hemoglobin, haemoglobin. The red pigment, composed of protein and iron, in red blood cells.

Hemorrhage, haemorrhage. A term for bleeding.

Hepatic. A term used to describe anything connected with the structure or function of the liver.

Hepatic duct. The channel that carries bile from the liver to the duodenum, having as a side branch the cystic duct, which leads to the gall bladder.

Hepatic portal vein. The vein that conducts digested food materials in the bloodstream from the intestines to the liver.

Heredity. The passage of family characteristics from parents to child.

Hiccup. An involuntary contraction of the diaphragm, the sheet of muscle between the chest and abdomen, which may result from irritation by overdistension of the stomach, a local infection or a nerve disorder. The sound of the hiccup is made by the sudden closing of the glottis, the entrance to the air passages leading to the lungs.

Hindbrain. The part of the brain joined directly to the spinal cord. The hindbrain contains the medulla, which is concerned with sleep and consciousness, breathing and blood circulation, and the cerebellum, which is concerned with muscular control.

Hip. The ball-and-socket joint between the nearly spherical head of the femur, or thighbone, and the three bones of the pelvis, where they meet to form the cuplike socket, or acetabulum.

Histamine. A substance released from injured tissues that initiates inflammation, the reddening and swelling caused by dilatation and increased permeability of the blood capillaries, intended to speed the healing processes. Many allergic reactions are caused by the histamine-producing mechanisms responding either excessively or to inappropriate triggers; these can be suppressed by antihistamine drugs.

Homeostasis. The term given to the maintenance by the body of a constant internal environment despite external changes.

Hormone. A chemical substance secreted into the bloodstream, usually by an endocrine gland, in order to act on an organ or organs in other parts of the body.

Human chorionic gonadotrophin (HCG). A hormone produced by the outer layer of cells of the fertilized egg, the trophoblast, which maintains the corpus luteum of the ovary and thus ensures that the wall of the uterus remains suitably prepared for the development of the implanted egg.

Humerus. The bone of the upper arm from the shoulder to the elbow.

Hyaline cartilage. Thin cartilage, glassy in appearance, which forms the embryonic skeleton, covers the ends of bones at freely movable joints and is found in the nose, windpipe and larynx.

Hyaloid canal. The canal running through the center of the vitreous humor of the eye. It is apparently functionless in adults, but in the developing embryo is filled with blood vessels.

Hymen. The thin membrane which partially closes the lower end of the vagina.

Hypothalamus. The region at the base of the forebrain between the cerebral hemispheres. It contains nerve centers for the regulation of certain of the body's vital processes, including the metabolism of fat and carbohydrate, sleep, temperature and sex drives. From it extends the pituitary gland, occupying a small hollow in the bone at the base of the skull.

I

Id. The term Sigmund Freud used to label the unconscious, instinctive and pleasure-seeking part of the personality.

Identical twins. Twins who grow from the splitting in half of a single fertilized ovum and so have the same genetic makeup. Such twins are also called monozygotic, or monovular, twins.

Ileum. The part of the small intestine which joins the jejunum to the cecum, the first part of the large intestine, at the ileocecal valve.

Illusion. An incorrect perception of something actually present, in contrast to a hallucination, which has no basis in external reality.

Immune response. The body's natural reactions to infection or the presence of foreign material, involving antibody production and the mobilization of cellular defenses.

Immunity. The body's ability to resist infection by disease-causing organisms. Natural immunity is acquired when the production of specific antibodies is stimulated by the invasion of the body by bacteria or other infective agents. Artificial immunity is induced by inoculating weakened or dead bacteria to stimulate antibody production and give protection.

Immunoglobulins. A group of proteins, including antibodies, produced by white blood cells to protect the body against disease.

Implantation. The embedding in the wall of the uterus of the zygote, the cell formed by the union of sperm and ovum. Implantation happens about a week after fertilization occurs.

Incisors. The four chisel-shaped front teeth of both upper and lower jaws, which have the function of biting off small pieces of food.

Incus. The anvil-shaped second bone of the three ossicles, tiny linked bones that transmit sound vibrations from the eardrum to the inner ear across the middle ear.

Infection. Invasion of the body by harmful bacteria, fungi or viruses, which are able to establish themselves and multiply.

Inflammation. The response of tissues to injury or infection, and the prelude to repair processes. The four cardinal signs of inflammation are heat, redness, swelling and pain.

Ingestion. The taking of food into the digestive tract.

Inhibition. The complete or partial reduction in the response to a stimulus, or the slowing down or prevention of a chemical reaction.

Inner ear. The innermost of the three parts of the ear, encased in bone and containing the cochlea, the organ of hearing, and the semicircular canals, responsible for balance.

Insulin. A hormone secreted by the pancreas which regulates the level of blood glucose, increasing its uptake by the tissues, and suppressing its formation from glycogen and animo acids by the liver. Insulin also reduces the mobilization of fats from storage for use as fuel. Lack of insulin results in excessively high levels of blood glucose, and its spillage into the urine. At the same time the tissues are unable to utilize the glucose as fuel.

Intelligence. An individual's capacity to act purposefully, to think rationally and to deal effectively with his environment.

Intelligence quotient (IQ). A measure of a person's ability to do intelligence tests which have been standardized so that the average success rate for the population is 100. On the Weschler Adult Scale 70 is defined as subnormal; while 150 and over is genius.

Intercostal muscles. Sheets of muscle between the ribs that move them upward and outward during inspiration, and downward and inward during expiration.

Interferon. A substance, produced by cells infected with viruses, which can interfere with the growth of infecting viruses.

Interneuron. A short neuron, or nerve cell, in the spinal cord, which links other neurons. The majority of nerve cells in the body are interneurons.

Interstitial. An adjective applied to material found between other structures. Thus interstitial fluid lies between the cells in tissues, and interstitial cells in the testes lie between the sperm-producing tubules.

Intestine. The digestive tract between the stomach and anus. The first part, the small intestine, is divided into duodenum, jejunum and ileum. The large intestine consists of the cecum (with the appendix attached), the ascending, transverse, descending and sigmoid portions of the colon and the rectum.

Iris. The colored circular diaphragm in the front of the eye which controls the size of the pupil, the hole in the center of the iris, and so regulates the amount of light entering the eye.

Islets of Langerhans. Groups of cells in the pancreas which secrete the hormones insulin and glucagon into the bloodstream.

Jaw. The upper jaw is formed by the cheekbones, or maxillae, and the lower jaw by the mandible, which hinges with the upper jaw at the two temporo-mandibular joints.

Jejunum. The middle portion of the small intestine, leading from the duodenum to the ileum. About eight feet long, it is concerned with the digestion and absorption of food materials.

Joint. Any meeting point between bones usually, but not necessarily, allowing a controlled amount of movement. The sutures of the skull, where its bones meet, are immovable joints, while at the other extreme are the hip and shoulder joints, which are freely movable. The joints between the vertebrae of the spinal column are slightly movable since the bones are united by disks of fibrocartilage. In a movable joint the ends of the bones are capped with articular cartilage, and lubrication is provided by synovial fluid in an enclosing synovial membrane. Surrounding the joint is a fibrous joint capsule, in which thickened bands form ligaments. Joints may be described in terms of their movement. Thus there are such joints as hinge joints, saddle joints and ball-and-socket joints.

Joule. The basic unit of energy in the standardized systems of metric measurement. Approximately four and one-fifth joules are equal to one Calorie.

Jugular veins. The four large veins in the neck which conduct blood from the head and neck and brain back toward the heart.

Keratin. The tough protein from which nails, hair and the surface layer of skin are formed.

Kidneys. The paired organs at the back of the abdomen that filter from the bloodstream waste products of metabolism and materials and fluid in excess of the body's needs. The resulting fluid, urine, is passed through the urethra to the bladder for eventual disposal to the exterior.

Knee. The large, freely movable joint, classified as a hinge joint, between the lower end of the femur and the upper end of the tibia.

L

Labia. The Latin term for lips, generally used to refer to the labia majora and labia minora, the outer and inner folds of skin around the vulva, the external female genitalia.

Labor. The intense and repeated contractions of the uterus which culminate in childbirth. In the first stage of labor the cervix, or neck of the uterus, dilates to open up the birth canal from uterus to vagina. The stronger contractions of the second stage, which is much shorter, cause the actual birth of the baby. In the third stage the placenta, or afterbirth, is expelled.

Labyrinth. The intricate system of fluid-filled canals of the inner ear which contain nerve cells concerned with balance.

Lacrimal glands. The glands beneath the eyelids which secrete tears, so moistening and lubricating the transparent cornea at the front of the eye. Tears contain lysozyme, a substance with antibacterial action.

Lactase. An enzyme, secreted by glands in the small intestine, which breaks down lactose, or milk sugar, which forms a component of human and cow's milk.

Lactation. The production of milk by the mammary glands after birth.

Lactic acid. A chemical produced when supplies of oxygen reaching muscle tissues are insufficient to satisfy demands for energy from the chemical burning of glucose and fats with oxygen.

Lactose. "Milk sugar," a carbohydrate composed of two simple sugars—glucose and galactose—joined together. It is broken down into these components by the enzyme lactase in the intestine.

Large intestine. The portion of the alimentary canal between the last part of the small intestine, the ileum, and the anus, about five feet long and two and a half inches in diameter, consisting of the cecum, colon and rectum. It receives material from the small intestine, reabsorbs much of the water this material contains and stores the semisolid waste until it can be ejected as feces.

Larynx. The voice box is positioned in the neck at the top of the windpipe, or trachea. It is a cylinder of muscle and cartilage about one inch across and one and a half inches long which houses the vocal cords, two folds of membrane that tighten to vibrate in the airstream from the lungs and so produce basic sounds.

Lens. The transparent body in the eye behind the iris. By changing shape under muscular control, the lens brings to a fine focus images cast onto the light-sensitive cells of the retina at the back of the eye.

Leucocytes. White blood cells, of which there are three main types—granulocytes, lymphocytes and monocytes. Leucocytes are able to pass into the body tissues and fight infection by engulfing bacteria and forming antibodies.

Ligaments. Bands of tough, fibrous and inelastic connective tissue that hold bones together and keep joints stable.

Limbic system. Nerve centers in the forebrain, within the temporal lobe and thalamus of the cerebral hemispheres, which are concerned with emotions.

Lipases. Enzymes in the pancreatic and intestinal juices which break down fats after the emulsifying action of bile. Other lipases aid the removal of triglyceride from blood for use in such tissues as muscle. A further lipase enzyme is found in fat cells to mobilize the stored fat when it is needed to provide energy.

Lipids. Another word for fats, a group of substances that provide energy in the body and contribute to cell structure.

Liver. The largest organ in the body, lying in the upper right-hand corner of the abdomen beneath the diaphragm. Its many functions include bile production, synthesis of proteins and carbohydrates, storage of glucose, minerals and vitamins and the breaking down of poisons—including alcohol.

Lumbar. The adjective applied to structures of the back between the ribs and the pelvis.

Lungs. The two baglike organs filling most of the chest which, in breathing, draw air in through the windpipe and distribute it through their complex system of air passages to thin-walled air sacs that are richly supplied with blood vessels. Here, where blood and air are in close contact, the exchange of carbon dioxide and oxygen occurs.

Luteinizing hormone (LH). The hormone secreted by the pituitary gland which causes ovulation.

Lymph. The watery fluid from body tissues which returns to the bloodstream through lymph vessels rather than directly into the capillaries. Lymph is somewhat similar in composition to blood plasma.

Lymphatic system. The network of vessels throughout the body which conducts fluid, the lymph, from the tissues toward the base of the neck, where the lymphatic system joins the blood circulation.

Lymphocyte. A type of white cell found in most of the body's tissues. Lymphocytes are a little larger than red blood cells and mainly concerned with the formation of antibodies to fight infection.

Lymphoid tissue. Tissue in the spleen and in the lymph nodes of the lymphatic system which is concerned with the formation of lymphocytes.

Macrophages. Large white blood cells that destroy bacteria by engulfing them.

Malleus. One of the three bones of the middle ear, shaped like a hammer, that transmits vibrations from the eardrum to the hearing organ of the inner ear.

Maltose. A sugar that is broken down in the small intestine by the enzyme maltase to the simple sugar glucose.

Mammary glands. The milk-producing glands of the breast which mature at puberty, under the influence of female hormones, into a system of lobes connected by ducts to the nipple. The secretion of milk is initiated shortly after birth by the baby's suckling. For the first three or four days after childbirth the mammary glands produce colostrum, a yellowish fluid rich in proteins, antibodies and minerals, but this is then replaced by milk secretion.

Mandible. The single bone of the lower jaw, which is hinged to the skull in front of the ears.

Marrow. The soft core substance of bones. Red marrow fills all bones early in life and manufactures both red and white blood cells. Later in life, fatty white or yellow marrow fills most of the long bones of the limbs.

Mastication. The biting and chewing of food.

Maturation. The attainment of full adulthood—or the completion within the body of such processes as bone growth. Physical maturity is attained in the late teens or early twenties.

Maxilla. The upper jaw, the bone of the skull which forms the main part of the face.

Meatus. A passage, or channel, such as the external auditory meatus, or ear hole.

Meconium. The green fluid, composed mainly of bile, present in a newborn baby's bowel and passed shortly after birth.

Medulla. The term for the central part of an organ such as the medulla, or middle, of the kidney.

Medulla oblongata. The part of the hindbrain which is joined to the spinal cord and governs sleep and consciousness, breathing and blood circulation.

Meiosis. The process of cell division that produces sperm and ova. In meiosis the chromosomes of the original cell, which contain the hereditary information, divide into two sets of twenty-three chromosomes, instead of duplicating to give forty-six chromosomes as in mitosis, normal cell division. When a sperm fertilizes an ovum, the resulting cell, or zygote, has a full set of forty-six chromosomes, half from each parent.

Melanin. The pigment that gives coloration to skin, hair and eyes.

Menarche. The onset of menstruation at puberty.

Meninges. The three layers of membranes surrounding the brain and spinal cord. The innermost layer, the pia mater, clings closely to the surface of the brain and spinal cord. The outermost layer, the dura mater, lines the skull and spinal canal. Attached between the two is the middle layer, the arachnoid mater.

Menopause. The cessation in middle-age of ovulation and of menstruation which accompanies alterations in the balance and amount of sex hormones.

Menstrual cycle. The sequence of changes in a woman's body, repeating approximately every twenty-eight days. Immediately after menstruation, the endometrium, the uterine lining, under the influence of hormone production triggered by the pituitary gland, begins to re-form, readying itself for the implantation of a fertilized ovum. A follicle works its way to the surface of one of the two ovaries and in the middle of the cycle releases a mature ovum into a Fallopian tube. If the ovum is fertilized by sperm passing up the Fallopian tube after coitus then pregnancy may occur. If the ovum is not fertilized, it passes out of the uterus at menstruation and the cycle is repeated.

Menstruation. The "period" of bleeding from the vagina, caused by the uterus shedding its lining, that occurs at approximately monthly intervals from puberty until the menopause.

Mesoderm. The middle of the three germ layers of the early embryo from which bones, connective tissues and muscles are formed.

Metabolic rate. The rate at which a person consumes energy in activities and bodily processes, reflected directly in the body's oxygen intake. There is a minimum basal energy requirement (basal metabolic rate), but the rate increases dramatically during exercise.

Metabolism. The processes of chemical change in the body consisting of the twin aspects of anabolism, the building up of the complex substances the body requires, and catabolism, the breaking down of complicated molecules to provide both the body's "basic building blocks" and the continuing requirements of the body for energy.

Metacarpals. The five bones that form the middle of the hand from wrist to knuckles.

Metatarsals. The five middle bones of the foot.

Microbe, microorganism. Living creatures, some of which can cause disease, that are too small to be seen by the eye. Microorganisms may be tiny animals, fungi, bacteria or viruses.

Micturition. The act of urination or passing water.

Midbrain. The part of the brain stem connecting the forebrain to the hindbrain, the midbrain is concerned to some extent with processing information from the eyes and ears and with the regulation of consciousness and sleep.

Middle ear. The mechanism consisting of the eardrum and three small bones, the incus, malleus and stapes, by which sound waves entering the outer ear are transformed into vibrations and transmitted to the oval window of the cochlea, the hearing organ of the inner ear.

Milk teeth. The twenty teeth that appear between the ages of six months and two years and are completely replaced, by about the age of twelve, with the permanent teeth of adulthood.

Mitosis. The normal process by which the body's tissues grow, mitosis is the duplication of a cell by splitting in two after the doubling of the chromosomes so that each "daughter" cell has a full set of twenty-three pairs identical to those of the "parent" cell.

Mitral valve. The valve that controls the flow of blood between the upper and lower chambers of the left side of the heart, the side that takes freshly oxygenated blood from the lungs and pumps it around the body.

Molars. The twelve flat-surfaced teeth, three at the back of each side of both upper and lower jaws, that grind food down to prepare it for swallowing.

Molecule. The smallest unit of a substance that retains its chemical identity.

Monocytes. Long-lived white blood cells that can engulf bacteria and foreign materials which invade the body's tissues.

Morula. The block of cells formed by the rapidly dividing fertilized ovum which quickly develops into the hollow blastocyst before implantation.

Motor area. The areas of the cerebral cortex toward the rear of the frontal lobe of each cerebral hemisphere which are concerned with the voluntary control of muscular activity.

Motor nerve. A nerve cell, or neuron, that carries impulses away from the brain via the spinal cord to initiate activity in a muscle or gland.

Mucous membrane. Thin, delicate tissue found lining many of the body's internal surfaces, such as the digestive tract and the respiratory tract, moistened with the secretion mucus, formed by the mucous glands that it contains.

Mucus. The viscous fluid secreted by glands in mucous membranes that lubricates and protects the surface of these tissues.

Muscles. Bundles of fibers which contract in response to stimulation, usually from the central nervous system. Striated or skeletal muscle is called voluntary muscle. Smooth muscle, or involuntary muscle, is found in such internal organs as the intestines and blood vessels. Cardiac muscle, the muscle of the heart, is the only muscle able under certain circumstances to initiate its own contraction.

Muscle tone. The slight tension giving a degree of preparedness for action, normally present in every muscle.

Mutation. A changed form of physiological development of functioning resulting from an alteration occurring in a chromosome in one of the germ cells.

Myelin. The white, fatty substance that covers and insulates some nerve fibers and plays a part in increasing the speed with which they can conduct nerve impulses.

Myosin. The thick protein fibrils in muscle that interact with thin fibrils of actin to produce contraction.

N

Nails. The hard covering of the ends of the fingers and toes, the nails which grow from a fold in the skin called the matrix, are formed mainly of keratin, the horn-like substance also present in hair and the outer layer of skin.

Nasopharynx. The space behind the nose and above the soft palate of the roof of the mouth.

Navel. The depression in the abdomen where the umbilical cord was attached.

Neck. The part of the anatomy joining the head to the trunk, containing the seven cervical vertebrae, that supports the head and encloses the spinal cord, the trachea, or windpipe, the esophagus, or gullet, the blood vessels supplying the head and brain and the thyroid and parathyroid glands.

Neocortex. A term for the cerebral cortex, the outer layer or gray matter of the cerebrum.

Nephron. The basic unit within the kidney, each nephron—there are more than a million in each kidney—starts in the outer layer, or cortex, of the kidney as a cuplike Bowman's capsule, which filters urine from blood vessels looped into the capsule, and continues first as the tortuous proximal tubule, then as the loop of Henle and finally as the distal tubule. In all three parts of the tubes, nutrients and excess water are reabsorbed before the urine finally drains into the collecting ducts and then to the ureter, which carries it to the bladder.

Nerve. A bundle of nerve fibers that can conduct information to or from the brain and any part of the body.

Nerve fiber. The extended processes leading from the cell body of a neuron, or nerve cell, along which nerve impulses can pass. The longest nerve fibers in the body may be several feet in length.

Neural coding. The transformation of sensory or motor information into a pattern of nerve impulses that can be passed through the nervous system.

Neuroglia. Supportive tissue in the central nervous system.

Neuromuscular junction. The junctions between a motoneuron, or motor nerve fiber, and a muscle.

Neuron. A nerve cell, consisting of the cell body, short filaments called dendrites, which exchange impulses with neighbouring cells, and an extension, the axon, which can be inches or feet long and which passes nerve impulses between other nerve cells, muscles and body organs.

Neutrophils. The most numerous white blood cells, capable of engulfing bacteria and other foreign material as part of the body's defense system.

Nipple. The protruding teat of the breast through which a baby can suck milk.

Noradrenaline. A chemical released at some nerve endings in the brain and in the sympathetic nervous system. Also a hormone released by the medulla of the adrenal glands to transmit nerve impulses.

Nose. The part of the face around the nostrils, supported by a central bone and pliable cartilage.

Nucleic acids. The vital substances concerned with the passing on of hereditary characteristics and the production of proteins within cells.

Nucleus. The dense center of a living cell containing proteins that control the processes of the cell.

O

Occipital lobes. The hindmost of the lobes of the cerebral hemisphere, mainly concerned with visual perception.

Odor. The sensation produced by chemical stimulation of the receptors in the nose and characteristic of the substance smelled.

Olfactory. A term meaning to do with the sense of smell.

Olfactory mucosa. Mucous membrane in the nasal cavities containing the smell receptors, cells with sensitive protruding hairlike processes that when stimulated pass impulses to the brain through the pair of olfactory nerves.

Oocyte. The primitive egg cell within each potential Graafian follicle in the ovaries that can develop into a mature ovum in the process of development called oogenesis.

Operant conditioning. By reinforcing some spontaneous activities with rewards, operant conditioning acts as a form of training and alters behavior.

Optic chiasma. The junction where the optic nerves coming from the eyes merge and redivide so that all impulses coming from the left visual field, roughly everything seen to the left, pass to the visual cortex of the right cerebral hemisphere and those from the right visual field go to the left.

Optic nerve. The bundle of nerve fibers coming from each eye which carry impulses from the light-sensitive cells of the retina to the brain.

Orbit. The socket in the skull in which the eyeball is held and moves.

Organ. A distinct structure of the body that has particular functions such as the brain, heart, lungs and liver.

Organic. A term meaning to do with organs or living organisms.

Organism. A living thing, whether animal, plant or microbe.

Organ of Corti. The part of the cochlea of the inner ear that responds to waves in the fluid filling the inner ear caused by sound waves.

Orgasm. The climax of sexual excitement. Psychologically, orgasm is the moment of release of sexual tension, physiologically it is the point at which involuntary muscular contractions produce ejaculation in the male and a "thrust of intense sexual awareness" in the female.

Osmoreceptors. Cells in the hypothalamus, the region at the base of the forebrain lying between the cerebral hemispheres, which monitor the osmotic pressure of the blood, a measure of its fluid content, and adjust the output of antidiuretic hormone (ADH) to control the amount of fluid passed from the body in urine.

Osmosis. The selective passage of water through a membrane between solutions of different concentrations of a substance that cannot itself pass through the membrane. The water passes from the weaker to the stronger solution to bring them both to the same concentration. Osmosis plays a large part in the passage of water through, for example, the walls of blood vessels and kidney tubules.

Osmotic pressure. The pressure required to stop pure water from passing through a membrane into a solution, a measure of the concentration of the solution. Osmotic pressure comes into play, for example, in forcing fluid back into the capillary system after it has been forced out into the body tissues by the blood pressure.

Osteoblasts. Bone-forming cells at the ends of the long bones and under the fibrous covering of all bone surfaces which act by producing enzymes that cause insoluble calcium phosphate to form with the bone matrix of protein fibers.

Osteoclasts. Bone-eroding cells which produce an enzyme to dissolve hard calcium phosphate and so allow reabsorption and modelling of bone that is already formed.

Osteocytes. Cells in bone which are stable, neither growing nor being broken down, which seem essential to its health.

Otolith organ. Part of the vestibular apparatus of the inner ear, which is concerned with balance and movement, the otolith organ contains cells that by sensing the pull of gravity assess the position and attitude of the head.

Outer ear. The external ear, called the pinna, and the auditory canal leading to the eardrum.

Oval window. The membrane stretched across the upper canal of the cochlea, the organ of hearing of the inner ear, is made to vibrate by the action of the stapes, one of the three ear bones of the middle ear which are set into movement by the eardrum and so produce waves in the fluid inside the cochlea.

Ovaries. The primary female sexual organs. The two ovaries, one on each side of the uterus in the abdomen, are about an inch and a half long. Throughout life the ovaries secrete hormones that regulate sexual development, and between puberty and the menopause they alternate in releasing a mature ovum at approximately monthly intervals in the middle of the menstrual cycle. The ovaries also produce, under the control of the pituitary gland, the hormones that affect the menstrual cycle, prepare the uterus for implantation of the fertilized ovum and adapt the body to pregnancy.

Ovulation. The release of a mature ovum from an ovary into the Fallopian tube, where it may be fertilized by sperm passing up the tube after coitus. Ovulation, which occurs at about the midpoint of the menstrual cycle, is triggered by hormones from the pituitary gland.

Ovum. The female egg, or germ cell, released from a mature Graafian follicle on the surface of an ovary at ovulation.

Oxygen. The gas essential to the energy-producing processes of the body by which food materials are "burned up" by oxidation. In the lungs, oxygen from the air combines easily with hemoglobin in red blood cells and so can be carried to all the body's tissues.

Oxygenated. The term meaning charged with oxygen, as when hemoglobin in red blood cells is converted to oxyhemoglobin, which gives arterial blood its bright red color, through contact with oxygen in the lungs. The opposite term, applied, for example, to dark venous blood, is deoxygenated and refers to blood which is relatively short of oxygen.

Oxytocin. The hormone produced by the pituitary gland in women which triggers contractions of the uterus during labour and initiates milk flow from the breasts when the baby suckles.

PQ

Pain. A symptom of physical hurt or mental distress, often providing a useful warning sign. Pain may be "superficial", caused by injury to the skin or organs near the surface of the body, or "deep," when associated with internal organs. Pain-receptor nerve cells pass impulses to the spinal cord, where a nerve tract carries these warning signs to the thalamus, which is concerned with the analysis of sensation.

Palate. The roof of the mouth, between the mouth and the nasal cavities. The hard palate, at the front, consists of bone covered with mucous membrane. In the soft palate, at the back of the mouth, the mucous membrane lies over muscular and connective tissue.

Pancreas. The gland just below the stomach, tucked into the loop formed by the duodenum, that pours several different digestive enzymes into the intestine and is also an endocrine gland, secreting into the bloodstream the hormones insulin and glucagon, which regulate glucose metabolism.

Papilla. A small bump or projection from any surface, such as one of those found on the surface of the tongue.

Paradoxical sleep. Periods during sleep, also known as rapid eye movement (REM) sleep, that last about twenty minutes and occur about every hour and a half. It is during these periods that dreaming takes place.

Parasympathetic nervous system. One part of the autonomic nervous system, responsible for unconscious control of glands and the smooth muscle of internal organs.

Parathyroid glands. Four tiny glands in the neck that secrete parathormone (PTH), the hormone that controls the level of calcium and phosphate in the blood.

Parietal lobe. The area to the rear of the deep central division of the cortex of each cerebral hemisphere that is concerned with sensory perception.

Parotid glands. The two salivary glands, situated just in front of each ear, that pour saliva into the mouth through a duct that opens on the inside of the cheek.

Patella. The kneecap, a small bone lying in front of the knee joint, in the tendon of the quadriceps femoris, the muscle that straightens the leg.

Pectoral girdle. The group of bones, consisting of the paired shoulder blades, or scapulae, and the collarbones, or clavicles, that supports the arms.

Pelvis. The ring of bones that supports the contents of the abdomen and transmits the body's weight through the hip joints to the legs. At the back is the group of five vertebrae called the sacrum, and on each side are the bones called the ischium, the ilium and the pubis. The ischial bones are those on which we sit, each hip bone is an ilium and the pubic bones meet in the front to form the pubic arch.

Penis. The male organ of sexual intercourse and urination. Through the penis runs the urethra, the tube opening at the glans at the end of the penis, which carries semen during ejaculation and urine during micturition. During sexual arousal the two corpora cavernosa, twin columns of erectile tissue lying side by side along the upper surface of the penis, and the corpus spongiosum below, engorge with blood and change the penis from its usual flaccid state, hanging down in front of the scrotum, to the larger, stiff and upright state of erection in which it can enter the vagina.

Pepsin. A protein-digesting enzyme formed in the stomach from pepsinogen and acid, both secreted by cells in the stomach wall.

Perception. The overall awareness of the environment resulting from interpretation of sensory information.

Pericardium. The double membrane surrounding the heart. The outer sheath protects the lubricated inner membrane that allows the heart to expand and contract with minimal friction.

Perineum. The area of the body between the thighs on each side, the genitals in front and the anus.

Periosteum. The thin layer of connective tissue, containing blood vessels and nerves, that covers the outside surfaces of all the bones in the body.

Peripheral nervous system. The nerves and nerve fibers connecting the brain and spinal column with all other parts of the body; all nervous tissue outside the central nervous system.

Peristalsis. The regular waves of muscular contraction that pass down the walls of the intestine and push the contents along.

Peritoneum. The lubricated membrane that lines the abdominal cavity.

pH. A measure of the acidity or alkalinity of a solution. A solution with pH 7 is neutral, one with a higher number is alkaline; acid solutions have low pH numbers.

Phagocyte. Any cell that is able to engulf bacteria and digest or destroy foreign material in the body, in the process called phagocytosis.

Pharynx. The cavity behind the nose and mouth containing the opening of the windpipe, or trachea, and the gullet or esophagus, which carries food to the stomach.

Pineal body. A small mass of tissue at the back of the brain which does not appear to have a specific function in man, but in certain lower animals, notably lizards, is capable of detecting light and may also influence pigmentation.

Pinna. The external cartilage flap of the ear.

Placenta. The disk-shaped attachment of the developing embryo to the lining of the uterus during pregnancy. Through this organ and the linking umbilical cord, oxygen, body-building and energy-supplying materials and waste products are interchanged between the bloodstreams of the embryo and the mother. The placenta develops from membranes surrounding the embryo. Following the birth of the baby, the placenta is expelled from the uterus as the afterbirth.

Plasma. The watery solution of salts and proteins that forms the fluid part of the blood. The plasma proteins are albumins, globulins and fibrinogen. Serum is plasma without the fibrinogen.

Platelets. Cell fragments in the blood that are essential to the clotting of blood.

Pleura. The double membrane surrounding the lungs. The outer layer lines the chest cavity and the inner layer covers the lung surfaces. The lubricated surfaces of the membrane allow movement of the lungs within the chest without friction.

Pons. Nerve fibers in the midbrain connected to the forebrain and the medulla oblongata of the hindbrain, concerned with hearing, the facial muscles and breathing.

Portal vein. The large blood vessel which carries nutrient-laden blood coming from the veins of the stomach and intestines to the liver, where the nutrients are processed.

Prefrontal lobes. The foremost lobes of the cerebral hemispheres. They appear to influence the personality.

Pregnancy. The period of development of a baby from fertilization until birth, lasting about thirty-eight weeks.

Premolars. The two teeth in each quarter of the jaws lying between the molars and the canine teeth.

Progesterone. One of the two female sex hormones produced by the ovaries. During the second half of the menstrual cycle, the corpus luteum, the yellowish mass of cells formed after a Graafian follicle releases a mature ovum, secretes progesterone to ready the lining of the uterus for implantation of a fertilized ovum. If implantation occurs, progesterone production, from both the corpus luteum and the placenta, continues throughout pregnancy to maintain the uterine wall. Synthetic progesterones are used in the contraceptive pill.

Prolactin. The hormone produced by the pituitary gland that stimulates milk production by the mammary glands.

Prostaglandins. Substances, first discovered in the fluid secreted by the prostate gland, that cause smooth muscle, in particular the muscle of the uterus, to contract strongly. Prostaglandins are now known to be present in many tissues and to have different effects.

Prostate gland. The mass of muscle fibers and glands about one and a half inches long that, in males, surrounds the urethra, or urinary passage, just below its exit from the bladder. The prostate produces some of the seminal fluid that, along with sperm from the testes, forms the semen ejaculated by muscular contractions at orgasm.

Protein. One of the three main components of the diet that, with fat and carbohydrate, are essential to tissue growth. All human proteins consist of long chains made up of various permutations of some twenty-one amino acids, the building blocks of living matter. Pepsin and other protein-digesting enzymes in the stomach and small intestine break proteins in food into their constituent amino acids for absorption into the bloodstream. These amino acids are then reassembled by the liver and other tissues into the proteins the body needs.

Puberty. The period during which the sexual organs attain maturity, leading to the formation of sperm and ejaculation in boys and menstruation and ovulation in girls, and the secondary sexual characteristics, including breast growth and body hair appear.

Pubis. One of the three bones on each side that, together with the sacrum, form the pelvis. The pubic bones meet in the center to form the pubic arch, beneath which are the external genital organs of both sexes.

Pulmonary circulation. Twin arteries carrying deoxygenated blood to the lungs from the chambers of the right side of the heart and veins returning oxygenated blood to the left side of the heart to be pumped around the body.

Pulse. The rhythmic beat, or expansion, in arteries caused by the waves of blood driven through the circulatory system by the heart.

Pupil. The opening in the center of the diaphragmlike iris of the eye. The size of the pupil regulates the amount of light falling on the retina, and also acts as a "stop," as in a camera, to help focus images on its light-receptive surface.

Purkinje fibers. Muscle fibers within the heart that conduct electrical impulses from the atria to the ventricles, so coordinating their contractions to pump blood through the pulmonary and systemic circulations.

Pylorus. The exit from the stomach to the duodenum, surrounded by a muscular, valvelike ring which is called the pyloric sphincter.

Quadriceps. The large muscle at the front of the thigh that straightens, or extends, the leg at the knee joint. The patella, the kneecap, is a bone within the quadriceps tendon, which is attached to the tibia.

Quickening. The movement of the fetus in the uterus.

R

Radius. The bone of the forearm on its outer, or thumb, side, parallel with the ulna, the inner bone. Both bones take part in the wrist and elbow joints.

Rapid eye movement (REM). Spasmodic movement of the eyeballs occurring during sleep in approximately twenty-minute periods every hour and a half. REM sleep is associated with dreaming.

Receptor. A cell or nerve ending sensitive to specific forms of stimulation, resulting in the passage of impulses in the nervous system.

Rectum. The final ten inches of the large intestine, between the colon and the anus.

Reflex. A response to a stimulus that occurs without conscious, voluntary control. A reflex arc consists of sensory receptors, afferent nerves carrying impulses inward to the central nervous system, connecting, or interneurons, efferent, or outgoing, nerves and effectors, which are muscles or glands.

Releasing factors. Substances released by the hypothalamus, a region at the base of the forebrain concerned with the emotions and regulation of the body's internal systems. The factors stimulate the hormonal output of the pituitary gland, and so control some of the other endocrine, or hormone-producing, glands.

Renal. A term relating to the kidneys. The renal arteries and veins, for example, are responsible for the circulation of blood within the kidneys.

Reproduction. The process of creating and producing offspring.

Reproductive system. The organs within the body concerned with creating and producing offspring. In men, these include the sperm-producing testes along with other glands which secrete seminal fluids, the ducts that carry and store semen and the penis which, when erect, can enter the vagina during coitus to deliver sperm at ejaculation. In women, the reproductive organs include the ova-producing ovaries, the Fallopian tubes, which collect the ovum, the uterus, which holds the developing fetus during pregnancy, and the vagina and clitoris. The breasts are regarded as accessory parts of the female reproductive system.

Respiration. The process by which oxygen is delivered to tissues and the waste products carbon dioxide and water are removed from them. In one of its senses, the term is used to describe the overall movements of the thorax during breathing—inspiration is breathing in, and expiration is breathing out. Cellular respiration, on the other hand, refers to the complex chemical changes within the tissues during the "burning" of food to release energy.

Respiratory system. The organs concerned with breathing. The nasal passages and mouth lead to the trachea, or windpipe, which splits into two bronchi leading to the lungs. Within the lungs the bronchi divide and subdivide into bronchioles, which culminate in air sacs, alveoli, that are richly supplied with blood vessels. When the diaphragm and muscles between the ribs contract, the chest expands and air passes into the lungs. When the muscles relax, air is pushed out, carrying with it waste carbon dioxide.

Resting potential. The difference in electrical potential between the surface and the interior of a nerve cell that is ready to carry an impulse.

Reticular formation. The network of cells deep within the midbrain and hindbrain that regulates sleep and consciousness. The reticular formation shares incoming sensory information with the cerebral cortex, and in response adjusts the overall activity of the brain.

Reticulocytes. New-formed red blood cells.

Reticuloendothelial cells. Phagocytic scavenger cells in the blood vessels of the bone marrow, spleen and liver that can engulf and destroy bacteria, worn-out body tissues and blood cells that have aged.

Retina. The lining at the back of each eye, composed of the light-sensitive rod and cone cells that transmit impulses through the two optic nerves to the brain.

Rhesus factor, Rh factor. A factor, or antigen, sometimes present in blood.

Rhodopsin. The light-sensitive pigment, sometimes called visual purple, found in the rod cells of the retina. Light breaks rhodopsin into opsin, which activates the nerve fibers from the rods, and retinene, which, so long as vitamin A is present, can re-form into rhodopsin during periods of darkness.

Ribonucleic acid (RNA). Large molecules in the nuclei and cytoplasm cells which are concerned with protein synthesis. These chemicals "read" the genetic code carried by the deoxyribonucleic acid (DNA) of the chromosomes and assemble appropriate combinations of amino acids into proteins.

Ribosomes. Bodies within the cytoplasm, or cell fluid, which consist of ribonucleic acid (RNA) and are concerned with the synthesis of proteins according to genetic instructions carried by the deoxyribonucleic acid (DNA) of the chromosomes.

Ribs. The twenty-four bones attached to the twelve thoracic vertebrae that form and protect the chest cavity. The top seven ribs on each side join directly to the sternum, or breastbone, the next three are attached by strips of cartilage and the last two, which are much shorter, float freely.

Rods. One of the two types of sensitive nerve endings in the retina of the eye. Rods do not distinguish between colors, but do respond in dim light and so provide colorless night vision.

Round window. The membrane stretched across the lower canal of the cochlea, or hearing organ, enclosing one end of the column of fluid, which is closed at the other end by the oval window.

Sac. A small cavity of tissue, such as an alveolus, one of the small air bags within the lungs.

Sacrum. The five vertebrae of the spine below the lumbar vertebrae which are fused into a single bone.

Salivary glands. The three pairs of glands in the mouth that secrete saliva, a watery fluid containing digestive enzymes that lubricates and cleans the mouth and the teeth.

Scalp. The tough and relatively inelastic skin over the skull carrying the hair.

Scapula. The shoulder blade, a flat bone of the back that forms part of the shoulder joint.

Sciatic nerve. The large nerve, formed from several spinal nerves, that runs down the back of the thigh to supply the leg. The term sciatic means of the ischium, one of the bones of the pelvis.

Sclera. The white of the eye.

Scrotum. The bag of skin, hanging between the thighs of the male, which contains the testes.

Sebaceous glands. Glands around hair roots in the skin that produce sebum, an oily substance that keeps hair and skin supple.

Secondary sexual characteristics. Male and female characteristics that develop at puberty, including body shape, fat and hair distribution, the breaking of the voice or the growth of breasts, but which are not directly essential to reproduction.

Secretion. A substance produced by a gland, whether internal, such as a hormone or digestive juices, or external, such as sweat.

Semen. The viscous fluid ejaculated through the penis at orgasm which is made up of sperm from the testes and secretions from the prostate and other glands.

Semicircular canals. Three canals in the inner ear which sense movement of the head.

Sensation. The experience of stimulation of a sensory nerve.

Sense. The body's faculties for perception and awareness. The five major senses are sight, hearing, touch, taste and smell.

Sensory cortex. The areas of the cortex, or surface layer, of the rear half of each cerebral hemisphere. The sensory cortex is concerned with analyzing sensations.

Sensory neurons. Nerve cells which consist of a receptor sensitive to a particular stimulus and a nerve fiber which carries impulses from the receptor to the central nervous system.

Septum. A division between two parts of the body such as the cartilage between the nostrils.

Serotonin. A substance in the brain that may be involved with transmitting nerve impulses.

Serum. The major component of blood fluid which separates out as a yellow liquid when blood clots.

Sex chromosomes. The pair of chromosomes among the twenty-three pairs of the newly fertilized zygote that influence the sex of the child-to-be. Two matched X chromosomes result in a girl, an X chromosome and a Y chromosome—so-called because this is the shape they appear under the microscope—produce a boy. A woman always passes on an X chromosome in her ovum, a man's sperm may contain either X or Y.

Sex glands. The sperm-producing testes in the male and the egg-producing ovaries in the female. The sex glands also produce hormones that influence sexual development and functioning.

Short-term memory. The aspect of memory that retains limited items of information for short periods without necessarily transferring them to the long-term memory. Remembering a telephone number long enough to dial it is one example of short-term memory.

Shoulder. The ball-and-socket joint formed between the head of the humerus, the bone of the upper arm, and the scapula, or shoulder blade.

Sinus. A narrow body cavity, or opening, such as the air sinuses of the head which open off the nasal passages.

Skeleton. The articulated bony framework of the body.

Skin. The body's protective external covering, skin has two main layers. The inner dermis, which is up to one-eighth of an inch thick, is made of elastic connective tissue, is well supplied with blood vessels and nerves and contains the sweat glands and hair roots. The outer epidermis contains a layer of growing cells which gradually migrate into the thin covering of hard, dead cells and are ultimately rubbed away.

Skull. The bony structure of the head and face. More than twenty bones are rigidly joined together to form a protective covering for the brain and to provide underlying support for the face.

Sleep. The period during which a person becomes unconscious, the electrical activity of the brain alters and many of the body's processes slow down. There is still, however, a level of selective awareness in which the person can react to significant stimuli. In the course of the night, the person moves through various stages of sleep, including several periods lasting twenty minutes or so of rapid eye movement (REM), or paradoxical sleep, which is associated with vivid dreams.

Smell. The sensing of odor in response to chemical stimulation of the olfactory receptors high in the nasal cavities.

Speech. The act of speaking, or talking, which involves the mental formulation of words and the physical control of the voice in pronouncing them.

Speech areas. The areas of the cerebral cortex in the frontal lobes of the brain which control speech. The speech areas are closely associated with areas for hearing, reading and writing and in most people are located in the left cerebral hemisphere.

Sperm, spermatozoa. The male germ cells produced in the testes and ejaculated at orgasm as part of the semen.

Spermatic cord. The bundle running from each testis to the urethra, the tube running through the penis. It includes the supporting muscle, blood vessels and the duct that carries sperm.

Sphincter. A ring of muscle that closes and opens. One example is the anal sphincter, which closes off the rectum.

Spinal column. The flexible backbone made up of a "column" of separate bones, the vertebrae.

Spinal cord. The column of nervous tissue occupying the upper two-thirds of the canal running through the spine. The spinal cord is an extension of the brain and contains nervous tissue able to initiate reflex responses, as well as nerve tracts carrying impulses to and from the brain.

Spinal nerves. The thirty-one pairs of nerves passing out from the spinal cord through the gaps between vertebrae to supply all areas of the body and limbs.

Spinal reflex. A response to a stimulus in which the sensory impulses initiate motor impulses from the spinal cord without passing to the brain. One example of a spinal reflex is the jerking of the leg when the tendon running down from the kneecap is tapped.

Spleen. The large fibrous organ, richly supplied with blood and lymphoid tissues, just below the ribs on the left side of the abdomen, which is concerned with the formation of blood cells and the removal and breakdown of worn out and imperfect red blood cells.

Stapes. One of the three small bones of the middle ear that transmit vibrations from the eardrum to the inner ear.

Starch. A complex carbohydrate, contained in such foods as cereals and potatoes, that is broken down into simple sugars in the digestive system.

Sternum. The flat breastbone to which the ribs are attached by cartilage.

Steroids. Chemical compounds sharing a common backbone of four linked rings of carbon atoms, steroids include such substances as vitamin D, cholesterol and hormones produced by the testes and ovaries. Often the word steroids is used to describe only the corticosteroids, hormones secreted by the cortex of the adrenal glands which influence secondary sexual characteristics, muscle growth, the balance of salts in the body and other physical processes.

Stimulus. Any experience that initiates a change, response or sensory activity in a person as a whole or in any part of the body's systems.

Stomach. The large muscular bag in the abdomen which takes food materials from the esophagus, or gullet, mixes them with protein-digesting enzymes and acids secreted by the stomach walls, churns and softens them and passes them on to the small intestine, where they are digested and absorbed.

Sugars. Simple carbohydrates such as glucose, lactose, maltose and sucrose.

Sulcus. A groove in the surface of any organ, particularly the grooves or fissures dividing the lobes of the cerebral hemispheres.

Superego. The term Sigmund Freud used to label the partially unconscious process of imposing moral constraints on the instinctive and pleasure-seeking aspects of personality.

Sweat glands. Glands in the skin that during the process of perspiration secrete sweat, a fluid that evaporates from the surface of the skin, cooling it.

Sympathetic nervous system. Part of the autonomic nervous system, responsible for unconscious control of blood supply, heart rate, breathing rate and volume and digestive activity. Sympathetic nerves connect nerves coming from the thoracic and lumbar regions of the spine directly to the organs they control.

Synapse. The junction between two nerve cells where nerve fibers mesh in a network of fine branches and nerve impulses are passed chemically from one nerve to the next. The synapse between a nerve and a muscle is called a neuromuscular junction.

Synovial membrane. The smooth membrane which lines the fibrous capsule around a joint and secretes the joint-lubricating synovial fluid.

T

Tarsus. The base of the foot, formed by the seven tarsal bones, connecting the bones of the leg in the ankle with the metatarsal bones which run to the base of each toe.

Taste. The sensation of flavour. Taste buds, small clusters of nerve cell receptors on the tongue, respond to the four basic sensations of sweet, salt, sour and bitter. Different combinations help to create the varying tastes we recognize, but awareness of smell and texture of food play a great part in the overall response.

Teeth. The adult has thirty-two teeth, eight in each half of each jaw. From back to front the eight are comprised of three molars, two premolars, one canine and two incisors. Children have twenty temporary, or milk, teeth. The white part of the tooth above the gum is the crown; the part set in the gum is the root. Each tooth consists of a layer of enamel over the crown—the root is covered with bony "cement"—a much thicker layer of the tooth's main structural material, dentin, and a central cavity containing pulp and a nerve ending.

Temporal lobe. One of the lobes of the cerebrum, low at the side of each cerebral hemisphere, mainly concerned with memory, the synthesis of emotion, hearing and, on the dominant side of the brain, speech.

Tendon. A sinew or strong fibrous bundle of strands of collagen which joins a muscle to a bone.

Testes, testicles. The male sex glands, each about an inch and a half long and held hanging behind the penis in the scrotum. The testes produce sperm, the male germ cells ejaculated in the seminal fluid at orgasm, and male sex hormones, including testosterone.

Testosterone. A male sex hormone, or androgen, produced by the testes, testosterone, together with the other sex hormones produced by the adrenal gland, plays a major part in the development of masculine secondary sexual characteristics at puberty.

Thalamus. The mass of nerve cells at the base of the forebrain between the two cerebral hemispheres, the thalamus is believed to be involved in the assessment of sensation, in control and in relaying nervous impulses to the cerebral cortex and other centers in the brain.

Thigh. The upper leg, the thigh is built around the largest of the body's bones, the femur, or thighbone, and contains some of the body's most powerful muscles. The femoral arteries and veins and the femoral and sciatic nerves run through the thigh to supply the leg.

Thorax. The chest compartment, or thorax, enclosed by the ribs, backbone and diaphragm, contains the lungs, the heart, major blood vessels and the esophagus.

Threshold. The critical level for a stimulus to produce a response. A nerve fiber, for example, will not "fire" when given a stimulus below the threshold value. Any stimulus above the threshold produces a full impulse.

Thrombin. The enzyme formed when blood comes into contact with damaged tissue or a wound, thrombin initiates the change of the blood protein fibrinogen to fibrin, producing a blood clot.

Thymus. A gland in the top of the chest cavity, particularly important in childhood for the development of the immune system. The thymus is responsible for the development of one type of lymphocyte (white blood cells which produce antibodies) but ceases to function by puberty.

Thyroid gland. Two lobes lying on either side of the windpipe in the front of the neck, the thyroid gland is stimulated by a hormone (thyroid-stimulating hormone (TSH)) from the pituitary gland to produce thyroxine, a hormone which increases the rate of activity of body cells.

Tibia. The shin bone, the tibia is the main bone of the lower leg.

Tidal volume. The volume of air moved in and out with each breath which, except during heavy exercise, is a small part of the lungs' total volume.

Tissue. Extended blocks of a similar variety of living cells performing a particular function. Epithelial tissue, for example, such as the skin, consists of cells packed closely together. In muscle tissue the cells are grouped into fibrils. In connective tissue, bone and cartilage the cells are spread through a matrix of ground substance.

Tissue fluid. The watery fluid that fills the spaces between cells and helps to transfer nutrients and waste products between cells and capillary blood vessels.

Tongue. The muscular structure attached to the floor of the mouth, the tongue helps to manipulate food into balls which can be easily swallowed and to shape speech sounds. Taste buds, set between the small projections of papillae on the upper surface of the tongue, respond to taste sensations. The tongue is also sensitive to touch and to temperature.

Tonsils. The two masses of lymphoid tissue which lie at each side of the throat and which play some part in the body's defenses against infection.

Touch. The sensation produced by pressure on the skin. Nerve endings which respond to pressure are most thickly clustered in such sensitive areas as the fingertips and lips.

Trachea. The windpipe, or trachea, runs from the larynx, or voice box, at the back of the mouth to just above the heart where it splits into the two bronchi going to the lungs.

Trophic hormones. The hormones produced by the pituitary gland that do not act directly, but influence other endocrine glands.

Tunica intima. The innermost lining of all blood vessels and the only component of the walls of capillary vessels, the tunica intima is only one cell thick and allows fluids, nutrients, waste products and oxygen to pass through between capillary and cells.

Twins. Two children born together from the same pregnancy, twins occur in about one pregnancy in ninety. Fraternal, or dizygotic, twins, which develop when the mother's ovaries release two ova at once and both are fertilized, are no more similar than other children sharing the same parents. Identical, or monozygotic, twins, which develop when a single fertilized ovum splits in two before growing into an embryo, inherit exactly the same genetic makeup and so are physically similar.

U

Ulna. The inner bone of the forearm.

Umbilical cord. The bundle containing blood vessels that runs between the fetus and the placenta, the organ that grows in close contact with the lining of the mother's uterus where interchange of oxygen, nutrients and waste occurs. After birth the cord is cut and withers away.

Unconditional response. A response to a stimulus that occurs naturally and without training.

Unconditional stimulus. A stimulus that produces a particular response naturally and without training.

Unconscious. Temporarily unaware, for example when asleep. According to Sigmund Freud's psychoanalytic theory, an active unconscious level of mind influences behavior and conscious thought.

Urea. The end breakdown product of the metabolism of nitrogen-containing compounds, such as proteins, which is excreted in the urine.

Ureters. The tubes that lead from the kidneys to the bladder.

Urethra. The tube that leads from the bladder to the tip of the penis in males, and to the opening in front of the vagina in females.

Uterus. The womb, the hollow, muscular organ in a woman's abdomen that holds the fetus and enlarges during pregnancy. At delivery, rhythmic contractions of the uterine muscle push the baby out through the birth canal.

Uvula. The small lump of tissue hanging from the soft palate at the back of the mouth.

V

Vagina. The channel running from the uterus to the vulva, the vagina receives the penis during coitus and forms the final part of the birth canal through which the baby passes from its mother's body at delivery.

Vagus nerve. The tenth of the cranial nerves, and an important part of the parasympathetic nervous system running down from the brain to control the lungs, the heart and the digestive organs.

Veins. Wide, thin-walled blood vessels that carry the blood back from the body's tissues to the heart. Veins are broken into sections by valves that prevent the blood flowing back toward the periphery and allow the general muscular activity of the body to pump it toward the heart.

Venule. The small veins that collect blood from the capillaries.

Vertebrae. The separate bones that make up the backbone, or spinal or vertebral column, which supports the body and protects the spinal cord within it.

Villi. Minute projections from the lining of the intestine which increase its surface area and aid the absorption of digested food.

Virus. Small infective agents, much smaller than bacteria, responsible for producing many diseases. Viruses are little more than packages of nucleic acids that, when they invade the body's tissues, turn cell processes to their own ends, reproducing further virus particles and causing illness.

Visual cortex. The areas of the cortex, or covering layer of the occipital lobes at the back of the brain concerned with vision.

Vitamins. Substances essential to the body's functioning, but which are required only in small quantities in the diet. The normal diet usually supplies more than enough for most people. The most important vitamins include: A, concerned with vision and healthy skin; B, a complex of different substances essential for the formation of new cells, such as red blood cells, and many of the body's chemical processes; C, necessary in forming connective tissue; D, vital to the formation of healthy bone.

Vitreous humor. The clear, jellylike substance that fills the section of the eyeball between the retina and lens.

Vocal cords. Two folds of membrane in the larynx, or voice box, at the top of the windpipe which can be tightened to vibrate in the airstream from the lungs and so produce the basic sounds of speech and song.

Vulva. The female external genitalia, or external reproductive organs, including the opening of the vagina, the clitoris and the inner and outer lips or folds of skin, the labia, that protect and enclose them.

Water. More than half the body weight is made up of water, which provides the basic fluid environment for many of the body's processes. The careful balance between water drunk, taken in food and formed in chemical reactions within the body and that lost in urine and sweat is important to life itself because the body begins to suffer if it loses as little as one-tenth of its water.

White blood cells. The leucocytes, or white blood cells, produced in the bone marrow, which are able to pass into the body's tissues and fight infection by forming antibodies and engulfing bacteria.

Windpipe. The trachea, or windpipe, running from the larynx to the lungs.

Wisdom teeth. The hindmost of the three molar teeth in each quarter of the jaws, the wisdom teeth often do not appear until early in adult life.

Womb. The uterus.

Wrist. The flexible joint between the forearm and the hand.

X

X chromosome. One of the two sex chromosomes that can be passed on in the sperm or ova. If the fertilized zygote has two X chromosomes it develops into a female, if it has an X chromosome and a Y chromosome then the child is male.

X rays. Electromagnetic radiation of much shorter wavelength than visible light which can pass through apparently opaque substances, such as the body's tissues.

YZ

Y chromosome. One of the two sex chromosomes that can be passed on in the sperm. If the fertilized zygote has an X chromosome and a Y chromosome it develops into a male; if two X chromosomes are present then the child is female.

Zygote. The cell formed when a sperm fertilizes an ovum.

Answers to tests on pages 127–135

Page 127 Visual memory problems

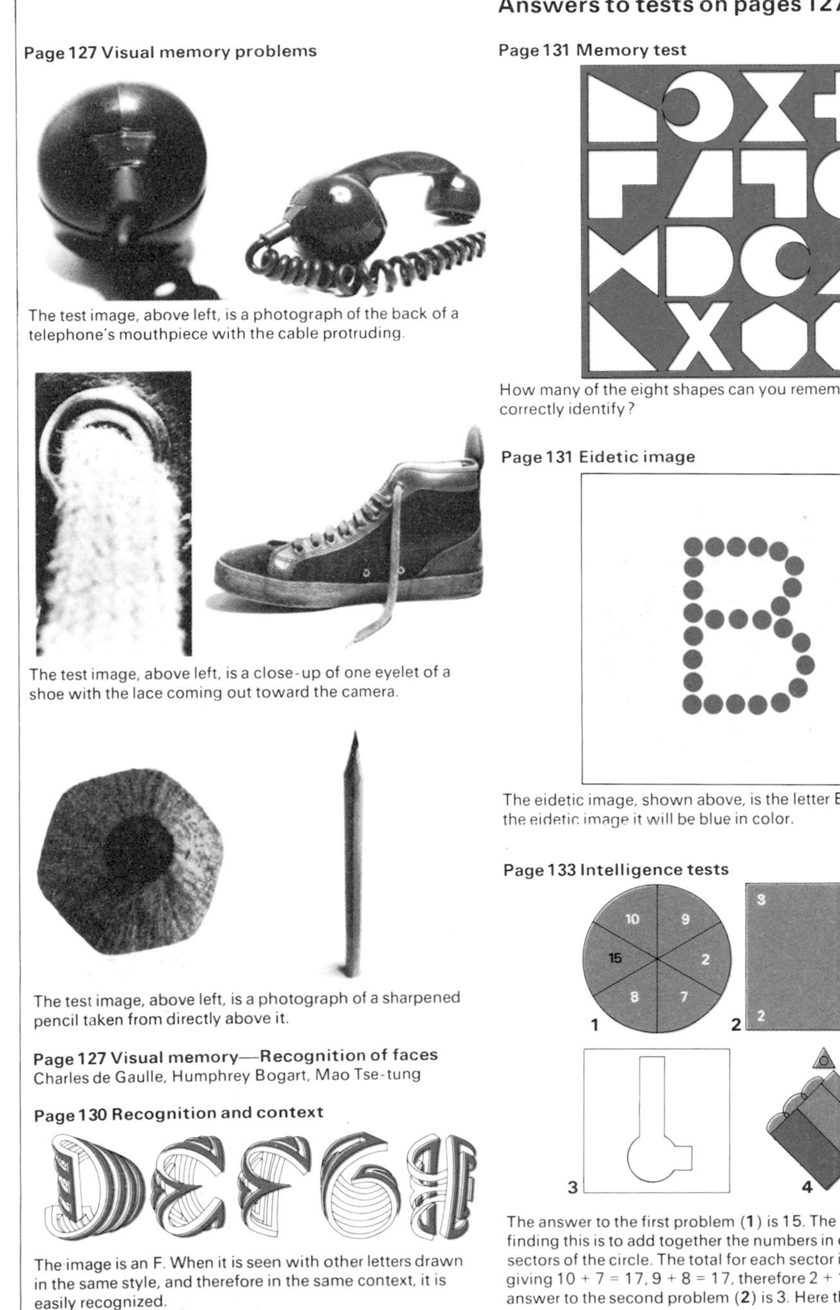

The test image, above left, is a photograph of the back of a telephone's mouthpiece with the cable protruding.

The test image, above left, is a close-up of one eyelet of a shoe with the lace coming out toward the camera.

The test image, above left, is a photograph of a sharpened pencil taken from directly above it.

Page 127 Visual memory—Recognition of faces

Charles de Gaulle, Humphrey Bogart, Mao Tse-tung

Page 130 Recognition and context

The image is an F. When it is seen with other letters drawn in the same style, and therefore in the same context, it is easily recognized.

Page 131 Memory test

How many of the eight shapes can you remember and correctly identify?

Page 131 Eidetic image

The eidetic image, shown above, is the letter B. If you form the eidetic image it will be blue in color.

Page 133 Intelligence tests

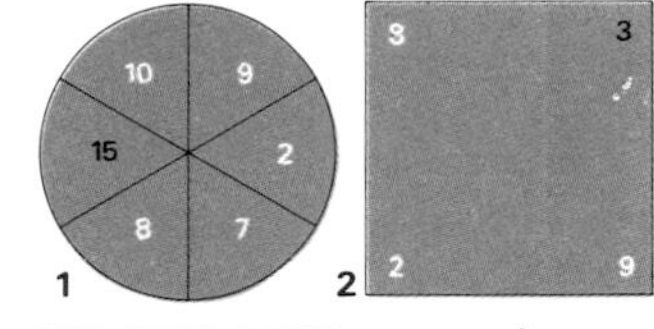

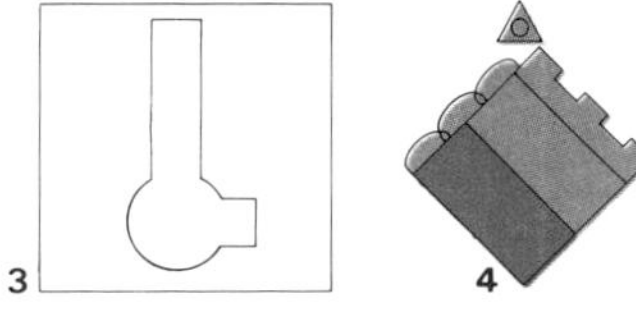

The answer to the first problem (**1**) is 15. The principle in finding this is to add together the numbers in opposite sectors of the circle. The total for each sector is constant, giving 10 + 7 = 17, 9 + 8 = 17, therefore 2 + 15 = 17. The answer to the second problem (**2**) is 3. Here the principle is that the sum at the base of each shape equals the sum at the top of each shape. Therefore 4 + 5 = 9, 8 = 6 + 2, and 2 + 9 = 8 + 3. The third problem (**3**) is visio-spatial. All the objects are rotated clockwise except the second, which is rotated counterclockwise, and is therefore the odd one out. The third alternative is the correct answer to the last problem (**4**), which is another visio-spatial test. It has the correct shape, with the patterns changed over on each side. The position of the colors is reversed and the main shape is reversed with that of the small shape, while the same inner shape remains.

Page 133 Intelligence test—Find the errors

1. June has twenty-eight days instead of thirty days.
2. No shadow is cast from the calendar onto the table.
3. The shoe has an uneven number of eyelets.
4. The clock has only eleven markings.
5. The plug has round pins while the socket has square holes.
6. The book has the title on the wrong cover.
7. Calendar has five-day week instead of seven-day week.

Page 135 Concept formation

The concept is the blue, spotted rabbit.

Page 135 Creativity test—Hidden objects

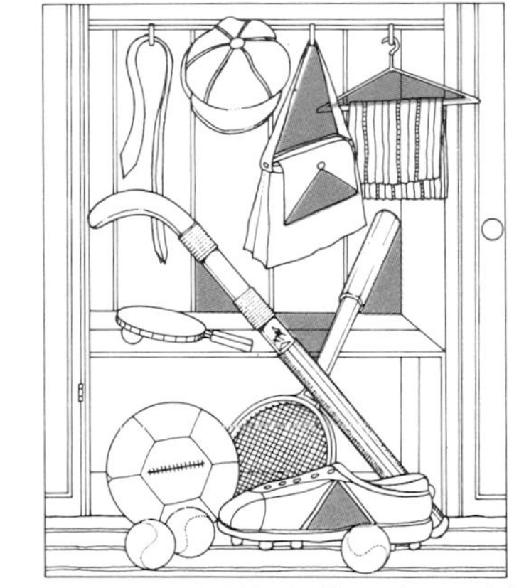

The seven triangles formed within the illustration are picked out above in solid color.

Page 135 Creative thinking—joining the dots

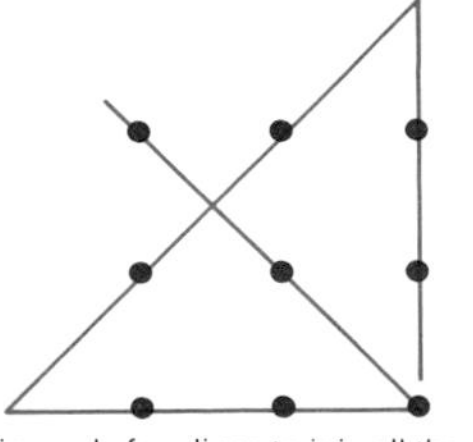

The solution, using only four lines to join all the dots together, is shown above. Three of the lines extend beyond the boundaries formed by the perceived square.

Index

italicized references denote illustrations

D

E

F

G

H

I

JK

L

M

R

S

T